PRAISE FOR

The Tangled Wing

(continued from the back cover)

"How wonderful to have a new *Tangled Wing* which incorporates the rich findings made in the last twenty years in the fields of evolutionary and behavioral biology. We find the same graceful writing as in the original classic and the same facility to clarify complex issues and to come to stimulating conclusions."

 —ERNST MAYR, professor emeritus, Harvard University

"The word 'masterpiece' almost never springs to mind about a work in biosocial science. It does about *The Tangled Wing*. This is a jewel of virtuoso scholarship written with lustrous and punctilious warmth. It's a remarkable accomplishment."

 —LIONEL TIGER, Charles Darwin Professor of Anthropology, Rutgers University, author of *The Decline of Males*

"In the great tradition of Charles Darwin and Thomas Henry Huxley, Konner updates the argument for a biological basis of the human mind and spirit with force, clarity, and eloquence."

 —J. ALLAN HOBSON, director of the Laboratory of Neurophysiology, Harvard Medical School, author of *Consciousness*

"*The Tangled Wing* is a magnificent synthesis of the latest knowledge of biology, psychology, and anthropology. No one is better qualified than Melvin Konner to illuminate the complex workings of the mind. This volume should be on the bookshelf of anyone interested in human nature—and who is not?"

 —AARON T. BECK, M.D., University Professor of Psychiatry, University of Pennsylvania

"If you read the breathtaking first edition of *The Tangled Wing* and were mesmerized, your pleasures have just begun. Now Konner does it again, with a vastly new account of our nature. Each sentence is a pleasure to read, and the powerful points he makes about the human condition can leave one sleepless."

 —MICHAEL GAZZANIGA, David T. McLaughlin, Distinguished Professor, director, Center for Cognitive Neuroscience, Dartmouth College

"In an age in which scientific opinion about violence and biology is overwhelmed by the political correctness of the left and the genetic determinism of the right, the reader can turn to Mel Konner as the sensible, sober, and authoritative voice."

 —MARTIN SELIGMAN, Fox Leadership Professor of Psychology, University of Pennsylvania

"Mel Konner has a breathtaking set of skills. I do not think anyone else could have put together this synthesis, ranging from rigorous medicine, to cutting-edge psychology, to some funky field anthropology. He also has a rare command of language and knows how to tell a story. *The Tangled Wing* is heavy-weight scholarship, rendered in an attractive and readable style."

— **MARK RIDLEY**, professor of zoology, Oxford University, author of *Mendel's Demon*

"Mel Konner has taken on the most fundamental question—what is our nature—and provided a beautifully written, original account, as broad as it is deep. It is rare for a book by a scientist to be a page-turner, but Konner keeps the reader ever alert to his rare insights and profound understanding of our nature. Konner joins Lewis, Gould, and Pinker—a scientist who can write, who teaches us without our realizing we are being taught."

— **PAUL EKMAN**, professor of psychology, University of California, San Francisco

"This new edition of *The Tangled Wing* updates and expands a seminal work on human nature. Mel Konner is a gifted synthesizer and a great teacher, who weaves together insights from many distant ports: anthropology, evolution, neuroscience, molecular genetics, and literature. But he is more than a reporter of diverse discoveries. In *The Tangled Wing* he brings a critical eye to the science of human nature. Whether describing studies of single gene mutations or hunter-gatherer societies, he tells us not only what is known but what is not yet known and what may not be knowable. With clarity, insight, and, at times, poetry, he simultaneously explores and demonstrates the wonders of human nature. This is a superb book."

— **THOMAS INSEL**, M.D., director, Center for Behavioral Neuroscience, Emory University

"In the upcoming postgenomic era, linking basic biology and behavior is likely to be the most important challenge of the life sciences, demanding an ability to move with poetic creativity yet with rigor from the humanities to molecules. *The Tangled Wing* is the finest such effort I have seen. Konner has penetrating insight into hitherto unappreciated difficulties of Darwinian and Skinnerian approaches to sociobiology. He comes up with novel, creative formulations. All of this in a volume that is as gripping as a mystery novel."

— **SOLOMON H. SNYDER**, M.D., director, Department of Neuroscience, School of Medicine, Johns Hopkins University

"In a time marked by increasingly complex, highly specialized disciplines, it is rare to find an author with a depth of knowledge in both the biological and social sciences, combined with superb writing skills sufficient to provide a reader with a breathtaking new understanding of human behavior. Mel Konner did that twenty years ago in the first edition of *The Tangled Wing*. The updated and revised edition of this book is an even greater achievement, culling major new findings from the neurosciences and behavorial biology that deepen our understanding of the intricate interdependence of genes and social behavior in our own and other species and cultures. In addition, Konner alerts us to the serious problems attending the expanding size of the human population against the shrinking resource base on our tiny 'blue dot' of a planet. Konner brings the eye of both a poet and a scientist to his work, artfully providing us with intellectual wings to soar across the abyss that has artificially separated the biological and social sciences, and reporting new research findings that pave the way to their integration. *The Tangled Wing* is sure to be among the most important books published in the first decade of the new millennium."

> —ALICE S. ROSSI, Harriet Martineau Professor Emerita of Sociology, University of Massachusetts, Amherst

"Discussions and arguments about how to describe, explain, and value human nature and activity pervade the natural sciences, behavioral and social sciences, literature and the arts, and humanistic scholarship, religion, and popular culture. Anyone, scholar or general reader, who is seriously interested in the human ought to engage this thorough revision of *The Tangled Wing: Biological Constraints on the Human Spirit*. Konner's general erudition, his comprehensive research, his combination of deeply moral and humanistic sensitivity with scientific rigor, his wonderful prose, his critical analytical prowess, but most important his brilliant synthetic capacities provide a powerful and distinctive interpretation. Its thick but lucid arguments ought to be taken into account by theologians, philosophers, and other humanists in their assessment of the human—our embedment in nature as well as our distinctive 'spirits,' our limitations as well as our creativity, our natural proclivity to evil as well as our natural orientation toward the good. And many general readers who reflect on their own life experiences, or on the diverse accounts of the human to which they are exposed, will find themselves refining and revising their own thinking. This new version will, like the first, become a point of orientation for many persons in the coming decades."

> —JAMES M. GUSTAFSON, former Henry R. Luce Professor of Humanities and Comparative Studies, Emory University

The Tangled Wing

BIOLOGICAL CONSTRAINTS
ON THE HUMAN SPIRIT

Second Edition

Revised and Updated

MELVIN KONNER

An Owl Book

HENRY HOLT AND COMPANY   New York

To Herbert Perluck
and Irven DeVore,
and to the memory of
Doctor Julian Gomez

Henry Holt and Company, LLC
Publishers since 1866
115 West 18th Street
New York, New York 10011

Henry Holt® is a registered trademark of
Henry Holt and Company, LLC.

Library of Congress Cataloging-in-Publication Data
Konner, Melvin.
   The tangled wing : biological constraints on the human
spirit / Melvin Konner.—2nd ed.
      p. cm.
   Includes bibliographical references and index.
   ISBN 0-7167-4602-6
   ISBN 0-8050-7279-9 (pbk.)
      1. Emotions—Physical aspects. 2. Human behavior. I. Title.
QP401. K6 2001
152.4—dc21   2001005720

Henry Holt books are available for special promotions
and premiums. For details contact: Director, Special Markets.

Second edition published in 2002 by Times Books

First Owl Books Edition 2003

An Owl Book

Designed by Diana Blume

Printed in the United States of America
10 9 8 7 6 5 4 3 2 1

Contents

Acknowledgments

Many an author, sitting with pencil stub in hand, bent over sprawled galleys, must have thought as I did of the first passage of *Don Quixote*: "Idle reader, you may believe me without any oath that I would want this book, the child of my brain, to be the most beautiful, the happiest, the most brilliant imaginable. But I could not contravene that law of nature according to which like begets like." If such can be said of Cervantes's brainchild, what can one possibly say about one's own? Only that one has done one's best. And yet, if the truth be told, that "best" may prove to belong as much to certain others as to oneself.

The new *Tangled Wing* owes most to John Michel, Executive Editor of W. H. Freeman's Trade Division. Undaunted by the complexity of the project, he supported and helped me at every stage in a three-year process, sometimes taking heat for decisions that favored me. His direct contributions to the book were invaluable, but his friendship has been equally important. Margaret Wimberger copyedited the manuscript and notes with a keen and highly intelligent eye, and wrote the phrases keying the notes to the text. Her hundreds of comments gave new depth to the concept of quality control. Scott Amerman proofread the manuscript and caught errors of substance far beyond the responsibility of a proofreader, in at least one instance saving me from great embarrassment. Diana Blume met the challenge of designing a book and jacket that would be beautiful without compromising its rather weighty substance. Vivien Weiss, Project Editor, and Susan Wein, Production Coordinator, coped gracefully with my obsessive concern for detail, even when it slowed production. Everyone at Freeman also had to deal with the fact that its Trade Division was sold to Holt/Times Books just as this book was going to press. Their support for it and for me during this difficult transition in their own lives will remain with me in grateful memory.

Judith Robertson has been my personal research assistant throughout the four years I have worked on the revision. Her calm, balanced, capable approach to every problem from initial library and internet explorations to the final crossing of t's and dotting of i's helped give me the equanimity to persist and finish. Angela von der Lippe was an early supporter of this project, and her confidence in it was important to me. My agent, Elaine Markson, has been a good friend through both editions and during the intervening years.

There would not be a new *Tangled Wing* if there had not been an old one, and I gratefully name a few of the people who made that book a success. Gerald Henderson of Brooklyn College first taught me to see the human psyche in the context of evolutionary anthropology, a visionary idea at the time. Alex Gold first encouraged me to write the book. Marion Wood, my editor at Holt, believed in it when there was little reason to, and applied her exceptional gifts to making that first edition work. Important and generous early influences that helped give rise to it came from Rabbi Bernard L. Berzon, Nicholas G. Blurton Jones, T. Berry Brazelton, Jane Chermayeff, Stephan Chorover, Victor Denenberg, Nancy DeVore, Howard Eichenbaum, Marjorie Elias, Michael Elias, Robin Fox, Sarah Blaffer Hrdy, Jerome Kagan, Marc Kaminsky, Lawrence Konner, Jane Lancaster, Richard Lee, Robert Liebman, Kopela Maswe, Myrtle McGraw, Richard Morris, Walle Nauta, Penelope Naylor, !Xoma N!aiba, Paul Pavel, John Pfeiffer, Leonard Rosenblum, Alice Rossi, Robert Sapolsky, Laura Smith, Stefan Stein, Charles Super, Lionel Tiger, Robert Trivers, Dora Venit, Eric Wanner, Beatrice Whiting, John Whiting, E. O. Wilson, Carol Worthman, Richard Wurtman, and Paul Yakovlev. The support and friendship of Nancy and Irven DeVore in those years made all things possible; Earl and Martha Kim also opened their hearts and their home to us. My parents, Hannah Levin Konner and Irving Konner, overcame both physical handicap and poverty to give me the desire and the opportunity for a lifelong dedication to learning. They did not fully understand this book, but they believed in it and in me.

For improvements in my writing since the first *Tangled Wing*, I thank the following gifted and dedicated editors, several of whom are superb writers themselves: Geoffrey Cowley, Robert Wright, Holcombe Noble, Elisabeth Sifton, and William Phillips.

For teaching me much more about human nature over the two decades since the first edition, and for valued encouragement, I thank Reverend Joanna Adams, George Armelagos, Michael Bailey, Roger Bakeman, H. Thomas Ballantine, Ronald Barr, Joseph Beck, Michael Cantor, Carl Degler, Mahlong DeLong, William Durham, S. Boyd Eaton, David Edwards, Irenäus Eibl-Eibesfeldt, Arthur Falek, Daniel Federman, Rabbi Emanuel Feldman, John Felstiner, Mari Fitzduff, James Flannery, John Gardner, Norman Geschwind, René Girard, Avram Goldstein, Rabbi Arnold Goodman, Sarah Gouzoules, Robert Green, Patricia Greenfield, James Gustafson, Reverend Robert Hamerton-Kelly, Jeffrey Houpt, Nancy Howell, Thomas Insel, Daniel Kevles, Harvey Klehr, Jane Lancaster, Deborah Lipstadt, Paul MacLean, Melissa Melby, Richard Michael, Kathy Mote, André Nahmias, Ulric Neisser, Charles Nemeroff, Robert Paul, Paul Pavel, Massimo Piattelli-Palmarini, Misha Pless, Gérard Prunier, Mark Ridley, James Rilling, Robert Rose, Alice Rossi, Wendy Roth, Sandra Scarr, Gary Scharnhorst, Wulf Schiefenhövel, Stuart Seidman, Martin Seligman, Mark Shell, Susan Shell, Bradd Shore, Katherine Sieck, Sarah Steinhardt, Judith Stern, John Stone, Sherry Turkle, George Vaillant, Patricia Whitten, Polly Wiessner, Reverend Alison Williams, Carol Worthman, Michael Zeiler, and Doris Zumpe. James Rilling read through the manuscript at an early stage of the revision, and his comments were very helpful.

Certain people at the institutions I have belonged to deserve special mention: At Brooklyn College, Herbert Perluck and Gerald Henderson; at Harvard University,

Irven DeVore, John Whiting, Beatrice Whiting, and Jerome Kagan; at the Massachusetts Institute of Technology, Richard Wurtman, Judith Wurtman, and Loy Lytle; at the Harvard Medical School, Walter Abelmann, Daniel Federman, Norman Geschwind, Joseph Lipinski, and Hans Bode; at Emory University, James T. Laney, David Minter, John Palms, Billy Frye, Robert Paul, Peter Brown, and Carol Worthman.

My wife, the late Marjorie Shostak, was at my side for thirty years. Her companionship, support, friendly criticism and intellectual stimulation were at the core of my life leading up to the first edition of this book and during most of the time since. The words "fought cancer" are often said, but never more truly than during her last eight years. Her love of life was intense, but her desire to remain as a force for good in our children's lives was almost fierce. Her two major books, *Nisa: The Life and Words of a !Kung Woman* and *Return to Nisa* strongly shaped my ideas about human nature. Her parents, Jerome and Edna Shostak, have helped me to care for our children. The late Lois Kasper, who when Marjorie died was already ill with the same cancer, never failed us. She was high-spirited, brave, and funny until her last day. Sarah Steinhardt, Bruce Tuchman, Maureen O'Toole, Betty Castellani, Polly Wiessner, Millie Broughton, Richard Leff, Julian Lokey, and Sidney Stapleton all helped care for her and for our family.

My brother, Lawrence Konner, and I have been exceptionally close, and have guided each other through some of the worst personal crises that occur in modern life. Certain friends were also vital: Herbert and Hazel Karp, David and Shlomit Ritz Finkelstein, and Boyd and Daphne Eaton provided a context of warmth and support—really the equivalent of an extended family—during the most difficult years. My friends Joe Beck, Dudley Clendinen, and Steve Berman have provided crucial personal support for me. Becky Perry, my "extra daughter," and Ronald Kirby, my "extra son," have helped me through the bad times. Ronnie Wenker Konner, during the fourteen years since her serious stroke, has inspired me and taught me to respect the resilience, strength, and humor in the human spirit.

Kathy Mote has been the mainstay of our family during fourteen years of tragedy and loss. As our nanny, she has been a second mother to our children. As my assistant, she has been creative, supportive, efficient, and fiercely loyal. I simply do not know what we would have done without her.

My children have long since passed the point at which they are only learning from me. Susanna Konner guided me on a two-week tour of Palestine, Jordan, and Egypt that transformed my thinking about the Middle East conflict. Her wisdom about that and other conflicts has helped to shape my view of human violence. Adam Konner supplied a pivotal observation for chapter 12, but more importantly, he is trying to teach me the art of happiness. Although this is an uphill battle in my case, it was greatly aided by a month we spent together in the south of France. Sarah is the light of my life, wise beyond her fourteen years, and the brightness of her spirit, the firmness of her character, and her steadily positive approach to life have made my losses bearable and my days rather beautiful.

Ann Cale Kruger is the person with whom I have found comfort, love, and tenderness. She has taught me a great deal about psychology, but much more about life. There is little in this long book that has not been enriched and deepened by my conversations with her, and by the bond we share; I am grateful.

The book is dedicated to three men who have been my mentors in every sense of the word. Herbert Perluck, Professor Emeritus and former Chair of English at Brooklyn College, CUNY, taught me to read and understand literature and shaped my sense of myself as a human being and as a man. Irven DeVore, Professor Emeritus of Anthropology at Harvard, taught me the ideas at the heart of this book, and gave me the crucial research and teaching opportunities that launched my career. Julian Gomez was my psychiatrist during the most difficult time of my life, and his wisdom about human experience, combined with exquisite clinical skill, was in no small part the reason for my survival. He taught me how to live with tragedy, and then, by example, taught me how to die. I feel his loss keenly.

These friends, and the many others mentioned above, were the good angels behind whatever is best in this book. With gratitude and affection I have tried to live up to their expectations. But, wistfully, I must concede that they do not share the burden of its many flaws and follies.

A Prefatory Inquiry

Unless we wilfully close our eyes, we may, with our present knowledge, approximately recognize our parentage; nor need we feel ashamed of it. The most humble organism is something much higher than the inorganic dust under our feet . . .

—CHARLES DARWIN, *The Descent of Man*

We must recollect that all our provisional ideas in psychology will presumably one day be based on an organic substructure.

—SIGMUND FREUD

Human nature exists.

To some this may be a modest claim, but to others it is a frontal assault on two common, deeply held beliefs. The first is that humans are not *in* nature, but somehow outside or above it, because we have souls. These souls may be encumbered with a pile of earthly junk, but we are soon to slough that off and ease into a transcendent dimension, a perfect world in which (if we have been good) we permanently overcome disappointment, pain, anguish, and that most egregious insult to human ambitions, death. The soul is our essence and therefore our true and eternal nature. It has little to do, really, with the sticky pulp in the skull or the organs dangling awkwardly from the skeleton. Most of all, it—and therefore *we*— have little to do with animals. Fewer than one in ten Americans accept evolution as a process that goes on without divine intervention.

This book not only endorses the last view, it also claims that every aspect of the human spirit—mind, thought, feeling, love, dreams, hope, admiration, decency, faith, and in general everything that the religious person takes as evidence for the soul—came from that same natural process, without need of divine assistance. The human spirit *is* made of the sticky, pulpy stuff and, except to the extent that some echoes of its throbbing may continue in other sticky pulps, becomes silent when that stuff stops throbbing.

Which brings us to the second deeply held belief: the human spirit, and therefore human society, is basically perfectable by the usual cultural means. We make a decision to bring humanity into line with some ideal, we figure out how to do it and—presto!—we have a new woman or man, and consequently a new social order. Centuries—no, millennia—of effort in this line have largely failed, but that does not seem to daunt the keepers of this buoyant faith. Ironically, many of the religions that would deny the notion of a noninterventionist evolution have a handle on something that *this* faith has pretty much missed: the human spirit is intractable in enduring ways, and you don't get it in line with any ideal unless you engage in relentless struggle with, or, more exactly, within it. So the central idea of this book, that of a real, defined human nature, is also incompatible with this second belief system.

Once upon a time I was being interviewed on camera for the BBC science program, *Nova*. The producer and I were chatting about the idea of family among the !Kung San, hunter-gatherers of Africa's Kalahari Desert. I was supposed to be revealing something about the biological limits on the natural variation of human behavior, one of the subjects I am paid (it still astounds me, too) to try to know about.

"We're almost finished," the producer said. He paused and looked at a piece of paper folded in his hands. "What," he went on, "can you suggest for the prevention of future global conflict?"

There was a long pause. I believe I grunted before I burst out laughing.

"Cut," he said good-naturedly. "You don't have to answer that one."

"You didn't warn me there were going to be questions like that."

"We can skip it," he said wistfully.

"Well, if you give me a minute to think, I might be able to come up with something that's not completely ridiculous."

What I came up with was that, if we are ever to control human violence, we must first appreciate that humans have a natural, biological tendency to react violently, as individuals or as groups, in certain situations. If we can understand this response and its determinants, we might begin to have some hope of prevention. For example, if, as now seems clear, women, like other primate females, are less aggressive than males for biological reasons, then an overall increase in the number of women in power would tend to buffer political systems against violence. But this, I said, was only one of many possibilities.

"Was that more or less what you wanted?" I asked.

It was. But in retrospect it was not at all what *I* wanted. As I walked away from the makeshift studio in Harvard's Museum of Comparative Zoology, I thought of everything else I should have added. Leaving out population growth was a serious error, since this is the gravest threat to world peace. We had been talking about San hunter-gatherers, yet I failed to mention the grossly unequal distribution of planetary wealth and power. The San provide this model: hoarding of wealth by some increases the risk of violence; sharing of wealth with others correspondingly reduces it. I might have mentioned nationalism, our vivid historical extension of tribalism into the twenty-first century, although that is more a synonym for global conflict than an explanation.

If we'd had time, we could have talked about physiology. Given the ways people (mainly men) rise to power in every known society, it is unlikely that the least aggressive, greedy, or competitive sets of genes in a population will come to control the deployment of its military might. And we might have explored the fact that most people who run the world are sleep-deprived, aging, deluged by interactions with strangers, and under the influence of fear—from fear of loss of influence to fear of assassination. These features of their *biological* lives have known effects on the *biological* tendency to violence.

So much for my second thoughts, too rambling for the film even if they had been timely. We need simple, clear explanations; we are busy. But the answers must also be comprehensive. Simple, clear, comprehensive explanations. We have no time for endless academic musings, or for a litany of facts. We need theories that are incisive and illuminating, that enable us to grasp the solution at once, that transcend the complexities, paring away all irrelevancies, leaving only the elegant, decisive beauty of a Euclidean proof, the first paradigm of our intellectual training.

I offer no such theories. It is my belief that the failure of behavioral science up to the present day results precisely from the pursuit of them. Classical economics, Marxism, psychoanalysis, learning theory, instinct theory, cognitive theory, structuralism, artificial intelligence, neural network modeling, chaos and complexity theory, sociobiology—not a single one false in its essence, but each one false in its ambitions and especially in its condemnation of the others. A good textbook of human behavioral biology, which we will not have for another fifty years, will look not like Euclid's geometry—a magnificent edifice of proven propositions derived from a set of simple assumptions—but more like a textbook of physiology or geology, each solution grounded in a separate body of facts and approached with a group of different theories, all the solutions connected in a great, complex web.

What follows is a sort of game plan for such a text. It deploys the problems, the playing pieces, and the rules, but not the solutions—least of all simple, clear, comprehensive ones. Yet I can say with confidence that the solutions to come in the next few decades are prefigured by these complexities. In the modern cardiology ward, professionals talk of everything from chemistry to anxiety, surgery to diet, bioelectricity to love; and they had better, since no one theory can save a failing heart. Such problems as psychosis, depression, addiction, sexual misery, and yes, global conflict are similarly complex. Just as cardiac ward professionals rely on a surprisingly high fraction of the totality of human knowledge to find their solutions, so will human behavioral biologists require a great breadth of knowledge to find ours.

For the lay person, the difficulty of integrating this breadth of knowledge into a larger picture of ourselves is daunting. Occasionally, some fragment of human behavior, normal or abnormal, is analyzed before our eyes; a shaft of light falls on it and for a while we see it clearly. But most often the new knowledge fades before it does us any good. Scientists, too, are daunted by the effort to keep up. In the old cliché, they find out more and more about less and less until they know every-thing about nothing. But today the problem is that we know more and more about more and more. And though we will never know everything about everything, the time will come when we know so much that no one person can hope to know all the essential facts about, say, violence or anxiety, needed to make a single wise decision. Knowledge becomes collective in the weakest sense, and scientists become like people in a crowd, looking for one another, each holding a single piece of a very expensive radio. The boy with the cutting-edge tuning capacitor can't find the girl who invented the world's best frequency amplifier, and neither of them knows about the astoundingly good demodulator sitting in the back room of a bril-liant, nerdy technician. How can they tune in to the wisdom of the ages when they can't assemble the radio?

Every so often we must stop and see what we know. Without such pauses, epistemology is a bargain basement of shoppers jostling and shouting as they grab at a garment that fits and that, momentarily, is in style. None of the knowledge we have gained about behavior is of use to us if we only get it piecemeal, if we can-not somehow contrive to make it whole. Scientists are always saying, "We must know more." True, but the unspoken conviction that this in itself will save us is false. Knowledge does not automatically order itself in human terms, and if this is true of science generally, it is all the more true of the science of human behavior.

More startling, knowledge does not even always accumulate. A colleague of mine came back from a conference on information retrieval where one of the speakers argued (using, of course, calculations) that, for some of the knowledge unearthed by scientists of the past, it would be cheaper to rediscover it than to retrieve it. Imagine, if you will, the wave of uneasy movement that might have rip-pled through the graves of dead scientists at that argument. For some, the work to which they devoted their lives was not merely wrongheaded (this they were surely prepared for) but, even though right, was never absorbed in the body of human knowledge.

For those who do not share the conviction that bits of knowledge laid end to end lead to wisdom, the articulation of the bits becomes a challenge separate from that of unearthing them. When the knowledge in question is knowledge about human behavior, the emergent image must bear a human face. If it doesn't, the mind goes blank when we look at it, irrespective of how detailed or precise it is. So there are two tasks, really: first, the assembly of the pieces, and second, the discernment of a human face.

But both these tasks are daunting, and lie largely ahead of us. How then, we may wonder, do we live with ourselves *now*? By being, like poet Robert Frost, cheerful pessimists. By believing that change can happen, although it is some-where between difficult and impossible. By learning as much about human nature as possible, to avoid the fate of Sysiphus, backsliding again and again to the same discouraging starting point. And by leaving to the next generation the legacy of what we have learned in our struggle, in the hope that they will have better ways to deal with the stuff of the human spirit than we did or even thought of. For those who find in such goals a sufficient and admirable ideal, this book is a suitable investment.

It is also for anyone who wants to understand what we do now know about human nature. I present this new version of *The Tangled Wing* with more trepidation than I did the original, despite being far more confident in the accuracy of its content. I was much younger then, and full of the brashness of youth; I should have realized that such a book was almost impossible. But I stumbled forward, the book was well received, and it remained in print for seventeen years.

Why keep the framework of that earlier book instead of starting from scratch? First, readers valued the book for its interleaving of science and literature, and this approach is still worthwhile. Second, however young and brash, I was right about quite a few things. This is largely because I was not a true believer, on either side of the controversy. In 1976, a year after the publication of E. O. Wilson's *Sociobiology*, I presented a paper to the American Anthropological Association predicting that the new field would gain a significant but small place in the spectrum of behavioral and social science. Nobody wanted to hear that. Enthusiasts thought that conventional behavioral and social science would be subsumed or swept away; critics thought the new approach pernicious and simpleminded and vowed to stamp it out as soon as possible. Except for a few diehards, members of both camps are now embarrassed by their former claims. Most knowledgeable people agree that evolutionary psychology—as sociobiology renamed itself in a vain effort to escape opprobrium—has gained a significant but limited place in the spectrum of behavioral and social science.

I wrote *The Tangled Wing* to show what an integration might look like. Nearly two decades later, that integration looks reasonable, and shows signs of looking better in the future. I don't take credit for it; now, as then, I'm mainly summarizing and reporting on it. But in some form this integration is bearing down like a bulldozer on theology, philosophy, and much of psychology, and it has undermined key theoretical pillars of social science. Yet the bulldozer is no set of simple neodarwinian ideas. Rather, it is behavioral science itself set in a biological context. But let me explain why it is not a compromise.

It has been said that nothing in biology makes sense except in the light of evolution. It can now also be said that nothing in human behavior makes sense except in that same light. Cast it on ourselves, and we find that human nature is real, definable, and to some extent predictable. It is also quite different from what the world's cultures have thought it to be. The evidence to support these claims is so much stronger, richer, and broader now than it was two decades ago that I have had to eliminate the tentative tone of the prior book. In the heady early days of sociobiology, my mentor at Harvard, Irven DeVore, would begin his response to legitimate demands for proof of the theory, "The data are sitting up and begging." Today the data are sitting up, begging, heeling, rolling over, fetching, and playing catch.

During the 1980s, prominent critics resolutely opposed all applications of biology to behavior. Since then they have ceded large tracts of the field of action.

New medications for emotional, behavioral, and sexual problems have worked so well that few critics would deny us the choices they afford. Brain images of ever-sharpening resolution light up the substance of mind with shimmering, blotchy maps and promise to solve age-old mysteries of intelligence, talent, choice, passion, motive, joy, and suffering. New approaches to genetics, including behavior genetics, are simply not subject to the old critiques, because they have too much technical and scientific power. Sociobiology and evolutionary psychology have become the main research paradigms for thousands of professionals studying animal and human behavior, and evolution itself rests on a new scientific foundation, thanks to advances in molecular genetics, geology, and electron microscopy. Some critics continue to rail against these trends, but they appear increasingly shrill and impotent. Sociobiology has triumphed.

Still, it needs to be said that the light of evolution is just that—a means of seeing better. It is not a description of all things human, nor is it a clear prediction of what will happen next. Nevertheless, the science of biology has become ever more concrete and specific as we have moved into the Gene Age. Several species have been cloned from adult cells, and despite technical and ethical barriers, attempts at human cloning have already begun. Embryos routinely result from in vitro fertilization, and some of these are selected for genetic "superiority," or at least for the absence of known genetic defects. Embryonic stem cells are cultivated with a view toward eventually breeding replacement organs for sick or damaged humans. The genetics of aging, too, is being unraveled, as readily as the string of DNA that youth depends on. Sex selection of sperm is advancing rapidly. And of course, the human genome has been sequenced, faster than anyone expected. That project—the first Big Science effort in biology, with funding and international cooperation to match—sets the stage for aggressive pursuit of the mechanisms not only of disease but of normal function. Steps are being taken to decode the molecular biology of development—which allows a single cell, a fertilized egg, through the turning on and off of genes, to become the assemblage of trillions of organized, specialized cells that make a breathing, thinking person. Deciphering this mosaic will make the genome project look like a preschool child's jigsaw puzzle, and it will dominate twenty-first-century biology.

But geneticists have not been waiting around for the sequencing of the genome, let alone for the mosaic of development. Among other achievements, they have located and sequenced the genes for cystic fibrosis, the lung disease that used to be the most common genetic killer of children, and for Huntington's disease, a devastating brain disorder whose early symptoms often look like emotional problems. In both cases, definitive diagnosis is now possible, and we know which proteins begin the devastation. These are two of over 3,000 different, *simply* genetic diseases, any of which might be diagnosed or solved in the early twenty-first century. Diseases that are only partly genetic, or result from the action of many genes, are not being neglected. Genes affecting some breast and colon cancers are now known, and those contributing to heart disease, diabetes, obesity, and other common illnesses are under active study. And the claims go beyond disease.

Newsmagazines more than occasionally report new "genes for" happiness, novelty-seeking, homosexuality, intelligence, and other traits that have nothing to do with disease. The gate is ajar to a realm of wonders.

And dangers. More than once in the era of modern science, biology has become a weapon of abuse. In its infancy, Darwin's theory was used by some to justify colonial exploitation, racism, slavery, and sickening poverty. Genetics applied to mind and behavior produced grotesque political monsters in the first half of the twentieth century. Genocide was perpetrated to "improve" the species. Useless "monkey-gland" operations misled many men into thinking they could postpone aging. Ill-advised brain surgery, done for mental illness, damaged the mental and emotional capacities of many thousands. Crude drugs were used to control behavior, not for the benefit of those who took them, but for the convenience of those who gave them.

And the abuses are not over. Today in Communist China, thousands are sterilized each year because they are judged to have inadequate mental capacity to reproduce.In Afghanistan, women are so stigmatized by a religious gloss of their biological status that they can no longer practice any profession or even leave their homes without a man. In the United States and France, race is regularly invoked in legitimate discourse to explain why different ethnic groups perform at different levels in certain kinds of tests or tasks—differences that can easily be explained in other ways. And some scientists take very public platforms to raise doubts about the efficacy of schools, parents, health advisors, and others who attempt to shape human behavior in the face of powerful and pervasive biological forces.

Genes are not the only objects of biological study. The 1990s were heralded as "the Decade of the Brain," and the main annual brain science convention now has upwards of 10,000 scientists in attendance, all reporting new discoveries each year about the workings of this least understood, most marvelous organ. We know a lot now, but it is only a fraction of what we need to know to encourage the brain to work consistently for us—instead of sometimes for, sometimes against. Among the facts established more strongly in the 1990s is that the brain's structure responds to experience, sometimes dramatically. Also, at the right age and in the right locations, you can lose large parts of your brain and still adapt and function surprisingly well. Such facts are a tribute to the brain's plasticity and suggest vast possibilities for influencing it through education, nurturance, and other cultural strategies.

Still, as is all too clear to anyone who has watched a loved one struggle to live and recover after a stroke—even with an ideal rehabilitation environment—the brain has severe and intractable limitations. This fact, too, suggests possibilities, but unfortunate ones, for stubborn resistance to all kinds of well-meaning interventions. The favorable effects of prescription drugs on the brain, still somewhat controversial two decades ago, are universally accepted as fact, although ethical questions continue to trouble many. Depression, delusion, obsession, compulsion, overeating, addiction, and attention deficit are all treated with medicines.

Yet one need only change this list to read, sadness, imagination, conscientious-ness, diligence, appetite, habit, and restlessness to realize how very close we are to managing human nature by prescription.

In time, we will have the power to manage it by gene. Those of us who are wary of all these interventions, and of the more general biologizing of psychology, social science, and even philosophy, have a responsibility to understand what is going on. Denial is no longer an option, any more than reflexive disgust at every intervention or theory that smacks of biology. Our species is at a crux in its earthly history, aggressively mastering techniques that will let us guide and change our own nature. The biological genies—whether medicines, hormones, recreational drugs, brain cells, neural circuits, clones, or genes themselves—cannot be put back into the bottle. If you have read this far, you are among the literate elite of our precarious species, and so you are one of the people most responsible for see-ing to it that these new genies are servants to, not masters of, our destiny. This book will have done its intended work if you find it a useful tool.

But it is not an all-powerful tool. In Bertolt Brecht's play *The Life of Galileo*, the great astronomer argues with a philosopher who serenely invokes Aristotelian theory against observations just made through a new telescope. Galileo says, "Truth is the child of time, not of authority. Our ignorance is infinite, let's whittle away just one cubic millimeter. Why should we still want to be so clever when at long last we have a chance of being a little less stupid?" Late in the play he says it more clearly: "The chief cause of poverty in science is imaginary wealth. The pur-pose of science is not to open the door to infinite wisdom, but to set some limit on infinite error." That is the motto and hope of this book. For those who find it insufficient to their purposes, I cannot in good conscience recommend a further investment, since there are so many books that will open the door to infinite wis-dom at a comparable cost. But for those who share my view that to set some limit on infinite error is ambition enough, already pressing the point of human pride, I offer the first steps on an immense journey.

PART ONE

Foundations of a Science of Human Nature

Thus, from the war of nature, from famine and death, the most exalted object which we are capable of conceiving, namely, the production of the higher animals, directly follows. There is grandeur in this view of life, with its several powers, having been originally breathed into a few forms or into one; and that, whilst this planet has gone cycling on according to the fixed law of gravity, from so simple a beginning endless forms most beautiful and most wonderful have been, and are being, evolved.

—CHARLES DARWIN, *The Origin of Species*, 1859

What a book a Devil's Chaplain might write on the clumsy, wasteful, blundering low and horribly cruel works of nature!

—CHARLES DARWIN to Joseph Hooker, July 1856

CHAPTER 1

The Quest for the Natural

And the Lord planted a garden eastward in Eden; and there he put the man whom he had formed.

—Genesis 2:8

Rousseau was not the first, nor even the most naïve. But he was the most famous in a line of credulous people, stretching as far back as thought and as far forward as our precarious species manages to survive, who believe that what we have left behind is better than what we have, and that the best way to solve our problems is to go backward as quickly as possible. In this view, what is past is natural, what is present, unnatural—as if the march of history, with its spreading plague of gadgets, had somehow distanced us from the bodies we inhabit, from the lives we live, biologically, each day.

The nostalgia is undiscriminating. The naïve romantics of an era may look back just a few decades to find their Eden, little realizing that the romantics of that era also looked back, and so on, in infinite regress. Rousseau himself fell in love with the Swiss peasantry, both in the person of his wife and in the object of his philosophy. He saw in their atomized social (asocial?) life the road to happiness. Society was the root of evil, and in this "natural" peasant condition of supposedly complete self-sufficiency, the human atoms avoided collision by avoiding contact—friction being impossible in a vacuum. Each person, or at least each family, kept to its own farm and went about the work of survival with efficient aplomb, neither asking nor giving anything, without snobbery and, of course, without gadgets.

If this sounds more like the dream of a bookish philosopher than the social life of peasants, it is. No known society has ever been as unsociable as Rousseau's peasants, and the absence of ethnographic substance in his descriptions makes us surmise that they are not in fact descriptions but imaginings. In fairness, Rousseau was simply trying to define human nature. He surmised correctly that the distant past informs us about our nature, and he saw among Swiss farmers a profound contrast with his own life as a scion of the upper bourgeoisie. But he didn't go nearly far enough. More ironic than his failure to *see* peasant social life is that he should settle on it as the natural, original human condition. For if such a thing as the natural, original human condition exists, peasant society is as much a distortion of it, and almost as much of a novelty, as industrial urban life itself.

So what kind of society does fill the bill? Candidates abound, but all the serious contenders have this in common: they gather wild plant foods and hunt wild game for a living. No other sorts of lifestyles are admissible, because the archaeological record speaks so plainly: for 95 or 99 percent of human generations on this planet—depending on what you want to call human—our ancestors lived in this way alone. If we could write a true "book of begats" to replace the biblical generations of Adam, the generational wheel would turn forty thousand times—hunters begetting hunters, gatherers, gatherers—before the first Abel or Cain appeared to plant a seed or husband a tamed animal. A million years would have gone by before a few begetters would begin to scrape the earth with a view toward planting. Though our 10,000 years of agriculture seems long to us, it is only a minute in the hour of human life, a moment in the history of mammals. Imagine, then, an archaeologist standing in a trench a million years from now, looking at the layers of human history, like a geologist eyeing the ancient, successive stripes in the wall of the Grand Canyon. There would be a great, thick stripe holding the record of hunting and gathering, and on top of it an equal-sized layer for industrial life. Only on closer inspection would there appear a slender lens signifying peasant agriculture, a mere ten-millennial transition.

Naturally, anthropologists absorbing these archaeological facts wanted to document hunter-gatherer life before its inevitable disappearance. This task joined biological and social anthropology intrinsically for the first time. Hunting and gathering was the crucible in which natural selection pounded at the grist for the human spirit as well as the body. So it is important to those of us who presume to know the species to study the forces that formed it.

There were many. Heat and cold, torrential rain and arid sand, tropical forest and grassy plain, windblown tundra and balmy lakeside, crowded valleys and open, sparsely populated spaces, hunger and plenty, fierce predators in some places and pussycats in others, war and vendetta alternating with *Pax Paleolithica*—all these were tossed into the crucible. Rick Potts of the Smithsonian Institution has shown that the climate of East Africa alone during the time we were evolving changed so often that the creatures slowly becoming us must have been designed mainly for adaptability to those changes, not for any particular climate or mix of plants and animals.

Excellent studies of hunters and gatherers in varied settings—the Inuit and other Eskimo groups of the Arctic Circle, the Ache of the Paraguayan forest, the Hadza of East Africa's Serengeti Plain, the Walbiri and other aboriginal peoples of Australia's outback, and even complex hunter-gatherer societies like the Northwest Coast Native Americans—all are considered possible models, exemplifying the lives *some* of our ancestors might have lived. The Hadza, for instance, hunted with large, powerful bows and arrows, gambled a lot as a pastime, made their children gather some of their own food, and gave birth less than three years apart. The Ache got most of their food from hunting, staged regular ritualized club fights between men of different bands, and at one time practiced ritual child sacrifice. There is every reason to believe that such idiosyncrasies of adaptation and culture appeared and disappeared again and again during our long and uncertain evolution. We are a very adaptable animal.

So no one hunter-gatherer group can encompass the social and ecological patterns of our Paleolithic ancestors. Yet generalizations are still possible. One source of some of them is the Kalahari San, specifically the !Kung San. Made famous in the 1950s by the Marshall family—who have studied and filmed them brilliantly for half a century—they went on to become one of the most extensively studied of all simple societies. Interdisciplinary fieldwork, led by Richard Lee and Irven DeVore, reached a pinnacle in this setting, as the San weathered wave after wave of research, persuaded that their own worth was ample cause for scrutiny.

In the end they offered up enough of their essence—and perhaps, that of life as our ancestors lived it for over a million years—to challenge and then shatter one myth after another. *The Harmless People* is what one evocative book called them. And so they are, harmless to oppose the advance of a technical civilization, dwarfed by the might of surrounding peoples, both black and white, who bore down on them for centuries. During the white colonization of southern Africa the San were massacred for convenience and hunted for sport. They resisted bravely but with little chance of success, ceding territory everywhere. At last, they remained as hunters and gatherers only in far pockets of land unwanted by others, and that is where ethnographers of the twentieth century met them. They described the traditional San as living in organic harmony with nature and with one another. The San's knowledge of wild plants and animals is thorough enough to surprise and inform botanists and zoologists. Their commerce with the wild earns their living, but their knowledge goes far beyond necessity. Necessity itself seems harmonious. Even though women work just three days a week at gathering nuts, fruits, and vegetables in the wild, this work provides three-fourths of the food their families need.

The wide-world-as-supermarket also provides fresh meat on the hoof, in the form of some of Africa's most regal game. Eland, oryx, kudu, wildebeest, duiker, steenbok, even giraffe fall prey to San hunters. The animals, such as oryx and warthog, that will stand and fight are hunted with dogs and spears by small groups of men. The rest are hunted by one or two men with small bows and slender, deadly arrows poisoned with toxins made from the thorax of the larva of a certain beetle found in the sandy soil among the roots of a certain bush, in a certain season of the year. How the intricacies of this process became known to the San is only one among many mysteries.

Men pursue game about as intensely as women gather vegetables; three to four days a week seems enough. They kill only what they eat, and have maintained a balance with game populations for millennia. They have a great respect for and fascination with the animals they prey upon, sometimes to the detriment of their proper business of killing—as with the man who came upon an oryx pair *in flagrante* and was so transfixed he forgot to shoot them. And while San men are far from foolhardy, extraordinary physical courage becomes ordinary for them Some carry scars of hand combat with leopards; others have met a grim end in the jaws of a lion or hyena.

Women, too, have their ordeal of physical courage: childbirth is supposed to be managed by a woman alone. The labor usually begins at night and she will often simply go out beyond the edge of the village-camp—that anthropological twilight zone between culture and nature. There she will deliver by herself, under a broad dome of stars and surrounded by any number of unknown dangers, cut the cord, and return quietly to her hut. Often the first inkling of her confinement will reach her friends and relatives in the harsh, awkward notes of her baby's cry.

This deep dwelling in nature is mirrored by the harmony of the human world. Each village-camp is a small circle of huts; each hut holds a family in a hemisphere of grass just large enough to lie down in. The camp includes perhaps thirty people but is flexible. It moves with the changes of the seasons, the vagaries of food and water availability. It may move because someone has died there, making the place unpleasant to stay in, even though the grave is some distance away. People leave groups from time to time—there are other bands and camps to join. Conflicts are often resolved by group fission. The fragments may coalesce again months later, bygones forgotten, or they may form the nuclei of nascent bands.

War is unknown to them. Before they cause a split, conflicts are often resolved by sharing food and other goods, and by talking, sometimes half or all the night, sometimes for weeks on end. After two years with the San, I came to think of the Pleistocene epoch of human history (the 3,000,000 years during which we evolved) as one interminable encounter group. When we slept in a grass hut in one of their villages, there were many nights when its flimsy walls leaked charged exchanges from the circle around the fire, frank expressions of feeling and contention beginning when the dusk fires were lit and running on until the dawn. A potent selection pressure for the evolution of language must have been the respect gained by one who commanded attention around the fires.

Equality is strongly valued. Few social or economic distinctions are known, nor could they be maintained, since the powerful ethic of sharing separates a person from any accumulated wealth as soon as it becomes visible. Stinginess is the chief sin, punished by social ostracism; where mutual aid is key, no one can tolerate ostracism for long. Only intractable violence is more repugnant to San than selfishness, and the former is so strange it is classified more as mental disorder than sin. The smallest guinea hen or tortoise brought back from the chase will be split into pieces as it comes out of the fire, and an infant in arms, its fist closed around its share, halfway to its mouth, will receive a first crucial lesson, as an adult reaches toward the morsel and speaks slowly and clearly: "Give."

This ethic of sharing, mutual dependence, and organic social harmony is nowhere more vividly expressed than in the trance-dance ritual, an impassioned drama of healing. Here the line between sacred and social blurs and the people test their unity because only unity can engender the energy needed by healers who have to step into death to importune the gods. As Lorna Marshall, the classic ethnographer of the !Kung, put it in her 1999 book, "People bind together subjectively against external forces of evil. . . . [W]hatever the state of their feelings, whether they are on good terms or bad terms with each other, they become a unit, singing, clapping, moving together in an extraordinary unison of stamping feet and clapping hands, swept along by the music. No words divide them; they act in concert for their spiritual and physical good and do something together that enlivens them and gives them pleasure."

The patient, feverish with malaria or pneumonia, lies half-conscious by the fire. All the friends she has in the world are near her and are needed. The women sit in a circle around the fire, preparing to sing the songs that will alter consciousness, to clap the complex rhythms that will change the shape of time. Behind them, walking in a circle, the men begin to dance, getting the feel of their feet, adjusting the dance rattles around their legs. Marjorie Shostak, writing about a 1989 trance-dance she participated in, described it this way:

> The quarter moon drops toward the horizon, and the stars brighten as it descends. The sounds of a healing dance flood the air. Complex clapped rhythms drive the women's songs, fragments of undulating and overlapping melodies. Each woman tilts her head toward her shoulder, trapping sound near her ear, the better to hear her part. The women's knees and legs, loosely describing a circle, fall carelessly against one another—an intertwining of bodies and song.
>
> In the center of the circle, a fire flares as it is stoked and whipped by human breath, soon to ebb again into glowing coals. Beyond the circle, men and boys-almost-men, their taut upper bodies hardly visible in the darkness, pound the cool sand with bare feet, blending new rhythms into the song. So forcefully do they dance that a deep circle forms in the sand beneath them, enclosing us, separating us from the profane, protecting us from the unknown.

Will the women sing and clap well enough to inspire them? Will the men dance well enough to give the women confidence? Will the women stay together, hear one another, find their harmony? Above all, will the difficult whole be whole enough so that some of the men, on the strength of it, can drop into trance, into half-death, traverse the road to the other world, and berate the gods for causing human illness? So that returning, they can lay hands on the sick person and, shrieking and shaking for dear life, drag the cause of the illness out of her? Nowhere more than in this ritual do the San symbolize to themselves how life and death hang upon mutuality. In the Tsodilo Hills of northern Botswana, in sight of exquisite rock paintings made by San in centuries past, there is a dance-circle etched on a smooth plateau of rock, made by countless generations of steadily dancing feet. This ritual has been a spiritual mainstay since time immemorial.

Finally, they have a rich oral tradition of stories that help them order the world for themselves and their children, a mythic philosophy of life that extends from hunting to sex, from human and animal origins to pregnancy and birth. This rich lode of folk narratives, mined and preserved by Lorna Marshall and Megan Biesele, makes a strange amalgam of sacred story, soap opera, and situation comedy that lets people know who they are and what they should do—just as the book of Genesis helps to do for many Westerners and Classical myths did for the ancient Greeks.

According to Hobbes, that dismal philosopher—based on as little evidence as Rousseau's opposite fantasy—the life of primitive people was "solitary, poor, nasty, brutish and short." Yet, in the San, we see the most basic of human societies operating as it must have for millennia. Archaeological evidence shows hunter-gatherers living around the San's present village locations for at least 11,000 years. Ancient hunting blinds have been unearthed beside seasonal pans where modern San still come and lie in wait for game. They had had contact with farmers and cattle people for centuries, yet until late in the twentieth century many still chose to hunt and gather for a living.

Plainly theirs is a life with a proven viability, much more ancient than our own. Far from solitary, it is above all mutual. Far from poor, it is adequately supplied and leisured; it has even been called "the original affluent society." Far from nasty, it is based on human decency, respect for others, sharing, and giving. Far from brutish, it is courageous, egalitarian, good-humored, philosophical—in a word, civilized—with an esthetic so fine its very music touches the gods. Although many die in early childhood, those who do not may live to a ripe old age, an old age not consigned to a ghetto or to a "home" that is not a home; rather, it is one embedded in that same intimate social world, surrounded by grandchildren full of delight, by grown, powerful children full of courtesy.

Now for the bad news.

First, there is no going back. Throughout the world the hunting-and-gathering way of life supports at most one person per square mile. Since the total land area of the planet is under 200,000,000 square miles, the human population turned its back on the hunting-and-gathering possibility around six doublings ago. We are committed to and dependent on technology.

Aside from being irrecoverable, some features of San life, and of hunting-and-gathering life in general, have been overlooked in a wave of romantic enthusiasm. For example, homicide rates are comparable to those in American cities. We hear much about San egalitarianism, but what of the accounts of wife beating and the obvious sexist ideology evident in all-male discussions? We talk about San sharing, but in fact, almost all sharing occurs among close relatives, and modest forms of status and prestige do exist. We admire the absence of war; but the San lack the manpower to mount even the simplest war, and when they talk about other people, even other groups of San, they are not above prejudice. If they could make a war, perhaps they would, and they may have done it in the past. We speak of their "affluence" without mentioning shortages, in which they may lose 10 percent of their body weight; of their kindness to children without mentioning illness, which kills half before they grow up; of their decency to the aged, without mentioning that only 20 percent of infants can expect to live to be sixty, leaving San society with an incomparably lighter burden than the elderly and infirm constitute in our own.

We should not be surprised to find that certain ethnological idylls are ill-founded. In 1748 the philosopher David Hume wrote, in *An Enquiry Concerning Human Understanding*, "Should a traveller, returning from a far country, bring us an account of men . . . wholly different from any with whom we were ever acquainted . . . who were entirely divested of avarice, ambition, or revenge; who knew no pleasure but friendship, generosity, and public spirit; we should immediately, from these circumstances, detect the falsehood, and prove him a liar, with the same certainty as if he had stuffed his narration with stories of centaurs and dragons, miracles and prodigies." Yet still ahead were two and a half centuries of anthropological writing that often succumbed to the temptation of romanticizing. Not only the !Kung and other hunter-gatherers but the Samoan islanders of the South Pacific, the Balinese, the Zuni Indians of the American Southwest, and even our nonhuman relatives the chimpanzees and bonobos have all served as canvases on which ethnographers and other travelers have sometimes projected their fondest wish—that somewhere, sometime, there must be (or have been) a really, truly *good* society.

The great French essayist Michel de Montaigne indulged in this kind of enthusiasm in 1580, after reading Jesuit accounts of the "wholly noble and generous" Brazilian Indians:

> [T]here is no sort of traffic, no knowledge of letters, no science of numbers, no name for a magistrate or for political superiority, no custom of servitude, no riches or poverty, no contracts, no successions, no partitions, no occupations but leisure ones, no care for any but common kinship, no clothes, no agriculture, no metal, no use of wine or wheat. The very words that signify lying, treachery, dissimulation, avarice, envy, belittling, pardon— unheard of.

Except for the last sentence (silly) and the bit about no occupations (didn't they have to eat?), most of these claims were true. Yet strangely enough, Montaigne thought that these same people were cannibals. Whether they were or not, we know from subsequent studies that they did indeed have lying, treachery, dissimulation, avarice, envy, belittling, and pardon and in addition probably had words for them as precise as our own. Further, we know in retrospect that riches, political superiority, servitude, partitions, and addictions were all present in incipient form, in effect just waiting for a chance to show themselves. In fact, hunter-gatherers of British Columbia and Washington—the Northwest Coast Indians—had a lush resource base in shellfish and salmon, and as their populations became denser, increasing competition and allowing subordination, they became fiercely hierarchical. Overall, recent research on primitive societies—Robert Edgerton's *Sick Societies*, or Lawrence Keeley's *War Before Civilization*, for example—have strongly confirmed Hume's intuition, not Montaigne's.

What is the truth, then? That the emperor has no clothes? That the idyllic picture of simple societies is just a Rousseauan ruse? Not quite. But the positive features of hunter-gatherer life have been oversold, and the negative features of our own life blamed too much on historical changes that took us away from our ancestors' way of life. The emperor has clothes, but the clothes show rents and tears. Through this far from perfect suit we glimpse our naked selves. Yet there is something in the cut of it that we would like to get, if only we could figure out just how it fits. We cannot step out of our own clothes into those, and yet . . .

In the pages that follow we will unceremoniously remove our new clothes, and the garments of other, older cultures as well. We will become intimate with the animal inside, and we will see how what is natural to that animal might be turned to our own benefit. But the way will not be obvious. When Picasso and others illustrated imagist poet Guillaume Apollinaire's play *Les Mamelles de Tirésias*, the poet wrote an insightful introduction. Here in the drawings were strange new shapes and views, and yet, to Apollinaire, they somehow revealed more of nature than had ever before emerged in two dimensions. He wrote, "I thought that one should return to nature itself, but not by imitating it in the manner of photographs. When man wanted to imitate walking, he invented the wheel, which does not resemble a leg."

This is an apt guide to those of us who want to learn from "primitive" societies. Perhaps we *should* return to nature, and for our kind—*Homo sapiens*—this must mean in some sense a return to life as it was when we were evolving. What better proposal do we have for muting the social chaos swirling around our hapless planet? But if we return, it should be to learn, not to mimic. The direct study of hunter-gatherers does not and cannot mean mere imitation of them, but it *can* lead to insights that must figure in the design of a workable world. We could see that an ethic of sharing is not unnatural and try to revive it even though we dwell among people not our relatives. We could try to relinquish our fierce hold on things and seek a better hold on people. We could give pride of place to social, not just economic, values. We could conclude that war is an absurd way to try to resolve conflicts. And we could search for some ritual, some sacred social symbol, that so transcends the field of conflict as to release us from our mundane hatreds and so, finally, to heal. But if we can make a workable social world (and it is not clear that we can), it will resemble hunter-gatherer life as little, or as much, as a wheel resembles a leg.

CHAPTER 2

Adaptation

It should by now be obvious that there is, indeed, a general theory of behavior and that the theory is evolution, to just the same extent and in almost exactly the same ways that evolution is the general theory of morphology.

—ANNE ROE AND G. G. SIMPSON, *Behavior and Evolution*

No one in the behavioral sciences today doubts that biology is making a profound impact on those fields. Our view of behavior, by the mid-twenty-first century, will be as changed as physics was after the early twentieth. Modern physicists did not invalidate Newton; they merely found the edge of Newton's universe. Beyond it, in the macro- or microcosm, their observations demanded new laws, and they invented them. Back in the middling universe, among dropping apples and banging billiard balls, Newton still served, but it was clear that even his world would never again look quite the same.

A parallel change in how we see human behavior is happening now. Biology is not necessarily wiping the old views away; in fact, it enhances many of them. But it is opening new frontiers where prebiological laws are useless, and as we chart these unknown reaches and turn to gaze back, the human spirit will never look the same. Biology has encroached on behavioral science along two broad frontiers: evolution and genetics on the one hand, anatomy and physiology on the other. These two frontiers have a natural bridge in the science of embryology, and the connection, developmental genetics, is the main task of twenty-first-century biology. Yet at every stage in the life of every organism a bridge already exists. That bridge is adaptation.

Adaptation is the process by which a creature adjusts to the world around it, including that of other creatures, hostile and distant or near and dear. This process includes both physical and behavioral change. For instance, if we insist on running around the park every evening, we feel gradual changes in the body's shape and weight. We know, too, that there are changes in coordinated patterns of nerve and muscle and in the balance of chemical pathways. Finally, there is a mental change: what was painful and exhausting becomes pleasurable though exhausting, partly because of changes in joint, muscle, and bone, partly because of endorphin release, and partly because we have gotten into the habit.

All such changes are deeply biological, yet they are set in motion by a choice. We choose to run around the park in the evenings, starting a causal chain in which behavior and biology act on each other until body and mind are both in a new set. "That's culture, not biology," says a traditional observer. "It shows you can do anything you and your culture decide." This is wrong on two counts. First, running is the most natural form of vigorous human exercise. Try gymnastics, pole-vaulting, or dancing en pointe to see the sorts of biological obstacles that can confront a human decision. Second—and this is the reason for the first—we are designed by nature to respond to daily running with just such changes. Actually, designed by evolution, or, more precisely, by natural selection, since evolution is mainly the result of such selection.

Darwin's 1859 book on the idea—he and Alfred Russel Wallace had outlined it concurrently a year earlier—set off a century-long stir, but not because *evolution* was new. Epicurus, the Roman poet, had seen it at least dimly. Goethe had studied it and Darwin's own grandfather had written an epic poem about it. Charles Lyell's masterpiece, *Principles of Geology*, describing the fossil record of evolution, was in Darwin's trunk when he embarked on the H.M.S. *Beagle* for his grand youthful voyage around the world.

The idea had been in the air, accepted by many despite vigorous opposition, for more than 100 years. Darwin's contribution, aside from a grand synthesis of the known relevant facts, was to set out in clear language the first convincing theory of how evolution could work: the theory of adaptation through natural selection. As a recent reviewer put it, Darwin's argument was "as simple and lifeless as a crystal." That is, it contained no mysticism or divine intervention, not even any mild to moderate mystery of life. It was just the story of life rewritten in the language of physical and chemical causes and effects.

It goes something like this. Certain elements of heredity, later called *genes* and now known to be mainly DNA, became organized from simpler chemicals early in Earth's history. They are ordered, and so should have been degraded, in accord with the second law of thermodynamics, which decrees a relentless progression toward disorder. But these complex molecules also contain information about the environment that helps them resist dissipation. Slowly they organized other molecules around themselves as a bigger and better shield against disorder. This higher level of organization we call an *organism*, but it is basically a gene's way of making another gene. So organisms tend to do what genes direct them to do, which tends to perpetuate those genes. This is always for one simple reason: the genes that couldn't get the clumsy biological machinery they rode around in to perpetuate them aren't around for us to observe.

The idea of natural selection was first fully presented by Charles Darwin in the 1859 book entitled, *On the Origin of Species by Means of Natural Selection or the Preservation of Favoured Races in the Struggle for Life*. Yet the title was misleading. By the custom of the age, it was long, with room for three misrepresentations, at least two of which Darwin was aware, as he showed in the book itself. He might not have realized that he was not solving the problem of the origin of species. This highly specific knot in evolutionary biology was not to be untied until Ernst Mayr's *Animal Species and Evolution* was published just over a century later.

But Darwin's title contained two other misleading phrases. First, it wrongly implied that evolution works through competition among races. This viewpoint has been discredited by modern science, replaced by the view that evolution works mainly through competition among individuals and their kin. The selection or elimination of larger population units is real but minor. The concept of group selection has many difficulties, the most embarrassing being that no one has ever observed an animal group that functioned so cohesively that in-group competition was successfully suppressed. In fact, despite his title, Darwin was no group selectionist and he emphasized individual selection throughout his life. The idea recurs so frequently, but recently it has been on the defensive since 1966, when George C. Williams published his classic *Adaptation and Natural Selection*. It baldly proclaimed that every creature is engaged in a constant struggle with every other creature in its environment, however related, however closely allied. A third of a century later, group selection has had something of a resurgence. It is mathematically plausible under certain conditions—as even Williams conceded in his 1992 book, *Natural Selection*—but it is of only minor importance in the evolution of complex animals.

The third error in Darwin's title is the catch-phrase "struggle for life." The goal of the struggle was never life, but reproductive success. This would not be news to Darwin, since he himself discovered sexual selection. The idea is basically this: while being better adapted for survival is certainly one way for an organism to be favored—individuals dying before breeding cannot ordinarily perpetuate their features—it is not the only way. Visualize a population in which all individuals die at the same age; evolution could still proceed rapidly through different rates of breeding. Imagine, for instance, an all-blonde population—how boring. But—mutation to the rescue—a brunette appears. Although he (like all the blondes) dies on his fiftieth birthday, he is much in demand as a breeder, and he leaves ten brats behind to the blondes' average of five. Keep that up for a few generations, and you'll be as bored with brunettes as you once were with blondes. But now imagine that our first brunette only lives till forty, leaving, say, eight offspring. He is so busy breeding that he burns himself out, yet he passes on his prodigal characteristics (along with his cute dark hair) more effectively than others do their calm, tempered, healthy ones.

Recognizing these possibilities in 1859, Darwin wrote "a few words on what I call Sexual Selection," words that echo to this day:

> This depends, not on a struggle for existence, but on a struggle between the males for possession of the females; the result is not death to the unsuccessful competitor, but few or no offspring. . . . Sexual selection by always allowing

the victor to breed might surely give indomitable courage, length to the spur, and strength to the wing to strike in the spurred leg, as well as the brutal cockfighter, who knows well that he can improve his breed by careful selection of the best cocks.

Amongst birds, the contest is often of a more peaceful character. All those who have attended to the subject, believe that there is the severest rivalry between the males of many species to attract by singing the females. . . .

Thus it is . . . that when the males and females of any animal have the same general habits of life, but differ in structure, colour, or ornament, such differences have been mainly caused by sexual selection; that is, individual males have had, in successive generations, some slight advantage over other males, in their weapons, means of defence, or charms; and have transmitted these advantages to their male offspring.

Thus, Darwin introduced not only sexual selection but the whole subject of behavioral evolution, and while he was at it he fielded a theory of sex differences.

The theory had problems, to be sure. For one thing, it was male-centered. There are species in which females fight over males—phalaropes, for instance, and jacanas, also known as "Jesus birds" because when they scoot over lily pads they seem to walk on water. Sexual selection in those species has made the female more spectacular than the male. But this only strengthens Darwin's point, while taking away its sexist sting, as he realized in his later writings on sexual selection. Moreover, even in species where male competition rules, there is still competition *for* males *among* females. Finally, although males usually vary more in reproductive success, females vary, too. Those that raise the most offspring to maturity, by means of, say, more efficient care, or by enlisting male help, will win in reproductive competition.

This argument puts us—as it indeed put Darwin—squarely in the realm of sociobiology. While some in the field prefer the term "evolutionary psychology," this narrows the subject matter too much. Sociobiology, as defined by Edward O. Wilson in his book of that title, subsumes evolutionary psychology, Darwinian anthropology, some aspects of behavioral ecology, and other paradigms. Despite this breadth, sociobiology has a quite specific meaning: the use of natural selection to explain reproduction. It attempts to see how far we can get with the idea that the goal of life is to maximize reproductive success. It also deals with structure, but is most useful with structures involved in mate selection and reproduction. The Darwin passage quoted above is a prime example—a century before its time— of sociobiological reasoning.

But here is what sociobiology is not: it is not the whole study of animal behavior, even behavior in the wild, nor does it look at all the effects of social life on biology. It is not an all-encompassing theory of animal and human nature, and it is not a pernicious, cynical, reactionary pack of lies about animal and human nature. Furthermore, it is not the same as behavioral biology, the subject of this book. The vast majority of behavioral biologists—ethologists, neuroethologists, psychobiologists, behavioral geneticists, neuropsychologists, behavioral neurologists, behavioral endocrinologists, biological psychiatrists, comparative psychologists, biological anthropologists, psychopharmacologists, and others not even anchored yet in the sea of subfield gobbledygook—would not call themselves sociobiologists.

G. G. Simpson, one of the twentieth century's greatest evolutionists, was perhaps a herald of the movement when he coauthored the claim at the head of this chapter. The best theory of behavior *also* Darwin's? It's a sweeping claim, but as much a restriction as an extension. When we ask, *Why does the male peacock have such extravagant feathers?* it is not very satisfying to talk about how they grow, or detail the chemistry of the pigments that make them blue or green, or even to describe their fractal branching patterns. Each of these lines of reasoning has its own fascination, but such structural and chemical answers leave the biggest puzzle nagging, and it is not until we say, *For eons peacock females thought they were pretty spiffy*, that we begin to feel the question being answered. In this case the ultimate cause, or teleological question—the *why* question—is too intriguing to let go of. But not every question in biology leads to the same balance of explanatory power.

Consider the question, *Why do long nerves in our bodies tend to have fatty sheaths?* The functional answer is, to make conduction faster and more efficient—obviously an advantage for any animal. Since the average sixth-grader can figure this out, it doesn't exactly compel our attention. But what mechanism *makes* conduction faster? Here the physics and chemistry of nerve pulses provide impressive answers, and if we listen for a few minutes to someone who knows the subject well enough to explain it in simple terms, we will see nerve-fiber function falling into place among the laws of high school science—a lovely illumination. Large-bore tubes bear faster flows; electricity jumps across small patches of insulation, charged particles try to flow down gradients, and so on. A naïve evolutionist, who may not have been listening very carefully, will now say, "I knew the reason all along. The nerves work better that way, otherwise they wouldn't have been selected for."

If we have not shared his parochial training, we are puzzled to what he thinks he knows. We know that we have come up against one of the innumerable questions in animal function that yields little to evolutionary theory and calls out for mechanistic explanation. It isn't that what the adaptationist thinks isn't true, but that it is true of everything, which is why it sometimes offers us so little. It's the cases like peacock fantails that cry out for adaptive explanation, since at first glance they seem so *non*functional.

Evolutionary biologists—including Darwin—rarely define adaptation, leaving us to surmise what they mean, as in this sentence from *The Origin of Species*:

Whatever the cause may be of each slight difference in the offspring from their parents—and a cause for each must exist—it is the steady accumulation, through natural selection, of such differences, when beneficial to the individual, that gives rise to all the more important modifications of structure, by which the innumerable beings on the face of this earth are enabled to struggle with each other, and the best adapted to survive.

The passage—strangely doughy for writerly Darwin—sets out the three key elements of the theory. For living things to evolve, they must have, first, a way of inheriting some features from their parents. Second, there must be a source of natural variation, renewable in every generation—grist for selection's mill. And third, some individuals must be more successful than others in giving rise to future generations. These are said to be the best adapted.

Fine. But in retrospect they are just more effective ancestors. When we define adaptation as the ability to procreate more successfully, and then go on to deduce that adaptation exists in every organism we observe in the natural world, we have indulged in circular reasoning. If we know a little about logic, we begin to wonder whether saying that the nerve fibers are adapted says anything at all about the world.

Stephen Jay Gould and Richard Lewontin, both distinguished evolutionists, have long claimed that it does not. Spurred by sociobiological excess, which seemed to see in every bump and blemish on an animal a cleverly concocted adaptation—the best, in this best of all possible worlds—they have argued that the concept of adaptation is just a one-size-fits-all garment. If it fits, wear it, the ultra-adaptationists seem to say; and if it doesn't fit, try it on a different way, stretch it, squash it, make it fit. It has to. It's the law of evolution: if it evolved, it must be an adaptation.

This kind of reasoning has produced some odd-sounding hypotheses: that women's breasts evolved to mimic their behinds, which our male apelike ancestors were already quite pleased with; that physical handicap is a fine thing for sexual competition, since it makes your particular heartthrob at least take notice of you, while proving you had the right stuff to survive despite the handicap; and that infertility—even permanent infertility—can be highly adaptive in reproductive competition if it's the wrong time and place to have offspring.

These theories may seem implausible, but they are taken seriously and have worked in mathematical models. Although devilishly difficult to prove in nature, they are also hard to dismiss. An organism has a trait; it must have been selected for, or it wouldn't be here now; and it must be adaptive, or it wouldn't have been selected for—a completed derivation, just like in Euclid. Now think up an adaptative function. But there's the rub: it *is* like Euclid—a logico-deductive proof that follows necessarily from certain assumptions. Just as the Pythagorean theorem follows from the definitions of a straight line, an angle, and a perpendicular, so the adaptationist theorem follows from the definitions of variation, inheritance, and natural selection. In a strict philosophical sense, neither theorem tells us anything about the real world, because either can be arrived at without reference to that world. Neither gives us real empirical knowledge.

Still, every day, carpenters and bricklayers go out into the world (or at least they long did) carrying in their pockets a three-four-five string—a string knotted to make lengths equal to three, four, and five units, a tool for producing a right angle. This bit of worldly activity uses the Pythagorean theorem, which holds that if a square on the longest side of a triangle is equal to the sum of the squares on the other two sides, it must contain a right angle—a theorem that, strictly speaking, says nothing about the world.

In the everyday life of the naturalist or anthropologist, the theorem that creatures are adapted to their environments has the same seat-of-the-pants usefulness. When I was investigating the neurological status of newborn infants among the !Kung San, I wanted to know whether these infants—born to parents very different from us in many ways—would exhibit the same basic nervous-system reflexes observed in our newborns. Contrary to previous claims about African infants, they did. But I was also interested, and only a bit more idly, in the possible functional value of the reflexes. For example, a brand new baby, fresh from its taxing journey, will, if held upright and carried, turn its eyes and head in the direction of movement, as if looking where it is going. If held against the body, it will push off with both hands, lift its head away, even make crawling movements. If support is released from its head, it may grab hold as if trying to keep from falling. In pediatrics these reflexes are described as signs of an immature nervous system (they are) and as vestiges of an evolutionary past when some of them may have had some function.

In the culture of the !Kung they were functioning still, because it was one of the many cultures in which the smallest infants are carried around all day in a sling. Perhaps in our advanced industrial state, where infants loll in cribs all day, the reflexes are useless; but where I was studying them—in context, used by the infant to adjust itself in the sling, to get comfortable, perhaps to avoid smothering—they had the look of adaptations. At a minimum, natural selection didn't need to get rid of them. Now, I wouldn't want my reputation to stand or fall on whether they are or are not adaptations. It is easy to argue that the reflexes are just signs of immaturity, which evolved for other reasons—that they have no function except to enlighten pediatricians as to the status of the nervous system at birth. But I spent two years watching infants in that African society, and the concept seemed to fit—to be useful, like the three-four-five string. In some cases this sort of reasoning seems more useful, in others, less.

But this adaptationist puzzle-solving is not a simple-minded exercise; there are criteria for judging usefulness, some of which are rigorous and quantitative. For instance, in the Puerto Rican frog, fathers guard eggs; remove the male and you reduce the number of eggs hatched into tadpoles, just in proportion to the guarding time lost. In sage grouse, at the annual breeding grounds, only a few males get sex, and the specifics of courtship displays predict who those lucky few will be. Among vervet monkeys, grandmothers often reside with their grown daughters; remove the grandmother, and the daughter will have fewer offspring and be a more likely target of attack. These kinds of studies measure outcomes, and so they provide good evidence that fathering, displaying, and grandmothering are adaptive in these species.

But adaptation is not always testable, and in that event either those who know the organism in question find the concept useful or they don't. Gould has often pointed out that not all features of organisms are adaptive. Some are simply the structural by-products of building a body for other purposes. Gould and Lewontin call these by-products *spandrels,* named for a nonfunctional architectural feature of cathedrals. A favorite biological example is nipples, which obviously serve a purpose for females but may be a mere structural by-product—a kind of spandrel—for males. Spandrels may later serve a function in some descendants of their builders, but natural selection didn't so much create them as it just left them lying around.

The kernel of truth here is that life is a crude mess. Natural selection, even with the best will (or greatest power) in the world, cannot hammer and weld that mess into a perfect machine. In fact, one reason evolutionists of all stripes are doubtful about God's design is that only a very inept creator god would have built such ugly, unreliable, and inefficient machinery. As anthropologist Irven DeVore pointed out at a 1999 conference on "Cosmic Questions," "If 97 percent of all creatures have gone extinct, some plan isn't working very well." Or we could follow historian and philosopher Isaiah Berlin in quoting Immanuel Kant, who said that of the crooked timber of humanity nothing straight was ever made. Applying this dictum to life itself, we could say that of the viscous goop of protoplasm nothing sharp was ever made.

Of course, some fairly pointy things have been made out of the viscous goop, and these suggest that selection has been a pretty serious shaper. Darwinians rightly say that if we don't look for adaptations, we won't find them, so why not err on the side of interpretation? In the end, the feuds get annoying, and we might be tempted to put a plague on both their houses, except that both are full of excellent scientists. Couldn't we stop the squabbling? Natural selection is the theory of behavior to the same extent and in the same ways that it is the theory of morphology; that is, not entirely. Neutral mutations hang around; developmental constraints slow evolution with organismal inertia; compromises are made between opposing semi-successful features; adaptations for one purpose get used for another; and spandrels probably abound.

Darwin himself wrote in *The Descent of Man* "that in my earlier editions of my 'Origin of Species' I probably attributed too much to the action of natural selection or the survival of the fittest. I have altered the fifth edition of the Origin. . . . I had not formerly sufficiently considered the existence of many structures which appear to be, as far as we can judge, neither beneficial nor injurious; and this I believe to be one of the greatest oversights as yet detected in my work." Darwin himself rejected ultra-Darwinism; he was no Panglossian adaptationist, and modern critics have underestimated his open-mindedness and profound insight into all aspects of evolution.

As for modern adaptationism, it is not an evil, ultraorthodox, neo-Darwinian plot. It is an effort, usually honest if sometimes too keen, to see how far the theory can go. In the late twentieth century many investigators—especially field biologists—

have gone farther with it than anyone has before, as Darwinian theory won the hearts and minds of naturalists. Increasingly, it works for them. And when they go overboard, critics rein them in. In the meantime adaptationism has stimulated countless excellent studies; it has enormous heuristic value. Adaptation is thus that most puzzling of scientific phenomena: a not entirely provable yet eminently usable idea.

Darwin certainly thought so: it is the key idea in *Origin*, the central fact of nature the book explains. It was what struck Darwin most vividly when, as a young man, he sailed around the world, and especially as he studied variation like that in the finches of the Galapagos: the remarkable fit, the suitedness, of creatures to their environments. This is part of what ecologists and evolutionary biologists mean by adaptation. Evidence for selection in the wild—ongoing, measurable selection, not just adaptedness—is anything but weak, despite the claims of naïve creationists. Back in 1986 John Endler, using very strict and very explicit criteria, compiled over eighty direct demonstrations of selection in the wild for animals and well over thirty for plants. Many examples have been added since, and if microbes are included, the number is much larger.

The classic case is the rise and fall of industrial melanism in the moth *Biston betularia*. Records in several Western European countries and Britain show that black moths largely replaced speckled ones over several decades, when industrial smoke first blackened tree trunks. Black moths had always been easier for birds to pick off trees, but after industrial pollution began *they* became the more camouflaged variant and speckled moths were eaten down to minority status. Then, when ecology-minded governments forced a cleanup in smokestack industries, trees were no longer sooty, speckled moths were more hidden, and selection made black-moth numbers dwindle again.

Today the most elegant instance of ongoing selection measured in the wild can be found in Darwin's own Galapagos finches. As Darwin knew, these birds have beak shapes well suited to the kinds of feeding they have to do, and he reasoned that a relatively new (in evolutionary terms) dispersal to the Galapagos Islands from the South American mainland had given the finches new worlds to conquer. They soon evolved the divergent beak shape and became the varied species he discovered. But this was explanation after the fact. Darwin could not see evolution going on; he could merely infer that it had happened from the end results he saw. Now, thanks to more than a quarter century of work by Peter and Rosemary Grant, the *process* has been seen.

The ability to crack seeds—a skill vital to survival—varies with beak shape both within and (more strongly) between species. Half a millimeter of beak length or depth determines whether or not a bird can crack certain widespread seeds. Raising finches with adoptive parents shows that beak shape is highly heritable, not a product of diet itself. Through decades of research, as drought or flood changed the need for certain hard-to-crack seeds, the Grants found measurable changes in average beak shape in several different finch populations. Over successive generations, natural selection changed beak shape before their eyes, proving that the adaptedness Darwin observed can and does rapidly evolve when survival is at stake. In addition, elegant experiments on mate choice showed that beak shape mattered to the birds themselves; thus, sexual selection—selection through mate preference—may well augment the effects of selection through mortality.

There is also evidence for ongoing evolution of *behavior*. Territorial aggression in fruit flies, cannibalism in flour beetles, and courtship in tiny fish called sailfin mollies have been shown to be not just genetically controlled, not just adaptive in the wild, and not just responsive to natural selection, but all three. These are the necessary and sufficient conditions for Darwinian evolution of behavior, a process that has been proven in a growing number of species. John Endler's research on Trinidadian guppies, for example, showed that body size, color patterns, color vision, and mate choice all are evolving in concert in the wild, right now.

Increasingly sophisticated studies demonstrate adaptedness at the species level and higher. Stephen Jay Gould and other skeptics of Darwin's theory have frequently claimed that adaptation is weak and evolution too indeterminate to replay "the tape" of life and expect the same outcome. That is, if we started evolution all over again, there is no reason to believe that the world would be the same as it is, or that we would be a part of it. At some level this is certainly true—the details of body structure and mental function would surely be quite different. But Darwin's critics really want us to think that the whole process of adaptation, the whole idea of directional trends in evolution, is a quasi-religious myth, and that indeterminacy is the fundamental feature of the history of life.

Not so. In fact, this claim has been debunked at every level from the species on up. For example, Jonathan Losos and his colleagues studied lizard varieties on four different islands in the Greater Antilles. On each island evolution has repeated itself; parallel sets of five or six ecologically specialized species—"ecomorphs"—occur on all four islands. Thus, one genus of lizard, *Anolis*, has diverged independently four times to produce closely parallel groups of species, due to parallel adaptations to parallel habitats. The treetop-dwelling lizards on each island all have short legs and large toe-pads, adapted for life on twigs; but their DNA shows they are closest cousins to the longer- and stronger-legged species on their own island, not to the short-legged ones on other islands. Adaptation, not kinship, caused anatomy to converge. This is just what happened with the marsupial mammals of Australia, independently generating most of the major adaptations achieved in parallel among placental mammals throughout the rest of the world. And at a broader level, the hydrodynamic sleekness of sharks, sturgeons, ichthyosaurs, dolphins, and penguins points to the power of selection to reproduce adaptations when the "swim well" or "back-to-the-sea" tapes are replayed, starting with different kinds of animals.

Simon Conway Morris, whose work on the great storehouse of fossils called the Burgess Shale sparked some of Gould's musings on evolution's indeterminacy, has firmly separated himself from this attack on Darwin. His important 1998 book on the history of life, *The Crucible of Creation*, argues that to refute indeterminacy, it is necessary to introduce the evolutionary phenomenon known as convergence. This is the phenomenon that animals often come to resemble each other despite having evolved from very different ancestors. Nearly all biologists agree that convergence is a ubiquitous feature of life. Convergence demonstrates that the possible types of organisms are not only limited but may in fact be severely constrained. The underlying reason for convergence seems to be that all organisms are under constant scrutiny of natural selection and are also subject to the constraints of the physical and chemical factors that severely limit the action of all inhabitants of the biosphere. Put simply, convergence shows that in a real world not all things are possible.

Morris fully admits, "So tortuous is the path from an animal like *Pikaia* [the 525-million-year-old ancestor of vertebrates] to a human via all sorts of fish, amphibians, reptiles, and mammals that there does indeed seem to be no predictability in the outcome." But convergence has shown again and again that evolution is in important ways predictable. The isolated, now-extinct marsupials of South America produced an animal that looked and worked uncannily like a sabre-toothed tiger but that evolved completely independently.

Consider animals that swim in water. It turns out that there are only a few fundamental methods of propulsion. It hardly matters if we choose to illustrate the method of swimming by reference to water beetles, pelagic snails, squid, fish, newts, ichthyosaurs, snakes, lizards, turtles, dugongs, or whales; we shall find that the style with which the given animal moves through the water will fall into one of only a few basic categories.

Whales were not inevitable, but "the evolution of some sort of fast, ocean-going animal that sieves sea-water for food is probably very likely and perhaps almost inevitable." Likewise, we can say with Gould that the evolution of human beings as we know them was not causally preordained. Yet some sort of land-living, large, intelligent, warm-blooded, social creature mobilizing oxygen efficiently from air to tissues and using ion conduction to make internal electrical signals respond adaptively to light and sound, and eventually think—this was inevitable, given the Earth's geochemistry and cosmological setting.

So, just as Darwin thought, adaptation is ubiquitous and real. But in different contexts, scientific or otherwise, the word *adapted* has a wide range of meanings, and using it loosely can cause a lot of confusion. We can spare ourselves some of that confusion by being very clear at the outset. Common-parlance adaptation— better called "adaptability"—is just flexibility in the face of change. Getting out of the hot sun, learning a new job, coping with and surviving grief—all show us adapting to the world's endless flux. As we have seen, body and mind changes caused by daily jogging are a virtual showcase of human adaptability. Yet such changes cannot in any way be passed on to offspring through the genes.

Adaptation in the evolutionary and genetic sense is the fit of a creature to its environment through traits widely shared within a species. Adaptation may or may not entail adaptability, and it includes much more besides. We humans are adapted *for* adaptability in jogging, because of the circumstances of our evolution; but there is more to it than flexibility. Go back a mere six or seven million years in our ancestral line, and you will find a creature incapable of jogging or even walking well on two legs; resurrect that creature and you will find it inflexibly unable to do so. It simply does not have the nerve, muscle, and bone for it—the structures that we *do* have, coded by our genes: our adaptive advantage. In other words, major parts of adaptation are not flexible at all and do not share that common-parlance meaning of the word *adapt*. Humans are not adapted by evolution to dwell in Antarctica but are so adaptable that some of us do; emperor penguins are adapted to live there but are not adaptable enough to move anywhere else.

Like most of the era's evolutionists, Darwin failed to separate evolutionary and common-parlance adaptation, as well as adaptation and adaptability. The confusion was indeed a centerpiece of the theory of evolution set forth by the French naturalist Jean Baptiste de Lamarck in 1809, exactly half a century before *Origin*. The idea, known as "use and disuse," or the inheritance of acquired characteristics, was widely believed by scientists who accepted evolution, and was explicitly—even warmly—embraced by Darwin. (Alfred Russel Wallace, the coauthor of the theory of natural selection, distinguished himself, among other ways, by avoiding this error.)

We now know that the hereditary material carried in the germ cells is subject to no such systematic change. Pump iron as you may, your offspring will be born with the same pec potential as if you had lain around eating bonbons. The kind of change that occurs in a lifetime cannot translate into evolutionary adaptation, which takes far longer. Still, there are three different, intriguing relations between the two.

First, the stricture only applies to genetic evolution, not to other kinds of transgenerational change. For instance, if you begin life speaking Italian, and later learn to speak English instead, your children may grow up speaking English, having learned it from you. Subtler changes, such as those that have altered English since Elizabethan times, occur in much the same way. Although sometimes called language evolution, this is really language history (we have enough confusion about evolution without giving a single name to both genetic and nongenetic processes). Still, generational change can occur without genetic change.

And not only in us. For instance, rhesus monkeys on the Caribbean island of Cayo Santiago seem to inherit their rank in the dominance hierarchy by cultural transmission from their mothers. In another primate example, a Japanese monkey genius invented potato washing, to get sand off the spuds she was provisioned with. Before long other monkeys in her troop had learned it, partly from watching her, and spud soaking became a permanent part of group behavior in subsequent generations.[38] Similarly, among birds, the song of the white-crowned sparrow, sung by males to attract females, is learned by the young from their fathers. Driving up the mid-California coast, you can hear different white-crowned sparrow songs; they are dialects, in effect, a kind of protoculture, carried on by learning through imitation. Chimpanzees pass on different modes of nest building, nut cracking, termite fishing, and other learned techniques of survival, depending on the habits—the customs—of their elders in the neighborhood.

There are countless other examples. Nongenetic transmission is a fact of generational change and stability in animals. Consider the fact that heroin addicts have children who are addicted at birth. The results of nongenetic transmission—and indeed the process itself—may reach much more deeply into the function of the organism than most of what we are pleased to call culture.

Second, as Ernst Mayr wrote in 1988, "Behavior is the pacemaker of evolution." Nongenetic change can lead to genetic evolution, although certainly not in Lamarck's mechanistic sense. In a process called the Baldwin effect, change within one generation creates conditions that in time select for related genetic features. Imagine a population of perching birds in which some individuals *learn* to like a new kind of berry—say, blueberries. These individuals start nesting in blueberry patches, and their offspring learn to like blueberries just as they did. Eventually, just randomly, the genetic shuffle produces a few individuals who like blueberries right off—they don't have to go through the process of learning. These birds may be favored by selection—blueberries are a readily available food, and a gene-coded taste for them means some nestlings start eating them sooner than others, in turn gaining weight and maturing faster, doing better in reproductive competition, and so on. Eventually we have a generation in which all, by genetic propensity, are very fond of blueberries *without* learning.

Meanwhile, selection proceeds in related areas. Enzymes to digest blueberries better or retinal cells more sensitive to blue may arise by chance and spread through the population gene by gene. But the *initial* conditions for this genetic change were created by learning within one life span. Behavior, in this theory, pioneers genetic change.

In a classic experiment, the English geneticist C. H. Waddington showed that it works even in anatomy. He noticed that some fruit flies changed the vein structure of their wings in response to a change in surrounding temperature. By selecting those that made the change most easily and breeding them, he eventually produced a population that had the new vein structure from birth. Although this was misconstrued by some as Lamarckian evolution, it is just another example of individual life-span adaptability later *mimicked* by genetic adaptation. The flies born with the new vein structure did not inherit it from their parents, who had acquired it. But now, a gene allowing switching of the wing vein pattern in warm conditions had been replaced by a gene that directly coded the second pattern—a kind of genetic progression. The new pattern was evidently a basic biological alternative for the species, reachable with or without complete gene coding.

This brings us to another kind of relationship between individual adaptive change and evolutionary genetic change, and it's perhaps the most important of the three. We saw it in the response of mind and body to daily running, but it has wider implications. It is that individual adaptability itself has a genetic basis and is subject, like any other feature of an organism, to standard evolutionary change. This fact has always been recognized by social scientists, but in a weak sense. In the past, they have stressed that the hallmark of our evolution is increased adaptability. Many behavioral features that in other animals are fixed by the genes are labile with the environment in human beings—certainly in a life span, often sooner. To be fair, they recognized that this flexibility has a genetic basis, a product of evolution during the last few million years.

But these statements simply don't go far enough. First, genes specify more about adaptability than just some general flexibility owed to a bulging brain. For example, we might be more malleable when young, or after giving birth, or while starving. Second, genes influence adaptability not only in degree but in kind—for example, exceptional ease in memorizing odors. But they can specify much more—the kinds of environmental change an animal will be able to respond to, the shape of the response, the nature of the links between stimulus and response, the speed of the change, the time in the life cycle when the change can be best effected, and the permanence of the change. The error was, to quote biologist Julian Huxley, to "forget that even the capacity to learn, to learn at all, to learn only at a definite stage of development, to learn one thing rather than another, to learn more or less quickly, must have some genetic basis." And through this error some were led to believe that the genes have receded into the distant background of behavior, while in fact they are too much with us, even now.

An example from the laboratory rat shows just how silly it is to view evolved, innate mechanisms as separate from or opposed to behavioral change. If there was one species learning psychologists thought they understood, the white Norway rat was surely it. By the 1950s there were quite a few major journals devoted to an ever greater refinement of the laws of rat conditioning. Under the now-tattered banner of the era's reigning psychologist, B. F. Skinner, hundreds of researchers showed how the rat could be made to acquire or lose certain responses (such as bar pressing) when presented with certain stimuli (such as flashing lights). From this it seemed possible to generate laws not only for the behavior of rats but for— as in the title of Skinner's major early work—"the behavior of organisms."

Consider this for a moment. Skinner's book consisted mainly of a description of how the white lab rat learns a few artificial behaviors, tied by reward and punishment to an equally small number of artificial stimuli, controlled by scientists. Any other extraneous stimuli were ruled out, and the rat's naturally occurring behavior was ignored. In fact, the book was about bar-pressing responses in a handful of situations where the rat was rewarded with "reinforcing stimuli"—for example, food pellets—so called because they increased the rate of any response just preceding them. The circularity was deliberate. Skinner and his disciples viewed the brain as a black box best left dark. They studiously ignored such questions as *What is a reinforcer?* and *Why does it reinforce?*

Those riddles would have led them to the boundaries of learning where they themselves would have learned some new tricks: that reinforcers are determined in the first place by genes; that they may differ from one species to another; that this kind of learning presumes some *existing* behavior that can be reinforced; and that different cases of learning follow very different laws, due to the special, inborn readiness of the brain. Yet Skinnerians viewed their laws of learning, established in one species, in a few controlled situations, as uniform and universal—"the behavior of organisms."

Wrong. In the 1960s a series of studies by John Garcia and his collaborators made even the white rat seem a much more interesting creature. Rejected by major journals, Garcia's papers appeared in relatively obscure ones. An editor, a distinguished learning theorist himself, who had rejected one paper later apologized for his failure to see the importance of the work. Despite the obscurity of the journals, the papers created a scientific revolution.

Here is the key experiment: Take four groups of genetically identical rats and give them classical training for avoidance. Have them drink water and punish them while they drink, but precede the punishment with a signal. Unless they are desperately thirsty, they will soon learn to stop drinking when the punishment occurs; but in addition they will gradually learn to stop drinking *before* the punishment, when the signal is given. They will have acquired an avoidance response.

So far this is all the standard stuff of the learning lab. But there is a twist: You give the four groups of rats different pairs of signals and punishments. One group gets a noise-and-light signal followed by an electric shock. The second group gets distinctively flavored water, followed by an artificial, X-ray-induced nausea. Up to this point the experiment and the results are conventional. The rats in Group 1 and Group 2 learn the avoidance—they stop drinking at the noise-and-light signal or the flavor after a typical number of repetitions, or trials—as in many prior studies. The surprise comes with Groups 3 and 4. In those, the pairings of signal and punishment are swapped. Group 3 gets the noise-and-light signal followed by the nausea, and Group 4 gets the flavor followed by the shock. Here is the kicker: These two groups don't learn.

In other words, it is easy to teach a rat to associate a taste or a smell with nausea, and it is easy to teach it to link a light or sound with a shock. But it is very difficult to make the rat learn the converse associations. It can't seem to get the idea that a taste or smell can signal shock, or that a light or sound can signal nausea. When the rat feels sick to its stomach, it apparently has a genetically coded hypothesis: "It must have been something I ate." The rat's expectation is strong enough so that it ignores the *real* signal of forthcoming nausea—in this case, sight and sound. Similarly, when it feels external physical pain, it is not prepared to think in terms of flavors.

The link between taste and nausea could be made in spite of long lapses—up to an hour and a quarter—between taste and queasiness. This was considered a very long wait between signal and punishment, yet the rat still learned avoidance. So the taste–nausea link broke another time-honored law, that lengthening the gap should systematically weaken the learned association. So much for general laws of learning.

To give the, uh, flavor of how these studies were received, one researcher who had worked on similar problems publicly declared, "Those findings are no more likely than birdshit in a cuckoo clock." They were not only likely, they were true, confirmed many times in several species. Hawks, for instance, normally eat white mice. Give them a black mouse—a sort of mickey, nefariously spiked in the lab to make them sick—and they will have trouble getting the idea that they should avoid it; but make that black mouse *taste* different, and they will back off strongly the next time they see it. Disgust at a taste that causes illness is a privileged, special connection; so is fear at a sight or sound that causes pain. These patterns of learning are separate because evolution has made them diverge.

While this special learning may have startled psychologists, it was no surprise to biologists. It is obviously adaptive. The ancestors of rats, in the wild, got into situations where tastes or smells led to nausea, and where sights and sounds led to external pain. But selection in the wild had little chance to favor rats who could associate lights and sounds with nausea or taste with shock and flinching. So it did not produce proper Skinnerian rats, ones that could link stimulus and response equally well across the spectrum of sensations and actions. Instead, it produced genetically limited rats, primed to learn some lessons better than others. Selective association, as it came to be called, has now been proved beyond doubt. To paraphrase William Timberlake, a leading current theorist, an organism comes loaded with systems "that precede, support, and constrain learning." But how could genes, mere swirls of organic acid, confer such specific and strong channeling of learning?

Experiments at the millennium began to show how genes can affect learning ability. In 1999 Joe Tsien of Princeton University inserted a gene into normal mice that made them smarter, by tinkering with the receptor for the neurotransmitter glutamine. The molecule in question is part of a channel in the membranes of certain brain cells; calcium flows through this channel, affecting the cell's activity. But the channel molecule comes in two forms. Young mice learn easily, and tend to have the B form of the molecule, while older, less trainable adult mice have the A form. Tsien used gene technology to preload ordinary mice with genes for the B form. These gene-altered mice, with a sort of fountain of youth in their brains, were smarter than controls in every one of six tests of mousy cleverness and learning. This was the first time learning was improved through gene technology.

Almost simultaneously a group in Bonn led by Markus Noethen confirmed the first human gene tied to dyslexia, on chromosome 15. Others have for years been searching for genes that affect general intelligence, but these have been elusive. Noethen's research took a different approach, exploring a syndrome that limits learning in one domain, that of reading. Reading impairments come from many sources, including purely psychological ones, and even the biological forms of dyslexia do not fit a single pattern. But some dyslexics have an obstacle to learning that involves a gene on chromosome 15. Thus, a general advantage in mouse learning and a quite specific deficit in human learning were both linked to specific genes.

For now we can only speculate about how the connections become biased for special learning. For instance, the B form could be expressed in some brain circuits but suppressed in others. Also, genes can direct how one group of cells in the brain gets wired up to another in the womb, by coding for proteins on the growth cones of pioneering axons. When certain proteins meet their destined matches on the surface of another cell, two cells get wired together. Selection has used such a mechanism to shape, over many generations, privileged connections between certain stimuli and certain outcomes.

So we understand that general laws of learning—laws that apply to all stimuli and all responses—are implausible not only ecologically but neurologically. The nerve path of taste leaving the tongue, for instance, has its first way station in a brain-stem center called the solitary nucleus; not coincidentally, visceral sensation from the stomach reports first to the same nucleus. Why should a sound stimulus, with a brain pathway quite removed from these, form an equally easy link with the visceral sensation of nausea? This could only have made sense in Skinner's world, where the brain was a black box, but it makes no sense today, when the sciences of brain and gene are rapidly converging. Tongue and stomach sensations privilege each other in anatomy and chemistry. Similarly, there is a close brain affinity between sound and skin-surface sensations. This link, in the thalamus and cerebral cortex, is far from the circuits where a mammal senses nausea. Such major structural features of the nervous system are determined by genes.

Study of other species' behavior confirms the principle. In the golden hamster, some complex actions, such as digging, rearing on the hind legs, washing, and scratching, are "instinctive" in the sense that they emerge without training, and yet modifiable in the sense that they can be trained to certain cues. But if you reward all four of these patterns with food, only digging and rearing will increase in frequency, because these are linked to the food quest in the wild. Worse for Skinner's theory, if you train pigs to deposit a coin for food, they later get "pigheaded" and do what comes naturally, forgetting the food reward and pushing the coin around on the floor with their snouts, as if they were rooting in the farmyard. Raccoons rub coins together, applying the obsessive food-cleaning actions *they* use in the wild. Chickens peck at coinlike tokens as if they were food, instead of sensibly trading them in for a real food reward. Such "instinctive drift" has interfered with learning experiments since they first began, and has aptly been called "the misbehavior of organisms."

Psychologist Martin Seligman coined the term "sauce béarnaise phenomenon," to explain a related experience of his own. One evening before the opera he had filet mignon with béarnaise sauce, a favorite dish at the time. Six hours later he became very sick. The cause turned out to be not food poisoning but stomach flu. Yet despite the hours that had elapsed and his knowing the sauce had nothing to do with his nausea, despite the many other stimuli—such as the dinner plates, *Tristan und Isolde*, or the route home from the opera house—that he might have become disgusted by, he acquired one and only one association: a permanent distaste for béarnaise sauce. In time he saw his experience as a clear human case of prepared learning.

Such learning goes way beyond béarnaise sauce. Philosophically, it is our best current path to a reconciliation between those ancient rivals, instinct and learning. Instinct is not pure instinct, of course, nor is learning pure learning; that has been known by intelligent people for centuries. But these experiments may provide a theory of the middle ground. Let us study (the theory says) the many kinds of prepared learning—describe them, distinguish them, try to explain them as nervous-system adaptations—and let instinct and learning go hang. Fortunately, some researchers have done just that.

The sauce béarnaise syndrome, for instance, has been studied in the setting of cancer treatment. People who eat ice cream before receiving nauseating chemotherapy can develop a strong aversion to ice cream, despite all previous happy tastes of it. But the implications go far beyond taste aversions. Seligman analyzed clinical phobias based on the same idea: evolution designed us to learn certain fears—of strangers, the dark, snakes, heights—much more easily than others. Even learning to jog supports the principle. For a million years or more, our ancestors had to adapt themselves—in a time frame of weeks to years—to running long distances; the result was a creature that has, in its gene code, an elaborate set of abilities to adapt in that manner—*adaptability* for running. Few people can run three miles in half an hour without practicing and getting into condition, but most people can do it after a few weeks of work. We are adapted, in the evolutionary sense, to adapt, in the individual sense, to running.

Jogging is a dull example; without risk of political offense, it shows how silly it often is to draw a hard line between instinct and learning; such a line will slice through and destroy some very interesting observations. Running does not provoke cries of indignation when we give it a biological analysis. But the reader who is inclined to find such analysis uncongenial should now feel warned: running is only a model for less dull things. Sex differences, sexuality, fear, violence, mental illness, and the shape of the human future are among the things we must set in the harsh light of evolution. Social conscience is a fine human quality, but we need to invoke it after we have gained understanding, not use it in advance to limit inquiry.

Since we rudely pointed out one of Darwin's major errors (these are not numerous), it is only fair to give him the last word on a subject so close to his heart, especially since he foresaw the subject of this book, and of evolutionary psychology in general: "In the distant future I see open fields for far more important researches. Psychology will be based on a new foundation, that of the necessary acquirement of each mental power and capacity by gradation. Light will be thrown on the origin of man and his history." These are the matters to which we now turn.

CHAPTER 3

The Crucible

Origin of man now proved. He who understands baboon would do more towards metaphysics than Locke.

—**CHARLES DARWIN**, *The M Notebook*, 1856

To be sure, San hunter-gatherers are not early humankind evolving. For one thing, biologically and psychologically, they are thoroughly similar to ourselves. Second, we have no direct way of studying the functions of brain and mind in our ancestors, even as recently as 50,000 years ago. It is only in gross anatomy (mostly bones), material culture (mostly stones), and some aspects of their ecology (pollen and other indicators of climate) that our ancestors are known to us. The San do not even represent in detail the way of life of living hunter-gatherers. From the arctic, igloo-dwelling Eskimo to the boomerang-throwing desert wanderers of aboriginal Australia, from the elephant-hunting Efe Pygmies of the African tropical forest to the longbow-stretching Siriono of the Amazon basin, hunter-gatherers are far too varied to be known from a few thousand Kalahari San.

Still, we can use our knowledge of these cultures to think more clearly about the past. We begin by ordering all the information we have about hunter-gatherers— the San among others—to arrive at general laws of their culture and ecology. Once such principles are established—for example, the fact that most hunter-gatherers are nomadic, whatever their continent or climate, or that they live in small face-to-face bands—they become reasonable candidates for facts about our preagricultural ancestors. But hunter-gatherers are only one source of data. Darwin, in his notebook, alludes to another. Of course he did not mean that the baboon provides all the answers. But since we are products of evolution, the study of other animals must help us understand ourselves. And if they are our close cousins, they have that much more to say—if we listen carefully.

In the 1960s, baboons were thought to represent our prehuman ancestors: they were monkeys, they were smart, they were social, and they lived on the African plains, facing many of the challenges our ancestors of, say, three to ten million years ago must have faced. But baboons went out of fashion as model ancestors, only to be replaced by chimpanzees. Chimps were bigger than baboons, smarter, and genetically even more closely related to humans; all of which made them dandy model ancestors, and they certainly did things differently from baboons. Then bonobos showed up on the guest-ancestor list. These gentle, supersexy, brilliant apes made us look better than either chimps or baboons had, so we started to gaze at them as if in a favorite mirror. Here at last were ancestors to be proud of.

The problem is not that these animals are dull or those who study them misguided. But to pin so much responsibility on any species is to miss their real value. Baboons, bonobos, or any other species can only yield one set of pictures in the panorama of primate behavior. For example, James Moore has set out a cogent set of criteria for judging the relevance of savanna-dwelling chimpanzees to human evolution, based on the known facts about both. His analysis provides a testable model. After the facts—about *many* relevant species—yield an overall structure, we can derive general principles, and from principles we can make good guesses about our ancestors. Baboons and bonobos, chimpanzees and people are what they are because of unique histories and specific evolutionary demands. Clarity will come not from choosing one species as the very best model but from finding the general laws.

Working backward from behavior is no different from studying soft body parts—the appendix, say, the chambers of the heart, or even proteins and DNA. These structures don't fossilize, but with today's comparative biology—*cladistics* is the technical name—we can reason backward from modern forms, through past branching points, to a plausible reconstruction of animal history. It works with the brain as well as the gut. Take the substantia nigra ("black substance"), an important brain-stem center. It helps regulate movement, and a defect in it is the cause of Parkinson's disease. A direct ancestor of ours of perhaps 5,000,000 years ago would, like ourselves, have had a substantia nigra larger, relative to brain size, than that of the average mammal. But how do I know this, when the substantia nigra has the consistency of jelly—a squishy part that, with the rest of the brain, is washed out of the skull soon after death and (unlike some structures on the brain's surface) does not even leave a mark on the skull?

This is a fair question about any soft body part, and the answer is usually this: all such claims come from comparisons of the anatomy and chemistry of living species, together with fossilized parts from the ancient creature itself. All living higher primates have large substantia nigras, and the ancestor in question was indisputably a higher primate—that we know from the skeleton, which we *can* see. It would have to have been a bizarre collage of an animal, built by very mysterious biological processes, *not* to have had a large substantia nigra, which is standard higher-primate stuff.

The same applies to molecular evolution. While we can hope to find ancestral blood cells in an ancient mosquito trapped in amber, or even to extract some DNA from fossil bone, a greater change in how we think about our origins has come from comparing large molecules of different *living* species. The DNA sequence of either the chimp or the bonobo is about 98.5 percent the same as ours. If we analyze the structure of certain genes—say, the gene for hemoglobin, which carries oxygen through the blood, or for a receptor for the neurotransmitter dopamine—we can guess pretty accurately the sequence of that gene in an ancestor that died millions of years ago.

Just as certainly, the ancestor in question must have had complex facial muscles and used them to communicate emotion. There are no fossils of facial muscles, although there may be traces of their insertions on the bones of the face. But complex facial expression communicates emotion in all Old World monkeys and apes, and it would have been too unlikely an evolutionary anomaly for the ancestor in question *not* to have had it. Finally—and here we seem to be on shakier ground but in fact are not—we can guess that the creature in question had, as part of its reproductive life, competition among males for access to females; that infants kept in contact with their mothers and freely nursed; and that males, after maturity, rarely attempted to mate with their own mothers. Why? Because if it had violated any of these generalizations, it would have had to be the most freakish primate that ever lived—so unusual, in fact, that we would doubt it had ever existed.

Such assurances come from fieldwork on almost all monkeys and apes, and from laboratory studies of brain and behavior. Some older claims have gone by the wayside. For example, it used to be thought that social structures of primate groups would fall in place much as brain size does: apes would have complex social structures, monkeys less so, lemurs and lorises less, and so on down the scale of nature. But in social complexity, Japanese monkeys may be closest to humans, followed by baboons, while chimpanzees and bonobos, our nearest relatives, seem almost disorderly. And orangutans, much closer to us than monkeys, have no obvious social structure at all—just a mother, her young, and an occasional male visitor, easing their way through the thick, wet Bornean jungle.

Another fond hope was that basic ecology would explain social complexity. There are a few such regularities—baboon groups do tend to be more organized on savannas than in forests—but we still have no good predictors of social structure. Karen Strier has found great variation in whether and how male or female primates disperse from their native groups; this behavior is central to social structure but difficult to predict. Except for the central tie between mother and young, aggressive competition (especially in males), and facial expressiveness, few rules about the social lives of monkeys and apes hold up. One species in different environments can vary almost as much as different species do. So there are no assurances about what our five-million-year-old ancestors' social structure was like.

Theoretically at least, they could have lived like orangutans in isolated, single-mother families; or like Japanese monkeys in hierarchical troops of over 100; or like mandrills in troops numbering up to 600, built around a core of females, with males peripheralized in combative bachelor bands. They could have lived like gorillas in calm harems, grazing languidly among dense jungle greens; or like chimpanzees in promiscuous groups full of conflict; or like bonobos in deliciously sexy, peacefully sociable bands; or like gibbons in mostly separate, mostly faithful pair bonds. Probabilities can be offered, but no assurances. Further knowledge is likely to come from the fossil record itself, the most direct evidence of our ancestors. Without it, behavioral observation means little, whereas with it the same observations take on great intellectual power.

It is difficult to overstate the excitement surrounding fossils that may be ancestral to ourselves and the exhilaration shared by those who devote their lives to the search for them. We begin with *Aegyptopithecus*, "the Egyptian ape," named for where its bones were found. It would be fun to reach down and down into evolutionary time, trolling for much earlier ancestors—into the Cretaceous, for instance, of 100,000,000 years ago, when flowers first appeared and spread, and when the major forms of modern mammals began to emerge—living, as one scientist put it, "under the feet of the dinosaurs." There we would look for the first primate, a furry runt that could sit in our hands, and that we might come upon at night, surreptitiously finding a meal in a broken dinosaur egg, an annoyance in the life of *Tyrannosaurus* much as mice or roaches are in ours.

Or we could delve deeper, among the Devonian seas, 400,000,000 years ago, where we'd look for a fish using its odd, strong, lobular fins to scuttle along the floors of shallow, muddy pools and, occasionally, to crawl through the mud from pool to pool. Or deeper still, in the exquisite fossil deposit known as the Burgess Shale, where early versions of all the main types of modern animals, and many more besides, were left behind for us, and where, 520,000,000 years ago, a small, darting swimmer called *Pikaia* was about to give rise to all animals with backbones. Or even deeper, into the Precambrian, *two billion* years ago, when the atmosphere of the planet was just beginning to grow rich in oxygen, and the dominant life-form, a kind of advanced bacteria called blue-green algae, was making that oxygen by aerobic photosynthesis.

But to dally in those epochs would make for too luxurious a journey. We must set out in the middle of things, and *Aegyptopithecus* makes as good an Adam as any. It is a small creature, more apelike than true ape, that Elwyn Simons found in 1966, in the Fayum Oligocene. *Fayum* places it in a shallow dip in the landscape sixty miles southwest of Cairo, once on the edge of the sea; *Oligocene* fixes it in its epoch, roughly 30,000,000 years ago. It was quite an idyllic spot then, heavily wooded in places and interlaced with rivers; water birds abounded. Now it is barren, but not to the fossil hunter's eye; it is one of the world's richest lodes of extinct primates.

The species name *zeuxis*—after the Greek father-god—gives some indication of its importance. Simons described it as "the skull of a monkey equipped with the teeth of an ape," and later, when a limb bone was found, the body too seemed monkeylike. This was the piece needed for its patch of the puzzle, since ape- and monkeylike creatures were then emerging. *Aegyptopithecus* was a beagle-sized tree-climber, something like the modern-day Central American howler monkey. Its eyes faced forward in a pair of nearly complete bony cases, and the arrangement and fit of its teeth were distinctly apelike. These are advanced features, with intriguing implications. The eyes recall, in their shape and disposition, 30,000,000 years of prior evolution. Through the Paleocene and Eocene—the dawn epochs of post–dinosaur history—the early primates diverged and multiplied, helping to fill a planet largely emptied by an asteroid. But they broke from the mammalian mold mainly in their eyes. Relentlessly arboreal, they won their place in nature by evolving the most impressive eyes—eyes that could lock on a bug on a twig in the dark, or judge the spring or sturdiness of a branch while their owner leaped and scrambled away from a famished raptor. These skills required depth perception, which means overlapping, binocular fields of vision streaming into forward-facing eyes, in eye sockets that can be readily seen in fossil skulls.

By this time, 40 or 50 million years ago, fossil brains, too, are committed to eyes. The apparatus of smell, which so many mammals were investing in, had already shrunk in our first monkeyish ancestors, while the visual brain, toward the rear of the skull, was larger. Ultimately, banking on vision would create features special to higher primates—not only stereoscopic but color vision, not to mention advanced focusing, eye-hand coordination, and visual learning. But even before *Aegyptopithecus*—from 60 to 30 million years ago—there was a shift of immense importance for the evolution of mind, a shift from the realm of unpatterned chemical signals—individually recognizable but then floating vaguely on the air—to the realm of light, where pattern is almost infinitely realized.

Even in its teeth *Aegyptopithecus* points to the future. It or a very similar creature gave rise, during the next 15,000,000 years, to an immense variety of monkey- and apelike creatures; from them came not only the modern apes but ourselves. The Miocene, between about 25 and 5 million years ago, produced all the major forms of modern apes, including those that somehow carried seeds of human potential. The path to gibbons—slim, long-limbed, mostly monogamous acrobats of the high jungle canopy—diverged first, followed by orangutans, then gorillas. Meanwhile we stayed primitive, small, and unspecialized, both anatomically and functionally, for millions of years more. Molecular evidence shows that humans, chimps, and bonobos are more closely related to one another than any of us is related to the gorilla, and that the orangutan is still farther out on a limb of the family tree. Some biologists think of us as just "the third chimpanzee."

This is not much of a stretch. We share around 98.7 percent of our genes with chimps, and since we only have perhaps 35,000 genes, we are not too many genes away from our closest ape relatives. Not surprisingly, gene expression has changed most in the brain, but we have only found two differences. The first, discovered by Ajit Varki, is a gene that apes have but we lost two or three million years ago, so that we lack a certain sugar on our cell surfaces. It *could* be a signal for wiring up the brain and a key to learning. The second, discovered by Wolfgang Enard and Svante Paabo, has a unique, more recently evolved form in humans. It is also a gene which, if mutated, can cause a rare defect in speech. Could it be a key to the evolution of language? Future research will clarify things. In addition to Varki and Paabo's teams, Japan's Genes and Mind Initiative (GEMINI) focuses on neurological gene expression in apes and humans, to find the genes that make the human brain unique.

But to reconstruct the history, we must continue to grope around in the Miocene for our own and the apes' ancestors, without always being sure which is which. The confusion comes from an embarrassment of riches—there are many different apelike fossils as well as increasing molecular evidence on the relatedness of modern forms. All these Miocene apes stemmed from *Aegyptopithecus*, or at least from a group to which that beast belonged. Some, like *Ramapithecus* and *Sivapithecus*, were found on the Indian peninsula and symbolically named for Hindu deities. Others, like *Dryopithecus*, named for the Druids' tree spirits, the dryads, had drifted gamely into a cooler Europe.

But it is an African species, *Proconsul*, that seems most godlike among the apes—at least to the extent that gods must be in our own image. Crucially, it had a small face. It also lacked a tail, and was apelike in other ways, but the face stands out because it was not snouty or doggish. *Proconsul*'s competitors, in that world of 18,000,000 years ago, whether armed with big, long canines for tearing or with massive molars for grinding, had all committed themselves too far in the direction of big-snoutedness to be candidates for our ancestral line. Peering into its face, after doing the same with the others, we might have felt a shock of recognition.

Proconsul-like fossils have been found in the middle Miocene of Europe, Asia, and Africa, a range suggesting wide success. Unlike many other apes of the era, *Proconsul* exhibits nothing that excludes it from human ancestry. It is not the only possibility, but it is the most likely one. It is also likely—from its jaws and teeth, which make for its modest face—that it was comfortable either in tropical forest or on the open savanna; it may have been making the transition. It could grind the small, tough seeds of grasses that blanket the open plains, and it may have relied on these for food. A seed-eating phase this early in human evolution could have preadapted us for upright posture. Like geladas, modern baboons that eat seeds, *Proconsul* may have spent time sitting upright in a grassy field—perhaps a meadow in an open woodland—grazing busily with its hands.

From indirect evidence, these hands were generalized—committed neither to the long-fingered, short-thumbed brachiating needs of the apes nor to a prominent human thumb, neatly opposed to the four fingers for precision grip. But if it was indeed spending most of its time searching for tiny seeds, grasping them, and bringing them to its mouth, then it might have been in a new phase of selection pressure for visual skill and eye-hand coordination, beyond that achieved by the early primates. This inadvertently would set the stage for what would follow. It was also smallish, even smaller than the pygmy chimpanzee, which is part of the reason we can accept it as ancestral to later, paltry protohumans no bigger than four feet tall. But its size made it very vulnerable to predation as it moved from protected forest into dangerous open savanna. When out in the open it may have moved in groups of substantial size—say, fifty or more—and these groups may have had a fair amount of hierarchy, especially among males, who would have had to cooperate smoothly in defense.

This was the model for one phase of our evolution that anthropologists Irven DeVore and Sherwood Washburn proposed in the 1960s, and that led them and scores of others since to study baboon social behavior. In baboon troops deployed on the African plains, stable social structure hangs on a core of related females. But day to day, the arbiters of power are dominant males. One of the interesting things about savanna-baboon troop structure is that if you match males on a one-to-one basis—as DeVore did, instigating fights over food—you do get a linear hierarchy of dominance. But it is not the natural hierarchy—it only appears in such forced conditions.

The discrepancy is explained by coalitions. At the top of a troop will be not a lone, enduring tough, but a sort of troika in which the members reliably come to one another's aid. This enables a politically astute elderly male to exercise dominance over other males who could easily best him in 'boon-to-'boon combat. Coalitions are also seen in defense against predation. Sometimes older males protect females and young at the center of the troop, while younger males array themselves strategically at the periphery. But sometimes females join actively in fights between groups, and spooked males head for the hills. The subtleties are endless, and individual variation can be dramatic. Robert Sapolsky's marvelous eye for baboon triumphs and foibles lights up the East African savanna and fills it with stunning social detail:

> Obadiah came of age and pushed off for parts unknown, never to be heard from again. Scratch, named for the deep gouge on his nose, was an awkward loopy subadult who joined the troop and went nowhere in the hierarchy, was even pushed around by the hapless Adam, and therefore contented himself to lethargically dominating Absolom and Limp. Jesse, another adolescent transfer, introduced a new behavior, giving credence to the notion of "cultural differences" among different populations of the same species of primates. Whether this was his own invention or a habit of everyone in his old troop . . . I never knew, but soon all the younger baboons were crossing streams bipedal.

These complex social groups contrast with the more rigidly linear ones in simpler primates. And they are not limited to baboons; other ground-living monkeys, like rhesus and Japanese macaques, have similar arrangements. So do our closest relatives, who inspired Frans de Waal's *Chimpanzee Politics*, a vivid account of strategic, hierarchical behavior in this extraordinary animal. In fact, this kind of "Machiavellian" intelligence might have been a key to primate braininess itself. Beyond brawn, political success requires intersubjectivity, the ability to see social dynamics from the viewpoint of the other. But unfortunately for simplicity, the same species that is an intricate piece of political clockwork on the savanna may seem a disorderly hodgepodge in the forest. Among forest baboons, the largest males may be the first to scramble up a tree when a leopard appears. This difference is presumably an instance of the species' adaptation for guided flexibility. The *range of reaction* of a baboon species' social order includes both these polar possibilities, and much else besides.

So *Proconsul's* social life remains a mystery, but it was a clear triumph. Other apes appeared in succession: *Equatorius* at 15,000,000 years, *Kenyapithecus* at 14,000,000 years, parts of a whole array of apes that spread throughout Africa, Europe, and Asia. The facts enable us to guess that our ancestor who pioneered life on the ground—something very like *Proconsul* or *Equatorius*—was perhaps capable of a baboonlike social order when on the savanna. But they also imply that the same creature, ranging into the forest, using powerful hands and feet to grasp branches and clamber in trees, could adopt a looser, less hierarchical structure. Clearly social hierarchy is not simply wired in. Still, a range of reaction is not ultimate freedom. Rather, it is an equation the genes provide to relate a creature's characteristics to its environment. The hardwired equipment includes if-then statements: "If the environment is rich, grow large; if poor, grow small." Similar if-then statements exist for behavior.

According to Smithsonian anthropologist Rick Potts, the kind of creatures that evolved—two-legged walkers collectively called *hominids*—had adaptability as their crucial adaptation. His studies of climate during human evolution show that variation was the hallmark of this period in Africa, and suggest that the search for a creature designed for one climate will lead nowhere. On the contrary, it is variation, the ebb and flow of wet and dry, hotter and cooler climates, with all their consequences for the spectrum of plants and animals, that our hominid ancestors adapted to. "The world is continually changing, and the extremes of sunlight and seasonality, ice ages and interglacials, intense aridity and precipitation, themselves change on cycles of tens of thousands of years."

For unknown reasons, the fossil record weakens between about eight and five million years ago for our ancestors and closely related animals. One would not want to call what is missing "the missing link," since that phrase implies a single crucial transition to human status. "Missing links" have appeared on the landscape for over a century (witness Darwin's remark at the opening of this chapter), and most of them are no longer missing. Yet each time a link is found, the gap is only partially closed, leaving new gaps on either side of the new fossil— room for more missing links. Today there are thousands of fossils relevant to human ancestry and at least nine different species of hominids alone, a densely crowded field compared with that in 1856 or even 1956. Yet the lineages remain incomplete and controversial. There are certainly some missing fossils between eight and five million years ago, because what we have at the end of that time is new indeed.

Molecular clocks, based on studies of proteins and DNA in ourselves and our closest relatives, suggest that our ancestors diverged from chimps and bonobos just about 6,000,000 years ago, so it is a particularly embarrassing time to have a gap. But just over 4,000,000 years ago two new species arose in East Africa. *Ardipithecus ramidus*, from the Ethiopian for "ground-living root ape," was unearthed in 1994 by a team led by Tim White of the University of California at Berkeley. It was decidedly apelike in its teeth and jaws, but erect in posture—the opening where the spinal cord enters the skull is set far forward. It was probably forest-dwelling, a largely upright ape that sometimes ambled bipedally on the ground and sometimes scrambled up and into the trees. It may be off the line of our evolution, but only a little.

The second new species was discovered the same year by Maeve Leakey and her colleagues at Kanapoi, southwest of Lake Turkana. They named it *Australopithecus anamensis*, for the lake—*anam* in Turkana—by which it had lain for more than 4,000,000 years. It was clearly more primitive than any hominids yet found in our own lineage. In both the upper and lower jaw—the tooth rows are straight and parallel, and the chin slopes and recedes—it resembles the apes, like *Proconsul* and *Equatorius*, that had roamed the neighborhood 10 or 15 million years earlier. Yet the bones of its lower leg made it unmistakably erect.

The next stage of our evolution, that between four and three million years ago, was already well charted during the 1970s. As if Mary Leakey had not already contributed enough to our knowledge of fossil humans, she and her research group, in 1977, made one of the most spectacular finds in the history of the field. At Laetoli, in Tanzania—not far from Olduvai Gorge, where she and her husband, Louis, had spent decades finding fossils—she found in hardened, ancient mud, dated at just over 3,500,000 years, footprints of a trio of upright-walking animals. Imagine what she must have felt, after a lifetime of searching—occasionally exhilarating, usually dull, with the biggest discoveries coming in the form of a rim of tooth or bone peeking out of a rock. Imagine strolling across some mud you know to be almost 4,000,000 years old, frozen all that time as rock, looking down, and seeing the imprint of an unmistakably human foot, and another, and another, making a passage across the mud, just like yours.

In fact there could have been three who crossed the mud that day in the Pliocene—an adult much smaller than ourselves, and two youngsters. One walked beside the grown-up and the other, even younger, seems to have stepped in the grown-up's footprints. The mud, moist volcanic ash, had perhaps been hot from the African sun, but was cooler where Mom or Dad had just stepped. As a human child on the beach at noon will stretch to put his or her next footfall right into a parent's cooler print, this australopithecine squirt may have toddled along using a similar strategy.

Many bones were also found in the vicinity, and in most respects they match the bones of that era elsewhere in Africa. Footprints and bones alike belong to *Australopithecus*, "the southern ape," named in the 1920s when it was first found in South Africa. But the footprints were made by a certain type of southern ape: *Australopithecus afarensis*. In 1974, in the Afar region of Ethiopia, Donald Johanson had found the greater part of a skeleton of the species, a small biped famously known as Lucy. The usual arguments over whether a separate species had been found soon ended. The little woman from Ethiopia and her cousins who made the Tanzanian footprints deserved all the distinction they got; they were stunning new evidence about our ancestry.

In 1999 a team led by Maeve Leakey found a new hominid skull that is causing experts to rethink part of our evolution. A contemporary of Lucy living in
Kenya, this protohuman had small teeth and a flat face, both of which make it
likely for our lineage. *Kenyanthropus platyops*—Kenyan protohuman with a flat
face—would be not just a new species but a new genus, taking a place beside
Lucy as a potential ancestor of ours from a time in Africa more than 3,000,000
years ago.

Like their (and our) *anamensis* ancestors, these creatures were upright walkers. Whatever the missing fossils between *Equatorius* and *Ardipithecus*, we know
that near the end of these 10,000,000 years, our line split from that of the other
two chimpanzees, and in ours upright walking emerged. The long, slender ape
began to give way to a short, sturdy, weight-bearing bowl of bone. The foot, too,
became unmistakably weight-bearing, although the stride may not have been fully
human—more like a partly running walk, with some persistent, occasional, four-
limbed clambering in trees. As we know from a recent discovery of a complete
Australopithecus hand, the long, curved, slender fingers would still have lent
themselves to grasping branches. But the use of the hands in earthbound locomotion had been decisively abandoned.

The reasons? No one is sure, but there have been many proposals. Freeing
the hands for carrying food or babies, seeing farther over the distances of the
savanna, even displaying the breasts or genitals in the service of sexual selection—
all have been proposed. Peter Rodman and Henry McHenry calculated that, compared with the chimpanzee's four-limbed knuckle walk, human progress on two
legs uses less energy. That improvement alone could have made the difference in
ultimate fitness. But whatever selective advantage made our ancestors bipeds, the
change was momentous. First, this creature clearly had a harder time with birth
than its Miocene forebears. Apes have much easier deliveries than we do. We
have "cephalopelvic disproportion"—heads too big for our mothers' hips. The
first cause of this rude squeeze was that fully erect posture and locomotion turned
the pelvis into a thick, bulky affair that holds up half the body—not at all the sort
of set-up that an infant would most like to sidle through. This problem was
already faced by Lucy and her children, and the consequent crunch was a pivotal
focus for natural selection. It is likely, too, that this change played a role in the
evolution of woman and in the emergence of physical courage, a peculiarly conscious form of animal mettle.

But one clear result was that Lucy's hands were free. Unneeded for walking,
they were liberated for carrying, a distinctively human act. Chimps and bonobos
walk on two legs for short distances with arms full of bananas, but for Lucy such a
move would have been far easier and more common. Recall Jesse, the young immi-
grant baboon in Robert Sapolsky's troop, who brought with him the habit of two-
legged walking across streams, and was soon copied by the young in his adoptive
troop. This one observation glimpses several kinds of potential in even the monkey
grade of primate evolution. As for the apes, orangutans and gorillas occasionally walk
bipedally, cradling their infants with their hands. This seemingly innocent posture

suggests both pressures and possibilities for Lucy. Did she have long body hair? If not, then she was one of the first primates whose infant could not cling to her as she walked. Even if it could, it is likely that, as in modern apes, the baby's clinging was weak at first, and some support from its mother's hands was needed. Among Lucy's contemporaries and descendants, then, the woman who thought of using an animal skin, tied around the infant and mother to give the needed support, would have gained great advantage.

Unfortunately, there is no evidence that Lucy and her contemporaries used tools. Much later, 2,000,000 years ago, we do have clear evidence of tool making: the pebble-tool collections of the next phase of human evolution. While evidence for tool making is usually clear, evidence for tool use is harder to come by—it is difficult to prove that an unworked, unmodified piece of rock has been used by a hominid, and not just damaged by natural causes. Nevertheless, there are persuasive indirect reasons for believing that Lucy belonged to a tool-using species. These reasons are derived from modern-day chimpanzees, first studied in the wild by Jane Goodall—at the instigation of fossil hunter Louis Leakey, who was convinced that they would throw light on human evolution.

Wild chimpanzees in the Gombe Stream Reserve in Tanzania and in the Tai Forest of Ivory Coast not only use tools but make them. Approaching a termite mound, Gombe chimps will strip the small branches from a twig and then use the twig to fish out termites, a coveted food. Young chimps closely observe their parents, and this facilitates learning. Also, adults will crumple leaves by moderate chewing and then use the resulting leafy mass as a sponge to extract water from the crux of a tree or even for cleaning out the inside of a baboon skull after a kill. In the Tai Forest, other tool-using and tool-making behaviors have been observed, including the use of a pair of stones to crack nuts. These, along with studies of chimps and bonobos in captivity, suggest that the two species have rudimentary cultural traditions in tool use and tool making.

In a pioneering experiment, ethologist Adrian Kortlandt recreated a kind of scene that might have been crucial to our evolution. He introduced a life-size model of a leopard—complete with roaring and head-turning—to a group of chimpanzees in the wild. The chimps made a group assault on the "leopard," using large sticks as weapons, attacking simultaneously and alternately, and throwing their arms around one another excitedly after scoring a blow. Kortlandt's study shows what the dawn of weapons might have looked like and suggests that our ancestors' capacities for tool use and cooperation were probably directed early on to aggressive, or at least defensive, purposes.

Although bonobos have not been seen using tools in the wild—puzzling, given their exceptional ability to absorb human culture in the lab—chimpanzees do it in all of the thirty-six wild populations that have been studied. William McGrew pioneered this work and his book, *Chimpanzee Material Culture*, is a cautious and convincing account. Gombe chimps, for example, show eleven different habitual tool-use patterns and, as among other populations, termite fishing is the most frequent. But this famous instance is only the beginning. Chimps in the wild use makeshift pestles, wave leafy twigs as fly-fans, aim and throw things, and remove bone marrow with sticks. A group led by Andrew Whiten compiled research from seven sites of chimp study in the wild, totaling 151 person-years of observation. The survey "turned up no fewer than 39 chimpanzee patterns of behavior that should be labeled as cultural variations, including numerous forms of tool use, grooming techniques and courtship gambits. . . . This cultural richness is far in excess of anything known for any other species of animal." The last statement is no doubt true, but whether this richness deserves the name "cultural" is something else again. The claim made by Whiten and Christophe Boesch, regarding a group of nut-cracking chimps in the Tai forest, that "in many ways this group could indeed be a family of foraging people," shows how naïve primatologists can be about human culture. The chimps lack language and symbol, virtually lack true teaching, and give no evidence of the sort of metacognition—self-awareness of mental process—that is the essence of human culture.

In addition, it has been very difficult to show consistent differences among populations in habitual patterns that serve similar purposes. Such differences have been avidly sought, because they would show true cultural variety. The research has become more convincing but remains open to the criticism that local "cultures" may simply represent idiosyncratic group responses to local conditions—something that would not qualify as culture in human beings. So both using and making tools among chimps may be merely a universal response to common environmental and social situations. Alternatively, the local varieties could be connected to genetic variation, which is known to be much greater in chimps than in ourselves. In fact, it would be surprising if true cultural variation existed in chimp tools, since it is minimal in human evolution until the emergence of our species, *Homo sapiens*.

Nevertheless, these chimp traditions, properly called protoculture, give us a glimpse of the first clumsy steps our ancestors must have taken into the dim light of simple culture. It is likely that Lucy and her species, *Australopithecus afarensis*, had a similar range of tool-making ability and tool use, and perhaps their ancestors *anamensis* and even *Ardipithecus* did as well. They did so with no more brain capacity, relative to body size, than the modern chimp. So although some of the most distinctive features of the human organism had already emerged by Lucy's day, the supreme feature—an immense brain—had not begun to evolve. We used to think that all the major human characteristics evolved in concert (Darwin's view), but this has been disproved—one among many proofs that evolutionary theories are indeed falsifiable. Erect posture preceded the start of brain expansion by at least 2,000,000 years.

What about Lucy's boyfriends? Some paleontologists believe that the early hominids are best understood as male and female of one species, with a substantial size difference between them. This fits the chimp model, as well as that of other sizeable ground-living primates like the baboon and the gorilla. If this is correct—or if indeed any of our ancestors had a lot more sexual dimorphism than we do—then we can guess some key features of their behavior. In a broad range of animals, a large sex difference usually reflects intense competition among males for females, which, over time, produces the size difference. Because pair-bonding species usually have minimal size difference between the sexes, substantial divergence makes the pair bond less likely and competition among males for females more likely.

Goodall's classic research opened one possible window to the social lives of our apelike ancestors. Among chimps, adolescent females rather than males change groups to seek mates. This gives their young lives an adventurous cast reserved for males in many species. It also means that the female kin group, often the core of primate social life, is usually not the core of chimp life—although, in one group, three generations of females managed to stay together, and this tendency is in a constant tug of war with female transfer. Monthly estrus gives females a cyclical, active role in soliciting sex. A female may go off with a single male during this period, by choice or through coercion, and these consortships—along with similar sexual friendships in baboons—have implications for the emergence of human pairing. After a period of adolescent infertility, females give birth every five years or so, and their subtle and enduring maternal capabilities go far beyond instinct.

Chimp males jointly hunt, kill, and eat small mammals—Thompson's gazelles, for instance, or infant baboons—a characteristic our apelike ancestors probably shared. Craig Stanford's research in the Tai Forest has shown that chimps and their main prey species, red colobus monkeys, have strongly influenced one another's evolution, and for these chimps hunting plays a very important role. Even among baboons, one group was observed to hunt systematically for a few years. With this and other evidence, some have advanced a "killer ape" hypothesis of human evolution, but hunting doesn't support it. Such arguments involve a false suggestion that killing one's own kind is a natural outcome of predation. In fact, there is little connection; many vegetarian animals show within-species violence, and the predatory antics of carnivores probably have more in common, psychologically, with play than with anger. Chimps are violent with each other both at the individual and group level, and this, not hunting, is the place to look for the origins of human violence.

But humans have another close ape relative, as close as the chimpanzee: the bonobo, or pygmy chimpanzee, which lives in a small population under the bend of the Zaire River in Central Africa. These have been studied in the wild for many years by Takayoshi Kano and other scientists, mainly from Japanese universities, and in captivity by Frans de Waal and his colleagues. Adrienne Zihlman, who has studied their anatomy, makes a good case for bonobos as our closest cousins on anatomical grounds alone. Behaviorally, bonobos resemble common chimps in some ways but not others: females transfer between groups and have long, tender mother-child relationships, but they also form exceptionally strong erotic friendships, something not seen in chimps. Male-female erotic relationships, too, are longer and friendlier in bonobos, and copulations are both more frequent and more promiscuous. Although violence occurs at both the individual and group levels, it is less frequent and less extreme than it is in chimps, and dominance hierarchies among males are much less important.

So which of these two ape cousins should we take as the model of our ancestors? We don't know. Ultimately we must learn from the fossil record and from the genes of living species, as we come to understand what those genes do. Meanwhile, we can probably assume that both species, and other apes as well, share some adaptations with early humans. A strong case can be made for savanna-dwelling chimps as a helpful model, but these wider-ranging, more ground-oriented groups have not been studied as well as their forest counterparts.

During the 1980s, it was fashionable to say that hunting was trivial as an influence in human evolution because our ancestors were mainly scavengers, stealing meat killed by more effective carnivores. But the fact is that most carnivores, including lions, hyenas, and human hunter-gatherers like the !Kung, do both. Hunting and scavenging are basically indistinguishable in the fossil record, so we may never know. But why rule out hunting in hominids when we know that even apes do it under natural conditions? It is highly likely that at some point, probably early on, our ancestors began eating the flesh of animals killed mainly by males. In addition, there is every evidence that it became a more elaborate, efficient, and central pursuit as our evolution continued.

After the spread of the first bipedal apes, our attention shifts to the eastern shore of Lake Turkana, in northern Kenya. Here Alan Walker and Richard Leakey have dug for decades, extending the work of Leakey's parents that began hundreds of miles away at Olduvai. Like the South African sites of Swartkrans and Sterkfontein before them, Olduvai and East Turkana have yielded to the patience of paleontologists a treasure of hominid bones—ancestral and otherwise—between three and one million years old. They range from a creature with great winglike cheekbones and enormous crushing, grinding molars—successful for over 1,000,000 years, yet ultimately an evolutionary dead end—to a relatively graceful creature with a big brain and delicate face that needed a new genus name: *Homo*.

One thing is certain about even the first members of this group: they made tools. The earliest pebble tools are more than 2,000,000 years old, and they are modest—no more than oval rocks with one rough edge made with two or three blows of another rock. Nevertheless, they are tools, and as Louis Leakey dramatically demonstrated in 1955, such bits of sharpened stone make butchering a carcass far easier. It is possible that several million years of chimplike, spur-of-the-moment hunting led very gradually, through selection for tool-making ability, to the pebble-tool tradition. They were perhaps first used in butchering, crucial to scavenging and hunting alike.

We need to remember, however, that none of the modern chimps' tools—termite sticks, leaf sponges, nut-cracking stones, or logs as clubs—would appear as fossils. In fact, some of the most important tools used by modern hunter-gatherers—the digging stick, for instance, no more than a sturdy, smoothed, pointed stick for getting at roots in the ground, or the sack for carrying vegetables, or the infant sling—can almost never come to archaeological light, since all are perishable. Such tools were at least as important in the first halting steps toward human culture as are the stone tools and weapons we find. The low profile of women in the study of human origins may be due as much to the fragility of their artifacts as to neglect by male scholars.

The notion that tool use and tool making are a cause more than a consequence of human brain evolution—that "tools make man"—is an old one. It was hinted at by Darwin in *The Descent of Man* and was made explicit, in a Lamarckian cast, by Karl Marx's collaborator Friedrich Engels in his essay "The Part Played by Labor in the Transition from Ape to Man." Many modern anthropologists have argued that the use, manufacture, and design of tools gave unprecedented advantage to the more intelligent, and so led to larger brains.

Immensely ingenious research by Nicholas Toth has opened new paths of study on stone tools and has shown that we can think more precisely about their effect on brains. Toth devoted years to mastering the skills of stone tool making and proceeded to study the tools left by hominids using his hard-won knowledge of their craft. He has been able to make important inferences about their mental abilities; they clearly had cognitive capacities greater than those of apes. This is underscored by Toth's work with Kanzi, an enculturated, language-using bonobo studied by Sue Savage-Rumbaugh at Georgia State University.

Despite the fact that bonobos have not been seen to make tools in the wild, or even to use them much, Kanzi learned to make stone flakes as he needed them and to use them instrumentally, cutting a rope to get to a banana. But Kanzi has not reproduced even the earliest pebble-tool assemblages, nor has he invented a stone-tool process, as our ancestors did; evidently, by 2,000,000 years ago, they were already smarter than apes. Glimpses into brain function are also possible; Toth showed that the earliest tool makers were mainly right-handed. But apes are equally likely to be right- or left-handed, so this means that by 2,000,000 years ago the human brain had already distinguished the two cerebral hemispheres, transcending the hints of lateralized function present in ape brains.

The only difficulty (and it may not be a serious one) is that the fossils of hominid bipeds from the next crucial horizon of human evolution are so numerous and so variable that it is very difficult to associate the known tool types with the known skeletal specimens. Walker and Leakey, looking only at the specimens from the eastern shore of Lake Turkana and at only one time depth (around 2,000,000 years), found enormous variation in skull form—variation that would not have been believed possible before the 1980s. Surveying these and other fossils in 1994, Leakey reasoned that four different species names were necessary, not counting the one that led directly to ours.

It is not surprising to find among them both *Australopithecus africanus* and *Australopithecus robustus*. These two species embraced much of the earlier evidence, except for an advanced form at Olduvai. They represented, respectively, the gracile (small and lithe) and robust (larger and thicker-boned) forms of the genus, thought to range over much of Africa before the rise of *Homo*. Like Lucy, both have small brains for their bodies (under half a liter of mental stuff), but the robust form has much more massive jaws and teeth for grinding and crushing, with a crest on the top of the head for muscle attachment. An even more thick-boned form, almost a caricature of robustness, deserved its own species: *Australopithecus boisei*. In addition, descendants of Lucy's species, *afarensis*, and of *Kenyanthropus*, might have persisted until this later time.

Yet even this great variation was not the key to our ancestry. In the earliest time depth at East Turkana, contemporary with the small-brained robust and lithe bipeds, there was another form with a small face but a much larger brain. This third form, represented by the now-famous catalog number 1470, had a cranial capacity estimated by physical anthropologist Ralph Holloway at 775 cubic centimeters. This ruled out membership in any of the species of *Australopithecus* and made it a better candidate for the maker of stone tools in the region, as well as for direct ancestry to us. It belonged to a species named by Louis Leakey at Olduvai: *Homo habilis*, or "capable man."

A half million years later at East Turkana we again have three forms. But while the two small-brained forms, one slender and one robust, have changed little, their brainier cousin is quite different. Brain size has increased only 75 cubic centimeters, less than a third of a cup, but the face is much more modern, and the species is now *Homo erectus*, the same as for "Java man" and "Peking man." This was the last species before *Homo sapiens*. It had a worldwide range, a brain approaching a liter in volume, and exceptional skill in making stone tools—no longer crude choppers but much more articulate hand axes, so good that they were destined to stay almost uniform in design throughout Africa, Europe, and Asia for the next 1,000,000 years.

One member of this species dominated discussion during the early 1990s. Discovered in 1984 by Kamoya Kimeu and analyzed by Walker, Leakey, and others, his skeleton was complete, allowing them to assess his age and likely growth trajectory. He was an eleven-year-old boy who, had he survived, would have been over six feet tall. This in itself was stunning, since it means that our average *Homo erectus* ancestor might have been bigger than most humans have been. It also seemed that he could have been speechless. This was cleverly inferred from his anatomy; specifically, from the size of the spinal canal behind his chest. We have a bulge in that part of our spinal cord, believed to help control speech, which is after all just an elaborate (all right, exquisite) modulation of the stream of exhaled air. The Turkana boy, like most nonhuman primates, lacked that bulge, so Ann McLarnon and Alan Walker surmised that speech evolved later on in the human story.

This specimen brings us to the verge of our own species. There are arguments with the scheme, of course, but there is agreement that an ape descendant of *Proconsul* (18 or 20 million years ago) gave rise to *Equatorius* (15,000,000 years ago), which led to *Australopithecus anamensis* (over 4,000,000 years ago), and then (by 3,500,000 years) to both Lucy and the new flat-faced hominid. One of them gave rise to *Homo habilis* (by 2,000,000 years ago), which in turn produced *Homo erectus* (by about 1,500,000 years ago). Some would like an extra species between Lucy and *habilis*; others would like *habilis* to be deemed an early *erec-*

tus, but these are quibbles. The known fossils may punctuate the continuum and make it seem segmented, but—at least for higher animals—it is still continuous. As the fossil record improves, we will find out whether the process was actually gradual. If so, the naming of intermediate forms will always be arbitrary. As the great evolutionist G. G. Simpson used to say, the transitional forms are argued over because they are transitional, and that is why they are fascinating. Echoing Simpson, evolutionary biologist Richard Dawkins has said, "We may believe that the genus *Homo* is descended from the genus *Australopithecus*. But it is ludicrous to suggest that there must once have been a *Homo* child at the breast of an *Australopithecus* mother. . . . Where we are dealing with fossils, such subjective judgments are inevitable."

The rest of the story is simpler. One million years ago there was only one species, one sole survivor of all the upright walking apes. The robust and gracile forms of small-brained hominids, whether three species, four, or more, are gone forever. There is only *Homo erectus*, with groups that range throughout the Old World, all similar in brain size, facial form, body size, and stone-tool culture. Foragers with a standard but powerful tool kit, they preyed upon big game, which they could only have done through sophisticated cooperation in hunting. They probably had a low population density with flexibly organized nomadic groups. Their young developed slowly and needed intensive care, almost undoubtedly a female responsibility. In all likelihood, women were also responsible for gathering wild fruits and vegetables. By this time in evolution, infant brains were big enough to fill childbirth with grave risk and hardship.

Over the next 1,000,000 years there was a steady, gradual increase in brain size to the modern average of 1,350 cubic centimeters. Why? Although in modern living humans the large range of normal brain sizes is difficult to correlate with measures of intelligence, predicting intelligence from brain size does make sense when comparing different species. This is especially true when the ratio of brain to body is examined, so that we focus on "extra" brain cells beyond what a big animal needs to manage a bigger body. More important, the brain did not just become larger, it was reorganized.

Ralph Holloway, who has devoted decades to brain evolution, believes the reorganization is fundamental. He studies the convolutions and blood vessels of hominid brains through the impressions they leave on the inside of skulls. Some of the distinctly human features of the surface—things not found on chimp brains—were emblematic of *Australopithecus*. These brains are more humanlike than apelike—even the one found by Raymond Dart at Taung, South Africa, in 1924. It was the first *Australopithecus* found, a three-year-old with a naturally formed and exposed stony cast of the inside of its skull, looking remarkably like a brain. The frontal, parietal, and temporal lobes of the cortex—serving association, delayed response, reasoning, decision making, language, and memory—are larger and more human in the Taung child's brain than they are in apes. So even the early hominids may have been smarter than apes; their brains were small but already being reorganized.

The Taung youngster also reminds us that our ancestors were evolving the long childhood of our species. Braininess in the animal world goes with lengthy growth, and both had to come into play before we could be human. As psychologist Jerome Bruner presciently said long ago, foreseeing a new approach to child psychology, "The nature and uses of immaturity are themselves subject to evolution, and their variations are subject to natural selection." Our ancestors were selected for longer childhoods in part so that they could play, observe, and learn more. They needed the time to grow larger brains, but they also needed to bend those brains to the purposes of culture. Controversy continues as to whether the australopithecines had an apelike or humanlike plan of childhood growth, but either way, their brighter descendants, up to and including us, were under selection for bigger brains with more complex functions. Comparison of the toothy australopithecine skull with the more globular first skulls of *Homo* led to an apt distinction: the "eating brain" and the "thinking brain."

But thinking about what? Among the more plausible claims are puzzle-solving ability, needed to track game without a decent sense of smell; information storage, for memory-guided gathering; language and planning, for group hunting; intersubjectivity, for teaching skills like toolmaking; foresight, for the protection and training of offspring during that same long childhood; "Machiavellian" intelligence, for reading the plans of rivals, to do unto others before they do unto you; and abstract thought, for reckoning complexities of kinship, marriage, and economic exchange. What is certain is that brain size and cultural complexity, as reflected in stone-tool traditions, increased in concert over the last 1,000,000 years.

There is no doubt that hunting played a role in this process, engaging some of the highest features of prehuman emotion and intelligence. Recent hunting-and-gathering cultures have elaborate and moving beliefs and rituals for hunting. Among the gentle Pygmies of the Ituri Forest, for example, young men learn to kill elephants, and a successful hunt of any kind promotes band solidarity and gives rise to a day of ritual and story. The Siriono of eastern Bolivia rub their bodies with the feathers of a harpy eagle they have killed, to take on some of the bird's awesome power. The simple, persuasive remark, "Women like meat," made by the !Kung of the Kalahari to anthropologist Lorna Marshall, became the title of Megan Biesele's book about their myths. With such a preference embedded in female choice, it is little wonder that even non–hunting-and-gathering cultures like the Sambia of New Guinea and the Masai of East Africa made hunting central to their rituals of manhood. And as shown by anthropologist Matt Cartmill in his profound meditation on the role of the hunt in human culture, *A View to Death in the Morning*, hunting is a transforming and passionate experience, celebrated in the classic art and literature of great civilizations. Despite all skepticism it is likely to have played a crucial, if still mysterious, part in our evolution.

Until the late 1990s it was thought that human control of fire first appeared at Zhoukoudian in China, about 500,000 years ago. That *Homo erectus* roasted meat seemed to be proven by the charred bones of various prey species littering the floor of the Zhoukoudian caves. But now it seems that the roasting there may not have been deliberate. Still, at some point in the early evolution of our species, fire did come under our ancestors' control. If their taste apparatus was anything like ours, such cooking must have greatly increased the appeal of the meat in their diet. It must have made certain parts of the carcass more digestible and relaxed selection pressure on the teeth, jaws, and face, allowing them to shrink even further. Fire would ultimately figure in the spread of agriculture and the founding of civilizations. But the dawn of controlled fire has deeper implications. If these near-human creatures reacted to fire the way we do—especially those of us in hunter-gatherer and other small-scale societies—then control of fire meant a quantum advance in human culture: rituals centered around a steady, managed blaze and, as language emerged, the chance for nightly discussion of each day's events, plans for the next day, important events in the personal and cultural past, myths and tales, and future possibilities for the people and the band.

Of course, such talk can go on during the day or in the dark. But the day must generally serve more urgent purposes, and anyway, there is something about the night—the fear of isolation, the deep need for social life, the longing for light and warmth, the soothing, even mesmerizing effect of the flickering burning. "Look not too long into the fire," says Melville's Ishmael in *Moby Dick*, a warning against the lapses of pragmatism—the lapses, even, of self—that such gazing may occasion. We can visualize a group of these early people, huddled together, touching, talking, and gazing into the fire, safe from the night behind, attaining perhaps a new plateau of human consciousness—a distinctively *cultural* consciousness—and perhaps too, the rise of an impulse that we might be inclined to call religious.

Human ingenuity soon created spaces. Four hundred thousand years ago, in what are now the environs of Nice, there were not only hearths but houses built around them. We may call them huts, but they are as much houses as many dwell in today. They are oval—twenty to fifty feet long and twelve to eighteen feet wide—and could shelter between ten and twenty people. They have postholes for large branches, a wall of stones, and odd slabs of limestone for sitting or working at. There is much debris on the floors, but immediately around the hearths places are cleared—perhaps for sleeping beside the fire, as native Australians used to do. There is clear evidence of repeated brief occupation. Pollen in coprolites—the fossils of feces—comes from spring and early-summer plants, showing that the occupation of these sites was seasonal, much in the style of modern, seminomadic, gathering-hunting peoples. Thus the French Riviera, 400,000 years ago.

At around the same time in Spain, at Torralba and Ambrona, are signs of collective and brutal big-game hunting. There archaeologists like F. Clark Howell and Richard Klein have found the bones of too many woodland elephants, accumulated in too short a time, probably associated with human killing and butchering. Together with evidence of grass fires, perhaps deliberately set, these characteristics of the Spanish sites may point to collective elephant-driving—a cruel, dangerous, and potentially wasteful hunting method. These facts undermine misconceptions about "natural conservation" practiced by primitive cultures; whole herds of elephants were apparently driven by grass fires to their deaths, careering over a cliff, just as the Plains Indians stampeded bison. Unless Ambrona had an implausible population explosion, a lot of elephant flesh was wasted. We humans may love nature, but unfortunately we are not quite natural conservationists.

In another 100,000 years, or at most twice that time, we have the earliest specimens most will agree to call *Homo sapiens*. At Swanscombe in England, Steinheim in West Germany, and elsewhere were skulls with shorter, more tucked-in faces, smaller brows, and a thinner but higher vault than *Homo erectus* has. While we might not have been overcome with an impulse to embrace them, they were members of our own species—people. But these European fossils led, in yet another 100,000 years, to the quite different Classic Neanderthals—subjects of controversy since the day the first specimen was found in 1856 by quarry workers, in the Neander Valley near Dusseldorf. It was not just the first Neanderthal, but the first humanlike fossil of any kind, and it provoked Darwin's remark that heads this chapter. We needn't dwell on the heated arguments of the day, made in the first flush of passion about evolution, nor on recent arguments about the Neanderthals' correct place in the family tree. What counts is that they were basically human. While a Neanderthal man dressed up in a business suit could probably not ride unnoticed on the subway, in face shape and brain volume, even the Classic Neanderthals were roughly like us, and some earlier Neanderthal-type skeletons from the Near East look even more modern.

As shown in superb studies by Erik Trinkhaus at the University of New Mexico, two things are distinctive about them. The first is their exceptional robustness. The diameters of their long bones and the insertion points for muscles indicate a body made to bear tremendous stresses. The legs are adapted for running, climbing, and weight bearing, and the arms for controlled vigorous activity, as in throwing or striking blows. The face is intermediate between the very robust form of *Homo erectus* and our far more delicate form. Neanderthal sturdiness is evident in children's skeletons, at least by age five, so it was not just adaptability in an individual sense, emerging during a strenuous life, but genetic adaptation, fixed in the body's plan. The only exception to this is the pelvis: in both sexes, these bones are delicate, probably because the female had to pass a large head during delivery, a problem that had already been worsening for three or four million years before the Neanderthals. The male's similarly delicate pelvis could have resulted from evolutionary economizing—building both genders on the same basic plan. The light-boned pelvic basin might have been the Neanderthal woman's answer to infant robustness, which would have made the treacherous passage through the birth canal even worse.

The sorts of activity that might account for this robust build include hunting big game, carrying heavy loads, walking steep terrain, and engaging in combat. Any of these activities could in turn account for the second highly distinctive feature of the Neanderthal people's skeletons: the exceptionally high frequency of injuries. This is especially true at Shanidar, a rich Neanderthal find in a cave in Iraq. There are simply too many healed fractures and unhealed broken bones to give us a picture of peaceful life. Although we cannot tell the source of most of the injuries, modern rodeo cowboys have similar patterns of injury, suggesting that close encounters with large, dangerous animals—encounters during hunting that could get you hooked, tossed, or trampled—might have done the damage. I saw such injuries and even risked them while hunting with the !Kung, and similar mishaps during hunting might have been frequent for Neanderthals.

One Shanidar man, however, does offer a likely sign of the cause of not just an injury but a death. A partially healed scar on the top of his left ninth rib was unmistakably caused by a sharp object thrust into the chest. This could have been accidental, but it also could have been a deliberately inflicted spear wound. The man might have suffered a collapsed lung, and it is evident from the limited healing around the wound that he lived no more than a few weeks. If it is evidence of combat, it is the only such evidence for Neanderthals; but, along with their high rate of injury, it raises the possibility that violence was part of their lives. And a discovery in 1999, clear proof of Neanderthal cannibalism, suggested that they were not sentimental about the body. In 100,000-year-old bones found in the cave of Moula-Guercy in France, Alban Defleur, Tim White, and others found evidence that the skilled techniques of butchering used by Neanderthals on deer and goats were also used on other Neanderthals. Still other investigators have suggested that cannibalism might be much older, and it has certainly persisted up to recent times.

Yet despite their grisly side, the Neanderthals had accomplished much. First, there is the Mousterian stone-tool industry, characteristic of all their kind and a quantum advance in complexity and refinement over the hand-held axes of *Homo erectus*. Advanced statistical techniques reveal the complexity of these tool assemblies, and the scanning electron microscope reveals exquisite invisible landscapes that show how the tools were used. Their skills were as advanced as those of many modern humans—indeed, the more we know about these people, the less strange they seem, except for their robustness and numerous injuries.

What is most intriguing about Neanderthal culture, however, is the possible evidence of ritual. Although subject to legitimate skepticism, archaeological evidence has suggested to some that these people had systematic cults. According to one view, in the Cave of Witches, west of Genoa in Italy, Neanderthal people penetrated more than a quarter mile into the murky stone depths, where they habitually threw pellets of clay at a certain stalagmite vaguely shaped like an animal. In the cave of Drachenloch—"the lair of dragons"—8,000 feet up in the Swiss Alps, in what seemed to some a deliberately built cubical stone chest, were seven skulls of bears larger than grizzlies; six more skulls were apparently set in niches along the walls. Since the Ainu of northern Japan and other hunting peoples of northern Asia have had bear cults in modern times, it did not seem far-fetched to propose a Neanderthal bear cult. In yet another cave in Lebanon, dated at 50,000 years, a dismembered fallow-deer carcass was apparently laid on a bed of pre- .

arranged stones and sprinkled with red ocher. Of course, these tantalizing hints merely establish a possibility. But taken together with another, better established set of facts, their significance is enhanced.

These are the facts of burial. Deliberate burial of the dead occurs for the first time in human history among the Neanderthals. In the cave of La Chapelle-aux-Saints, a Neanderthal hunter was laid out in a shallow trench, a bison leg on his chest, and the trench strewn with tools and animal bones. The similarity to many recent burial rites, speeding the soul of the deceased on its way in the next world, was unmistakable. At La Ferrassie, also in France, archaeologists found a rock shelter that had been used as a family cemetery. Sixty thousand years ago a man, a woman, two children about five years old, and two infants were buried here, with ceremony. The man and woman were laid head to head, the children were neatly set at the man's feet, and one of the infants, newly born, was interred with three beautiful and valuable flint tools. The baby was in a small earth mound adjacent to eight other similar mounds of unknown significance. In the Crimea, in Israel, and elsewhere in the Near East are burials with the legs pulled up tightly to the body, with one intriguing exception: a forty-five-year-old man with the jaw-bone of a boar held in his arms. At still another Near Eastern site, a boy's grave is ringed by pairs of goat horns.

But the most astonishing burial is at Shanidar. Here in the 60,000-year-old grave of a hunter with a fatally crushed skull, are the fossil pollen remains of a multitude of flowers. Analysis shows that they were brightly colored, related to hollyhock, bachelor's button, groundsel, and hyacinth. The man had been laid on a bed of blossoms, perhaps woven together with soft pine boughs; more flowers were probably strewn over him. There are doubts, as always, and there are specu-lations by experts as to what the flower burial, if it indeed occurred, may mean. But such speculations offer no more, and perhaps less, than what each of us may privately do with this knowledge.

What the Neanderthals did not have, despite their advanced stone technol-ogy, was any significant plastic art. We cannot accuse them of lacking a musical culture or a (spoken) literary one. In fact, a fossil bone found in 1997 has holes that suggest it could have been a flute, and in 1998 research at Arcy showed that body ornaments made from animal teeth and ivory were a cultural pattern inde-pendently invented by Neanderthals. But the best inroad we have into the aes-thetic world of any of our ancestors, the main reflection we have of their sense of beauty, is in painting, carving, and sculpture. Of these the Neanderthals left noth-ing, except for the ornaments. By about 30,000 years ago the Neanderthals disap-peared from the earth. By evolution's standards, they went quickly. They evolved gradually, both in physique and culture, over about 50,000 years; but they were replaced, in Europe at least, in only about five or ten thousand years. The specter of mass slaughter by a technologically superior race, was often raised by early the-orists, but there is no need for such a process. Five thousand years is still quite a long time—though not long enough for the Neanderthal people of Europe to have *evolved* into later Europeans.

The most widely accepted theory has later Europeans evolving in Africa, migrating to Europe, and gradually overtaking—absorbing, conquering, squeezing out, and perhaps killing—the technologically more primitive Neanderthals. This "out of Africa" model, based on two decades of research on mitochondrial DNA, has given us a new respect for the unity of humankind. In fact, it has suggested that (since mitochondria are only transmitted through females) all humans alive today can trace our ancestry back to one woman, often called "Mitochondrial Eve," who lived perhaps 100,000 years ago in Africa. Certainly the genetic data support the unity model. As Svante Pääbo put it, "The genetic variation found outside of Africa represents only a subset of that found within the African continent. From a genetic perspective, all humans are therefore Africans, either residing in Africa or in recent exile." In effect, we are all brothers and sisters under the skin, and this is why genetic research has consistently shown that the vast majority—more than 90 percent—of all human variation is contained within any one race. Differences between races are very small by comparison. It is also why gene sequencers have found no racial differences. As Craig Venter put it, "We have sequenced the genome of three females and two males, who have identified themselves as Hispanic, Asian, Caucasian, or African-American. . . . In the five . . . genomes, there is no way to tell one ethnicity from another."

So the European story is one instance of the human colonization of the world, from a fairly recent starting point in Africa. Those sons and daughters of Eve who migrated into Europe beginning around 40,000 years ago are called Cro-Magnon, after a hermit who lived in a rock shelter high in the limestone cliffs of the Dordogne. This shelter housed a few of these new people about 30,000 years before the hermit's birth. The migration and replacement—partly brutal, partly merely selfish, energetic, inadvertent—seems very easy to understand, and indeed very human, if we compare it to the likely earlier replacement of *Homo erectus* by *Homo sapiens* in East Asia and to the almost continual sequence of similar processes up to the present day.

What the Cro-Magnons carried was simply the most impressive array of stone tools and weapons—not to mention others of bone, antler, and ivory—that had ever been brought together by any human culture: spear points, lances, knives, chisels, needles; tools for scraping, perforating, sawing, whittling, pounding; even iron pyrite for making fire—and much more. At this, the most human level of human evolution, any attempt to minimize the importance of hunting would be naïve indeed. But while this superior technology must have seemed the central fact about them to the people who became their victims, it cannot seem so to us. What we see as so distinctive are a few items of small practical value that they left on the walls of certain caves. John Pfeiffer, in his book *The Emergence of Culture*, convincingly argued that this signaled the final evolution of the modern human mind, and he called it "the creative explosion."

At Altamira, a cave in northern Spain, a raucous herd of polychrome bison tumbles around the ceiling of the gallery—walking, standing, crouched, charging in warm, rich reds and browns, with a horse, two hinds, and a few wild boar tossed in for good measure. At Font de Gaume in France, a pair of exquisite reindeer grace the rock, the hart almost tenderly sniffing the head of the kneeling hind. At Trois Frères, we have a collection of puzzles: an utter chaos of masterfully engraved big game superimposed on one another on a series of rock panels, a pair of sketchy snow owls guard their chick, and a poignant, eerie figure called "the Sorcerer" presides over the animals, with antlers, staglike ears and body, human hands and feet, and a prominently human male sex beneath the horse's tail—none of which match the impression made by the all-too-human eyes, shocked, plaintive, sad, set in the midst of a mouthless, bearded face.

And then there is Chauvet. In 1994, in the last decade of a century of discovery, three explorers found a stunning painted cave in southern France that turned out to be by far the oldest known. The paintings are reliably dated to between thirty and thirty-three thousand years ago. They included lionesses, bison, and rhinoceroses of exquisite realism; shading to show the nuances of animal coats is rare in other cave art but common in Chauvet. A preponderance of dangerous animals links it thematically to cave art at Arcy in northern France. There are some sketchier, more stereotyped depictions of wooly mammoths and parts of humans—including two large pubic triangles complete with vulvas. Stripes, circles of dots, and other signs of unknown meaning are also present. Chauvet thus takes its place among the great ancient galleries of art that decisively show the appearance of a fully human consciousness.

But none of these finds can really hold a candle to Lascaux. Dubbed "a Paleolithic Sistine Chapel" by an early student, it was unknown until 1940 when a pair of boys discovered it and thought to alert an expert. It impresses the observer—religious or otherwise, expert or not—with a strong sense of the sacred. Its halls, naves, and galleries—as they are justly called, twisted though they be—have walls and ceilings covered with gorgeous animals, masterpieces of realism, although by no means merely zoological. A bistre yellow and black horse surrounded by flying arrows, a herd of ornately antlered, graceful little stags, a horse with a fluffy mane, a "Tibetan" antelope, a pair of bison tail-to-tail, a big red dappled cow, a fire-eyed five-meter-long black bull. Among these and other works at Lascaux are several that art historians with no particular brief for prehistory readily accept as some of the greatest art of all time.

We do not know what these works of art mean or what they were for. We do know that many of the friezes in these caves were physically accessed with only the greatest difficulty; they are in the bowels of the earth. They can only have been approached with torches, some only by crawling through cramped spaces. Pfeiffer, a distinguished student of cave paintings, who devoted years to viewing and thinking about them, proposed that they were part of a kind of theater, perhaps religious in intent, in which the torch-lit, strained, obstacle-filled approach was the spellbound prelude to a drama. Among other explanations are hunting magic for propitiation or restitution, totemism, shamanism, initiation ritual, and decoration. As Ann Sieveking said of Paleolithic paintings and sculpture, "They are simply a language for which we have no vocabulary." And it is quite possible that we never will.

Suppose for a moment the simplest interpretation—that they are merely decorative. Auguste Renoir said, "The purpose of a painting is to make a wall pretty." Coming from him, those words give dignity enough to the merely decorative. In the long trek of human progress on this planet, these are the first "objets" for which the question *But is it art?* must be answered unequivocally in the affirmative. Fernand Windels dedicated his book about Lascaux "To our distant ancestors who worked in the silence of the caves some two hundred centuries ago in belated honor of genius never surpassed." It is significant that we begin to speak of centuries, as if welcoming these women and men at the boundary of the realm of the recent past. That some of them could execute such paintings, engravings, and carvings is impressive enough; but more impressive still is the majority who could not but who evidently could appreciate and support those who could. There is no more significant single advance in human history, and no more convincing demonstration of the final, decisive emergence of the utterly, distinctively human brain.

CHAPTER 4

The Fabric of Meaning

After every foolish day we sleep off the fumes and furies of its hours;
and though we are always engaged with particulars, and often enslaved
to them, we bring with us to every experiment the innate universal laws.
—RALPH WALDO EMERSON, "Nature"

Anatomy may not be destiny, but it is all we bring into the world and all we take with us. And while we are not endowed at birth with a fixed and changeless structure, since it is dying and being reborn continually, a whole does arise from the sum of our parts. The love and hate, joy and grief, hardheaded analysis and excited imagination we experience during our stay on this planet all lie in the intricate, dense, wet web of cells we carry around with us, and not in any airy thing attached to it. It is only because certain cells make signals, chemical and electrical, that we are able to think and feel at all.

Take, for instance, that crown of evolution, the pyramidal cell of the new, outer crust of the brain. Named for its shape, it sits like an ancient symbol in a gelled sea of cells, tentacles almost motionless, poised for messages. It has millions of incarnations all over the cortex, each a centerpiece of higher brain functions, swallowing signals into its steady, electric rhythm, modifying in its tiny, crucial way what it says to other cells. As the incoming messages flow around its skin, coursing together, they may add up to a single trigger impulse capable of affecting remote parts of the brain and body. Just to move a muscle, this impulse may have to run a course from the brain's outer layer all the way to the base of the spine. In the giraffe this is no small distance; in the blue whale it's a far cry indeed.

For us humans, more important even than motion-controlling signals that leave the brain by the billions is the vastly larger number of within-the-brain signals. It is mainly because pyramidal cells are pestering one another over a great web of circuits that we can have these lofty thoughts about them. This web was woven over hundreds of millions of years. Entrusted as it is with much of what we need to get through life, even to have sex, in our brain-ridden species, its assembly during growth could not be left to the mere vagaries of experience. Growth cones of axon trunks—the tips of nerve cells in the embryo—grope like snakes inching through jelly, sniffing at chemical gradients and climbing along them blindly. They reach their private, ultimate, anatomical destinies while the infant is in the womb, before the lure of bright lights or the warmth of mother love can make any impact. Once near the cells they are to bind to, axons make synapses—direct chemical bridges to the tentaclelike dendrites of the next receiving cell.

So far the process is mainly run by genes. Nutrients and energy are supplied from outside, minimal conditions of warmth and protection must be met, and chaos in the formal sense—extreme sensitivity to initial conditions—introduces an unpredictable element. Development is not random, yet it is impossible to predict from any data that could possibly be collected. So by mid-pregnancy, even the brains of identical twins are different; genetic control is imperfect and, well, stuff happens. Tiny differences in the way cells move and grow in earliest embryonic life are amplified with time. Yet the genes keep tabs on things, continually calling the process back to its general plan. It is fast, it is orderly, it is amazingly complex, and on the whole it is really rather miraculous.

Critics are right in saying that little is now known about how the genes get the job done, but new research makes them less right every day, and they will be just about completely wrong in a few decades. The human genome's sequence is nearly complete, its basic internal structure revealed. The structure is hierarchical, so that a small number of genes can interact to produce astounding complexity. Regulatory genes, growth-factor genes, cell-adhesion genes, and especially homeotic genes—the governors of growth that encode animal body plans—are taking the mystery out of development. What is already clear is that, except for some unpredictability introduced by chaos, the genes are in charge at this stage. When a growth cone is recruited, the sign says NO EXPERIENCE NECESSARY. Brain growth after birth is also embryological, in the sense that it is largely inner-directed. But where experience counted for very little in the womb, now the brain must swallow huge doses of it, in addition to doses of oxygen and milk sugar. In a classic experiment on experience and the brain, rat pups were raised in rich or poor environments. The favored pups, selected at random, grew up in a world full of toys and other pups. A control group grew up under ordinary lab conditions. A third group was impoverished, not even getting the low level of stimulation available in the typical lab.

These conditions stamped differences in the brain. In the visual part of the cerebral cortex, where patterns from the eye are built into usable thought, those same pyramidal cells so crucial to higher mental life were changed. Not in their basic placement or overall structure—in rats as in ourselves these are formed by genes, mostly before birth—but in the fine structure, experience is manifest and can be seen under the cold eye of the microscope. Animals raised in a better environment have more small branches, far out along the main trunks of the dendrites. On these branches, countless tiny spines, contact points for incoming messages, crowd in more numerously in rats that grew up with more stimulation. Everything from learning ability to brain chemistry is changed by experience. Yet somehow it is anatomy—that most ancient biologcal subject—seen and drawn or photographed through the classic tool of the microscope, that persuades at last. It is something we so want to believe that we have to see structure with our eyes, mistrusting every other form of evidence. "Look," I thought when I first saw the photographs, "see for yourself. *Experience really does change the brain.*"

Yet those same spines can be stripped away by a gene. In Tay-Sachs disease, a dreaded form of mental retardation that strikes selectively at the children of Eastern European Jews, the dendrite branches lose all their spines over the first year of life. To compound the tragedy, the child is normal at birth. The Tay-Sachs gene causes an enzyme deficiency that allows a usually harmless chemical to build to poisonous levels in the brain. This in time distends the dendrites and denudes them of precious spines. Since being bare of spines is tantamount to being deaf to incoming messages, the cells become slowly functionless, while parents and doctors stand by helplessly.

There is no known treatment, and death is inevitable, usually before age two. The speed and certainty with which the Tay-Sachs gene destroys brain cells could never be opposed by mere experience, and when there is a treatment it will be molecular, probably genetic. Yet it is illuminating to know that there is a visible, real structure in the brain—the spines along the dendrites—that can be changed *either by genes or by experience.* When you listen to barren arguments about nature and nurture, heredity and environment, think of the spines on the dendrites. Remember the rat pups when you hear the stubborn claim that most mental function is determined by genes; and remember the tragedy of the Tay-Sachs infant when someone says that genes have no known effects on behavior.

Other studies of experience and the brain supply intriguing details. For instance, age is no obstacle. While it made sense to experiment first with very young rats, even elderly rat brains change with experience; old rats *can* learn new tricks, and the tricks change their brains. Then, since most of the changes were in the visual cortex, researchers began to wonder if merely seeing could make the difference. To test this possibility, some rats were kept in small cages within the larger "enriched" cages, from which they could watch the toys and other rats but not participate. Their brains showed no changes. Evidently, we must grapple with the world to change the brain, not just sit back passively watching. It is conventional to say that if rat brains change with experience, this must be all the more true in humans. Aren't we far brainier, plainly capable of many times the learning known in rats? If environmental enrichment can cause new spines to sprout on a rat's dendrites, imagine what it must do in the human brain, where dendrites are thousands of times more numerous. Learning must be thousands of times more possible.

This argument has truth in it, but it is also treacherous. It can lead us into temptation, into believing that such enrichment liberates the brain from the genes. No one, of course, believes that genes do nothing; but many psychologists and social scientists want to believe that the known facts of brain modification, multiplied to the human scale, push the effects of the genes back to the crudest aspects of structure. The genes start the ball rolling in the womb during the early weeks of pregnancy, the thinking goes. Then, after some unspecified early point, the tiny human brain somehow gels into a rough outline of its destined ultimate form, and presto!—beginning with the pregnant mother's diet, drugs, habits, and moods, experience takes over brain-building. Michelangelo said of stone carving that it was merely liberating the figure from the marble that imprisoned it—that he had merely to chip away the excess stone. But this poetic, modest view, if unrealistic for the struggling apprentice sculptor, is actually appropriate for the brain. Indeed, the massive new burden of human brain size and complexity lies more with the genes and their guidance of maturation than it does with the forces of nurture—even while it offers vast new fields for nurture to plow and seed.

Consider what has happened in evolution. Encephalization (from the Greek *enkephalon*, meaning "brain"), the ratio of brain to body size that shows how brainy a species is, has not changed in reptiles in a good 200,000,000 years. Yet even the initial transition to mammals, just about that long ago, caused this index of braininess to quadruple, requiring major genetic change. In the last 50,000,000 years of mammal evolution, another four- or fivefold increase has occurred—in all, a twentyfold increase for the *average* mammal, compared with the average reptile. Although evolution affects all parts of the brain, the varied sizes of mammal brains are due overwhelmingly to differences in its most advanced part, the cerebral cortex. Even the lowliest mammal is a far cry from a reptile; yet our earliest primate ancestors were already ahead of the mammal pack. *Necrolemur*, not even much of a primate but very near our ancestral line, had a brain-to-body-weight ratio of 1 to 35 as early as the Eocene, 60,000,000 years ago. Meanwhile, one counterpart, an ancestor to the rhinoceros, had in the same era a brain-to-body-weight ratio of 1 to 2,000. The disparity has greatly increased since.

The initial surge in brain size from reptiles to mammals meant a major shift from a mainly visual world view to one also dependent on hearing and smell. Harry Jerison, a leading theorist of brain evolution, argued that the first mammalian brain expansion was basically a solution to a packaging problem. In vision you package many of the basic neural cells you need in the retina itself. But similar processing in hearing and smell has to be done at higher levels—perhaps because it is integration in time, not just space. Research in the late 1990s confirmed Jerison's insight: the evolution of the tiny middle-ear bones—a mammalian hallmark—from precursors in the angle of the reptilian jaw goes along with brain expansion. Archaic mammals, driven from daylight niches by the more successful reptiles, needed better hearing and smell to adapt to the night or twilight world.

Although the basic change was genetic, experience could have played an initiating role. Imagine an early mammal—a shrew- or volelike thing—living in a twilight niche, with a brain slightly changeable by experience, just like that of the enriched rats. Suppose that during its one and only life this creature finds itself forced to rely on hearing to find food and elude predators, more so than another member of its species—say, a few hundred miles south, where the daylight hours are longer and vision remains more useful. In the animal that relies more on sound, we would expect a change in the number and shape of the dendritic spines in the hearing part of the brain.

Of course, this more hearing-reliant animal cannot pass this hard-won advantage on to its offspring—there is no inheritance of acquired characteristics. But it can rear those offspring in an environment where the same brain change is likely to happen to them, because of exposure to a similar balance of sight and sound. And in the long run, differences in reproduction will favor those with genes for better hearing. This would be a case of what is known as the Baldwin effect: adaptability during the individual life cycle brings a population into contact with new selective forces. These forces can in turn produce genetic change, in the same direction as the original environmental change. But change through experience merely *pilots* a change in gene frequency. In Ernst Mayr's apt phrase, behavior is the pacemaker of evolution. Meanwhile, the evolutionary transition, over millions of years, is relentlessly genetic. No experience in one individual's life, however powerful, could result in the sorts of changes that make the brain of the simplest mammal so different from that of the reptile.

The same holds for subsequent evolution. In the next phase, the impetus came from the disappearance of the dinosaurs—done in by a bursting, climate-shifting asteroid—and the reinvasion of the daylight niche by mammals. The brain change, for the most part, was a return to the visual sense but at a higher level. These new mammals evolved a visual system that was pivotal in later brain evolution. Compared with the reptile's, it was highly patterned in space and time, integrative, analytic, responsive to environmental influence, and, above all, fully open to the other sensory systems. This is what ultimately built the great association systems—the real circuits of thought—of the newest mammalian brains. These trends were most impressive in the primates, where analytic vision and brain expansion set the pace. In the end they took brain evolution farther than any other lineage, leading to tool using, then to tool making, and finally to language, the last step toward culture. All of this required expansive reordering of the brain—a change in genes and in the way they interact to generate structure.

The outcome, which we carry around in our heads, is for all its flaws impressive. It weighs more than three pounds, a big load for a midsize mammal; sometimes it seems hard to hold it up. It contains 10 to 100 billion nerve cells. The uncertainty is due to the granule cells, tiny components of certain brain structures. There are billions, and they are hard to count. Of the 100 billion overall, only a few million are sense cells—the points of input in the eyes, ears, nose, tongue, and skin—and only about 1 or 2 million are motor neurons, the final common pathways directly controlling muscles and glands. They are "final common pathways," because there are roughly 10,000 nonmotor, nonsensory nerve cells for every motor neuron; 10,000 cells converging on one.

The integration is enormous. Known as "the great intermediate net," these billions of central neurons account for the mental, emotional, and spiritual functions of the brain. Seventy percent are in the cerebral cortex, an intricately patterned, six-layered structure made for integration. Vast, sheetlike, and of roughly constant thickness, the cerebral cortex folds and refolds on itself according to genetically guided, species-specific patterns. Wrinkled and squeezed into a small, round skull, it occupies most of the space given up to the brain. Without it we cannot speak, comprehend, see patterns, learn associations, look forward, remember, think, or—at least in the human sense—feel. It is by far the newest, largest, most evolved, most interesting, and most ordered brain structure.

At the other extreme is a conglomeration of neurons (to call it a structure might dignify it too much) called the "reticular activating system," or RAS. Set in the core of the brain stem, it is as old and messy as the cortex is ordered and new. Yet silence it and we die, whereas removal of the entire cerebral cortex is perfectly compatible with survival, assuming some buffers against sheer stupidity. The RAS's role is to regulate arousal, or in the words of neuroanatomist Walle Nauta, "the posture of the internal milieu." From the core, through short connections of unspecialized, asterisklike neurons made for listening—as well as by longer direct connections—it can communicate, if slowly, with all higher and lower parts of the nervous system. It governs sleep and waking, as well as the briefer cycles within each. It modulates alertness and arousal, two kinds of wakefulness. It is pivotal, anatomically, for balance between the emotional work of the limbic system—"the feeling brain," cradled in or near the fringes of the cortex—and the physical functions of the body's nerve nets. Impulses travel both ways, upstream and down, meandering slowly but surely from one multipolar neuron to another. Because of their starlike shape, these cells can listen widely and pass on an integrated message. Collectively they belong to a vast population of similar starlike neurons deep in the core of the brain, a continuous Milky Way of weakly structured gray matter running up from the spinal cord, through the brain stem, and into the hub of the brain.

This order recapitulates evolution. In the brains of the shark or lamprey—close to the ancestral forms of the vertebrates—just such a stream of small, relatively unstructured neurons plays a dominant role. Their brains consist largely of what is in us the reticular core. Throughout 500,000,000 years we have made many new structures, most of them more orderly, attached to this ancient system, but the core still regulates. It seems at times almost to be looking on skeptically at the presumptions of the parvenu higher brain. The great neuropathologist Paul Yakovlev said it best: "Out of the swamp of the reticular formation the cerebral cortex arose, like a sinful orchid, beautiful and guilty."

Of course, the brain does not spring into being, least of all orderly being, by instant genetic magic; this absurd idea would indeed leave us groping for some environmental-determinist guidance. Both the ornate cortex and its ancient swampy counterpart in the brain stem are the products of millions of years of natural selection. The crucial events in this process took place in the chemistry of the genes that control structure. But whether you are a brain-burdened human or a neurologically slim little lamprey, your nervous system stems from a course of growth and, wondrous as it sometimes seems, it no longer looks magical. Indeed, it now seems miraculous only in the sense that much of the natural world does—the colors of dawn, say, or autumn leaves—miracles that are now largely explainable. Not that the answers are all in, or even imminent. Brain growth holds some of the most awkward unsolved problems in biology. Still, some answers are at hand.

Crudely, much has been known for over a century. In the late 1800s descriptive embryology occupied many biologists, who used the light microscope to describe how organs emerge in embryos. Such description still goes on today. The studies of Paul Flechsig, J. Leroy Conel, and Paul Yakovlev throughout the twentieth century have been carried forward by Peter Huttenlocher, Harry Chugani, and others into the twenty-first. They have vindicated description by continually refining it, and by seeing it through the latest stages of fetal development and, ultimately, through postnatal life. The electron microscope has made sense of finer structure and, with the scanning electron microscope—which revealed in three dimensions what we had long seen through flatly—the descriptive process yielded even more. Today, on the advancing frontier of imaging, under alphabet banners like CAT, PET, SPECT, and fMRI, we are starting to watch the breathing, growing brain.

Still, the field of brain development has shifted away from description. As adaptation ponders why, description asks simply, *What?* But there is an intervening question: *How?* Until the late twentieth century the how question—the central, vital question of developmental biology—remained an unscalable height. Today we are finally beginning to understand how the genes actually guide the miracle. Consider what has to happen as the brain starts to grow. In a few months the work of eons of evolution must be redone, yet not in the same way. Billions of neurons must be born, over 100,000 a minute, by the division of parent cells. These new cells must find their way to their anatomical destinations, sometimes large distances away, in an embryo that is constantly changing form. Half these cells must then die at predestined times, leaving other billions behind, like Michelangelo's hidden sculpture. Each must develop its own shape, including a well-built set of outgrowths: dendrites, generally shorter, the receivers of information; and axons, usually longer—sometimes very long—the main senders.

Once the cell is in place, the axon must find its own destination. The organs of the nervous system are unique in this regard; specific connections among cells are the basis of function. Even the dendrites are not still and we can visualize the cell as slightly motile in its place—especially its limbs—doing a long, slow, undulating, lifelong dance. But the axons bear the burden of making connections, sometimes traversing a great expanse of brain. They must not only get where they are going and connect, but they must avoid any number of other matches they might wrongly make with passing cellular strangers.

Each neuron must make one or more of some dozens of neurotransmitters—the chemicals used to signal neighboring cells. Thousands of properly placed synapses—the not-quite-touching interfaces between one nerve cell and the next—must correctly form. Cells must awaken electrically, making circuits work.

Hard-won connections in parts of the brain must die, streamlining the system, just as billions of hard-won cells have died earlier. Finally, support systems must develop around nerve cells. For example, the myelin sheath, a fatty wrapping around the axon, must form in many nerves. Without that sheath, made by nearby nonneural cells, the nerve would never function maturely, in terms of conduction speed, consistency, and the intricate timing of circuit. A growing understanding of the molecular genetics of myelination leaves no doubt that the process is largely under the genes' control. Yet multiple sclerosis unravels the same sheaths, tragically giving us some idea of what was first gained by their development.

Since genes are mere chemicals, and since all they can do is make more chemicals, our understanding must also be chemical; we can't just describe the look of things under the microscope. The overarching question is *how* the fabulous intricacies of the brain's trillions of contact points get themselves arrayed in orderly circuits. Although still relatively new, molecular genetics has gone beyond mere description and we will consider its findings, but description comes first. Around 1960, Richard Sidman, a neuropathologist at the Harvard Medical School, began to apply the technique of autoradiography to the study of how nerve cells are born. This technique, which involves injecting a radioactive label, revealed a new dimension of brain growth: the sequence of nerve-cell "birthdays." This sequence is set by regulator genes, in response signals from neighboring cells. These genes tell their own cells when to divide, and sequence in turn determines structure. For example, in the cerebral cortex, earlier-born cells stop in the first, inner layer, and latecomers sidle past them to form the outer rows; elsewhere in the tube-shaped early brain the reverse is true—early-born cells sensibly migrate farthest, with later ones settling in behind them. But evidently the less logical, upside-down pattern makes the cortex what it is, because any violation means disaster.

It turns out that such a disaster occurs in a mutant mouse, named "the reeler" for the sad dance its damaged brain causes. The reeler's defect is a recessive gene on chromosome 21, a tiny chemical change in an otherwise normal mouse that results in terrible flaws in the cerebral or cerebellar cortex. But how does the elegantly layered structure of a normal cortex get turned upside down? Researchers are closing in on the answer. First thoughts pointed to cell birthdays. Could the gene just alter the timing of cell birth? Autoradiography proved this hypothesis wrong. Cell birthdays in the mutant are normal, and the earliest-born cells are the size and shape they should be; the problem is that they are in quite the wrong place—on top instead of on bottom. This means that the genetic defect cannot be mainly affecting the timing of cell birth; rather, it must distort the migration path itself or, perhaps, the stop signal.

Like other genetically simple malformations, the reeler's defects can lead to insights into inherited brain problems in human beings—some of which are not in principle more complicated than the disorder of this mouse. For example, the brains of some (only some) dyslexics also show abnormal cell layering in the cortex. These cases appear to result from an abnormal sequence of cell births in one region of the brain—the one where language circuits and visual circuits converge.

Albert Galaburda, the neurologist who described these cases, worked with psychologist Victor Denenberg and others to understand similar cell-layering problems in mice and rats, whether due to genes or environment. Although the evidence is weaker, some theories of schizophrenia also invoke cell migration. Such insights have not yet solved these problems, but they may help. At a minimum, mouse mutants will help us learn how the genes of mammals, us included, build brains.

How do the cells of the early brain find their way in the morass of the neural matrix? Here, too, the answers are dimly visible, thanks in part to insights from immunology. That cells can tell friend from foe to destroy them or leave them alone suggests a whole range of hypotheses about the embryo. If cells can be programmed to recognize foreign tissue, then why not one another? In the embryo such recognition would mean not destruction, but rejection, attraction, or adhesion. Dramatic cell movements—the orderly migrations that build brain and body—could be ruled by a matrix of plus and minus valences. As in immune reactions, the cells could attract or repel one another by recognizing chemical markers on surface membranes, where they make contact. Such markers are made by genes.

Many laboratories deal with this process. With the genome sequenced—really an easy problem—the question of how they build a person will be the driving force of twenty-first-century biology. Central to the answer will be brain building, an even more difficult problem. French neuroscientist Jean-Pierre Changeux pointed out that there are about 10^5 genes in the human genome, but about 10^{11} neurons and 10^{15} connections in the human brain. Today we know that 10^5 genes is too many—we have only a third that number. How can so few genes specify so much brain? They don't, of course. Emergence, activity, and experience all play important roles. But the quantitative discrepancy between genes and connections doesn't mean that there are really billions of degrees of freedom in brain development. It means that in mammals, unlike in simple invertebrates, the genes must economize. For instance, by having one axon pioneer what will be the destiny of thousands, the genes get thousands of contacts for the price of one. The process is much more flexible in mice and men than, say, in grasshoppers; but in mammals, too, the genes govern.

Before these particulars can be worked out, however, developmental biologists have a deeper question to answer: How can the cells of an embryo produce such a variety of cell markers, enough to produce adhesion, repulsion, and everything in between, when all the cells are genetically identical? Indeed, this flaw in our understanding has plagued developmental biologists since the early twentieth century. The main job of biologists during the current century will be to figure out how different genes get switched on and off in genetically identical cells. About all that can be said now is that something about the physical situation of cells, even after very few divisions of the fertilized egg, makes them express different aspects of their identical sets of genes. Once divergent, they are committed to distinctness and can chemically change one another in a myriad of ways.

While we hear a lot about the effects of genes on behavior, intelligence, and emotions, it is in earliest brain development that we see the clearest effects of genes on design. To avoid the hand waving that often accompanies discussion of genes and large-scale human traits, it is worth focusing for a while on the physical and chemical events of those early stages, the first points at which genes do their work. Thanks to Pasko Rakic (rah-KEECH), Gerald Edelman, Thomas Jessell, and others, we know a bit about how these early cells get around. Cells divide frequently, perhaps a bit faster at one end of the embryo than the other. The shape of the embryo changes, with much of the movement caused by plain old mechanical shoving. Some movement occurs in blocks, but within these block movements, chemical markers that cause cells to stick, slip, or repel can direct the shoving. Although whole populations shift, each cell moves on its own, and the movement is essentially amoeboid. Since a cell is jellylike, part of it will slip in a certain direction, stick, and pull until the rest of the cell follows.

Some cells are also guided physically, and not just by being shoved en masse; these inch along structural tracks already laid down. This happens in the cortex of both the cerebrum and cerebellum, where newborn nerve cells cling to long threads called "radial fibers." Pasko Rakic described the long, slow climb of such cells along the radial fibers late in the rhesus monkey's embryonic life. These non-neural cells radiate through the cortex from the inner margin where neurons are born to the outer surface where they must finally find their way. It turns out that another mouse mutant, the weaver, has genetic abnormalities in these guide fibers that explain its wobbly gait and woeful memory. One bad gene reduces the size of the cerebellum, but a double dose produces severely abnormal movement, as well as a damaged hippocampus, a brain region vital to memory. Even without a gene sequence or metabolic pathway, the gene-dosage effect makes the pattern plain and real: the single defective gene impairs the radial fibers, but not too much, while the double dose makes the brain grossly abnormal because those same fibers don't work at all. As for the normal versions of the genes, you could say—paraphrasing the old ad—it's a case of better cells for better brains, through chemistry.

But where there are no obvious tracks to follow, what causes migrating nerve cells to go in one direction or another? Experiments are cracking the chemical code by which cells find their destinations. And when that code is cracked, it will be a short step to the genes that specify the signals in the code. This embryonic language—actually a vocabulary of secret handshakes—uses large proteins or sugar-protein combinations to form molecules on the skin of cells, stuck into or through that oily cell membrane. These chunks of chemical gook, uniquely able to clasp their own mirror images, are called "cell-adhesion molecules" (CAMs) if they make one cell embrace another; they're called "surface-adhesion molecules" (SAMs) if they merely make a cell stick to other stuff. Through the molecular sleight-of-hand now under way in many laboratories, we are learning to see cells and their extensions sidle and slip calmly through crowds of other cells.

Of course, beyond being drawn in a certain direction, the cells must actually make their connections. Some trail a long extension behind them as the cell body moves, so that the backward connection is already made. Others form purely local connections, so cell processes don't have far to go. But some must put out finger-like extensions that become long axons. Because the embryo is so small, these will not have to cover a distance as long as axons in the adult, yet they can extend thousands of times the length of the cell itself. Here the challenge to neuroem-bryologists is comparable to that of the pattern of cell migration: how does the growing axon move, find its way, and know when to stop?

This was a central argument in brain science in the mid-twentieth century, centering around chemospecificity—the idea that the guidance of growing axon tips is driven by specific chemical gradients. An older view was that the main forces were mechanical, due to the orienting effects of physical stress lines, a sort of continental-plate movement of microscopic structures, like the shoving referred to earlier. The newer view (really a revival of the classic view from the early years of the century) is that of the late Nobel laureate Roger Sperry. He insisted, and gradually proved, that chemical affinities guide growth cones far more than physical stress does.

Sperry worked on frogs and salamanders, animals simple enough to regener-ate nerves yet enough like mammals to be relevant. In many experiments, repeat-edly challenged by the mechanical-guidance school and redesigned to meet the challenges, Sperry and his colleagues detached the amphibian eye from the brain and scrambled the connections. Successive experiments made it ever harder for regrowing axons to find their way—even to the point of jumbling the retina like a jigsaw puzzle. Yet the fibers unerringly found their previous points of contact, even though the eye then functioned wrongly—for instance, a flying insect crossed one part of the visual field, but the frog's tongue flicked out to another. This was strong evidence for specific chemical labels, unperturbed by the scrambled cues of either anatomy or experience.

These studies do not rule out a role for mechanical forces. One way axons find their path is by following other axons. In some parts of the brain—the memory-forming hippocampus, for example—the timing of growth is critical. Axon tips can approach their hippocampal links but not reach the exact local destination—unless they arrive at just the right time. Change arrival time and you change the ultimate circuitry.

Another major strategy is trial and error; some connections are made but later die off, leaving the more active and useful ones. From the earliest moments of embryonic development, cells crucially influence one another. Some cells in the early brain can nourish a cell that contacts it, others cannot. These contacts help determine which cells live and which die. Genes do guide from within, but cells are also keenly responsive to other cells. These cues may be surface markers coded by genes or purely mechanical forces like the geographic layout and the movements entailed in growth. Some of the most fundamental events depend entirely on a passing kiss between certain cells at a special moment in growth.

Take an example familiar from basic biology, one of the oldest bits of knowledge in this field. The optic cup, destined to be the eye, is lensless in the early embryo. As it touches the epidermis—the outer shell of embryonic cells—it loosens some epidermal cells, which form the seed of the lens. Any epidermal cells will do; an optic cup planted under the skin of the future foot will also produce a lens, and without such epidermal contact, the lens will fail to form anywhere. Before this, the eye grows more or less under its own steam; after it, moving the organ virtually anywhere in the body will still give you a more or less normal-looking eye—though functionless for lack of the right connections. So the eye is set up by evolution and the genes to develop mostly on its own. Yet the process needs, at a key moment, a strategic intervention by an outside tissue.

It turns out that nerve cells can even build themselves into elaborate networks with fully mature connections between cells, *in the absence of any use.* Stanley Crain and others at the Albert Einstein College of Medicine grew tiny fragments of the cerebral cortex of the early fetal mouse, beginning long before functional links, or synapses, were made. Crain and his colleagues applied the local anesthetic Xylocaine—you've had it at the dentist's—to the fragments, suppressing electrical function without causing permanent damage. Despite the total absence of either outside stimulation or spontaneous activity, the tissue continued to develop under the Xylocaine. In the end it produced organized nerve-cell assemblies—the basis of circuits—including the hallmarks of synapses seen under the electron microscope. Supporting Sperry, this was strong evidence that nerve networks "are formed in forward reference to their ultimate function"—before they know what sort of work they are destined for.

Although these experiments were with mice, the process occurs in many species. For example, Corey Goodman and others showed that in both the grasshopper and the rat some neurons send axons across the long axis of body or brain. First they are attracted to nearby cells in the midline, but they don't get stuck there because an ingenious feat of genetic programming makes them lose that attraction as it is exercised. The meeting with cells in the midline changes a molecule on the axon's growth cone and the roving tip loses its taste for those cells. Tempted anew, it now moves on to the other side of the nervous system, makes a ninety-degree turn, and sidles up or down to its destined place. So the long journey is broken down into less ambitious legs, and at each crossroads guidepost cells point the way. In the paltry nervous system of the grasshopper it is possible for each neuron to be controlled by its own gene, whereas in the rat brain each gene must speak for thousands. Yet the mechanisms of attraction are similar.

Still, at a certain point simple and complex brains part company. The redoubtable roundworm, a see-through, millimeter-long, squiggly creature, has long been a favorite of biologists. These little snippets of protoplasmic thread have a total of 959 cells and about 20,000 genes. For them, genetic control is no problem: an average of twenty genes per cell gives the genes a good handle on things. For the fruit fly and the grasshopper, also favorites of the developmental lab, the situation is more complex, with less absolute determination of the cells' fates.

But even for simple mammals like the mouse, cells outnumber genes by several orders of magnitude, and for humans, matters are worse. We have at most a few percent more genes than the mouse, with brains many times larger. As for nerve-cell *connections*, we must multiply by 10 several times more. Thus Changeux's paradox: how can so few genes specify so many connections? Several answers have been fielded, each with a piece of the truth. The obvious one mentioned earlier is that a cabal of controlling genes can get thousands of cells or connections for the price of one. Mass-produce a given type of cell, and you can direct a small army toward a target as easily as you could a single soldier. But this is not the only answer.

First, there is great overproduction of brain cells. Half of those formed will die, many for lack of connections. This is the trophic theory of Dale Purves: early cells depend on other cells for their nourishment, and this makes for a certain genetic indeterminacy, a dependence of cellular destiny on neighboring cells' nurturance. *Only connect*, this theory says, and you will be much more likely to survive. Neurons growing in tissue-culture dishes may form surprisingly well without active connections, but in the rough world of the real embryo, where half the neurons born are doomed to die, these nourishing relationships may make all the difference.

Second, there is Changeux's own theory that synapses stabilize selectively. We know that extensions of cells and connections between them are greatly overproduced, and that many or most are pruned back. Changeux showed that, despite the substantial genetic independence of neuronal form, connections survive best if they are most active—a finding often confirmed since. Once some synapses are functional, the cells they link show electrical activity, and this gives them an edge in the cellular struggle for survival. We have long known that embryos, including human ones, are active very early; human fetuses flex their limbs, react to noise, and even suck their thumbs all within the first three months of pregnancy. Now we know that this activity shapes brain circuitry. The cells that fire together survive together, keeping their connections alive.

Third, in the late 1980s and early 1990s, an important refinement was added, an even better solution to Changeux's paradox. Gerald Edelman, a Nobel laureate in medicine for his work on the immune system, had turned to the nervous system because of the similarities we have mentioned. Immune-system cells must recognize foreign cells—bacteria or parasites, for example—to contain and destroy them. They do it by forming highly specific proteins on their surfaces, which recognize threats, or antigens. Edelman showed that the wiring up of the nervous system in the embryo depends on similar molecules—the CAMs and SAMs—but the result is cooperation, not destruction. Edelman's solution to how the brain's excesses are cut back and stabilized is different from Changeux's, however. Edelman's model, neuronal group selection, holds that functioning groups of nerve cells—simple circuits—form at the behest of genes, with little reference to activity or experience. This is not just possible but necessary, because eons of natural selection produced some fairly fixed, highly adaptive little modules for perception and action. In this view, the competition inevitable in the crowded developing brain occurs mainly among these preadapted neuronal groups. Experience plays a crucial role, but only by hooking up innate bits of circuitry.

Finally, the new sciences of chaos and complexity have begun to help explain growth in the womb. Brian Goodwin and Stuart Kauffman, among others, have pointed to two very interesting possibilities, using computer models of cells in embryos. First, exceedingly small differences at the start of a process as complex as this can multiply exponentially and cause large, unpredictable differences. This is chaos in the formal sense. It is what makes the weather unpredictable more than a few days ahead despite enormous amounts of data crunched by the biggest, fastest supercomputers. The same chaotic indeterminacy gives *identical* twins different brains by three months of pregnancy. Second, and much more strangely, computer models of embryonic cells as they divide, change, and affect one another show that chaotic unpredictability does not fan out and expand indefinitely. Patterns emerge, analogous to the patterns of turbulence in the air that eventually coalesce out of chaotic weather. Thus, in a process called "emergence," chaos in the developing brain ends of its own accord.

These models of chaos, emergence, and complexity pride themselves on needing no external constraints—genetic, environmental, or otherwise. But recall that they are only models; they happen for certain only in computers. And the fact that they *can* happen there, interesting as it is, is no proof that they happen that way in the growing brain. Chaotic and emergent effects, like activity-dependent ones, probably occur throughout development but are constantly pushed, pulled, shaped, and molded by the genes.

Yet, for all the importance of internal structures in brain development, later experience still has major effects, as we have seen in the classic experiments of Mark Rosenzweig, Marian Diamond, and Edward Bennett at the University of California at Berkeley. Impoverishing or enriching the environment of rats, even aged rats, affects the weight and thickness of the cerebral cortex, the ratio of supporting cells to nerve cells, the number and size of synapses, the amount of neurotransmitters and their enzymes, the complexity of higher-order branching of dendrites, and the number of spines crowded onto a given length of dendrite. It was later shown that the shape of the spines in jewel fish differs for those reared in isolation and those reared in communities. The latter, receiving much more stimulation, have spines that in cross section look like racquetball paddles—large, long heads with shorter stems—while cross sections of the isolates' spines look more like badminton paddles, with small heads and long stems.

This effect does not apply just to jewel fish. In a similar study, spines from mice reared either in light or in darkness showed similar shape changes. And in neurophysiological studies, after intensive short-term stimulation of single cells, such morphing of shape accompanies functional changes. Why? Because short-stemmed, fat-headed dendritic spines provide less electrical resistance to incoming signals than long-stemmed, small-headed ones do (this follows from basic physics), so changing the structure changes the action of cell, circuit, and brain.

We have long known that the brain is in some ways enormously plastic. Decades ago Nobel laureates David Hubel and Torsten Wiesel, neuroscientists then at the Harvard Medical School, together with Simon LeVay, showed that closing one eye of a young monkey during a sensitive early period allows the nerves from the other eye to compete unfairly. These competing nerves completely take over cells that normally would have become responsive to both eyes. No recovery is possible, and these monkeys never develop depth perception, which depends on binocular responses in those same cells. In other classic studies, Thomas Woolsey and his colleagues showed that removing a few whiskers from the snouts of newborn mice eliminates from the cortex areas called "barrels" that would normally have represented those whiskers, while nerves from surviving whiskers bulge their own barrels, taking over the empty spots. Fascinatingly, breeding mice for extra whiskers will in itself produce extra barrels in the cortex. This shows that genes are driving the cortical barrels indirectly, not by controlling the brain but by controlling whiskers, which in turn shape the brain.

Finally, this system can be placed in an evolutionary framework. In mice, rats, rabbits, cats, and seals, whisker length and density follow adaptive function, as their role as distance detectors for objects is enhanced. In some species—inveterate diggers like the star-nosed mole, for example—a vast cortical territory evolved just for the snout. Because the star-nosed mole burrows in soft, moist soil, it has an exquisite sensory system in its snout that is reflected in its brain. Its cousin, the common eastern mole, adapted for burrowing in dry, rocky soil, lacks these adaptations. At least some such differences are present before birth, coded by the genes. Yet evolution also builds in plasticity; selection doesn't bother to make the genes do things the environment can reliably do. In fact, it can't, since as long as the environment is doing the same job, selection lacks the cutting edge it needs to change the genes.

In the last two decades of the twentieth century it became clear that the brain can respond to experience even more dramatically than many had supposed. In experiments by Michael Merzenich, an *adult* owl monkey with a finger removed lost the brain representation of that finger, while nerves from adjacent fingers took over the territory. Another team in Merzenich's lab, led by William Jenkins, found that finger stimulation alone, with no surgery, can make lasting changes in how the fingers are represented in the cortex. In 1998 it was discovered that these changes are guided in part by a center at the base of the brain, the nucleus basalis, one of the key regions attacked by Alzheimer's disease. Thus, the tremendous intellectual losses of Alzheimer's may come in part from the brain's inability to rearrange its own structure with experience.

Human studies also show plasticity. A group using PET scanners to examine the brains of professional string players showed greatly expanded regions for the fingers of the musicians' left hands. Meanwhile, Helen Neville, a Salk Institute neuroscientist studying electrical activity in the brain, showed that people who have been deaf from birth but are fluent in American Sign Language have expanded visual regions. The left hemisphere, usually dominant for language, showed an especially expanded *visual* domain in the deaf signers. So human brains, too, are changed by experience, as existing cells expand their territories to take in new information.

But what about growing new *cells*? In the 1980s it was found that songbirds generate millions of new neurons every spring in response to testosterone, the better to sing with. Since that discovery there has been a search for functionally meaningful new neurons in mammals—whether for normal brain growth after birth, behavioral change, or recovery from injury. We have long known that, after birth, rats continue to make new granule cells, that small, numerous, densely packed class of neurons in the hippocampus. Since the hippocampus is involved in learning and memory, this finding intrigued everyone, and the search was on for something similar in primates. But for a long time it did not seem promising. In 1985 Pasko Rakic used radioactive labeling of DNA to search for evidence of cell division producing new neurons in rhesus monkeys at various ages. He found none, and the issue seemed closed.

But in the late nineties, using a more sensitive technique, a group led by Elizabeth Gould of Princeton University reversed this finding, showing that small South American monkeys called marmosets make thousands of new neurons every day. As in the rat, the neurogenesis was first seen in the hippocampus. Furthermore, the process was functionally relevant. For a male, the stress of being placed in a strange cage with the resident male reduced the production of new neurons by 30 percent, while environmental enrichment increased production. Similar effects were found in male tree shrews, whose brains are particularly responsive to subordination stress. By century's end, adult neurogenesis had been shown in rats, mice, guinea pigs, tree shrews, marmosets, owl monkeys, and rhesus monkeys, in various regions of the brain, including the lofty prefrontal cortex. In a separate line of study, Marla Luskin found an entire zone of the rat's embryonic brain, which supplies cells to the olfactory system, that continues to produce new neurons throughout life. The no-new-neurons doctrine, which had reigned for decades, was dead.

Most dramatically, in 1998 Peter Eriksson and his colleagues used an ingenious technique to detect new neuron formation in the adult human brain. Cancer patients had for clinical reasons received a radioactive substance that labeled DNA in dividing cells. When their donated brains came to autopsy, the label had been built into dividing cells—cells that also had the distinct chemical signature of neurons. Even Rakic accepted this proof of the birth of neurons in the human adult, again in the hippocampus. But there were two major caveats. First, very few cells were dividing; at the estimated 500 a day, they would only augment the brain's population of neurons by something on the order of one part per thousand in a decade. Even at thousands per day, neurogenesis would not add up to much in a brain with an estimated 10^{11} neurons. Second, we don't know whether the new neurons are incorporated into usable circuits in *any* of the species studied. Still, growing evidence proves that with or without new neurons, significant change in the brain is not just possible but routine.

How can change due to experience be reconciled with the evidence for circuit-building *without* function? One way is that embryonic events might carve out crude circuits in the absence of stimulation, leaving refinement to later experience. Unfortunately for this neat theory, we know that stimulation can play a major role in some aspects of embryonic brain growth. And conversely, some major events of postnatal brain development proceed very nicely without relevant experience.

The fact is that no simple construct will ever explain how the disparate tasks of brain building are shared between genes and environment. Talk of heredity and environment has transcended the "versus," passed beyond the "which" and the only slightly more useful "how much," to the mature question of "how." Now we know that this is not one question but thousands. For each system at each moment in development, we may have on our hands a different balance, a different division of labor, a different integration of the genes and the world. People being what we are, the torrent of argument between hereditarians and environmentalists, bigots of different stripes, will foolishly continue. For the unsuspecting listener, it will obscure subtle issues and sabotage understanding. Meanwhile, scientists will push the frontier of the field. Things have come far enough, though, to say that any analysis of human nature that tends to ignore either the genes or the environment can be decisively discarded.

CHAPTER 5

The Several Humours

The thoughts to which I am now giving utterance and your thoughts regarding them are the expression of molecular changes in that matter of life which is the source of our other vital phenomena.

—THOMAS HUXLEY, c. 1870, quoted by Charles Sherrington,
Man on His Nature

Whenever I hear an intelligent person deny the role of genes in complex behavior, I have an impulse to take him firmly by the elbow and escort him to Las Ventas, the main *plaza de toros* in Madrid. After the opening trumpets, the deliberately arch parade in the sun, and the few quiet moments staring down at the empty ring, a gate swings open and back and there prances, stumbles, or plummets into view about four hundred kilograms of herbivorous mammal, masculine gender, not in a very pleasant frame of mind. Ideally, holding the elbow of the cultural determinist would be not a geneticist, not even a scientist, but a seasoned follower of *toreo*.

Who would deny that the scores, possibly hundreds, of generations of careful breeding that have preceded our afternoon in the sun have shaped a set of genes that is somehow "behind" the behavior of the fighting bull? Each breeding ranch produces a strain with known tendencies: "For a century, the Cabreras, an Andalusian strain . . . were noted for their size and bravery. They also became noted for *sentido*, the ability to distinguish the *torero* from the lure, thus diminishing their popularity. In 1852, Don Juan Miura bought the Cabrera stud and combined it with other Andalusian strains to produce the magnificent and murderous Miuras that we still see occasionally today."

"Magnificent and murderous" does have a certain ring to it, but can details of behavior be specified? Indeed. The knowledgeable gentleman beside us at the *barrera* will check the ranch and strain of the six bulls on the card and try to predict their size, strength, and even their behavior—including not only "bravery" and *sentido* but, for example, how straight and far they will charge under the lure before turning, how tightly they will turn, how inclined they will be to charge the horses, how many times they will accept the picador's lance, how tired they will be after fifteen minutes in the ring, how they will carry their heads after six *banderillas* have been placed in their neck muscles, how sharply they will hook with their horns, how much they will tend to hook upward when the matador goes over their horns with the sword—even how easily they will die.

Our guide won't be able to predict all this about each ranch or strain, and in any case these features are not independent of one another, or of the size and shape of body and horn and the lie of muscle over bone. No prediction will apply to every specimen of a strain, however famous it may be for the trait. And no scientific studies have tested these predictions. Still, the expert follower of the bulls will be right more often than not, and will feel justly disappointed if expectations are not met. The next day's paper may even criticize the practices of the ranch, warning that the breed may be in danger.

This predictability comes from the genes. Bulls are not trained for the plaza, merely bred. Meticulous precautions are taken so that they will rarely, if ever, have seen a man on foot before they enter the ring. Growing up they see ranch hands on horses, but there are no unfriendly exchanges. There may be playful or serious locking of horns with other bulls, but at no time are there experiences of the sort that the animal will have during the last fifteen minutes of its life, when it will exhibit the behaviors that must uphold or detract from the strain's reputation. At no time will he have had an opportunity to observe and copy a model of action that might help him in that mortal situation.

It is conceivable that intrauterine effects—chemicals acting on the fetus, created from the mother's hormones or from hormonelike substances in grass—play a role in the development of some behavioral tendencies. But such effects are poorly understood and so are much more speculative than genetic effects. Learning must play *some* role; aggressive play with other bulls on the ranch helps behavior mature. But even learning certain moves at a certain pace in a certain way is itself genetically guided.

So *sentido* is something that a fighting bull learns quickly at the end of his life, in his only arena episode. That is why it is essential to prevent young bulls from seeing a man on foot. Such experience would increase the tendency to ignore the lure and go for the man, making *toreo* impossible; despite the English misnomer, "bullfighting," it is not supposed to be a fight, but a skillful, graceful, courageous ritual slaughter. What *sentido* describes, then, is the ease with which various strains carry out this learning under conditions of stress and pain; the differences in *sentido* among the strains are genetic differences in learning, specific to a narrow range of behavior and a spectacularly peculiar situation.

Of course, this is not science, it is only common sense; and at that, the common sense of members of one particular subculture to whom much of the rest of the world—or at least the Anglo-Saxon world—feels superior. But in the behavior of dog breeds, common sense and science converge. Take, for example, the cocker spaniel and the English bulldog. We accept that the distinctive stamp of face and body on these animals results from the arrangement of genes, created by hundreds of generations of careful, restrictive breeding. But we have more trouble admitting that the equally distinctive personalities of spaniels and bulldogs are also the product of genes, working through nerves and glands rather than bone, muscle, and skin.

In the 1950s and 1960s John Paul Scott and John L. Fuller took this idea seriously. At their kennel laboratories in Bar Harbor, Maine, using then-standard methods, they studied the genetic control of behavior in five pure dog breeds: the wirehaired fox terrier, the American cocker spaniel, the African basenji, the Shetland sheepdog, and the beagle. Each breed has special behaviors, and all were analyzed. But the work focused on two breeds, cocker and basenji, and on hybrids of the two. Lacking molecular methods, Scott and Fuller proceeded much as Gregor Mendel had in his classic experiments on the height, color, and texture of pea plants.

The cocker is America's darling family dog, good with children, friendly, tame. It was first bred for the hunt, but for retrieving, not assault or chase. The basenji is the hunting dog of the Central African Pygmies. Although we have no details of its breeding history, the shorthaired, short-eared dog—it resembles a small boxer—has long been used by Africans as a general-purpose hunting dog. Expectations for it included chase and attack, either alone or in groups, and the chance that it received or expected the sort of kindness that we lavish on cocker spaniels, given the attitude of rural Africans toward dogs, is nil. Basenjis that lived to breed were tough survivors, able to cooperate with humans without depending on, or offering, affection. In the lab they are obviously wilder, struggling against restraint, vocalizing and avoiding more in response to handling. They are timid and fearful of humans at five weeks of age; they yelp and snap when cornered and run away, generally acting like wild animals. At thirteen to fifteen weeks they show more playful aggression toward handlers than cockers do and throughout life are less likely to quiet when handled.

From what is known of the cocker spaniel breeding tradition, their behavior results from selection for two key criteria: crouching in response to an upraised hand, and the highly restrained "soft-mouth" bite of a good retriever. For at least dozens of spaniel generations, owners preferentially bred the dogs that showed those behaviors most clearly. Basenjis also bark much less, producing instead a wailing or yodeling sound, and breed only once a year, whereas cockers breed twice a year. Play fighting in the breeds also differs, and breeds true. Tested at fourteen weeks of age, basenji pups are much more likely to leap playfully at the tester, nipping at or wrestling with his hand. First-generation hybrids of the breeds are intermediate and unpredictable in play fighting. Second-generation hybrids (from matings between first-generation hybrids) are, like their parents, very variable. But the backcrosses—first-generation mongrels bred with pure basenjis or cockers—tend strongly back to the traits of the pure breeds. All these findings pointed toward one hypothesis: play fighting in dogs is controlled by two genes that somehow regulate the threshold for aggression, with neither gene showing much dominance or recessiveness; in crosses, the basenji and cocker traits coexist, neither dominating the other. This was the simplest model consistent with the facts.

Other studies with these dogs tested fearfulness—avoidance and vocalizing to human handling—and confirmed a simpler hypothesis. At five weeks most cockers showed no fear, while all basenjis showed some. However, the first-generation hybrids, with equal doses of basenji and cocker genes, were like basenjis, suggesting one or more genes for which basenjis carry dominant forms. This was narrowed down further by backcrosses of these hybrids to cockers. The offspring of such backcrosses were half fearful and half tame, a classic Mendelian ratio suggesting a single gene locus. Second-generation hybrids turned out to be three-fourths fearful, one-fourth tame, also a classic ratio. The data supported a very simple claim: unlike play fighting, fear in the dog is controlled by a single gene locus, and basenjis carry a variant that is dominant.

How could such a simple chemical change—perhaps a substitution of only one base among thousands in a long chain of DNA—effect a specific change in such a complex behavior? Little is known about how a minor change in DNA produces changes in chemical reactions in the brain or glands that could account for something like play fighting or skittishness in dogs. But there is relevant evidence from other species, including humans, to give us some clues to how the process works. At present it applies mainly to abnormal behaviors, including mental retardation, disorders of gender, Huntington's disease, and Alzheimer's dementia. Science begins with these abnormalities for humanitarian reasons: research on them will lead to improved treatment or prevention. Rapid progress is being made in describing the genetics of neurological syndromes in particular. This, along with progress in general medical genetics, will be accelerated by the genome sequence completed in 2000. But almost as rapidly the more subtle genetics of the normal range of human and animal behavior are also coming to light, initiating the era of behavioral genomics.

Nevertheless, in animals and humans, it is easiest to study the grosser abnormalities. Apart from neurologically mutant mice like the reeler and the weaver, dozens of other mutants are under study, not just in rodents but in roundworms, sea slugs, fruit flies, and other species. Several laboratories can now give a complete account of the chemical reactions that cause the abnormal behaviors, beginning with a simple single-gene mutation.

This kind of genetic disease tracing is also possible for some human mental disorders. In phenylketonuria (PKU), a severe but treatable form of mental retardation, much of the chemical story is in place. A defective enzyme, it turns out, is caused by one of several different mutations in the same vital gene. Researchers have figured out what chromosome the gene is on (12) and have sequenced it. We also know that a baby must have a double dose—two bad versions of the gene, one from each parent—to have the disorder. This tells us that the normal variant, if paired with the abnormal one, will chemically dominate, and no disease will appear—the bad genes are recessive. New research has shown that different errors in this same gene produce different degrees of impairment. But even before the precise structure of the enzyme or the gene that codes it was known, we knew much about its functions and how a simple, uncompensated defect in it can deprive a child of the most precious possession, a normal brain.

The defective enzyme is phenylalanine hydroxylase. Like all enzymes it is known by what it does—it changes a small molecule called phenylalanine to tyrosine, which has a slightly different structure but very different functions. Phenylalanine is an amino acid that we get by eating and breaking down protein. Its conversion to tyrosine (also an amino acid) and then to other useful things (thyroid hormone or adrenaline, for instance) is perfectly simple, except in the affected child. In that child, the pair of bad genes causes all molecules of the critical enzyme to be abnormal in chemical sequence and so in shape.

A defective enzyme means that a mutation in DNA sequence has altered the code for the molecule. An enzyme is a protein that scrunches into an intricate three-dimensional shape of coils and folds that decide its function. Enzymes catalyze (speed up) chemical reactions in the body, such as the conversion of dopamine to norepinephrine, by attracting the key reactants to their complicated surfaces and urging them into position to work on each other. Changes in shape can block the function of an enzyme, changes in amino-acid sequence change shape, and changes in genes change that sequence. So even without a major structural effect like damage to a nerve cell or gland, a gene can change a molecule that affects the way we act, think, or feel.

In the case of the PKU enzyme, the shape change cripples it; it can no longer speed the reaction that transforms phenylalanine. The amino acid, plentiful in a normal diet, piles up behind the enzyme block. Tragically, it is poisonous to nerve cells and just kills them. Yet it can be treated with almost complete success by keeping phenylalanine out of the diet. (Since tyrosine is also a common amino acid in food, there is no problem supplying the rest of the needed metabolic pathways—phenylalanine just gets skipped.) Universal screening of newborn babies for high levels of the chemical—a tip-off that the enzyme is botched—now allows the diet to be adjusted almost at birth.

In another form of mental retardation, an enzyme called beta-galactosidase is defective, also due to a recessive single-gene mutation. This causes a different metabolic blockade, with a parallel result: a compound normally carried away builds up to toxic levels. Dominick Purpura of the Albert Einstein College of Medicine described the neuropathology of this syndrome, right down to the nerve cells seen with the electron microscope. Ultimately, the toxin causes cell extensions to take on an abnormal shape; this distorts electrical transmission, and mental abilities are lost.

In effect, the chain from gene to enzyme defect to toxin to abnormal cell has been sketched in these two illnesses. In both cases, the final problem in how the brain works is due to the cell structure; because of the laws of electric conduction, such cells must fail. And while it may all seem obvious in retrospect, most of these links rest on recent discoveries. But given enough research on the genetic materials and their chemical mechanisms, it will only be a matter of time before every detail of the process is filled in, not only for these but for diseases like Huntington's,

Alzheimer's, and others. We are on the verge of complete explanation of certain disorders of the human mind, beginning in each case with a chemical breach in a gene.

But we can't get away from the environment so easily. PKU, the best understood of such conditions, is finally caused not by a bad gene but by a poison in the diet. Remove the poison, no disorder, despite the bad genes. Of course, the "poison" is no poison to most babies, only to those with the damaged gene. Strangely, this wound of the mind is all genes *and* all environment. No more impressive evidence can be brought against the folly of partitioning mind and behavior into proportions of genetic and environmental causes. Nor is there better proof that finding a genetic cause for a disorder may lead to an environmental change that largely cures it. Knowing that the disorder is genetic allowed us to identify those babies that phenylalanine would poison; screening then led to a change in their environment, which saves them.

In this era of molecular engineering, genetics holds out another hope: intervening in the machinery. Genes are regularly deleted from and added to strains of mice, and these deletions, known as knockouts, are a staple of genetic research. To placate those who claim we can't learn much from mice would be sheer folly. The fact is that (thanks to a common ancestor scrounging around among the dinosaurs) 97 percent of human DNA corresponds to that of mice. So if a human gene or protein has a puzzling function, making a knockout mouse and figuring out what it *can't* do will likely point to the human molecule's role.

Today genetic studies of severe mental disorders are accepted even by opponents of behavioral biology, since to resist would simply mark the critics as inhumane. But they still claim that this has nothing to do with normal behavior or even with less severe abnormalities such as emotional disturbances and mental illnesses. They would never concede that the perfectly clear research we now have on severe disorders of mice and men provides suggestive evidence on lesser abnormalities. And they continue to insist that there is little or no evidence of similar metabolic effects on normal behaviors.

For example, in the apparent single-gene control of the fearful reaction to handling in two dog breeds, skeptics will wait to see a metabolic chain leading to a stable difference in the level or rate of production of behavioral molecules—hormones, neurotransmitters, and their related enzymes. They'll want to see a difference in the sequence of an enzyme, receptor, or membrane channel between cockers and basenjis, proven to result from the proposed gene difference and to cause the structural and behavioral difference. That would show continuity between severe disorders and normal differences, at the level of gene control. No *complete* account exists for fearful behavior or anything similarly complex. But the extreme environmentalists are being shaken from their complacency by a good deal of research on other animals, from the tiny fruit fly *Drosophila melanogaster* to the bipedal primate *Homo sapiens*.

When Seymour Benzer, a geneticist at the California Institute of Technology, became interested in brain function, he took a crash course in neuroanatomy. He was already famous for his role in breaking the code that translates DNA into protein. But from his study of the mammal brain, he concluded that "everything in the brain is connected to everything else," and such a messy system might make it impossible to tease out the impact of the genes. The sleuth who had cracked the code of codes could not get bogged down in so inelegant a morass. He wanted a simpler, more tractable system and he found it in the common, well studied fruit fly. Not one to think small, he and his colleagues took on the task of completely describing—from gene to action—how behavior unfolds in a wide variety of fruit fly mutants. As a result, for many behavioral mutants, today we know the chromosomal location of the gene, where on the map of the early embryo the gene first finds expression, what structural change in nerve, sense organ, or muscle explains the behavior, and how, chemically, the gene makes the structural change. Such are the advantages of simple systems, as Benzer knew when he passed on mammals. A group of scientists, in about a lifetime, can hope to arrive at a global understanding of how the fruit fly system works. In mice, this would take generations; in human beings, centuries.

Yet as different as simple systems are, they have dividends of their own, because evolution links us not only to the rat but also to the slimy slug, the lowly worm, and the pesky fruit fly. And the paltry increase in genome size during evolution—a mere doubling of gene number since our common ancestor with the fly—makes the study of simple systems all the more compelling. At least 10 percent of our genes are clearly related to those of the fly.

Some fly mutants, like the mice with brain mutations, have grave disorders. The drop-dead mutant, for example, emerges and acts normal for a day or two then becomes less active and uncoordinated. Within hours, it falls on its back and dies. The syndrome begins with a recessive gene on the X chromosome and can be traced through a chunk of the early embryo destined to become the brain. As the disorder unfolds, the brain becomes riddled with holes, due to some intrinsic gene-determined factor. Of course, this drop-dead mutant is no more interesting from the viewpoint of normal behavior than are the severest brain mutants among mice. But it is only an extreme case, and there are much subtler mutants. These include changes in instinctive behaviors such as approaching light, climbing against gravity, and the twenty-four-hour behavioral cycle. Mutants with nineteen-hour and twenty-eight-hour cycles have been found, with possible implications for human "early birds" and "night owls."

One fly mutant is sluggish in its movements; another is speeded up. The "easily shocked" mutant has a seizure in response to a mechanical jolt, then recovers after a few minutes to go about its business. Another mutant is "paralyzed" when the temperature warms uncomfortably; it promptly revives as soon as the mercury drops. The "comatose" mutant is also normal, except at high temperatures; but it takes much longer to recover, depending on the length of the heat exposure. This is because electrical function in the nerves is depressed and revives only slowly. Most intriguing are the mutations affecting the fly's highly stereotyped courtship and sex: one type of mutant male courts, but with less-than-normal vigor; another pursues males as eagerly as females; still another, called "stuck," fails to disengage after intercourse. The one called "coitus interruptus" needs no explanation.

In other laboratories, research has traced the effects of gene mutations on the basic biology of nerve transmission, a process far more likely to be similar in fruit flies and humans. Some fly mutations have been linked to abnormal structures in the channels of nerve-cell membranes. These channels, which sit astride the membrane and have pores of varying size to carry ions, are made from globs of protein stuck together during development. They tend to be relatively conserved in evolution, so studying fly-membrane channels is going to be more relevant to humans than, say, the fly eye would be. Specific genes code for these proteins, and defective genes yield defective channels. Bad channels retard the flow of ions, and even a reduced flow can drastically alter nerve firing. For example, the behavior of a mutant known as "shaker" is due to changes in the proteins of the potassium channel, especially in nerve cells in the abdomen. These in turn have been traced to a mutation in one of a family of potassium-channel genes.

The links to human behavior are emerging. In the late 1990s, a group at the University of Bonn, led by Christian Biervert, found a human gene closely related to the fruit fly's shaker gene. It is a dominant gene on the short arm of chromosome 20 that is altered in one inherited form of epilepsy. In this relatively benign disorder, young infants have unprovoked seizures but usually outgrow them within a few months of birth. The gene responsible is similar in sequence to the fruit fly's potassium-channel gene. In an elegant though now common strategy, Biervert and his colleagues made RNA from the defective gene and injected it into frog eggs where, they knew, the channel would be expressed. Not surprisingly, Biervert found that with the defective gene the channel did not function—but the normal version of the same gene gave all the signs of potassium flow. So research on the lowly fruit fly, via the not-much-more-impressive common frog, ultimately shed light on a human brain problem.

Yet, as in PKU, genes are not the only things that can affect potassium channels. At almost the same time that Biervert and his colleagues made their findings, Jiuyong Xie and David McCobb showed that stress hormones affect how the products of potassium-channel genes are spliced. This means that in an animal under stress, despite normal genes, hormonal changes will produce abnormal channels, by causing abnormal splicing of messenger RNA. This alters the protein, which changes the shape and function of the channel. So modified splicing is yet another of the many now proven ways experience reaches down to affect the genes.

Finally, the fruit fly contributed even further by unveiling the genetics of a normal mutation known to be adaptive in the wild. Marla Sokolowski, of York University in Toronto, had spent twenty years trying to figure out the genetics of laziness in flies. Some fly larvae roam widely in their search for food; others are, well, slugs. Separate the two types, mate the grown-up flies with only their own kind, and sluggishness breeds true. Furthermore, a poky larva becomes a poky adult, while the energetic ones remain that way for life, despite being transformed by metamorphosis. Among flies from Toronto backyards, around 70 percent were "rovers," the rest "sitters." Sokolowski's search was rewarded when she and her colleagues showed that the gene in question, on chromosome 2, was the same that codes for an enzyme that mobilizes energy in cells. This enzyme's activity predicted behavior. More impressive, putting the rover genes into sitter eggs made larvae break faith with their heritage; *these* sitter eggs grew up to be rovers.

Then, to clinch the adaptationist explanation, an evolutionary experiment was done. The sitter-rover distinction becomes apparent only after feeding, so the scientists reasoned that sitting would be adaptive where a fly could be sure of its next meal (why waste energy?). But in a situation where food was scarce, eating a bit should not shut off the food quest. Taking wild populations, the group kept them in mixed jars of either 50 or 1,000 flies. The numbers were kept steady for seventy-three generations. Sure enough, the crowded jars now had a larger proportion of rovers, while in the roomy jars sitters thrived. Thus, a clearly adaptive, single-gene trait affecting the normal range of behavior was located, sequenced, cloned, used to modify behavior through gene transfer, shown to work through a known enzyme, and made to evolve through a fair simulation of natural selection. In short, everything we could ask for, including confident predictions that the findings will be relevant to mammals, possibly us.

The mutants described above and many others under study run from severe abnormalities to individual differences that would be normal in the wild. They include behavioral syndromes that appear only under certain environmental conditions, and they extend even to genetic differences in learning. Yet all are single-gene changes, similar in principle. In this simple insect, at least, science has established an explanatory continuum for both severe behavioral disorders and at least some more subtle behaviors. This is not to say that all fruit fly behaviors are controlled in the same way. A behavior may result from many genes rather than from one acting alone; it may depend heavily on diet or temperature during growth; or it may be shaped by learning. Even simply controlled behaviors can be subject to many influences. What we know is that some normal behaviors of great interest to natural history can result from the simplest genetic causes.

What about the laboratory mouse, a creature with thousands of times the brainpower of the fruit fly? Can we extend the spectrum of genetically influenced behaviors toward the normal range for this much more complex animal? Clearly, yes. In classic studies, genetically different mouse strains showed differences in behavior that in turn were linked to differences in metabolism. Thyroid hormones were related to activity level, serotonin and other neurotransmitters to emotionality and reaction to shock, liver enzymes to alcohol preference, ATP and glutamate to noise-induced seizures, and androgen sensitivity to sexual intercourse. The strains of mice in these experiments were raised in the same environments and differ only in genes, something like dog breeds or fighting bulls but with much greater scientific control. So some things were clear by the 1970s. One did not need to accept every item on this list to conclude that in the mouse, as in the fly, some subtle behavioral differences are under the control of plausible paths from genes to behavior. And normal mice offered a much more subtle behavioral biology than the neurological mutants did. Far from gross corruption of brain anatomy, investigators had found chemical differences: the turnover of a neurotransmitter, the shape of a membrane channel, the level of a hormone, the activity of an enzyme, the availability of an energy-generating molecule, the sensitivity of a receptor.

But that was only the beginning. By the year 2001, with the genome sequence in hand and human cloning an emerging reality, we have advanced to a level that few classic behavior geneticists had dreamed of. Even when measuring body chemistry, they were still always dealing with correlations. Today geneticists not only study genes directly, they manipulate genes to change behavior. Consider just the cases mentioned above.

Thyroid hormones were linked to activity in some strains, but today we know the genetics of several different thyroid-hormone receptors, each with several variants due to differences in RNA splicing. These are distributed strategically in the brain and expressed at crucial moments of development, to determine branching patterns of neurons both early and late in life, with profound consequences not just for activity but for learning.

We knew that serotonin was correlated with emotionality, but today we have knockout mice in which the gene for a particular type of serotonin receptor has been deliberately removed. This receptor (the $5HT1_B$) transports serotonin back into the cells that released it, and the mice that lack it have a shorter fuse and a higher frequency of attack against an intruder.

Liver-enzyme activity was known to relate to alcoholism, but today we have far more evidence of the genetic bases of alcohol susceptibility in mice and humans. Alcohol dehydrogenase converts ethanol to aldehyde and aldehyde dehydrogenase further converts aldehyde to acetate. In humans, the latter enzyme has two genes controlling it and each has several variants. In any of the three genes, a difference in a single amino acid changes the metabolism of alcohol and predicts different risks of dependency.

Noise-induced seizures had been related to brain chemistry in otherwise normal mice, but today glutamate is implicated by more than a mere correlation. Some knockout mice have altered glutamate transmission at nerve endings. Knock out Grik2, and you alter one type of glutamate receptor, making the mouse more resistant to certain seizures. Knock out Gria2, on the other hand, and you get a lax receptor that lets calcium in, leading to more seizures. Generally, glutamate receptors regulate ion flow; change the genes that make them and you change ion flow, causing greater or lesser seizure susceptibility.

And as for the old role of androgen sensitivity in sexual intercourse, well, today we recognize a broad spectrum of actions. Since androgen, ironically, often works on the brain by being converted to estrogen first, we know that the estrogen receptor plays a key role in male sex. Disrupting the gene for that receptor affects male mounting and intromission and completely prevents ejaculation. The effect of ER knockout on male courtship behavior is minimal, however. But a growing array of genetic manipulations is tracing the power of genes over the enzymes, receptors, and hormones that control sexual acts.

Close study sometimes reveals obvious structural pathology due to such genes, but other effects remain in the realm of the messily, actively chemical — the realm where altered *tendencies* of nerve and muscle, due to molecular changes not seen under the microscope, are the basis of gene action. Subtle behavior, subtle physiology — yet still partly or even decisively genetic. Thus, in a creature as complex as the mouse, our close cousin as animals go, we have sketches of genetically caused behavioral physiology, without even leaving the realm of single genes. In fact, we have the basis for learning.

Joe Tsien of Princeton University manipulated the genes for a membrane receptor in mice and made them learn better — effectively, he made them smarter and he thinks he can do the same for us. Ordinary mice have hippocampuses (that memory organ again) that change biochemically as they grow. Young mice express a certain type of receptor for the neurotransmitter glutamate, which may be why they learn better than old mice, which don't express it. Tsien and his colleagues managed to put multiple copies of the youthful receptor type into otherwise normal mice. Measured six different ways, they learned better than unimproved mice. In essence, the window of opportunity for youthful learning had been kept open longer. So much, again, for the Berlin Wall dividing nature and nurture. Learning genes will continue to open windows in that wall and ultimately may bring it crashing down.

Strangely enough, we know less about genes and behavior in dogs, bulls, thoroughbreds, and other working mammals than we do in mice and fruit flies. So the next higher creature for which knowledge becomes impressive, for which it gets harder every day to deny the power of genes, is none other than us. Not surprisingly, human behavioral genetics is the most controversial pursuit in behavioral biology, even more so than the study of sex differences. Skeptics justly point to past failings and excesses. Most tragic are the ignorant, twisted theories that have been used to support political repression and even mass extermination, from the distant past (when it was about "blood" instead of genes) to the present day. Less dramatic is the sloppiness of what used to pass for respectable research in this field. In older twin studies, the statistics were often inappropriate; given small samples, findings could have been due to chance. Moreover, some famous studies of identical twins reared apart did not ensure against separate but highly similar rearing, and in some cases "separated" twins went to the same school and were in frequent contact.

Studies of heritability often make assumptions that apply to animals but certainly do not apply to humans, yet these procedures have been relied on as if they did. For instance, in mouse studies animals are randomly assigned to different environments that have no relation to their genes, whereas in humans environments are almost always correlated with genes, making it much harder to separate the two. Not surprisingly, conventional estimates of IQ heritability, for instance, ranged from about 45 to about 80 percent, which leaves a lot of room for error.

These estimates are also based on the assumption that two given genetic endow-
ments—say, extroversion and introversion—can be compared in a consistent way
regardless of the environment—an assumption now known to be false. In reality,
simple linear predictions may turn out to be quite wrong, because the rank order
of genotypes can change in different environments.

Worst of all, some of the most famous twin data on the inheritance of IQ
were faked. As shown by L. S. Hearnshaw, a reluctant historian of British psychology, the famous IQ expert Sir Cyril Burt either made up twin correlations or, at
the very least, repeated the same ones several different times, claiming they were
from different studies (exact to three decimal places). Astoundingly, he also
invented coworkers and coauthors for papers written and "researched" entirely by
him alone. It is difficult to explain this sorry history in any way but the long-standing
bias that blood is destiny.

Yet these criticisms, strong as they are, apply mainly to older studies. Work in
human behavioral genetics has become so good and has accelerated so much that
it is possible to base all generalizations on research published since 1980. Modern
studies have surveyed whole nations' worth, including thousands or even tens of
thousands of twin pairs. Overall, the results are decisive and by 2001 it was possible to say that "nearly all behaviors that have been studied show moderate to high
heritability—usually, to a somewhat greater degree than do many common physical diseases."

But excellent work began much earlier. Consider schizophrenia, a psychosis
marked by emotional withdrawal and thought disorder. Because of its great and
tragic human cost, it has been an intense focus of scientific psychiatry that holds
many lessons for those of us who want to understand behavior in general. The
same techniques that have been applied to schizophrenia will soon lead to similar
investigations of less serious problems—mood swings and flagging attention, for
example. Eventually, they will help solve ancient conundrums like the origins of
love, fear, and war. But as always, science begins with what is knowable and
known.

Before the development of antipsychotic drugs, the chronic form of schizophrenia often required permanent hospitalization. An acute form allows long
periods of remission between attacks and there are syndromes that may be
related but are much less serious, known as the schizophrenia-spectrum disorders. The identical twin of a schizophrenic has about a 50 percent probability of
becoming schizophrenic. But a randomly chosen unrelated person of the same
age and sex has a probability of less than 1 percent of having this illness, and a
nonidentical twin has a probability around 15 percent. These numbers suggested
strong heritability. Still, dissatisfied with twin research, Seymour Kety, a physician-
scientist at Maclean Hospital in Massachusetts, led an elegant study showing that
schizophrenia is to an important degree genetic. He and his colleagues hit on the
now common strategy of studying adopted children in a large population. They
had access to excellent Danish records for the whole adoptive population twenty
to forty-five years old in greater Copenhagen—more than 5,000 people. Of these,

33 were diagnosed as schizophrenic (a typical incidence); 28 of these had been adopted away from their biological parents before the age of six months. The 33 were carefully matched with 33 normal adoptees for demographic and economic characteristics. Several hundred relatives (adoptive and biological) of the two groups were found and interviewed.

Several different analyses led to the same conclusion. For individuals adopted before one month of age, schizophrenia in the *biological* relatives of *normal* adoptees occurred at the nonexistent rate of 0 in 92; in the *adoptive* relatives of the same people, it was 1 in 51. In the *adoptive* relatives of *schizophrenic* adoptees, 2 in 45 were schizophrenic. Statistically speaking, none of these three rates differed significantly from the others. But the incidence in the *biological* relatives of *schizophrenic* adoptees was 9 in 93, almost 10 percent—much higher than in the other groups of relatives or the general population, resembling levels in the biological relatives of schizophrenics who were *not* adopted away. This classic study, using every control lacking in earlier studies, proved beyond a doubt that schizophrenia is partly genetic. Subsequent large studies have had similar results. The risk of the disorder is around ten times higher in the primary relatives of schizophrenics than in other people. And since most studies have found little or no overlap between schizophrenia and manic-depressive illness within families, the inheritance seems fairly specific. One study in Finland failed to find identical twins more similar than fraternal twins, but it reversed itself in the 1990s as previously normal identical twins of schizophrenics became ill themselves—so the Finland study became the exception that proves the rule. The case for genes in schizophrenia is very strong.

These studies, then, tell us that there are between three and five gene loci that influence schizophrenia, along with some other genes that make minor contributions. But in a sense they are all ensemble players, since even the major genes increase risk by only two- to threefold. This is substantial, but recall that we must account for about a tenfold greater risk in primary relatives, suggesting that in schizophrenia-prone families several risky genes converge. To complicate matters, the various genes might play starring or supporting roles, or even bit parts. In the model presented by Irving Gottesman and James Shields, there is a continuum of vulnerability for schizophrenia. Some rare cases may be completely genetic, perhaps even due to a single rare gene. These would produce schizophrenia in almost any environment. At the other extreme, however, there may be environmental conditions—such as being reared in a closet or exposed to some devastating brain virus—that would produce schizophrenia regardless of genetic vulnerability. But most cases would be in between, along a continuum where various combinations of genes with small effects could accumulate to make someone slightly, moderately, or very vulnerable.

It remains for the future to find a convincing account, a step-by-step process from a gene or genes to the disturbance of feeling, thought, and behavior—links that must be chemical, physiological, and perhaps structural. So far, despite years of effort prior to and during the Human Genome Project, chromosomal linkages for schizophrenia have eluded investigators. As two authorities said recently, "The nomination of 'candidate' genes is on shaky grounds when the pathophysiology of a disorder is completely unknown." Yet there are promising leads.

First, the classic drugs that work against schizophrenia—drugs that do not just sedate but specifically improve disordered thought—share one action in common: all go to dopamine synapses, where they block that neurotransmitter's receptors. This dampens signals between dopamine neurons and the next cells in their circuits. Amphetamine and cocaine, which worsen schizophrenia and which in large doses can produce a full-blown psychosis resembling it, do just the opposite, by *stimulating* dopamine receptors. This and other evidence led to a focus on dopamine, and thousands of studies have confirmed this lead.

Advanced imaging techniques reveal other clues: both structural and chemical anomalies in the brains of living schizophrenics. Several studies have shown reduced brain mass, with enlargement of the ventricles and their watery, noncellular content. There are structural abnormalities in the hippocampus and decreased activity in the frontal lobes. Hippocampal imaging studies have been confirmed by microscopic research on schizophrenic brains in Arnold Scheibel's laboratory at UCLA; these brains have disorganized neurons in this structure vital to memory. But such evidence does not necessarily imply a genetic foundation. In one imaging study of identical-twin pairs where only one twin is schizophrenic, the well twin did not show weakened frontal-lobe activity. Whatever caused this difference, it wasn't genes.

Still, the specificity of the brain regions involved can be remarkable. David Silbersweig, Emily Stern, and their colleagues showed that hallucinations in schizophrenic patients activate the hippocampus and closely related limbic-system cortex, as well as dopamine-rich subcortical structures. But *neurological* patients with hallucinations—who, unlike schizophrenics, *know* that their visions or voices are not real—have different patterns of brain activation. Although none of the anatomical theories is proven, each is plausible and under study. In addition, other neurotransmitters, hormones, and brain chemicals modulate nervous activity and have been implicated in schizophrenia in even more complex ways. Whenever a molecule is implicated—its level, rate of production or removal, or the sensitivity of receptors for it—there lies an opportunity for one or more genes.

This possibility has not been lost on researchers. For example, the enzyme monoamine oxidase (MAO) oxidizes and removes several brain transmitters, including dopamine, norepinephrine, and serotonin. An early study showed that the enzyme's activity was reduced in the blood of schizophrenics compared with control subjects—although its brain level has been harder to measure. Slowed removal might leave an excess of dopamine or another neurotransmitter, which might in turn cause excess stimulation or be converted to something toxic. A defective version of MAO could be produced by a single gene leading to reduced MAO activity. Research continues to explore a possible role for genetic MAO variations in schizophrenia.

Another early candidate was dopamine beta hydroxylase (DBH), the enzyme that converts dopamine to norepinephrine. In pioneering studies, DBH activity was lower at autopsy in several areas of schizophrenics' brains than in normal control subjects. And in acute attacks of schizophrenia, patients had higher DBH activity than after the psychotic episodes had passed. In addition, DBH activity seems to be heritable. People with very low blood levels have siblings and other close relatives whose levels are lower than average, and identical twins have more similar levels than do nonidentical twins. Although more complex explanations are possible, a simple one consistent with the facts was that very low serum DBH activity stemmed from a single, incompletely recessive gene—that is, one usually masked by a normal gene in the pair, but only partly. Family pedigrees and other lines of investigation supported this analysis. Unfortunately, modern gene-linkage studies have so far failed to find an association between the disorder and the chromosomal locale of the DBH gene.

But perhaps the best hypotheses of this kind involve the genes for dopamine receptors. If a receptor is defective, or just different in some way, that may mean there is an altered gene, a different code for the molecule. Like an enzyme, a receptor is a protein and its three-dimensional shape is its signature. Just as the shape of an enzyme allows it to catalyze a reaction, attracting two molecules to pockets on its surface, a receptor of a certain shape engages a neurotransmitter or hormone as a lock takes a key.

Several different dopamine receptors have been cloned and their distribution in the brain is consistent with what we know about how antischizophrenic drugs work. Of at least five different dopamine receptors, the one known as D2 is most implicated. But D2 can reside before or after the synapse and blocking it would have very different effects in the two locations; it could either turn off the next neuron in the circuit or suppress feedback to the neuron that released the dopamine. One laboratory also found a specific mutation in the D2 receptor, the replacement of the amino acid serine with cysteine, due to an apparent mutation more frequent in schizophrenics. But other research failed to confirm any linkage of the disease to chromosome 11, where the D2 gene resides. Nevertheless, this kind of research, combining state-of-the-art genetics and brain imaging, should make us hopeful about an account of the disease in the near future, at least for some schizophrenics.

I have dwelt on schizophrenia because its tragic human dimensions have brought it exceptional scientific attention and talent, so that it serves as a model for analyzing other human behaviors. We are beginning to have accounts of such processes, aside from the dramatically abnormal ones. One begins with a recessive single-gene defect that alters the enzyme steroid 21-hydroxylase. The steroids, small but intricate molecules, are hormones vital to behavior. Made from cholesterol, all have an elegant four-ringed structure with some side chains attached. The main links are carbon atoms and, depending on the side chains, the molecule can be testosterone, the male sex hormone; estradiol, the main female hormone; progesterone, the hormone of female fertility and pregnancy; or cortisol, the main human stress hormone. All processing starts by clipping the tail off cholesterol to make progesterone. Then, different paths in the testis, ovary, placenta, and adrenal cortex make the other hormones. These varied metabolic machines depend on different enzymes to tack on or snip off various steroid side chains.

The enzyme that interests us is one of three that change progesterone to cortisol. This transformation takes place in the cortex of the adrenal gland. Because of the gene defect, the pathway of cortisol production is blocked and progesterone is forced into a different production path—one that normally leads to male hormones. If the baby is a girl, she will have higher levels of testosterone during early development than other girls. This can increase aggressive behavior and reduce doll play in childhood, even if corrected at birth. It can also dampen the fantasies of marriage and motherhood that most girls have. This was one of the first single-gene effects in the normal range of human behavior, and not a frank pathology like schizophrenia.

Advances in genetics have enabled us to observe even subtler effects. Marvin Zuckerman of the University of Delaware has for decades studied "sensation seeking." This includes physical risk taking, as in skydiving and mountain climbing; a relish for new experiences, whether in the arts, travel, friendship, or even drugs; hedonistic pursuits like partying, sex, and gambling; and an aversion to routine work or dull people. Many studies in the United States and England show consistent findings: males always exceed females, sensation seeking declines with age, and genes play a strong role. In one study, 233 pairs of identical twins had a correlation for sensation seeking roughly three times as large (.60 versus .21) as fraternal twins. Physiologically, sensation seekers have a bigger bounce in heart rate in reaction to novelty and bigger brain-wave responses to intensifying stimulation. Both main sex hormones, testosterone and estrogen, are higher in seekers, and monoamine oxidase (MAO) activity is lower, which may mean that certain neurotransmitters—notably norepinephrine, dopamine, and serotonin—are removed more slowly.

The questionnaires in most studies rely on self-description, but to the extent that behavior can be measured directly, the results are consistent. High scorers engage in more frequent, more unusual sex, with more partners; use more alcohol, cigarettes, drugs; eat spicier food; volunteer more for experiments; gamble more; and court more physical danger. If they become mentally ill, sensation seekers are more likely than others to suffer from hypomanic or sociopathic disorders. Studies at the University of Wisconsin, using a somewhat different measure called thrill seeking, supported the concept. Their measure predicted fights and escape attempts in prison inmates but also goes with the extroverted bent in ordinary people that can make them natural leaders.

Research had long suggested that these characteristics had a genetic component, but in the mid-1990s an Israeli group led by Richard Ebstein and Robert Belmaker stumbled on a possible *specific* gene for novelty seeking. It coded another dopamine receptor, the D4, which is heavily distributed in the limbic system. The gene's special feature is a "hypervariable" region, a central stretch of 48 DNA base pairs that can be repeated the same way up to 11 times without making the gene useless. Repeats make the protein longer, which makes the receptor weaker, so that it binds less dopamine. The key fact here is that all this variation is normal. That makes the D4 a good candidate for explaining some aspect of normal variation.

In Ebstein and Belmaker's study, a true-false test of personality, including novelty seeking, was given to 124 Jerusalem residents. Those having long D4 genes, with more repeats of the variable stretch, scored higher on the desire for new and exciting experiences. No other personality trait was associated with the D4 gene, but the relationship to novelty seeking occurred regardless of gender, age, educational level, or ethnic group. A research team led by Dean Hamer at the National Institutes of Health repeated the study with 315 ordinary American subjects. The finding was the same. Furthermore, when brothers or sisters who differed in novelty seeking were compared, the adventurous sibling was more likely to have a long D4 gene.

But is this *the* gene for novelty seeking? Statistical analysis showed that its effect is significant but small. It accounts for about 4 percent of the normal variation in the character trait. This is not much, but it is only a bit less than the percentage of breast-cancer variation attributable to BRCA1, a gene of tremendous interest for cancer research. Like breast cancer, heart disease, and schizophrenia, novelty seeking will turn out to have multiple determinants. In analyses of the overall contribution of genes and environment, the trait is estimated to be about 40 percent genetic. That means the D4 gene may explain as much as 10 percent of the *genetic* contribution to the trait; there may be nine or more others. And of course, since 60 percent of the variation is due to the environment, changing the environment could easily swamp the effects of having a long, repetitive D4. But, all other things being equal, such a gene could make you more of a risk taker.

Jerome Kagan, Nathan Fox, Steven Reznick, and other psychologists have taken a different tack in their study of timidity, shyness, and behavioral inhibition—traits that are something like the opposite of risk taking. In children as young as fourteen months they can identify a subgroup that is unusually bold and an opposite subgroup that is very timid. Those who at the earlier age are inhibited in a novel situation grow up to be four-, five-, and seven-year-olds who are restrained and socially avoidant with unfamiliar children and adults. They are also more likely to show unusual fears—fear of television violence, for instance, or of being alone in their rooms at night. Signs of the trait can be found in infancy and persist into the teenage years. At four months of age, the timid children were more reactive to stimulation in a laboratory test and fussed or cried more easily in response to a colored mobile. Between ages twelve and fourteen, during interviews with a child psychiatrist, children who had been judged timid in infancy were less likely to speak spontaneously, smile, or laugh than were their bolder counterparts in the same pubertal stage. Such stability throughout infancy and childhood, measured in various forms of actual behavior, is in some ways more convincing than the more common behavior-genetic approach of self-report through paper-and-pencil testing.

Not only is timidity very stable, it is also associated with biological measures that may link genes to behavior. Several measures suggest a higher tension or tone in the sympathetic nervous system. These include a higher resting heart rate, with comparatively little variation from moment to moment; a greater increase in heart rate in strange situations; higher blood pressure in the same situations; greater dilation of the pupils during questioning; greater tension in the vocal cords; higher levels of the stress hormone cortisol; and more manufacture and processing of norepinephrine, the principal chemical produced by nerve cells in the "fight-or-flight" system. A composite index of all these factors showed an even stronger link to timidity than each did alone. It has also been suggested that the set point for fear in the amygdala (the brain's fear center) is altered during development by these children's different levels of cortisol.

Timid infants can also be distinguished from bold ones through direct studies of brain function with electroencephalograms, or EEGs. Nathan Fox and his team at the University of Maryland found that at nine months of age, the reactive or fearful babies had more activity in the right frontal lobe than in the left, while for the uninhibited ones the reverse was true. This was consistent with decades of research by Richard Davidson and others showing that negative emotions tend to come more from the right hemisphere, positive ones from the left. Kagan and Fox believe that the amygdala, not only a center for fear but a key switching point of the limbic system, may mediate between the frontal lobes and the sympathetic nervous system in these children and may have a different set point or baseline in the different temperaments.

Animal studies suggest that the limbic system may indeed have a lower threshold for arousal—something more like a hair trigger—in inhibited individuals. "High-strung" would be one way to put it. Perhaps genes alter the enzymes that process norepinephrine, the main neurotransmitter in the fight-or-flight system. In any case, this subgroup of young children—perfectly normal, yet distinctive in behavior and biology—certainly exists. And it is probably only the tip of an iceberg of biologically stable traits in adults and children, most of which await discovery. Although claims have been made for specific genes for happiness, criminality, anger, homosexuality, anxiety, depression, and other traits, they should be taken with a grain (or maybe a sack) of salt. In 1987, based on excellent studies of large extended families among the Amish, a gene marker for manic-depressive illness, or bipolar mood disorder, was announced with great fanfare. It was located on chromosome 11 and was expected to lead quickly to an understanding of this dreadful disorder. Within a year or two these hopes were dashed and the claim withdrawn, because one or two individuals who had not been affected by the illness broke down and became ill in the interim. Since those crucial patients did not have the marker, their wellness had helped to confirm the gene's role in explaining the disorder. Now that they were ill, *without the suspect gene*, they had the opposite effect, destroying the validity of the original statistics and embarrassing the investigators. It is certain that manic-depressive illness is heavily influenced by genes, but this gene was not one of them, and so far the culprit genes have not been found. Genes for subtler human conditions, like happiness or sexual attraction, will be even harder to find with certainty. They just don't stand out enough in the great sea of genetic and environmental forces that constantly wash over us. Nevertheless, they are there and some day they will be found. The search for quantitative trait loci to explain the genetics of subtle human behavioral traits is well under way, greatly aided by genome sequencing. As Peter McGuffin and his colleagues have said, "Ultimately, the human genome sequence will revolutionize psychology and psychiatry."

Of course, biological variations in character have been described for centuries. The Elizabethans, like the ancients, thought that human dispositions came from four body fluids, called "humours." Each man or woman had a unique blend of humours; some had so great a predominance of one bodily fluid as to cause nasty behavior and mood. A sanguine temperament, reflecting a predominance of blood, might in moderation make a person warm, cheerful, lively, and hopeful, even ruddy in complexion. But carried to excess, the result would be far too animated and passionate. The phlegmatic, in whom phlegm held sway, could be sluggish, apathetic, and dull, although phlegm might make a normal person admirably cool and calm. Two types of bile, one producing the choleric or angry temperament, the other the melancholic or depressed, rounded out the four types. A subtler mix of the four fluids could give rise to intermediate temperaments, but only a perfect balance could make an ideal person with ideal moods.

We smile at these notions today, but they are useful in two ways. The first has to do with motivation. For generations psychology was driven by "drive theories," according to which behaviors are energized by "forces," innate or acquired, that push man or beast to eat, have sex, or explore. These forces were analogous to the forces in a hydraulic pressure system, and the consummatory acts—say, drinking or copulating—were thought to resemble letting off steam. The subjective sense we have of being "driven" to eat, have sex, or sleep reinforced the analogy. But these were only analogies, and nothing in the body corresponds to them.

Indeed, the humours metaphor of the Elizabethans is closer to the truth than is the drive metaphor. The brain is not a hydraulic system. No fluid in it builds up under pressure, urging us to do this or that action, relenting after we "let off steam." Of course, no fluids corresponding to the Elizabethan humours exist either, but the concept is nevertheless closer to current thinking. Behavioral scientists prefer words like *state, arousal,* and *excitation,* which can be specific or general. Rather than building up pressure until release occurs, action tendencies rise and fall with internal and external causes, often without deprivation or release. The terms we use to describe them are nonmetaphoric and epistemologically cautious, merely describing what can be seen and measured. But they have one great advantage: a plausible relationship to what goes on in brain and body.

Just as behavioral arousal can vary with the time of day, day of the month, season, and year of life—as well as respond to many more specific cues—neural or hormonal arousal varies, too. The day may come when we realize the Elizabethan aspiration, characterizing not only stable temperaments but also short-lived changes and long-term personal growth by reference to a subtle mixing of fluids. We will have to give up, of course, the simple calculus of blood, phlegm, and bile in favor of dozens of hormones and neurotransmitters, but perhaps the Elizabethans will forgive us, since we are closer than ever to their way of thinking. In the words of neuroanatomist Walle Nauta, "Etymologically at least, motivation is what moves us, and I believe that what moves us is moods." When we can describe the biology of moods we will understand some major forces behind behavior.

But the greater piece of foresight in the humours theory goes beyond fleeting motives to account for stable temperaments. The Elizabethans saw the temperaments as innate, due to this or that inborn balance or imbalance of fluid. Today we emphasize dimensions of personality rather than types, as in the five-factor model of Paul Costa and Robert McCrae. Thus, someone can be introverted and neurotic or extroverted and neurotic, and so on in endless combinations. But good studies supporting the heritability of normal temperaments are fairly recent. In the meantime, we have animal research and some knowledge of the effects of single genes on human behavior. These work through chemical pathways affecting hormones and the brain, but they are only the start of a hugely complex matrix.

Also, we must emphatically add that every genetic alteration is subject to reversal by an appropriate environmental change—one that either is known or can be found out. Nongenetic change can also mimic in various ways the effects of a bad gene. The contrary belief, which holds that genes are destiny and that proving a genetic effect is tantamount to throwing up our hands, would be silly if it were not both persistent and pernicious. It is false, of course, as eyeglass wearers and insulin takers can testify, and in the new era of human dominance over genes it is particularly shortsighted. The proof of a genetic effect says nothing at all about possible subsequent change, except that knowing the genetic, metabolic path can only help identify and treat the problem. Genetic analysis of behavior can lead to an increase not only in human welfare but in freedom.

What can be said now is that two individuals who differ genetically, in whatever particular chemistry controls behavioral tendencies, will, *if raised in identical environments*, grow up to act, think, and feel quite differently. The effort to ignore these facts and to trace every aspect of character, personality, or behavioral tendency, desirable or otherwise, to some feature of our experience—especially early experience in the intimacy of the family—is vanity. This approach can explain some things, but far from everything. There is in each of us a residue of traits of heart and mind that we brought with us not just when we left the womb but indeed when we entered it. The denial of this, as liberal as it sounds, is really a denial of individuality, every bit as dangerous as the most rigid genetic determinism. Extreme environmentalism had its heyday when the excesses of genetics threatened, for decades, not only human dignity but survival. It served a noble purpose and was for a time an objectively respectable position. But the sands of scientific time have shifted beneath that position, and those who tend it had best look to a foothold.

CHAPTER 6

The Beast with Two Backs

The beast with two backs is a single beast . . .
—ROBERT GRAVES, "Seaside"

Laura Allen, Camilla Benbow, Laura Betzig, Monique Borgerhoff-Mulder, Patricia Draper, Anke Ehrhardt, Helen Fisher, Patricia Goldman-Rakic, Kristen Hawkes, Sarah Blaffer Hrdy, Melissa Hines, Corinne Hutt, Julianne Imperato-McGinley, Carol Nagy Jacklin, Alison Jolly, Doreen Kimura, Annelise Korner, Marie-Christine de Lacoste, Jane Lancaster, Jerre Levy, Bobby Low, Eleanor Emmons Maccoby, Diane McGuinness, Alice Rossi, Meredith Small, Barbara Smuts, Judith Stern, Dominique Toran-Allerand, Beatrice Blyth Whiting, Patricia Whitten, Sandra Witelson, Carol Worthman. These are the names of some distinguished women scientists who have devoted their lives to research on the brain, hormones, or behavior. Each is considered a leader in her discipline, highly competent, rigorous, and original. Each has devoted many years of thought to the question of whether the behavioral sex differences she has observed—in the field, clinic, or laboratory—have a basis that is in part biological.

Without exception, they have answered this question in the affirmative. But one cannot imagine that they did so without difficulty. Each must have suffered, personally and professionally, from the discrimination against women that is common outside the academy and within it. Each must have worked with some man who envisioned her—in his heart of hearts—barefoot, meek, and pregnant in the kitchen. Each has sacrificed more than her typical male colleague has to get into position to work on a problem that interests her, and paying that price makes the truth more compelling than the politically correct. Yet each is wise enough to know that the very sorts of oppression she has experienced could be bulwarked by her own theories of natural gender differences.

Such scientists must perform a formidable balancing act. In their very achievements they exemplify the struggle for equal rights. Yet at the same time they uncover and report evidence that the sexes are irrevocably different—that after sexism is wholly stripped away, after differences in training have gone the way of the whalebone corset, there will still be something different, something grounded in biology. They have to endure the smirks of the "I knew it all along" sort of men—and they are not people who suffer fools gladly. And they have to contend with feminist critics who, in many cases, cannot or will not comprehend their research. We cannot know how they reconcile these conflicts. But we can do them the courtesy, before turning to their discoveries, of acknowledging their persistence and courage.

We can begin with Margaret Mead, one of the greatest of all social scientists. If she had become the first such scientist to win the Nobel Prize in medicine— she could have been cited, for instance, for her contributions to pediatrics and psychiatry, as well as for profoundly shaping our present concept of human nature—the choice would have done credit to the prize. In a world in which all odds were against it, she established a concept of human differences as more flexible, more malleable, more bent by the winds of life experience, as delivered by our very different cultures, than anybody thought possible at the time.

No question so engaged her as the role of gender. In trip after stubborn trip to the South Seas, enduring hardships rare for a woman of her culture, she gathered information impossible to come by otherwise. Among headhunters and fishermen, medicine men and costumed dancers; in steamy jungles, on mountaintops, and on vivid white beaches; in bamboo huts, in meeting houses on stilts high above lakes; and in shaky seagoing bark canoes, she took out her ubiquitous notebook and recorded the behavior and beliefs of men and women who had no awareness of American-style sex roles. By 1949, when *Male and Female* was published, she had done this in seven exotic, remote societies. She knew more about the variety of gender roles than any person ever had. And with that knowledge, which has stood the test of time, she surprised every smug male in America:

> The Tchambuli people, who number only six hundred in all, have built their houses along the edge of one of the loveliest of New Guinea lakes, which gleams like polished ebony, with a back-drop of the distant hills behind which the Arapesh live. In the lake are purple lotus and great pink and white water lilies, white osprey and blue heron. Here the Tchambuli women, brisk, unadorned, managing and industrious, fish and go to market; the men, decorative and adorned, carve and paint and practice dance-steps, their headhunting tradition replaced by the simpler practice of buying victims to validate their manhood.

And among the Mundugumor, river-dwelling cannibals of New Guinea, men and women seemed to be equally masculine:

> These robust, restive people live on the banks of a swiftly flowing river. . . . They trade with and prey upon the miserable, underfed bush-peoples who live on poorer land, devote their time to quarreling and headhunting, and have developed a form of social organization in which every man's hand is

against every other man. The women are as assertive and vigorous as the men; they detest bearing and rearing children, and provide most of the food, leaving the men free to plot and fight.

Imagine a startled readership in the United States in the wake of the world war. Fresh from victory, buoyed by a sense of the rightness of the American way of life, and experiencing that vivid apparent proof of the gap between the sexes that only triumph in war can bring, Mead's readers were treated to men who primped, gossiped, and danced all day while their down-to-earth female counterparts, reluctant mothers at best, took care of business. Moreover, these ways were perfectly workable, and some had gone on for 1,000 years.

Yet there was a flaw in Mead's argument. In all her cultures there was homicidal violence and, in all, that violence was done by men. Tchambuli men, as the passage above indicates, may have been effeminate in terms of certain American conventions, but they were still very devoted to taking victims—and, more traditionally, to hunting heads. Mundugumor men were evidently not threatened by the practice of their women providing for them, but that is because it freed them to plot and fight.

This part of the pattern can be traced through the world's thousands of cultures. In every culture there is at least some homicide—whether in the context of war, vendetta, ritual, or daily life—and in every culture, men are mainly responsible for it. Among the !Kung San of Botswana, noted for their pacifism as well as for equality of the sexes, the perpetrators in twenty-two documented homicides were all men. Fights over adultery or presumed adultery were involved in several cases, and a majority of the others were retaliations for previous homicides. These two themes of jealousy and vendetta pervade the cross-cultural homicide literature.

The perpetrators are overwhelmingly men. In fact, every measure devised to reflect participation in physical aggression favors men at every age, in every culture. In a sample of 122 distinct societies in the ethnographic range, distributed around the world, weapons were made by men in all of them. There are many exceptions at the individual level and in rare cases—such as modern Israel and Eritrea, or nineteenth-century Dahomey in West Africa—partial exceptions at the group level. What we are dealing with, to be sure, is a difference in degree, but one so large that it may as well be qualitative: men are far more violent than women. Even in dreams the distinction holds. In seventy-five tribal societies on all continents, men were more likely to dream of coitus, wife, weapon, animal, death, red, vehicle, hit, ineffectual attempt, and grass, while women were more likely to dream of husband, clothes, mother, father, child, home, female figure, cry, and male figure.

Recent research by Bobbi Low of the University of Michigan has revisited the relationship between gender and early warfare, and it has strongly confirmed what Mead had inadvertently discovered. Others, too, have repeatedly found that men account for the overwhelming majority of warriors in nonindustrial societies and that the capture of women is both a cause and a consequence of war in as many as half these societies. Literary sources, including Homer and the Bible, confirm the central role of young women as a goal or perquisite of ancient wars. And as Laura Betzig has shown in her fine historical survey, despotic empires carry this pattern to an extreme in which large numbers of women consistently flow upward to the beds of powerful men. Men have always made wars and they have often made them over women.

But what makes Mead's finding compelling is that it was the opposite of what she was looking for, indeed hidden in the context of her pervasive emphasis on the flexibility of sex roles. The women among Tchambuli and Mundugumor might have been disgusted with child bearing and rearing, but they did it. So do women everywhere in the preindustrial world. There are individual exceptions, but there is no society in the ethnographic or historical record in which men do nearly as much child care as women. This is not to say anything, yet, about capacity; it is merely a statement of plain, observable fact: men are more violent than women, and women are more nurturing, at least toward infants and children, than men. This is a cliché, and a politically incorrect one, but that does not make it less factual. For the moment, at least, there are no inferences to be made from it. It is nothing but plain, dull fact.

But of course, it is ethnographic fact and that raises some eyebrows. Although the cross-cultural surveys quoted above are quantitative, they are based on studies consisting mainly of mere description. As such, they are the victims of ill-founded "hard science" snobbery. Ethnology is in its earliest phase as a science. As in botany, zoology, anatomy, histology, geology, and astronomy, that phase must be descriptive. Indeed, to this day the pages of professional journals in histology and anatomy publish nonquantitative papers. Just as the "mere" description of a new brain nucleus or a type of bacterial infiltration is a first step on a new path in science, so, equally, is the description of a culture. Description, using the human eye, ear, and mind, given, at least at first, without statistics.

Yet numbers are needed, too, and are more usual in psychology than in anthropology. Psychologists in the Western world have now studied gender differences with an exactness that is hard to match in the tropical jungle. Eleanor Maccoby, a respected leader of American psychology, and Carol Jacklin, a younger scientist partly trained by Maccoby, wrote a major book, *The Psychology of Sex Differences*, that has stood the test of time. It systematically reviewed and tabulated hundreds of careful studies of sex differences in tactile sensitivity, vision, discrimination learning, social memory, general intellectual abilities, achievement striving, self-esteem, crying, fear and timidity, helping behavior, competition, conformity, and imitation, to name a few.

The main thrust of the book was to demolish clichés about the differences between boys and girls, women and men. There was no evidence that girls and women are more social, more suggestible, or have less achievement motivation than boys and men, or that boys and men are more analytic. On the measures of tactile sensitivity and fear and timidity there was weak evidence of a gender difference—girls show more of each. There was also weak evidence that girls are more compliant than boys and less involved in assertions of dominance. In the realm of cognitive abilities, there was some evidence for superiority of girls and women in verbal ability, and of boys and men in spatial and quantitative ability. A subtler but substantially similar case can be made today. Some theorists have linked these differences to tentative findings about the size of the corpus callosum and the degree of interconnectedness of the cerebral hemispheres. Others have pointed out that in species in which males have greater spatial ability (like humans), males also have larger hippocampuses, whereas in species with the reverse order of spatial ability, the female hippocampus is enlarged.

But the strongest case for gender difference remains in aggressive behavior. Out of 94 comparisons in 67 different studies compiled by Maccoby and Jacklin, 57 showed statistically significant sex differences; of those, only 5 showed the difference in favor of girls. In the other 52, boys were more aggressive than girls. The subjects ranged from age two to adulthood, and the specific measures ranged from hitting, kicking, and throwing rocks to scores on a hostility scale and included such things as fantasy and dream material, verbal aggression, and aggression against dolls. Of 6 studies that measured actual physical aggression, 5 found that boys exceeded girls, with 1 showing no difference. In an earlier book, Maccoby had summarized 52 studies of "nurturance and affiliation"; in 45 studies girls and women showed more of it than boys and men, whereas in only 2 did males score higher, with 5 showing no difference. As in the case of aggression, subsequent cross-cultural analyses have extended and supported this generalization.

While it is difficult to get accurate information for nonindustrial cultures on such measures as verbal and spatial ability, excellent work has been done on child behavior, using rigorous measurement and analysis. Beatrice Whiting has been a leader in this field; she originated techniques of study and sent students out to remote corners of the earth (as well as made field trips herself) to bring back accurate knowledge about behavior. She is one of the most quantitatively oriented cultural anthropologists and has built an edifice of exactitude on the foundation laid by Margaret Mead.

In research known as the Six Cultures Study, Whiting, together with John Whiting and other colleagues, organized teams to observe children's behavior in standard settings in a New England town ("Orchard Town") and in five farming and herding villages throughout the world. In Mexico, Kenya, India, Japan, and the Philippines, as well as in New England, hundreds of hours of observations were made on children of all ages, using uniform methods. Children were scored on twelve small units of behavior—things like "seeks help," "offers support," "touches," "reprimands," and "assaults." All the data were analyzed in a statistical procedure called multidimensional scaling. The children varied on two overarching dimensions: "egoism versus altruism" and "aggressiveness versus nurturance."

In all six cultures, boys showed greater egoism, greater aggressiveness, or (usually) both. The *difference* varied from culture to culture, presumably due to differing inculcation of gender roles. More interesting was the finding that although girls in one culture may be more aggressive than boys in another, the direction of the difference *within* any one culture is always and tediously the same. These studies of children too young to be fully socialized to their cultures, using more exact methods than most anthropologists have used, underscore rather than jeopardize the hypothesized gender difference. The data have also been extended to include five other quantitatively studied cultures, with similar conclusions. Under natural conditions, boys are more aggressive and less nurturing than girls.

But what are "natural conditions" for humans? Cultural ones, of course. The children in Whiting's studies, who ranged in age from three to twelve years, had been culturally trained and all of these cultures may be sexist. Cultures around the world are in fact quite consistent in their efforts to *train* gender roles: 82 percent of a sample of 33 cultures try to get more nurturance out of girls than boys and none attempt the reverse, according to ethnographers; furthermore, 85 percent of 82 cultures give boys more training in self-reliance than girls, again with no reversals. Couldn't the universal difference in aggressive and nurturing behavior stem from an equally universal (although unexplained) sex-role training pattern, rather than from biology? This in effect was the theory of Anne Fausto-Sterling, who proposed that because all modern humans are descended from a small population that lived around 100,000 years ago, arbitrary cultural patterns in that little founder group could have persisted to the present day.

To dig below the level of cultural influence, it helps to look at younger children. In Maccoby and Jacklin's list, 27 of the 94 studies involved children under age six. Of these, 14 found boys more aggressive, and only 2 found girls more aggressive; in one of those two the difference reversed when the sample was expanded. In a cross-cultural study not part of Whiting's analysis, three- to five-year-old children were observed in social interaction in London and among the !Kung—noted, as mentioned before, for egalitarian sexual politics. Two separate observers using different techniques—one concentrating on facial expressions, the other on physical acts—found boys more aggressive in both cultures. Other studies of three-year-olds have measured biological variables, such as heart rate, and related them to later aggressive behavior. Eleven-year-olds who'd had higher resting heart rates at age three were more aggressive than those whose heart rates had been lower. Such studies suggest that even when we don't readily find sex differences in aggression at early ages, there may be other differences that form a foundation for them.

It is harder to find aggressiveness in infancy, but we can certainly look at sex differences in behavior. For instance, at three weeks of age, in middle-class homes in the United States, boy babies are more active, cry more, look at the mother more, and sleep and vocalize less than girl babies—all oft-repeated observations. Unfortunately for simplicity, mothers of three-week-old boys hold, burp, rock, stimulate, arouse, stress, look at, talk to, and even smile at them more than do mothers of girls, and they have been doing so since the boys were born. What is the first question we ask at the birth of a baby? Precisely. We need that piece of information to know how to act, how to feel, even toward a newborn babe in arms.

Gender is a key to proper behavior. Among our closest animal relatives, the first reaction to a birth in many species is that adults show up and inspect the infant's genitals. Knowledge of gender shapes adults' behavior toward infants, even in monkeys and apes.

As for us, the same baby will produce very different responses from adults, depending on whether it is dressed in pink or blue; and a tape of a child's mischievous utterances will draw amusement and encouragement from adults who are told that the child is a boy but negative reactions from matched adults who think the child is a girl. This has been shown many times in many settings and there are now systematic rating scales for mothers' and fathers' gender stereotypes of newborn babies, complete with an acronym: the PASTON scale, short for "parental sex typing of newborns." Fathers stereotype more than mothers do, but both parents do it, with their own and others' infants.

These facts, along with much other evidence that stereotypes operate throughout childhood, make it harder to argue that boys and girls differ so much in their behavior that they make their parents treat them differently. At six months of age boys show more emotional distress than girls in face-to-face interactions with their mothers, including when the mother keeps a still face. But is that because the sexes are inherently different, or only because the mothers have had six months to shape theirs sons and daughters? Even at three weeks we have the chicken-and-egg problem; we can't prove that the infants' differences come first. Still, we can go younger. Annelise Korner spent decades studying newborns, and one of her interests was sex differences. Like others, she found reproducible differences within days or even hours of delivery. Boys at birth showed more muscle strength—more head lifting when prone, for example—while girls showed greater skin sensitivity, more reflex smiles, more taste sensitivity, more searching movements with their mouths, and faster response to a flash of light. The last response closely reflects brain function, being measured by tiny evoked electrical potentials recorded over the visual part of the brain.

In other studies, newborn boys are more irritable than girls, cry and grimace more, have more widely fluctuating emotional states, and smile and comfort themselves less. It is not easy to relate such differences to aggressiveness, and perhaps it should not be tried. But emotional instability and negative expressions might well be relevant. And babies with greater skin sensitivity could be conditioned by life experience to be less combative. In one good study following children from birth to age five, those with less skin sensitivity at birth were more likely to attack and surmount an inanimate barrier at five, and this correlation held *within* each sex, not just between sexes. Other studies have found sex differences in barrier behavior as early as age one, when boys are more likely to break through an obstacle and girls more likely to go around it or stop trying. Muscle strength at birth is also another plausible precursor of aggressiveness.

But before we resort to such indirect effects, consider another kind of evidence: studies of hormones, behavior, and the brain. The idea that humoral factors influence male and female behavior is very old; castration has long been used

to reduce aggressiveness in animals and men, and systematic experiments showing that this works began by 1849. So many studies have been done in so many
species—including humans—that the case is closed. The question is no longer
whether hormones secreted by the testes promote or enable aggressive behavior,
but *how*.

The main male hormone is testosterone. It belongs to a chemical class
known as steroids, which includes the justly famous cortisone—a nearly natural
compound with widespread medical uses. The two main female hormones are
also steroids: estradiol—the key estrogen in humans—and progesterone, the pregnancy enhancer made in massive quantities by the placenta and in lesser quantities,
in nonpregnant women, by the ovaries. Estradiol, progesterone, and the pituitary
hormones that regulate them produce the monthly cycle, a spectacular system
whose dynamics are not yet fully understood. Although nothing so sophisticated
as that exists in males, there is common ground between testosterone's mode of
action and that of the two female sex steroids.

Steroids are small, as biological molecules go, and they have many important actions on nonneural organs—testosterone, for example, promotes muscle
growth in teenage human males, an effect with clear importance for behavior.
But the most direct, and probably the overwhelmingly important, route by which
a steroid hormone (or any substance in the blood) can affect behavior is by transfer from blood to nerve and brain. The brain is the main organ of behavior; for a
molecule to affect behavior it must generally first affect the brain, or at least the
nerves.

Sex steroids are no exception. But in addition to the more typical means of
influencing nerve cells—direct, immediate action on neural activity—shared by
drugs, diet, and neurotransmitters, steroid hormones have a special access route.
Being fat-soluble, they dissolve right through cell membranes, which are basically
two rows of fatty molecules, and then through the same sort of membrane into the
nucleus. There they combine with receptor molecules, and the steroid-receptor
pair goes to the genes—literally. Steroid hormones work by directing DNA and
regulating how it makes RNA and protein, affecting the most basic functions of
the cell. Steroids may not interfere with the machinery of heredity, but they do
regulate its expression, in the most intimate example of gene-environment interaction. Here are genes whose effect on the organism—on the very cell they
inhabit—is so far from fixed that it is vulnerable to the merest winds of bloodborne humoral factors. That means anything in the environment that can influence, say, testosterone—diet, stress, temperature, seduction, even fantasy—can
potentially toss a molecular wrench among the delicate cogs of the gene machinery. So much for the fixed effects of genes.

To be fair, this conclusion is a bit improper; the genes fully expect such interventions. Still, this mechanism added a new dimension to our concept of gene-
environment interaction. Nerve-cell genes are no exception to this fifth-column
infiltration, although so far we know little of how they respond to it. One thing we
do know is that it is slow and long-lasting. Unlike the mechanism of a com-

pound—say, amphetamine—that acts quickly and, for the most part, transiently on nerve cells, steroid hormones bound to their receptors may land in the nuclei of their target cells and get in contact with DNA hours before they alter function. For example, giving a rat a shot of estradiol, radioactively labeled for tracing, produces a high concentration of the hormone in certain brain cells—specifically, in their nuclei—within two hours. Twenty-two hours after that, there is a correspondingly massive increase in the tendency of the rat—if female—to respond to stimulation with sexual posturing. What happens in those twenty-two hours tells a tale not only about hormone action but about gene action that is changing the way we look at cell biology.

Meanwhile, we know that steroids change behavior and that they get around quite well in the brain. Radioactive labeling shows that they pass from blood to brain selectively, concentrating in regions that serve reproductive behavior. Density is highest in the hypothalamus, at the base of the brain, and in other limbic, or emotional, areas—just where theory would like them to be. Concentrations occur in circuits active in courtship, sex, mothering, and aggression—the behaviors in which the sexes differ most and the ones most influenced by testosterone, estrogen, and progesterone. Although we don't yet understand how the system works, there are clues. For instance, injecting testosterone lowers the firing threshold for fibers in the stria terminalis; this pathway leads from the amygdala—the "almond" nucleus that is a switching center for fear—to the hypothalamus. As such, it helps the limbic system enable sex and aggression and gives substance to testosterone's role in behavior. It is one thing to say that a hormone probably influences sex and aggression by acting on the brain; it is quite another to find a major nerve bundle deep in the brain, implicated in emotion, that can fire more easily after testosterone seeps into it.

But we don't need to reach so deeply, since peripheral nerves also concentrate these hormones. In songbirds, the male is usually the singer, and testosterone targets the nerves to his voice box, or syrinx. This is part of how testosterone promotes song—not, alas, an expression of pure joy but more like the bombast of a self-important suitor. Yet males are not alone in these hormonal susceptibilities. Injecting female rats with estradiol enlarges the sensitive region of the pelvic nerve, even when that nerve is detached from the brain. This makes the female more susceptible to intimate touch. In these and many other ways hormonal flux affects nerve function and, through it, behavior.

Such is the the physiologist's view, which is, not surprisingly, pretty mechanical. More surprising is that someone like Alice Rossi embraced it. Rossi is a family sociologist and respected feminist. After years of distinguished research she became dissatisfied with Durkheim's dictum that only social facts can explain social facts and so began to think that some social facts might be partly explained by biological ones. Well versed in biology, she explained it to her sociologist colleagues and did not conceal her belief that some gender differences in behavior—in sex and parenting, for example—are attributable to biological causes.

Rossi (in collaboration with Peter Rossi) did classic research on women's behavior and mood during the menstrual cycle. Using precise measures and an econometric model, she showed that mood cycles in college women can be partly predicted from hormonal ones. For example, women had more negative feelings during the luteal phase of their cycles—beginning four or five days after ovulation, when progesterone peaks. More interesting, and rare in such studies, a group of men were included for comparison. While there was no evidence of a male cycle—despite some tantalizing clues, no one really has found one—men had the same number of days per month of physical discomfort as women. But at least some of the women's bad days were predictable from the cycle, usually coming during menstruation. Some men claim that menstrual distress disqualifies women from such positions as airline pilot and president. But do we want our plane—or our country—piloted by someone who has a few days a month of distress that come around like clockwork, or by someone with the same number of down days arriving randomly?

Also, cyclic hormone changes do affect women's mood and behavior but may work partly through fluctuating discomfort, not subtle hormonal effects. Other studies have shown increased irritability or depressed mood in the last few days before menses. Several studies have shown a small increase in sexual interest or sexual activity around ovulation, possibly due to fluctuations in testosterone rather than estrogen or progesterone. In research among the !Kung San of Botswana, Carol Worthman and Marjorie Shostak showed that women in that very different culture also report a significant increase of sexual desire at midcycle, in association with hormonal changes.

Thanks to research by Martha McClintock, it has long been known that women living in close association tend to synchronize their menstrual cycles and there is now definitive evidence that this occurs through pheromones—body secretions detected by smell. Still, these and other intriguing cycle-related social changes are weak effects for most women. As shown by Kim Wallen and his colleagues at the Yerkes Primate Research Center, even rhesus monkey females have far more complex social and psychological determinants of desire than rats do. If even monkeys aren't mechanical—*push-pull, click-click*—when it comes to hormones and desire, surely women are even less so.

But the issue of negative mood before and during menses is a different kind of question. Premenstrual syndrome (PMS) is a serious problem for some women, and claims that it doesn't exist or is purely cultural are insensitive and ill-informed. Experiments show that giving gonadal steroids to women with and without PMS has different behavioral effects, proving that the connection cannot be psychological. Both animal and human studies have confirmed that mental functions, activity, seizure susceptibility, and pain sensitivity, as well as symptoms of brain disorders like Parkinson's and Alzheimer's, are affected by menstrual cycle hormones.

But how important are these influences? In Rossi's classic study of average young women, workdays and weekends had a much larger cyclical effect on mood than hormones did. Another study, by Diane Ruble, dramatically demonstrated the power of suggestion: misleading women about what stage of the cycle they were in led to reports of mood more synchronized with the *false* cycle than with the true, hormonal one. Yet another showed that environmental factors are stronger than genetic ones in determining the severity of premenstrual symptoms. Such findings should signal caution to anyone who thinks these matters are purely biological. We know from ethnographic research that menstruation lives a vivid symbolic life in many cultures. Any process with so strong a grip on the human imagination, giving rise to countless myths, taboos, and rituals, can surely on occasion mislead scientists.

Cycles aside, Rossi has been broadly interested in sex differences in behavior and social roles. In reviewing the sex difference in nurturing—present in most families in all cultures—Rossi has accepted the possibility that its roots may be partly hormonal. She defended this viewpoint—a courageous one for a cofounder of the National Organization for Women—in many articles and in her books on sexuality and parenting. She knows that modern women are doing a complicated balancing act among economic, social, and cultural forces, but she also believes that a biological component unique to women must be brought into the balance.

In the tradition of Rossi's pioneering contributions, three works by women scientists, published at the millennium, laid a new evolutionary foundation for women's roles. Alison Jolly, a distinguished primatologist, wrote *Lucy's Legacy: Sex and Intelligence in Human Evolution*, a profound yet lively and often funny account of our evolution and destiny that accepts the inevitability of gender differences. Concentrating on behavior, Jolly weaves a tale of instinct and passion, intelligence and purpose, past and future that grants the greater aggressiveness of males and greater nurturance of females as part of the basis of who we are. To those who belittle these biological forces, she gives the back of her deft and steady hand. In her view, the past half-century of women's achievement and liberation will lead to and nurture a better world in part because of, not despite, gender biology.

Sarah Blaffer Hrdy, a leading theorist and observer of behavior, contributed an evolutionary account of mothering in *Mother Nature: A History of Mothers, Infants, and Natural Selection*. Her view is far more complex than the romantic ideal of motherhood, but it is frank about the differences between mothers and fathers. In a magisterial account of the relevant natural and human history, she establishes women as evolutionary strategists who, far from being passive vessels of male reproductive dreams, constantly weigh the costs of motherhood against its future benefits. Compelling as infants and children are, women are designed by nature to refuse and reject them at times, and to succumb to their appeal at others. This work extends Hrdy's earlier classic, *The Woman That Never Evolved*, which showed that in contrast to the male idealization, primate females are formidable competitors whose evolutionary legacy made women a force to contend with.

And Bobbi Low, a leading behavioral ecologist and historical demographer, published *Why Sex Matters: A Darwinian Look at Human Behavior*. After numerous accounts by men of the evolutionary psychology of gender—many useful but none completely convincing—Low's careful exposition destroys the claim that this line of thought is a sexist plot. She begins with the origin of sex itself: "One answer, then, to Professor Higgins's sad cry, 'Why can't a woman be more like a man?' is that once there are two sexes, with different paths to success in reproduction, the strategies that work for each are likely to be very different." After more than 1,000,000,000 years of sexual reproduction, the differences in both aggression and nurturance, although they vary marvelously among modern species, nevertheless continue to reflect the power of engrained, gender-specific goals.

But how do evolutionary aims play out on the field of mechanistic biology? Hormonally, nurturance itself has not been as easily studied as aggression, which in some ways at least is its antithesis. In many studies of humans and animals, testosterone enables and might even cause aggressiveness. While no one with any knowledge thinks that the causal link between testosterone and aggression is simple, a link does exist. Not only does castration reduce aggression and testosterone restore it in many animals, but there are fascinating correlations between aggression and testosterone level in group-living, normal animals. For example, when two groups of monkeys fight, the winners show a rise in hormone levels, the losers a fall.

Studies of aggression and testosterone in prison inmates do not paint a simple picture, but some findings stand out. Among male prison inmates, the higher the adult testosterone level, the earlier the age of the first arrest. That is, the men who had the highest levels had been arrested youngest, in early adolescence. In another study, the level of testosterone in male juvenile delinquents was correlated with their observed aggressive behavior. Recent research on large samples of prison inmates by James Dabbs and others during the 1990s confirmed the association. Dabbs has devoted many years to testing this hypothesis, working in a maximum-security prison. In a sample of 700 young male inmates, those whose crimes were violent—rape, child molesting, homicide, assault, and robbery—had testosterone levels more than 10 percent higher than those convicted for nonviolent crimes like burglary and drug offenses. Those who had violated prison rules had about 25 percent more testosterone than those who had not, and this difference was even stronger for those whose rule-breaking had been disruptive or combative. Testosterone was also tied to rank in the inmates' pecking order: "bo-hogs," the tough men at the top, had high testosterone, while "scrubs" did not.

So in this real world of very real violence, the main male hormone had some predictive value. One of Dabbs's collaborators, Marian Hargrove, had been an assistant to the attorney general of Alabama, where she learned all about murder and mayhem. She said that "when a high-testosterone man kills you, you are *very* dead." They had done things like stabbing a victim twenty-eight times, shooting a person twice in the front and five times in the back, or burning the dead body.

Women commit crimes, too, of course, and they too have testosterone, so Dabbs and his colleagues wondered whether they would show the same relationship. In 180 women in two prisons, murderers had more of the hormone than did other inmates and women who were aggressive and dominant in prison had higher levels than did those who were not.

Dabbs and his group have extended this work to many walks of life. With Robin Morris he examined Veterans Administration records of more than 4,000 men who had fought in the Vietnam War. They set aside the 10 percent with the highest testosterone levels and compared them with all the rest. The high group had been more likely to misbehave in school as children, get into trouble with the law as adults, go AWOL in the army, use drugs and alcohol, and have ten or more sex partners in one year. This was a lifelong pattern. Recall that the highest 10 percent were compared not with the lowest but with all the others; still, on average they had *double* the risk of these behaviors. Another research team, led by sociologists Alan Booth and Cynthia Gimbel, studied the combat history of the same group of men. A combat-exposure score was devised from questions about receiving and returning enemy fire, encountering mines or booby traps, being ambushed, and killing or seeing killings. Those who had no exposure had lower testosterone levels than those who had.

It needs to be stressed at this point that in most studies these differences are small. They are too large to have occurred by chance but too small to explain more than a fraction of the behavioral differences. We also know, from decades of research by Allan Mazur and his colleagues—in a sense the mirror image of the Dabbs group's research—that testosterone is not just a cause but also a *function* of experience, influenced by the stress of winning or losing. We are learning much more about the determination of violence. Still, too few ongoing investigations are focused on this important biological phenomenon.

One clear fact about sex hormones is that they surge at adolescence. From very low levels during early and middle childhood, testosterone (mainly but not only in males) and both estradiol and progesterone (mainly in females) rise to adult levels within a few years, and the female cycle starts. These massive changes influence behavior. The indirect effects start with physique. Male size, shoulder breadth, muscle mass, and deepening voice, all of which influence behavior, are determined almost completely by testosterone. Meanwhile, estrogen mainly accounts for the very different shape changes in girls' bodies, as well as fat deposition and breast development.

But it was surprisingly late in the history of this science, during the 1980s and 1990s, that the first really good studies were done attempting to link hormones to behavior without the intermediate link of changing physique and the resulting altered self-image. The physiological changes are not subtle or small: testosterone levels rise about eighteenfold in boys and twofold in girls, leaving boys with about ten times as much of the hormone as girls. Estrogen levels rise eightfold in girls and twofold in boys. The *rate* of change may be crucial, and a given boy's more than tenfold increase in testosterone may occur in a single year. Yet girls, despite lower levels, may be sensitive to testosterone, and small increases may affect them markedly. In boys, genital size, pubic hair, and total body growth are testosterone-driven; in girls, estrogens, progesterone, and androgens all contribute. The same hormones that speed growth later stop it, by shutting down the growth plates near the ends of the long bones.

Yet these events do not occur in lockstep fashion. Carol Worthman, who is both a field anthropologist and a world-class endocrinologist, has shown that the order and timing differ in different cultures. Still, the pace of puberty and the amount of problem behavior are heavily influenced by genes. For example, identical twins have higher correlations than nonidentical twins for the age at which growth is fastest (.80 versus .40), the age at first sexual intercourse (.41 versus .18), and the amount of delinquent behavior (.70 versus .48). Despite the great changes, correlations between identical twins, or between adopted children and their biological mothers, hold steady or increase as the child grows. Moreover, the pacing of puberty often runs in families, further supporting the influence of genes.

Emotions often track these growth changes. In some studies, depressive moods increase in both sexes, at least during puberty itself; but self-esteem declines in teenage girls while it rises in boys, and girls are depressed more often. Despite the conventional wisdom about teenage mood swings and hormones, it is not clear how these changes relate to body chemistry, as opposed to their relation to things like role change, mental maturation, and a new self-concept based on visible body changes. If acne and sullenness emerge together, were both caused by chemical turmoil, or was the mood just due to a glance in the mirror? Still, growing evidence suggests that the hormones of puberty can directly produce mental changes. Studies have shed new light on mood changes in teenage girls. Worthman, working with Adrian Angold and others, followed almost 1,300 girls between nine and sixteen years of age in the Great Smoky Mountains Study, one of the most comprehensive projects of its kind. The risk of depression for girls increases during a well-defined period at mid-puberty, and both estradiol and testosterone are involved. Other factors are statistically less important and these authors conclude that changing body image and other psychological factors are not as influential as androgens and estrogens in producing the characteristically higher prevalence of depression in adult women.

Studies of adults and of animals, where no external body changes confound the issue, show relationships in both directions between hormones and behavior. For instance, as we've seen, testosterone at least *enables* aggression and dominance in many species, including humans, and in clinical settings it stimulates desire in *both* sexes. Other studies link the hormone to aggression in teenage boys. Dan Olweus did long-term research on the total national population of bullies in Norway's schools. The overwhelming majority of these threatening children were male, and there was a strong relationship between testosterone and bullying. Another study looked at hormones, personality, and sex in 200 eighth- to tenth-grade boys and girls, who were asked to check which of 300 adjectives applied to them. Higher-testosterone boys were more likely to check *ambitious, cynical, dominant, original, persistent, pessimistic, robust, sarcastic, severe, showoff, spontaneous, stingy, temperamental, and uninhibited.* Higher-testosterone girls checked a list that was only slightly overlapping: *charming, cynical, discreet, disorderly, dominant, enterprising, frivolous, initiative, original,* and *pleasant.* These findings strongly suggest that the hormone makes a difference, but it is not so easy to label it.

In the late 1990s a growing body of evidence linked testosterone levels to the onset of sexual activity for both boys and girls and to the subsequent frequency of sexual intercourse in boys. This is not to say that cultural and social influences do not matter—on the contrary, factors such as attendance at religious services, media influence, and the sexual activity of best same-sex friends are also important. It is also difficult, as we have seen, to separate direct hormonal effects from bodily change. For example, boys whose faces are rated as more "dominant-looking" by other teenagers were found by Allan Mazur and his colleagues to have a distinctive sexual development. In the blunt terms of their title, "Dominant-looking male teenagers copulate earlier." But is that because testosterone has made their faces look more dominant, which in turn led to early sex? Or is it that testosterone independently influenced both facial form and sexual behavior?

In girls the drama of first menstruation makes it a little easier to separate the two types of influence. Most psychologists agree with Elisa Koff of Wellesley College that "menarche is a pivotal event for reorganization of the adolescent girl's body image and sexual identity"—even in our culture where little is done to mark it. Furthermore, the impact is typically negative. Girls who have been favorably prepared for their first monthly period have an easier time with it than girls who are taken by surprise; but for girls in general in our culture, first menstruation is experienced and remembered negatively. As Koff put it, "Menarche may have a positive personal significance" by association with adulthood, "but this . . . is overshadowed by a negative interpersonal significance as girls become increasingly self-conscious, embarrassed, and secretive." And that's without even mentioning the physical discomfort. We can't separate the effects of menarche on emotions from those of hormones, since hormones bring on menses, and bad feelings come in part from rising steroid levels. But self-image plays a key role. Regardless of hormones, the child, girl or boy, is an increasingly intelligent spirit aware of physical change.

Cultural influences matter as well. Consider a !Kung girl having her first menstruation, at sixteen. Wherever she is, whatever time of day or night, as soon as she notices blood she must sit down, keep absolute silence, and wait. Knowing that she has been ripe for this transition, the women in her village—mostly her relatives—will find her. They go to her, lift her up—her feet are not allowed to touch the ground—and carry her all the way back near the village. There they build a tiny grass seclusion hut—it may only take an hour—and place her in it. Soon a dance begins, for women only. All the adult women join in and as the day, the night, and the next day wear on, the dance becomes raucous and ribald. Women shed all clothing except a small leather pubic apron, which they toss wildly around as they dance—they even pull it up to show off their genitals, provoking uproarious laughter. Singing, dancing, and clapping go on continuously, and no men are allowed anywhere near the place. Meanwhile, the menstruating girl sits alone and sober-faced in the nearby seclusion hut, not allowed to laugh or speak, only to watch and listen. Yet the message cannot be lost on her: this is a spectacular celebration of womanhood.

Western cultures do nothing comparable. But a girl who chats openly about it with her parents in advance and then receives a dozen roses after the event will have a different experience from the girl who, completely unprepared, bleeds through her jeans on the baseball field. Ideally, there should also be a sense of pride in the growth of feminine curves, but if a girl overvalues slimness, she may resist the change in shape, which results in part from strategic deposits of fat. In contrast to preferences in the nonindustrial world and indeed to those of the West until the twentieth century, many girls hate normal fat levels and some end up with eating disorders. Boys, on the other hand, tend to be happier with their bodies; they want more muscles, and they get them.

A friend of mine, age 8, was asked at school if he thought life was harder for a girl or a boy. He wrote, "The same because they both need babies. They go to school. They go to college. They both work. They both learn. They both have feelings. From Andrew." Andrew's claims are deeply true and speak volumes about our shared humanity. One could leave it there—allude to the Graves epigraph, stress the overwhelming similarity between the sexes in brain and behavior, and suggest that evolution made "a single beast" with a single twist—a burst of different hormones from the gonads just before reproduction starts. In other words, just when the genders most need to be different. The trouble with that plot line is the overwhelming evidence that the sexes differ long before puberty, when there are not enough sex hormones to explain the difference. For example, studies throughout the world show that preschool-age boys are more aggressive than girls. Still, this would be fairly easy to explain without invoking biology. Parents behave in a gender-guided way, and small children identify with and model themselves on the same-sex parent. That could be all we need to get cross-cultural regularity.

But consider this experiment. Rhesus monkeys, those favorite objects of laboratory scrutiny, were raised in total social isolation. Subjects, male and female, had no sex-role training and no chance to identify with a parent—in fact, they had no social experience of any kind. At about age three—roughly our age ten—each monkey was put in a room with an infant monkey of randomly chosen sex. The result? Females cradled and cuddled the infant more, while males hit the infant more, and the difference was highly significant statistically. With no divergence as yet in hormones, and no differences in rearing, what could account for such a large gender difference?

Growing evidence puts the cause deep within the brain. In 1973 it was shown for the first time that there are structural differences between male and female brains. Near the front of the hypothalamus, one of the oldest parts of the brain, male and female rats differed in the density of connections among local nerve cells. Not only that, but castrating males just after birth left them with the female brain pattern, and injection of testosterone into females—likewise just after birth—gave them the male pattern. This discovery permanently changed the study of gender. For one thing, it was the first proof that the brains of the sexes differ in any animal. For another, the difference was where it should have been—in a center for the brain's regulation of sex hormones. But most impressive was the proof that sex hormones circulating at birth could change the brain.

For years it was clear that in mice, rats, dogs, monkeys, and other animals, testosterone and related hormones given to female young at or before birth suppressed their normal female sexual postures and in some species abolished sexual cycles. In males, castration or an antitestosterone drug at birth suppressed normal adult sexuality, *despite* replacement therapy with testosterone later on. One of the most impressive experiments produced "pseudohermaphrodite" monkeys by administering male hormones to female fetuses before birth. As they grew, these females showed neither the characteristic low female level of aggressive play nor the usual high male level but a level precisely in between.

For these reasons, investigators had already begun to talk about "androgenization of the brain"—a change in the brain made by male sex hormones at or before birth. It was already established that the basic mammalian body plan is female and stays that way unless diverted by male hormones. The new claim was that the brain, too, was being masculinized, but the evidence remained indirect and the claim speculative until the 1973 Raisman and Field study. They gave the phrase its first genuine meaning and also gave real credence to the view that *pre*adolescent gender differences in aggressiveness were as biological in origin as the more easily understood postadolescent ones.

But these discoveries were just the beginning. A few years later, Dominique Toran-Allerand of the Columbia College of Physicians and Surgeons did a tissue-culture experiment in which she was able to watch the process in action. She made thin slices of the hypothalamus of newborn mice of both sexes and kept them alive in glass dishes long enough to treat them with sex hormones, including testosterone. Her brief initial paper, published in *Brain Research*, showed the stunning results in photomicrographs. Many cells in the mouse hypothalamus are growing at that age—putting out neural tentacles that will finally form connections to other cells; but the cells in her hormone-treated slices showed more and faster-growing extensions than those treated just with the oil the hormone was dissolved in.

The only difference was the hormone. In effect, Toran-Allerand was able to watch as steroid hormones changed the newborn brain. This in no way implied that faster growth made the masculinized hypothalamus better, but it did make it different. Her work has continued for decades, confirming the discovery and revealing a chemical mechanism. First, paradoxically, testosterone had to be converted to estrogen in brain cells before it could have its effects. So why aren't female fetuses *estrogen*ized? Because proteins in the blood bind estrogen and prevent it from entering the brain. Testosterone, on the other hand, gets into the brain easily, is changed to estrogen, dissolves through the cell's oily membranes into the nucleus, and goes to work. Once *inside* the neuron, the hormone pairs with a new protein, the estrogen receptor, and the resulting gorgeously structured glob clasps a key stretch of DNA, turning a gene on.

This much was known by the mid-1980s. Toran-Allerand then discovered that the genes turned on by the hormone make proteins for nerve-cell growth. These proteins, including nerve-growth factor, or NGF, direct the cell in building extensions of itself, the same protruding tentacles that become axons and dendrites. Finally, steroid hormones like estrogen and growth promoters like NGF are cross-coupled. That is, they work together for maximum efficiency, promoting growth in specific parts of the brain. As mentioned, we knew that the basic mammal body plan is female. That this was *not* a necessary arrangement was shown by bird embryos, where the opposite is true; the basic plan is male, and added female hormones produce females. But the mammal story is clear: the genetic signal for maleness, on the Y chromosome, transforms a basically female structural plan, through male hormones. Gene signals cause the female organs to recede and the testes to be made, and the testes make hormones that do the rest. Furthermore, these male hormones masculinize parts of the brain later in development—often around the time of birth.

Are such generalizations applicable to that most puzzling of mammals, the one that studies its own nature? Doubts about this—formidable at first—were largely dispelled by the studies of Anke Ehrhardt and her colleagues, first at the Johns Hopkins School of Medicine (especially John Money), and later at the Columbia College of Physicians and Surgeons. Ehrhardt spent decades studying the condition and clinical treatment of certain unfortunate "experiments in nature"—people with anomalies of sexual and gender development. One such condition, the adrenogenital syndrome, was briefly discussed in the last chapter. A genetic misstep results in the absence of one enzyme in the adrenal cortex—the outer adrenal gland. As a result, instead of normal amounts of the stress steroid cortisol, it produces abnormally large quantities of the sex steroid testosterone. For girls with the syndrome, male levels of the hormone float around in the blood throughout gestation until birth. After birth the condition can be corrected, through surgery and medicine. If this is done, the hormone's effects are purely prenatal.

Yet at age ten and in adulthood these girls are psychologically different from their sisters and from unrelated controls. By their own and their mothers' accounts, they play with dolls less, are more tomboyish, and express less desire to be married and have children when they grow up. Studies in the 1990s by Melissa Hines, Gerianne Alexander, and others, on new samples of girls with adrenal hyperplasia corrected at birth, showed similar results on toy preference, rough-and-tumble play, and preference for playing with boys. June Reinisch, formerly head of the Kinsey Foundation for Sex Research, studied girls and women exposed in the womb to progestational hormones, given to their mothers for pregnancy maintenance. Such hormones also have androgenlike properties, with some parallel effects on later behavior.

Whatever value judgment we place on these differences, they seem to be real, confirmed by different investigators with different samples and even with different syndromes that have comparable hormonal effects. But they are mainly social differences, not cognitive ones. A careful review of available studies of clinical syndromes and drug effects that could correspond to early masculinization of the brain concluded that "evidence is most consistent for a developmental influence of androgens on sex-typical play. There also is some evidence supporting a role for androgens in the development of tendencies toward aggression and for androgens, or estrogens derived from them, in the development of sexual orientation." But quite categorically, "Suggestions that IQ, attainment of developmental milestones, and academic performance are enhanced after prenatal exposure to high levels of androgens or progestins have been refuted." Still, combined with increasing animal evidence, these findings suggested that humans, too, could show psychosexual divergence, affecting both brain and behavior but probably not intelligence, due to prenatal masculinizing hormones.

This would explain an extraordinary case that came to light in 1997. A baby born in the 1960s had a tragic surgical accident: his penis was destroyed by a botched circumcision. At the time medical wisdom held that gender identity was formed gradually in a child's mind, over the first three years of life, from appropriate models, cultural guidance, movies and television, and the like. For this reason, the unlucky baby was given the sex-assignment "female" shortly after the accident. Surgical construction of a vagina in childhood and female hormones given in adolescence seemed to complete the identity-forming process that, along with countless pervasive family and cultural influences, would make this baby an infertile but otherwise normal woman.

Or would it? Actually, the young woman this child became was relentlessly restive in her gender role. Eventually she began to search for clues about her early development and for the first time discovered her medical history. Poignantly, she experienced this discovery as a great relief. It appeared to explain a lifetime of feeling like a misfit, and she immediately began to take steps to reorient her gender identity. As "Joan" had once had hormone and surgical treatments to become more female, "John" now reversed the process with male hormones and genital reconstruction. "He got himself a van, with a bar in it," one of his doctors said. "He wanted to lasso some ladies."

Other cases of ambiguous gender at birth grow up in more ambiguous ways, traveling a slow path of identity growth in adulthood. For example, some XX-chromosomal females with congenital adrenal hyperplasia or other prenatal androgen exposure feel that they should be men, but most, despite their tomboyish natures, do not. This is not surprising, since they had not had full prenatal exposure to masculinizing hormones, as John/Joan had. In a detailed study of four such women who did decide to become men, the transition was possible but difficult. Three had had their ambiguous genitalia corrected back to the female form in infancy, and all were designated as girls within a few weeks of birth. But because of their somewhat masculine psychological tendencies, we may say that their gender identity was ambiguous in childhood. As they grew up and became sexually active, all were exclusively interested in women. We might say that at that point they were just normal lesbians. But they gradually formed a determination to undergo sex change. Supportive romantic and sexual relationships with women played an important role in several of these cases, the last step in the gradual process of forming a masculine identity.

Of course, as Anne Fausto-Sterling and others have argued, it might be better for a person like this if we recognized more than two sexes: males, females, and several intermediates. Why not, critics say, just let these rare babies be themselves, and stop trying to force them into our rigid two-sex mold? Michel Foucault, the French philosopher, reviewed the life of Herculine Barbin, an early nineteenth-century person of ambiguous gender who wrote a memoir suggesting that all was well until doctors began to exercise their curiosity. Thereupon a forced identity choice ended what Foucault called "the happy limbo of a nonidentity" and tragically led to suicide.

Nearly two centuries later the two-sex model still reigns in Western culture, but its tyranny is weakening. In the last decades of the twentieth century, gay men and lesbians ceased to hide behind conventional gender masks and brought about legal and social changes that allowed them to be openly different. Does that make them culturally designated third and fourth sexes? They seem to prefer to be seen as special categories of men and women, and indeed to classify them otherwise would be to reinforce stereotypes. Gay men as a group share only one characteristic, really: attraction to other men. Lesbians likewise, to other women. Modes of work, speech, entertainment, dress, aggressiveness, and nurturance vary independently of sexual orientation.

Indeed, one of the great advances in our thinking on this subject was to get beyond the stereotypes, as in endocrinologist Daniel Federman's classic model of the independent psychological and biological dimensions of gender. This model acknowledges that I can feel like a man but desire men; feel like a woman although I have male genitals; be a woman in every meaningful way despite having a Y chromosome; be a man in every way except that I like to wear women's clothing on weekends; be a hairdresser with a swishy manner who is a manic heterosexual; or be a solid family man, or a dozen other combinations we can think of, *if we get beyond stereotypes*. Some of these unexpected combinations result from genetic chances like androgen insensitivity, others from accidental hormone exposure with unpredictable results, still others from divergent experience in childhood, adolescence, or adulthood. People who live and often love these mix-and-match lives need the freedom they are gaining from increasingly flexible legal, medical, and psychological ideas about sex and gender. The third-sex notion could be a new kind of stereotype, and it could hurt them badly.

On the other hand, in cross-cultural research, there is ample precedent for the concept. Pueblo Indians of the American Southwest, like the Zuni and Hopi, recognize the *berdache*, or man-woman, a designated role for a man who not only feels and dresses like a woman but becomes the wife of another man. Far more numerous are the *hijras* of India, a collection of sects totaling thousands of individuals who are either sexually ambiguous from birth or (more commonly) transsexual men who become women through ritual emasculation. Because India is a vast and ancient civilization that has put sex and its variations in the center of myth, art, and literature, the *hijras'* ambiguous, sacred, and powerful sexuality will likely give them a permanent role in a culture that holds one-sixth of the human species.

But these traditions do not necessarily point the way for societies like those of the West that do not have a ready-made third-gender solution. Recent evidence suggests not only that transsexualism is becoming more common in the West but also that individuals who take this path are less committed to the two-sex model. If this is so, then an argument can be made for allowing infants to grow up as they are and decide later what they want to be. Organizations like the Intersex Society of North America strongly recommend against genital surgery in infants. Meanwhile, in the real world where parents have dreams and children have to fit in, decent pediatric endocrinologists and surgeons struggle to do the right thing every time a child is born for whom the question *Is it a boy or a girl?* does not have an easy answer. This debate will go on, but the theory that gender identity is wholly determined by upbringing is dead. What is certain is that human beings unsure about what sex they are will increasingly use the knowledge and power of biology and medicine to gain control over their own lives. And it is clear that there are limitations on how greatly culture determines gender identity and that hormones affect the brain before birth.

A brain masculinized at the start was probably the best explanation for an extraordinary discovery by Julianne Imperato-McGinley, an endocrinologist at the New York Hospital–Cornell Medical Center. She studied a syndrome of abnormal gender development that defied all the rules. Most cases occurred in three intermarrying rural villages in the southwestern Dominican Republic and, over a period of four generations, affected thirty-eight known individuals from twenty-three interrelated families. The condition is clearly genetic—several defective genes with simple changes in base sequence have been identified, not only in the Dominican villages but in Mexico, Turkey, and elsewhere. But it is the Dominican case that has been most carefully studied.

Nineteen people appeared at birth to be unambiguously female and were reared as normal girls. At puberty they first failed to develop breasts and then underwent a masculine pubertal transformation, including growth of a phallus, descent of testes previously hidden in the abdomen, deepening of the voice, and development of a muscular masculine physique. Physically and psychologically they became men, with normal or sometimes excessive sexual desire for women and with a complete range of male sexual functions, except they were infertile, due to abnormal ejaculation through a channel at the base of the penis. After many years of experience with such individuals, the villagers identified them as a separate group, called *guevedoce* (penis at twelve) or *machihembra* (man-woman), but they could not usually predict in childhood which girls would fall into this group.

Medical study showed that these people are genetically male—they have one X and one Y chromosome—but lack just one enzyme of male hormone synthesis, due to a defective gene. The enzyme, 5-alpha-reductase, changes testosterone into another male hormone, dihydrotestosterone. Although they lack "dihydro" almost completely, they have normal levels of testosterone itself. These two hormones are respectively responsible for male external sex characteristics at birth (dihydrotestosterone) and at puberty (testosterone itself). Despite normal testosterone, lack of the dihydro form makes for a female-looking newborn and prepubertal child. Yet the presence of testosterone itself produces more or less normal male puberty.

But the most stunning fact is that these people become, completely and securely, men of their culture. After a dozen years of girlhood, with all the psychological influences shaping that gender role in a fairly sexist society, they transform themselves into almost typical men—with male family, sexual, vocational, and avocational roles. Of the eighteen subjects for whom data were available, seventeen made this transformation. The last retained a female role and gender identity. The seventeen did *not* make the change with ease. It cost some of them years of confusion and anguish. But they made it, without special training or therapy. Imperato-McGinley reasoned that the testosterone circulating before these men were born masculinized their brains—and that this "appears to contribute substantially to the formation of male gender-identity." She went on to conclude, "These cases demonstrate that in the absence of sociocultural factors that could interrupt the natural sequence of events, the effect of testosterone predominates, overriding the effect of rearing as girls. . . . Our data show that environmental or sociocultural factors are not solely responsible for the formation of a male-gender identity. Androgens make a strong and definite contribution."

Studies by Gilbert Herdt and others, of the same enzyme deficiency in New Guinea, support this model strongly. The Sambia of New Guinea recognize these individuals as a third sex and label them with a term that means "becoming male" or the pidgin phrase "turnim man." This appears to be a stronger category than the Dominican words for these people and in Herdt's mind raises questions about whether even the Dominican case qualifies as a complete natural sex change. But as can be seen from the above quotes, they were never put forward as such. Imperato-McGinley and her colleagues merely claimed that the prenatal hormone "appears to contribute substantially to the formation of male gender identity" and that "environmental and sociocultural factors are not solely responsible." Herdt agrees. As he wrote in 1994, "The 5-alpha reductase deficiency syndrome clearly creates extraordinary prenatal hormonal effects in gender development. . . . Gender identity is not entirely a social construction, and sexual variations are not merely an illusion of culture." But here as well the standard two-sex world is too constraining, even in biological terms. As Herdt wisely concludes, "In some places and times, a third sex has emerged as a part of human nature; and in this way, it is not merely an illusion of culture, although cultures may go to extreme lengths to make this seem so."

One of the great social experiments with ordinary men and women has been the kibbutz movement in Israel, which after more than a century is one of the longest-lasting and most successful projects in the history of social engineering. But its success is due to its adaptability, and what started as a passionately communal system with many features of utopian socialism has become a cooperative, still communal effort, but with many features of enterprise capitalism. One of its most fascinating aspects, however, was the attempt to engineer sex roles. For several reasons, children were reared communally in houses separate from their parents. Women and men were freed to work as equals in parallel roles, and the ideology of gender equality was pervasive and strong. But by the 1970s it was evident that sex roles had become more conventional, reverting to classic bourgeois patterns. Tough, egalitarian pioneers had granddaughters who wanted their children to live with them and who were willing to let men play conventional male roles.

Although culture is not a factor, even animal studies are full of complexities. For instance, in songbirds in which the male is the singer, there is a sex difference in the area of the brain controlling song, a difference so large that it is readily seen under the microscope without subtle statistical techniques. Some songbird males have brain regions that grow seasonally, multiplying the cells for song every spring, all because of a surge in male hormone. While seasonal birth of neurons may not occur in animals like us, the bird brain findings show the power evolution has to deploy neural systems that both respond to the environment and produce adaptive, gender-specific behavior.

Closer to home, in mammals, the hypothalamus exhibits a much more striking size difference. This is the same part of the rat brain in which Raisman and Field first showed a sex difference in the number of synapses. The region—the "sexually dimorphic nucleus of the preoptic area"—is 3 to 6 times larger in male rats than in females, and this difference is, once again, a function of the presence or absence of testosterone around the time of birth. It is so striking that the seasoned observer holding the microscope slide up to the light can tell the sex of the brain with the naked eye. In the past two decades evidence has accumulated for structural sex differences, expressed in overall size or microscopic structure, in many areas of the rat brain. Yet the evidence remains best for the hypothalamus.

Does the human brain show a corresponding sex difference? The question has not been neglected. There is a group of four clusters of cells, the "interstitial nuclei of the anterior hypothalamus," or INAHs (rhymes with Dinah's). INAH-3 is triple the size in heterosexual men than it is in women, but in gay men it is woman-sized. Before we can appreciate this remarkable finding, however, we need to visualize the brain itself.

Cup your hands so that your fingertips meet and your thumbs dangle in space over the empty cup. You are looking at a roughly life-size model of your brain, seen from below. Your fingers are the frontal lobes, their tips the frontal pole. Your hands behind the fingers are the parietal lobes, except for the heels of your hands, which are the main part of the visual brain, the occipital lobes. Your thumbs, which should be comfortably an inch or so apart, are the temporal lobes. Thus, your hands—they would have to be a bit thicker than they are—correspond to the whole cerebral cortex, the newest, most advanced part of the brain. The space of the cup is in reality filled with a core of ancient structures inherited from our reptilian and early mammalian ancestors. Imagine the brain stem and spinal cord coming out from between your thumbs toward your chin. And between the tips of your thumbs, at the base of the brain, would be a grape-size blob of cells, the hypothalamus.

Small as it is, the hypothalamus holds many millions of cells. It is no blob really, but devilishly complex, subdivided in regions affecting body temperature, daily rhythms, eating, drinking, fighting, hormone making, and sex. If it seems marvelous that so much is packed into a grape, recall that it was a large part of the brainpower of our amphibian ancestors. They had to make do with it, and they made it do a lot. We added a large load of computational power around it, and those upgrades have been good for us. But the hypothalamus remains central, basic, complex, and powerful. The tail of higher brain function is weighty and big, but it still does not wag the hypothalamic dog.

So when the interstitial hypothalamic nuclei, tiny as they are, show sex differences, we pay attention. Studying such small centers is difficult. A Dutch team in the mid-1980s reported a sex difference in INAH-1 that did not hold up to further scrutiny. But several teams in the 1990s—including one in Roger Gorski's lab led by Laura Allen—have found large, consistent differences in INAH-3. A surge in credibility for this finding came when Simon LeVay, a neuroanatomist at UCLA, confirmed and extended it. LeVay was already renowned for his work with David Hubel and Torsten Wiesel, contributing to their Nobel Prize research on the visual system, the most elegant and well-understood part of the brain. His work in this area was world-class, precise, textbook-quality science accepted by brain researchers everywhere.

So when LeVay announced his finding about sex and the brain, the claim was taken seriously. More intriguing, he not only confirmed the INAH-3 sex difference but also found that gay men had an INAH-3 in the female range. At the same time LeVay made it widely known that he himself was gay, which added further interest to the finding. Some liberals take the view that a biological basis for sexual orientation can only increase persecution of homosexuals—not to mention reinforcing the conservative thrust of behavior genetics in general. Yet many others, including many gay men and lesbians, find the idea of a biological basis for sexual orientation liberating, because it meshes with their deeply felt subjective sense that being gay is no choice for them, moral or otherwise. It's just the way they are.

In any case, LeVay was too good a neuroanatomist for anyone to dismiss his discovery without further study. He assessed the size of INAH-3 in all the brains in his study blind to their group assignment, so while measuring the center in a given brain he had no way of knowing whether it came from a gay man, a straight man, or a woman. In retrospect, LeVay considered it "very likely that there are fewer neurons in INAH-3 of gay men (and women) than in straight men. To put an absurdly facile spin on it, gay men simply don't have the brain cells to be attracted to women." Being gay himself, obviously he wasn't suggesting that this was a deficit, and in his book *Queer Science*, he traced the fascinating history and human implications of such research. He also was not proposing that attraction to women can really be located in a group of cells. But somehow these cells participate in a circuit that enables straight men to become aroused by women—whether by smell, sight, interaction, emotion, or learning, we just do not know.

Melissa Hines, working with Allen and Gorski, also found sex differences in the amygdala and in the part of the hypothalamus that the amygdala connects to—just as Raisman and Field had found two decades earlier. Other scientists have repeatedly found sex differences in the part of the spinal cord controlling the pelvic muscles that move the penis. This difference appears prenatally in humans; in rats, at least, it depends on androgens circulating early in development. Interestingly, the hormones also prevent some of the usual cell death in the relevant muscles themselves, and more muscle cells ensure more spinal neurons. There is also a large environmental component. Mother rats lick the genitals of male pups more than females, and this stimulation helps keep the male spinal cells alive. Block the nostrils of the mothers and you will get males who grow up to have fewer neurons, just because mothers that can't smell lick their babies' genitals less.

So even in rats, part of the reason males and females have different sexual circuits is due to the fact that their mothers treated them differently. This indirect path to sex differences still serves the purposes of natural selection, even while partly circumventing genes. If this is true in rats, how much more must it be true of human beings? Think what we will of Freud's mistakes, we are indebted to him for focusing attention on the power of early experience in the growth of gender identity. "Anatomy is destiny," he famously said, alluding to his exaggerated conclusion about the importance of penises. Still, penises and testicles differentiate boys from girls; girls in preschool sometimes wish for penises, if only because they want to pee standing up, and boys do sooner or later discover the vulnerability of their genitals in rough-and-tumble play. Then, too, boys must envy girls their power of childbirth, something a lot more special than bipedal urination.

But there are more significant psychodynamics. So far in history, almost all children have had a woman as their first attachment object. That means girls have their first relationship with someone like themselves, someone with whom they will soon begin to identify. But boys have this primary attachment with someone quite different, indeed someone like the person they will one day lust after and love. Nancy Chodorow, in her contribution to psychodynamic theory, sees the process girls go through as formative for gender identity; most girls are destined in each generation to experience "a reproduction of mothering."

Although there are many other, later influences, mothers continue to shape girls in their own image, whether they want to or not. Carol Gilligan's study of moral development, *In a Different Voice*, showed that young women tend to solve problems of justice and punishment by emphasizing relationships, whereas young men solve the same problems by emphasizing rules. This overturned some naïve notions about universal moral judgments. Gilligan proposed that because girls identify so strongly with their mothers—the more nurturing parent—they cannot simply invoke harsh rules in making judgments but must also take love into account. Ann Kruger, in a powerful test of this idea, showed that eight-year-old girls benefit more from peers in making moral judgments involving fairness but benefit more from their mothers in making judgments involving care. The reproduction of mothering in growing girls extends even into ethical reasoning.

And of course, the cross-cultural evidence points to the power of socialization as strongly as it does to the underlying biological realities. A glance back at Margaret Mead's ethnographic descriptions demonstrates immense variety in the roles and behavior of women and men in different cultures, despite important constancies. Businesslike, unadorned women and men who primp all day may be taken for granted today, but in Mead's day, just a half century ago, her descriptions were a major challenge to what were then thought *essential* male and female natures. One of the main reasons for cross-cultural consistency in sex roles is that most cultures exaggerate rather than mute the biological differences. This is true in terms of childrearing practices and task assignment in childhood or adulthood, and it is even more true in view of the constraints, customs, and sometimes dramatic rituals that mark life-cycle events for both sexes. These cultural constructions give biological differences like menses and puberty a deep and strong symbolic meaning.

Even in the Six Cultures Study, which showed consistent sex differences across cultures, some cultures widened the differences while others narrowed them, using all the power of training, imitation, reward, punishment, ritual, and symbol. A classic study among the Luo of Kenya by anthropologist Carol Ember showed how this might work. The Luo, like most cultures, assign boys and girls different chores. But if families lack older girls, they will often assign boys to household tasks usually given to girls. Ember carefully showed that boys so assigned, especially if they become involved in baby care, change their behavior in other contexts. They are even less aggressive in settings outside the home, far from the pressures of household work. We might answer that this kind of feminizing task assignment does not stop Dominican or New Guinea boys with 5-alpha reductase deficiency from turning into men in adolescence. But in fact we don't yet know what residue their girlish childhoods may leave in their adult-male behavior.

Then, too, we grow up bathed in language; and not only gender labels and words of approval or criticism but grammar itself shape our feeling about what is possible. I have purged this book of the generic male pronoun and have never used *man* or *men* when I meant *human* or *people*. But if it were written in Italian, German, or Hebrew, gender would be everywhere and inescapable. And there are powerful forces in the realm of what are called pragmatic aspects of language— the often unconscious rules governing how, how often, and when we speak, depending on our social status, age, or sex. Thanks to linguist Deborah Tannen and others, we know that men and women converse very differently and that, in particular, men interrupt women far more frequently than the reverse. Tannen tends to see this more as a difference in conversational style (such as might be seen between Jewish women from New York and Episcopalian women from Virginia) than as yet another exercise of raw male power. But whether a matter of style or power and whether or not biology plays a role in men's interrupting, it creates a reality to which girls and women must adapt, and it is full of messages.

Still, despite the proven power of the environment, we can no longer deny that the genders differ in their reproductive and aggressive behavior for reasons that are to some extent built in. And there may be other differences. Growing evidence points to sex differences in the corpus callosum, the great arch of nerve fibers joining the two halves of the brain. Laura Allen's group, like that of Marie-Christine de Lacoste and Ralph Holloway at Columbia University, has made this finding, which extends to images of living human brains. If confirmed, the effect would be to put women's left and right brains in more effective communication, whereas in men the two halves would work more separately. This could help explain sex differences in verbal and spatial ability, as well as the greater ability women have to put feelings into words.

But still, the largest proven sex difference in behavior, outside the reproductive realm, is that men are more violent than women—and we are still trying to understand how this comes about. Eleanor Maccoby, probably the world's leading expert on sex differences in children's behavior, has proposed an elegant explanation for boy-girl contrasts, starting with small initial brain differences. In her 1998 book, *The Two Sexes: Growing Up Apart, Coming Together*, she summarizes a lifetime of research in a coherent model. First, she believes, comes the inborn difference in play pattern, with boys being more physically aggressive. Partly for this reason, the sexes begin to segregate in their play and friendships before age three and become increasingly separate throughout childhood. By age three they also classify themselves as belonging inflexibly to one sex or the other.

Cultural factors, especially adult stereotypes and *adult* segregation, clearly influence how children think and how much *they* segregate. But even in the most egalitarian environments the segregation tendency is very strong. Parents and teachers can force boys and girls to play together, but only a complete, top-down control of children's choices can effectively prevent voluntary sex segregation. In cultures where there is no prescribed sex segregation at home or in school, children still freely choose it in play and friendships. They shape one another much more than adults shape them. Within same-sex groups, boys and girls reinforce one another's "boyness" and "girlness," increasing the divergence that biology alone would create. When they meet again in adolescence and young adulthood, they are long since foreign to each other culturally as well as biologically. And prominent among their deep, enduring differences is a different male attitude and tendency toward violence.

One policy implication is plausible: serious disarmament will ultimately require an increased proportion of women in government. As it is, too many international confrontations turn into contests of honor and prowess among men. Add to their short male fuses their attraction to risk and their proclivity to have fun both building and wrecking complex toys, and you have a recipe for disaster. As novelist John Updike put it, "Nobody wants war, but men don't want only peace either." Countering with the record of past women rulers in this connection is a useless exercise. They got where they were by being unrepresentative of their gender, able to scramble up an all-male political pyramid. *Some* women are, of course, as violent as any man. But we can fairly guess that we would be safer if the world's weapon systems were controlled by *average* women rather than by average men.

Of course, as we know from Darwinian theory, there is not some ultimate essence of maleness or femaleness that guarantees the particular contrasts in our species. If we were jacanas or seahorses, we would have mild, nurturant males doing most of the parenting and competitive, arrogant females doing the nasty work of the world. There are even incipient signs that we could evolve in that direction. In *The Decline of Males*, anthropologist Lionel Tiger has argued that the ascent of women is already assured. Chemical contraception has placed the control of reproduction entirely in their hands, backed up by legal, safe abortion, often performed by women doctors. Fifty-five percent of college entrants are women, and that number continues to rise. Women still do not occupy the highest-paid leadership positions, but they also do not descend to the lowest. In the ranks of middle management, women's numbers are expanding.

As for reproduction, birth rates are falling in all industrial countries and the family as we thought we knew it—a woman, a man, and two or more children—now makes up just a fraction of households. Poorer women with children either juggle jobs and motherhood or embed themselves in multi-generation female families. Richer women hire the poorer ones to care for their children while they pursue much higher-paid activities. Men are, in Tiger's words, alienated from the means of reproduction through divorce, separation, or their own lack of responsibility. Their physical strength, their aggressiveness, and their bread-winning power all are increasingly irrelevant, and Tiger predicts that they will drift into cold, lonely, and possibly dangerous lives as mothers and children return to an ancient primate pattern in which their bond with one another is primary and male involvement is catch-as-catch-can.

But for a significant minority of future men, there is another possibility: gender-role reversal. The industrial world now has millions of families headed by single fathers, a rising number. As for our increasingly subtle corporate and bureaucratic world, neither violence nor brawn is any sort of asset. The skills of Machiavelli's wily prince are worth more than those of the warrior Achilles, and women's minds are probably as Machiavellian as men's. Many successful women have househusbands, and the human species could evolve toward a point where current sex roles are largely reversed, as in jacanas. Dominant women would compete for meek, sweet, generous men, mate with several, keep them in line, and deliver babies to them in more ways than one, while continuing to butt heads with other hardened women over corporate assets and raw political power.

However, to complete this reversal would take, at a minimum, scores of thousands of years. For the present, we are stuck with the facts of our own evolutionary inertia, the genetic history of human, not seahorse, males and females. So, for the foreseeable future, boys and men will continue to be, on average, more violent and less nurturing than girls and women. Men will also, on average, take more risks, seek more sexual partners, be less tolerant of emotional intimacy, have less interest in talking about feelings, and tend to come alive while watching boxers or football players pound one another's bodies. These are facts of male behavior in our species, arising in large part from the action of male genes on male brains. I think of Andrew's view of whose life is harder and it seems today more deeply true than ever: *The same because they both need babies. . . . They both work. They both learn. They both have feelings.* Our future seems to call for mutual sympathy.

Let us end where we began, contemplating the women who have helped unearth these facts about similarity as well as difference. Visualize them in their offices and laboratories, trying to sort out what it all means, to hear some pure tone of truth amid the cacophony of strident ideologies and the noisy, grinding machinery of relentless social change. How do they keep at it? Not by trying to second-guess the social future nor to press the point of some preconceived belief. Life, they know, is too short to do anything other than try to figure out what's true. And how do they sort out their own beliefs in the face of such truths? I suspect they do it by making a reconciliation—not a compromise, certainly not that, but a complex, difficult reconciliation between the idea of human difference and the ideal of human equality. It is one that we must all make soon.

CHAPTER 7

The Well of Feeling

I wish that I could be a thinking stone.
The sea of spuming thought foists up again
The radiant bubble that she was. And then
A deep up-pouring from some saltier well
Within me, bursts its watery syllable.

— **WALLACE STEVENS,** "Le Monocle de Mon Oncle"

What is the bodily source of grief? Which organ gives the warmth we feel at the touch of a loved child? Where is pain? Although such questions may be too simple, they are not meaningless, and for millennia philosophers have mused their way toward answers. But most of them, however brilliant and wise, were burdened to bafflement by the notion that feelings are not to be found in the body. Human feelings, the most faultless and fine indicators of the state of the human spirit, must be sought, they believed, in the soul.

By soul they meant no metaphor for qualities of heart and mind but a separate entity, completely insubstantial, coterminous with the body and linked to it during life but surviving it quite effectively after death. In Dante the souls of the dead exhibit almost every nuance of feeling they could have experienced during life. To suffer damnation, one must be sensible of pain; to reap the reward of heaven, feel pleasure. Purgatory and limbo depend upon anxiety and longing, and the bemused poet's interlocutors on every platform of heaven and hell give vivid testimony of the exquisite, wide-ranging sensibility of figures presumed to be made of airy nothing.

The mind-body problem, whether as a philosophical or biological issue, is not yet resolved, but it is fair to say that philosophers of the nineteenth century— at least those independent of theology—were already taking what biologists would consider a more useful approach to these questions. And by the twentieth century, particularly in the movement known as philosophic analysis, the mind-body problem came under the influence of biological and psychological research. This is not to say that philosophers were merely reacting to such research. Rather, they were engaged in clarifying the use of language—an innocent-sounding occupation that has undermined the basis of religion and metaphysics—and in the precise description of subjective experience, an exercise behavioral and biological scientists are either disinclined toward or do badly. Still, some of the best philosophers of the twentieth century—Bertrand Russell, G. E. Moore, A. J. Ayer, Gilbert Ryle, Paul Churchland, Daniel Dennett, and others—have also tried to explain modern science to their more metaphysical colleagues. Notable at the turn of the millennium are Paul Griffiths's *What Emotions Really Are*, the first philosophic account of emotion to be grounded in evolution, Aaron Ben-Ze'ev's *The Subtlety of the Emotions*, and the revised edition of John Rawls's *A Theory of Justice*, which incorporates the facts of economics and psychology in an analysis of moral choice.

To be sure, science alone does not produce philosophy, and some philosophers were empiricists long before the dawn of modern science. Aristotle denied the insubstantiality and immortality of what is usually translated as "soul," although he granted both to "mind." This stance, however ambiguous, would have made Dante's spirit world impossible. Epicurus, a younger contemporary of Aristotle's, carried skepticism farther. In a letter dated around 300 B.C., he wrote: "The soul experiences sensation only when enclosed in the body; and the body receives from the soul a share in this sensation. Sensation may survive the loss of parts of the body, but it ceases with the destruction of the soul or of the whole body."

This statement was made in the context of a wide-ranging natural science, including a theory of evolution, and it seems uncannily modern. But although its aim was to give comfort through acceptance of the inevitable and transcendence of irrational religious fears, it was widely viewed as a comfortless philosophy. In an ancient world charged with religious conviction, it languished, despite the eloquent discipleship of the Roman poet Lucretius, whose epic *On the Nature of Things* expressed similar views two centuries later. It is only at the end of the eighteenth century—around the time William Blake wrote the verse set in the flyleaf of this book—that we see the start of a modern kind of skepticism.

In philosophy the thread can be traced back to David Hume. Like Blake's, Hume's was a lonely voice. His major early work, *Treatise of Human Nature*, attacked all the reigning philosophies of his day. It was ignored. His *Dialogues Concerning Natural Religion* he considered too dangerous to publish in his lifetime. But advances over the next century formed the basis of future treatises on human nature, making it clear that a nonmetaphysical account would be possible— and, incidentally, fleshing out the sketch made by Lucretius two thousand years before.

Just a few of the names from that century—Charles Lyell in earth history; Darwin in organic evolution; Claude Bernard in general physiology; Louis Pasteur in microbiology; Karl Marx, Herbert Spencer, and Lewis Henry Morgan in social science; Hughlings Jackson and Charles Sherrington in neurophysiology; Paul Broca and Carl Wernicke in neuroanatomy; Santiago Ramón y Cajal and Camillo Golgi in cellular neuroscience; and William James and Sigmund Freud in psychology—

convey the changed climate of thought. With so much of "the nature of things," previously best explained by religion and metaphysics, now brought under the banner of the observable, the soul was ripe to be colonized by science.

As A. J. Ayer put it, speaking for many twentieth-century philosophers, a "prudent theory" would be "one that does not attempt to explain away the occurrence of experiences, or to maintain that our descriptions of them are logically equivalent to descriptions of physical events, but still claims that they can be factually identified with states of the central nervous system." In other words, our experience of our own brain's workings cannot be satisfactorily explained even by a full objective description of them, because the language of brain science cannot describe subjective experience. No description of brain function remotely expresses what we feel. Soul, at best, is a metaphor for that subjective experience and for what we guess is the inner life of others. Some such metaphor is needed—hence the word *spirit* on the cover of this book. But the subjective sense, whatever we call it, cannot transcend in space or time the body's mundane functions. It can only transcend them in perspective.

The great early-modern theorist of the soul was Sigmund Freud, a neurologist by training and, until his middle years, by practice. In the 1880s he was involved in brain research, including studies of the mental and medical effects of cocaine. It was only in the next decade that he began to write on psychology; by this time he was in his late thirties and a well-known medical scientist. His papers of the early 1890 s show that he knew the neurobiology of his day and seemed satisfied with the possibility of accounting for much of the human mind by reference to it. Interestingly, the other great psychologist of the period, William James, was also a physician expert in brain science and put it to use in his own psychology.

Two of Freud's papers from this period show what he expected to gain from neurology, as well as his growing frustration with it. "On Aphasia: A Critical Study," published in 1891, deals with the loss of speech, comprehension, or both, and with related mental disorders caused by brain damage. It offered Freud's views on the neural basis of language function and criticized the then reigning (and still largely accepted) model of language localization, established by Paul Broca and Carl Wernicke. This theory holds that the cortex has two major regions vital for language, one for speech production and one for comprehension, since these two functions can be deleted by different injuries, yielding one or another form of aphasia. Other disorders come from damage to the nerve pathways that link the speech-reception and speech-production areas ("conduction aphasias"). Still other circuits control reading and writing. These are not self-contained centers, but they *participate* in *circuits* handling language functions.

Freud reviewed the published cases and the state of the art in brain anatomy. But his main objections came from neurophysiology, the functional side, then being shaped by John Hughlings Jackson and Charles Sherrington. Freud underestimated Broca and Wernicke, but others' claims for brain localization had indeed been extreme. Also, the 1890s were a heady time for neurobiology, when the function of individual nerves was being organized into larger units in the first real theories of brain function. These developments inspired the young neurologist to try to transform the subject he had given his youth to learn. Freud knew that aphasics might retain the ability to swear, sing, or repeat a phrase spoken just before their injury. He discussed the pattern of recovery of mental functions after epileptic seizures, which sometimes includes a phase of transient word deafness. He knew of individual variability in the loss: "Different amounts of nervous arrangements in different positions are destroyed with different rapidity in different persons," a quotation he gives from Hughlings Jackson, corresponds to modern findings. Following his own teacher, the great French neurologist Charcot, Freud attributed this variation not to inborn differences in anatomy but to the vagaries of personal history. To take an obvious example, a literate and a nonliterate person would suffer different losses from a blow separating the visual areas of the cortex from the speech-perception areas.

Freud inferred that losses came from differences of function, not structure, that language is based not just on anatomy but also on the neural energy expressed in brain systems. He accepted Jackson's view that loss of function reversed the sequence of development in normal childhood and that recovery retraced the same sequence. He concluded that the retention of swearing, of simple, oft repeated words like *yes* and *no*, of singing, and of phrases learned in extreme stress all pointed to a powerful role for neural excitability in language — and, with that, a role for emotion. This marked the beginning of a transition that, in less than a decade, was to leave Freud studying not brain but mind. In his most famous works there is almost no reference to the anatomy and physiology that had once preoccupied him. In most of Freud's work the only reference to structure is to purely metaphoric ones: unconscious, preconscious, conscious; id, ego, and superego. While these are related to his earlier physiological notions, he was trying to distance himself from biology, and there is little evidence in his later writings that he followed the growth of brain science.

Yet much in the aphasia study prefigured his later work: the emphasis on the role of emotion in thought, the long-term impact of trauma, the significance of nervous-system excitation, and the idea that recovery is somehow parallel to the course of normal childhood. But it was only four years later, with the completion in 1895 of *Project for a Scientific Psychology*, that the transition to psychoanalysis really began. Freud's tenets of biopsychology as of 1895 were assembled from then-conventional knowledge of brain function, combined with a few newer, bolder ideas. The principles are in essence still valid:

- First, the central nervous system can be divided into two basic sectors: one containing "projection" systems — long tracts carrying impulses like pain from the body to higher brain centers such as the cortex, where the periphery is somehow represented; and a second containing "nuclear" systems, in the core of the brain, made up of short neural elements with many connections, monitoring the body's internal milieu — the sleep-wake region of the brain stem, for instance.

- Second, there are neurosecretory elements in the nervous system producing chemicals that circulate in the body. These chemicals can excite the neural elements of the brain, giving rise to a feedback cycle. (This remains true and is the essence of hypothalamic and pituitary function, but today we know there is crucial feedback *within* brain circuits.)

- Third, brain function consists of electrical activity of the neural elements, which, when excited enough, discharge. (Today we accept this as the "all-or-none principle" of neuron function, which gives the brain its basically digital character.)

- Fourth, the neural elements are separated from one another by "contact barriers," and in order for one in a circuit to excite the next, the contact barrier must be crossed. (We know this vital gap as the synapse, and it is the point of action for most psychiatric medications.)

- Finally, the neural elements can have a level of excitation *below* that required for discharge and transmission across the gap. (We now know that the neuron's ability to accumulate such lower levels of excitation until they reach a threshold for discharge is crucial to the brain's integrative action.)

Thus, Freud accepted and built upon the concepts of neuron and synapse at a time when this view of the nervous system—"the neuron doctrine"—was still hotly contested, dividing followers of the great microscopist Santiago Ramón y Cajal, who believed in it, from those of the equally respected Camillo Golgi, who wrongly believed that the nervous system was one vast uninterrupted web. A decade after Freud's acceptance of the neuron doctrine, Cajal had won wider acceptance, and Sherrington, in his great work *The Integrative Action of the Nervous System*, gave the synapse a central place in brain function. These features of brain function—some standard neurology, some educated guesses—were developed in Freud's *Project* into a stunningly original theory of the mind.

In this theory, exchanges with the outside world, such as sensation, pain, and muscle action, are made by the long, fast-acting projection systems. The core responds more gradually, influenced by bloodborne factors; it also influences those factors, playing a key role in what we now call homeostasis, the body's stable balance. The core's diffuse connections—complex or messy, depending on how one looks at them—can be excited by external stimuli or from internal spontaneous activity. If this excitation is mild and widespread, diffuse firing can occur, which, in Freud's view, might be maladaptive. In either system, repeated discharge of impulses could be expected to result in easier future transit over that pathway—learning. So far, there is little here that would raise a twenty-first-century neurobiologist's eyebrow. As for the rest, some of it would be viewed as pretty speculative, but none of it could be simply dismissed.

Subthreshold excitation of the core, Freud believed, could cause strain (in German, *unlust*, or unpleasure). Such excitement could come from positive feedback between the nerve cells and bloodborne factors, as might occur during hunger or sexual need. Pain would be a larger, more sudden increase in both cortical and subcortical excitation. The cortex, sensing strain, would animate acts like eating or sex, to bring the bloodborne factors and nerve cells back into balance, causing a decrease in strain, or pleasure. Pleasure tends to reinforce by lowering the resistance of the contact barriers to discharge along the same path. When this prevents built-up strain in the core, learning has happened. The core's structure is no longer random. The pattern of more and less reinforced pathways within it constitutes the ego structure or personality, which *is* these memory traces, selected by experience from all possible pathways. They include *motives* (pathways of core excitation), *wishes* (how the cortex sees motives), and *mechanisms of defense*—literally defense against excessive strain, by dispersal of excitation into pathways not obviously relevant. In this model, emotion is patterned change in the cortical tensions, and thought is comparison of the cortical patterns coming from the core (wishes) with those coming over the long projection systems from the periphery (perceptions). In plain words, thought is a mismatch between the way things are and the way we want them to be.

It is a pity that Freud did not continue to follow developments in neurobiology in an attempt to relate them to his theory. If he had, it would be easier to bridge the gap today, and he might not have wandered into certain speculations for which he is now heavily criticized. But perhaps cutting loose from his neurological moorings was a necessary step toward his insights into emotional life. William James, Freud's fellow physician and admiring older contemporary, had also mastered the elements of brain structure and function. But in place of Freud's training in neurology, he educated himself, during extensive travels in Europe, in the experimental psychology of the day, which was good—a rigorous field that measured human sensation and perception—and in the philosophy of the day, which was in many ways vague and bad. Philosophy distracted James in his effort, yet during many years teaching physiology at Harvard, he managed to found a comprehensive scientific psychology.

His remarkable achievement, the two-volume 1890 *Principles of Psychology*, is the model for all modern textbooks in the field. Compared with Freud's works, it shows greater balance and modesty, greater knowledge of experimental evidence, and greater grasp of the structure and function of sense organs. The reader does not feel the grip of a powerful personality, and genius does not intrude, either to inspire or dismay. Philosophy does intrude—in a didactic, distracting way that makes it difficult to separate psychology (which for James was based on evidence) from philosophical musing (which seemed to come from his own rather commonplace self-observations). He later abandoned science altogether for pragmatist philosophy, but he left behind a theory of the emotions that held sway for many years and was partly revived in the 1990s in a far more sophisticated form.

Named also for its Danish codiscoverer, Carl Lange, the James–Lange theory was embedded in physiology, and it recognized the role of bodily organs in both the subjective experience and the expression of the emotions. Our "natural way of thinking" tells us that what we see or hear causes emotions, which in turn cause physical arousal. James demurs:

> My theory, on the contrary, is that the bodily changes follow directly the perception of the exciting fact, and that our feeling of the same changes as they occur is the emotion. Common-sense says, we lose our fortune, are sorry and weep; we meet a bear, are frightened and run; we are insulted by a rival, are angry and strike. The hypothesis here to be defended says that this order of sequence is incorrect, that the one mental state is not immediately induced by the other, that the bodily manifestations must first be interposed between, and that the more rational statement is that we feel sorry because we cry, angry because we strike, afraid because we tremble.

The obvious objection that we may feel those things without performing the corresponding actions is met with the answer that while we are restraining ourselves from these actions, they still exist as tendencies and create sensations in the body perceived by our brains. James realized that, "stated in this crude way the hypothesis is pretty sure to meet with immediate disbelief." But the main efforts he made to dispel that disbelief were by way of argument, not evidence.

Although information about the brain and emotion made possible Freud's very different account a decade later, the James–Lange theory remained the accepted outlook in the United States. So much so that when Walter B. Cannon, the leading American physiologist, proposed an alternative in the *American Journal of Psychology* in 1927, he had to begin it "with some trepidation," although his critique was based on decisive evidence.

Referring to Charles Sherrington and other leading neurologists, Cannon methodically marshaled the points against James and Lange. First, total separation of the viscera from the brain by severing neural connections has little or no effect on emotional behavior, despite the absence of the visceral reactions believed by James and Lange to *cause* emotion. Second, similar visceral reactions occur in very different emotions. Third, the viscera are relatively insensitive organs, and we have a lot of trouble feeling what is going on in them, making them poor candidates for a medium of emotional sensitivity. Fourth, visceral responses are too slow for the task; they *lag* the emotional reactions rather than lead them. Finally, artificial induction of those same visceral changes generally produces irritation or malaise and rarely causes any recognizable emotion. "The processes going on in the thoracic and abdominal organs," Cannon concluded, "are truly remarkable and various; their value to the organism, however, is not to add richness and flavor to experience but rather to adapt the internal economy so that in spite of shifts of outer circumstance the even tenor of the inner life will not be profoundly affected."

Today those "truly remarkable and various" gut feelings are up for debate again. Considerable excitement followed the 1983 demonstration, by Paul Ekman and his colleagues, that measures of autonomic nervous system function, such as heart rate and breathing, actually discriminate among different emotions. When the same associations between emotions and gut reactions were found a decade later among the Minangkabau, a Muslim culture of Sumatra with a strong tradition of emotional reserve, it began to seem that investigators were onto a physiological account of universal emotions. But many of Cannon's objections still held up. Both Americans and Minangkabau had different visceral reactions to positive and negative emotions, but two primary negative emotions, rage and fear, could not be distinguished except by what happened between the ears.

What of James's argument that there were "no special brain centres for emotion"? Cannon countered by proposing a plausible linked group of such centers. In doing so, although he was wrong in some specifics, he emphasized brain control of the emotions. It is difficult to overestimate the importance of this step. It not only transcended James–Lange by locating the emotions in the brain; it also transcended Freud, whose appropriately brain-based theory of the emotions was anatomically vague, except for a sketchy two-fold division. Where James and Lange had located emotion in the periphery, Cannon brought it back into the brain. Where Freud, who had long before accepted the brain's primacy, based his theory on process more than on structure, Cannon took a chance on a functional theory packed with anatomical detail.

Recall that in the 1890s thoughtful people had had their fill of brain centers, of exaggerated, even ludicrous claims about the geography of specific mental functions and tendencies. The extreme version was in the short-lived but popular pseudoscience of phrenology—reading character from bumps on the head, which were supposed to reflect brain development. People even designed their parenting methods to counteract undesirable tendencies in the child that they believed they had "read" phrenologically—such as "amativeness," a polite word for sex drive. Small wonder, then, that Freud's growing skepticism led him to abandon anatomical detail. This healthy skepticism persists, as in a statement quoted by one brain scientist from another: "I find singularly little in all the vast volume of stimulation studies of subcortical structures that would promote the notion of 'centers.' There seems to be the need to introduce the notion of complex cortical-subcortical interrelations, rather than the autonomy of activity in the sense of centers."

In other words, emotions, like other mental and behavioral phenomena, must find their brain location in not a static geography of locations but a dynamic commerce of circuits. So when Cannon located the seat of the emotions in the thalamus, a pair of egg-shaped neural organs lying deep along the brain's midline, he was proposing too limited a model—but he did have evidence. Animals with the thalamus intact but with all higher brain centers removed showed much of normal emotional expression; removal of the thalamus abolished it. More interesting, some human victims of tumors on one side of the thalamus were perfectly capable of assuming two-sided smiles voluntarily, on command, but in normal situations of involuntary, spontaneous smiling and laughter, their faces responded only on the side without the tumor. This disjunction has often been confirmed. It suggested to Cannon that on-command smiling was run from the higher cortical centers, true emotional smiles from the thalamus itself; thus, the tumor victims were emotionless on one side of the face and body.

Cannon did nod toward circuitry by speculating that emotions come through a thalamic relay. Either impulses from the sense organs passed through the thalamus on the way to the cortex or they went to the cortex first and then down to the thalamus. In either case, "the peculiar quality of the emotion is added to simple sensation when the thalamic processes are roused." This put emotion *inside* the brain. Today we might say that the tumor's damage interrupted a complex circuit for emotions—a circuit passing through the thalamus among other brain organs.

Such a complex emotional circuit was proposed in 1937 by James Papez (it rhymes with *tapes*), an obscure physician-neuroanatomist working on his own in Ithaca, New York. His paper, "A Proposed Mechanism of Emotion," presented a disarmingly simple abstraction of a wealth of anatomical detail; it was as revolutionary as it was elegant, and it set the tone for research on emotional physiology from then on. Moreover, it had to be done by an anatomist. Neither a physician-psychologist like James, a former neurologist like Freud, nor even an outstanding physiologist like Cannon would have enough mastery of anatomical detail to see how the brain might handle emotion.

Papez located the emotions in circuits, not centers, sketching a set of interrelated structures we now call the limbic system. This system, situated where Freud would have wanted it—close to the core of the brain—included parts of the thalamus but also parts of the cortex and, most important, the hypothalamus. Through this region "below the thalamus," just about in the center of the head, the brain regulates the pituitary gland, and so the body's hormonal order. These circuits had once been thought to mainly process odors, but animals with little or no sense of smell, like dolphins, still have prominent limbic systems. By 1937 there was evidence—from lesion studies in animals and tumors in people—to support a strong relationship between this circuitry and the emotions. Papez proposed three "streams" of impulses corresponding to three broad domains of psychic life. All began with the sense organs, and each involved part of Cannon's thalamus.

The first, "the stream of movement," relayed sensations through the thalamus to the corpus striatum, or "striped body," a central structure in each half of the brain needed for normal movement. Today, precise surgery in this circuit can make major improvements in the stiffness and tremor of Parkinson's disease. The second, "the stream of thought," relayed sensations through the thalamus to the cerebral cortex. Much of the brain science of the last half of the twentieth century was devoted to filling in the details and cortical destinations of this broad second stream. The third, "the stream of feeling," relayed sensations through still other thalamic areas to the hypothalamus and parts of the cortex. These were midline structures, in the middle of the head, between the two hemispheres, near the core of the brain. "In this way," Papez believed, "the sensory excitations . . . receive their emotional coloring." He asked rhetorically, "Is emotion a magic product or is it a physiologic process which depends on an anatomic mechanism?" The evidence was "suggestive of such a mechanism."

These notions have largely survived the twin tests of time and increasing complexity. Papez's contribution, as large an advance beyond Cannon as Cannon's was beyond Freud, stands today as an outline of much of what we know about our emotional anatomy. Indeed, the most sophisticated current models seem eerily similar to the one Papez fielded long ago. Garrett Alexander, Mahlon DeLong, and others have advanced a model of reverberating, "parallel, segregated circuits" that separately mediate thought, emotion, body movement, and eye movement by taking different paths through the basal ganglia and thalamus and engaging different parts of the cortex and limbic system. Constant, rhythmic reverberations in these circuits, some due to feedback among neurons and some to intrinsic properties of membranes, ensure that they are almost always available for activation.

But this elegant modern vindication of Papez needs to be placed in an evolutionary context. As the great twentieth-century evolutionist Ernst Mayr put it, biology has thrown down two separate reductionist gauntlets before the figure of modern psychology. The first, physiological and biochemical, seeks to replace the language of act, thought, and feeling with another, entirely mechanical vocabulary. The second, evolutionary and adaptive, proposes a historical and functional account that seems more powerful than psychological process. To put it crudely, biologists pulled the soul down to earth, showing, first, how much can be explained by chemical reactions and electrical equations, and second, how much we share with other animals.

Freud, Cannon, and Papez had all felt Darwin's influence, but none of them took him seriously enough. In the *Origin* and *The Descent of Man* he had set in motion the study of social behavior as adaptation. This approach spawned the vigorous new fields of sociobiology and evolutionary psychology. To the extent that social behavior and emotion are bound together, this was also a contribution to the study of emotion. As E. O. Wilson, John Tooby, Leda Cosmides, and others have pointed out, our environment of evolutionary adaptedness must have given us brain systems designed for such functions as sexual jealousy, detection of cheating in exchanges, and parental concern. Only incidentally did it give us circuits able to do algebra.

But Darwin also addressed emotion directly, in his 1872 book *The Expression of the Emotions in Man and Animals*. Natural selection was mainly background music in this work, a muffled drumbeat for comparisons of humans with other animals and for theorizing about the functions of expressive acts. Darwin did not seem to want arguments about selection to be an issue; he wanted the book on emotional expression, published more than a dozen years after the *Origin*, to stand on its own. If he had written nothing else, it would have secured his place in the history of science. It is strange, then, that—with the important exception of Paul Ekman, a brilliant experimenter who has given his life to reading faces—it has had so little impact on psychology. But *The Expression of the Emotions* has been taken up by another research tradition, ethology. Rooted firmly in the soil of zoology, not psychology, ethology arose in Europe and England during the early twentieth century and now circles the globe. Three of the field's leaders won a Nobel Prize in 1973, and along the way it evoked from several major American students of animal psychology papers that were at once both capitulations to it and attempts to incorporate it into the comparatively impoverished American research tradition.

Ethology, the comparative study of animal behavior, established certain principles. First, studies of animal behavior must be carried out across a broad range of species. Second, the behavior of each species must be studied first under natural conditions, in the wild. Third, many aspects of the behavior of a species are as fixed as its anatomy and are equally attributable to genes. Fourth, fixed and flexible behaviors can be distinguished through deprivation experiments, withholding relevant aspects of experience. Fifth, once established, the fixed components of a species' behavioral repertoire are so reliable that they become, along with anatomy, a way to trace its evolutionary history. Finally, behavior evolves in response to selection in the natural environment, and such selection can be studied. These rules now seem obvious, but none of them played a significant role in the heyday of American animal psychology.

If psychology was guilty of ignoring ethology, anthropology was more so. Since Darwin's founding contribution was published in 1872, lack of opportunity for influence cannot be the reason. Although Darwin did not do formal ethnography, he made many observations during his youthful voyage around the world, and he referred systematically to observations by others. He canvassed many missionaries and amateur anthropologists for firsthand descriptions of emotional expression in remote societies throughout the world and proposed universals of human emotional expression. He compared these in detail to emotional expressions in animal species, especially mammals, observed in zoos, as pets, and in the wild. And of course, he studied the facial expressions and gestures of people and animals around him.

Darwin considered the role of experience by comparing expressive movements in infants and adults of the same species—including ours. In fact, part of the evidence was his detailed diary of his own firstborn infant. He speculated on the adaptive significance of many movements, from deceptive (as in the erection of the hair to make the body seem larger) to physiological (the widening of blood vessels in muscles preparing for action). He even proposed a crude neurophysiology of emotional expressions and actions. Peripheral organs were controlled by the nerves from the spinal cord, and emotional expression was the result of an "overflow of nerve-force" from the central nervous system, with the path of the overflow to be determined by nature and habit combined—a simple form of the theory Freud would embrace twenty years later. And he concluded very soundly "that the chief expressive actions, exhibited by man and by the lower animals, are now innate or inherited—that is, have not been learnt by the individual," although he went on to say, too optimistically, that this conclusion "is admitted by every one."

Ethology has vindicated Darwin. Concepts like the fixed action pattern (an inherited movement sequence, such as a dog's growl and bark at an intruder), the innate releasing stimulus (that causes predictable responses *without* prior experience, such as a frog's tongue-dart at a black speck in the air), and the ethogram (a catalog of a species' inborn behavior) are all current. And these concepts apply to humans as well as to other species. Cultural anthropologists routinely question the universality of emotions, citing exotic cultures as evidence. For example, studies of emotional reserve in South Seas cultures like Ifaluk and Bali have been viewed as showing that emotions are fundamentally cultural, not universal or biological. But research by other anthropologists and even careful reading of the ethnographic material on which the claims are based shows how exaggerated they are.

Marjorie Shostak's ethnographic classic, *Nisa: The Life and Words of a !Kung Woman,* is a case in point. It describes in intimate and psychologically rich detail, in autobiographical narrative and commentary, the life history of a woman in her mid-fifties in northwestern Botswana. She remembered intense sibling rivalry with her younger brother and attributed her small stature and other problems to allegedly early weaning. Her father fought violently with her mother, but they remained together until Nisa's adolescence. She was married several times in her teens and (despite a culturally typical pattern of sex play throughout childhood) had a stormy introduction to adult sexuality, but her parents tolerated her flight from her husbands.

She remained with her fourth husband and eventually had four children; two died in early childhood, one died of illness in his youth, and a fourth was tragically killed by her own bridegroom shortly after marriage. These losses, and being widowed shortly after the birth of her third child in her late twenties, shaped Nisa's adulthood. She had occasional contacts with lovers both before and after his death. This habit continued into her fifties, despite two further marriages, the latest one quite stable. Menopause began a time of sadness and self-assessment, but at fifty-five she had accepted her childlessness and was bringing up her younger brother's two children. She was vibrant, mildly eccentric with an at-times bawdy sense of humor, eloquent in discussing both her own life and her culture, open to new relationships (including the interview process with its probing self-exploration), and proud of having surmounted difficulty and tragedy with a willingness to go forward and a continuing joy in life.

Recall that this was a woman who lived among !Kung San hunter-gatherers, sleeping in a small grass hut, living on wild game and plant foods in the Kalahari Desert, practicing a pagan religion, with zero modern conveniences and almost no contact with Western culture throughout her life. Yet when the manuscript of the first book about her was circulated to publishers, one of them rejected it on the grounds that Nisa sounded too familiar, too much like an American woman, in her relationships, feelings, and foibles. On Shostak's final trip, the basis of the book *Return to Nisa,* Nisa was about seventy, and Shostak had in the interim become the mother of three children. She had also developed breast cancer and felt mortally threatened, as Nisa had in her own life previously. In addition to con-

firming the findings and feelings of many years earlier, the 1989 trip allowed the
development of a relationship between the two women based on far more shared
experience and the associated emotions. Consequently, the evidence of the universality of basic emotions is, if anything, stronger in the second volume than it
had been in the first. In 1997 Nisa, learning of Shostak's death, had this to say
about Hwantla (Shostak's !Kung name) and Shostak's husband, Tashay: "Strong.
Hwantla held me strongly. We held each other as if we were one. Hwantla, I greet
you in the sand where you are sleeping now. I don't know why the spirits have
taken you away. . . . Tashay is not alone in his sorrow."

But the Nisa volumes are only one example. One of the best ethnographies of
emotion is Unni Wikan's *Managing Turbulent Hearts: A Balinese Formula for Living*. It uses the method of "thick description" to explore instances in Bali in which
emotional expression seems far from our experience. Balinese are required to suppress sadness and anger and, in the words of a song, put on a happy face. The culture insists on suppression and masking not only on the grounds that it is better
for the individual, but through the more intimidating, shaming tactic of invoking
public mental health. The sufferer is repeatedly reminded that expressing sadness
stirs up sadness in others, and the result is ultimate damage to what we would call
community mental health. Yet the Balinese insist that our conventional Western
distinction between thought and feeling often misleads us miserably. "Stop thinking!" they would admonish the ethnographer. "You'll never understand what we
mean if you only use your thinking!" Influenced by her Balinese friends, Wikan
makes a plea for "resonance," her translation of their phrase *ngelah keneh*. It is a
deep understanding, based on what she calls "feeling-thought." To achieve resonance, she and the Balinese urge us to stop making false distinctions between
thought and feeling. As we will see, this advice can now be placed on a solid neurological foundation.

Astonishingly, ethnographies such as this one have been distorted to attack
the claim that there are universal emotions. For example, at a 1993 conference a
leading anthropologist cited Wikan's work as disproving the hypothesis that grief is
universal. Study of the work itself reveals how misleading this use of it is. When a
young woman, Suriati, received a telegram announcing the death of her fiancé,
she interpreted it as a joke. But when the truth was confirmed, "tears streamed
down her face. Whereupon people leapt at her, laughing, 'What's the matter with
you, are you crazy [*gila*]!' Such are characterizations that stick in Bali. Suriati
gathered herself together, composed her face into an exuberant expression, and
smiled left and right to passersby as she walked the long way home."

Yet she borrowed money to travel to her fiancé's memorial and had herself
photographed prostrate on his grave. Three months later she did cry openly, once.
Still, she was widely admired by her community, and this in itself was a testimony
to the difficulty of what she did. Her remarkable composure, exceptional even in
Bali, appeared to confirm the reputation Bali has among anthropologists of being
the only culture in which death does not call forth tears. As Suriati explained, "I
am afraid to think of it, that I might go mad, so I try to be cheerful always that I

may forget my sadness." It is absurd, then, to claim that the Balinese case disproves the universality of grief. Throughout the account, Suriati's grief is palpable, and there are many evidences of it, including at least two episodes of crying. Wikan's book is called *Managing Turbulent Hearts* because Balinese hearts churn, too. Emotions there are *managed*, not abolished or replaced by some exotic local cultural concoction. Of course, culture strongly shapes the *expression* of the emotions, but that is not news. Just as the stiff-upper-lip, upper-crust English superimpose calm and fortitude on grief, the Balinese choose cheerfulness. Canio, the clown in *I Pagliacci*, must take the stage and laugh, but we are not so foolish as to doubt that he is grieving. On the contrary, we feel that his grief may be worsened, and is certainly made more poignant, by the consciously forced disjunction between his feelings and his public face.

These are culturally mandated performances. They must influence underlying emotions in some way, of course, but the claim that they disprove the universality of emotions is silly. One might as well point to the cultural variety of cuisine as evidence that there are no universal processes of hunger or digestion. These distinctions are vital because psychologists, psychiatrists, and diplomats have often been misled by anthropologists' claims. Ethnography can challenge ethnocentric prejudices about human mind and behavior, but exaggerations of cross-cultural variation are misleading as well. Catherine Lutz, a postmodernist ethnographer, titled her book about Ifaluk, a Polynesian island, *Unnatural Emotions*; but there are no unnatural emotions to be found in it, merely culturally guided expression of the emotions that all human beings share. As Antonio Damasio, the leading authority on emotion and the brain, wrote in 1999, "The thing to marvel at, as you fly high above the planet, is the similarity, not the difference." It makes sense, given our species' recency of origin, and the unity of the genome that underlies our brain.

Another common error is the confusion of emotion words with emotions themselves. No one denies that languages label emotions in interesting and different ways. English distinguishes between liking and loving, but the French say only *aimer*. Do we think for a moment that the French have missed the difference? *Au contraire*. In fact we reach out to French for the noun *amour*, when we want to provide connotations no English word quite has. We look to German for the word *schadenfreude*, the joy we feel at the suffering of a rival or an enemy. But we do not imagine that the presence of such a word in German reflects a fundamentally different emotional makeup. Lutz finds it difficult to translate the Ifaluk word *fago*, which contains elements of compassion, love, sympathy, and perhaps even what Wikan would call resonance. But to conclude from this marvelous word that the Ifaluk have completely different feelings is to mistake the word for the thing.

The solution is found in Karl Heider's *Landscapes of Emotion: Mapping Three Cultures of Emotion in Indonesia*. This pathbreaking study used the statistical technique of multidimensional scaling to generate maps of emotion words in three different Indonesian cultures and languages: the Minangkabau language of

West Sumatra, the Indonesian language spoken in the same region, and the Indonesian of central Java. Based on similar work in European cultures, Heider proposed eight "basic pan-cultural emotions": sadness, anger, happiness, surprise, love, fear, disgust, and contempt. These were then tested against the clusters emerging from Indonesian emotion words.

The result was strong confirmation of the universality of sadness, anger, happiness, and surprise, and confirmation of the other four in the list with decreasing strength in that order. That is, words in the Indonesian languages clearly cluster in confirmation of the first four; but for love, fear, disgust, and contempt, boundaries are drawn in different ways. For instance, in European usage *love* loads with components of happiness, while in Indonesia it loads more with components of sadness. This research strongly confirms the universality of certain emotions and also points to the different ways in which cultures use emotion words. But it does not prove that Indonesians experience the emotions we call love, fear, disgust, and contempt in different ways, only that when asked to describe these feelings, they use different words.

Unfortunately, most cross-cultural work on emotion is less scientific than Heider's. The writings in this field do make passing reference to the biological underpinnings of human mental life. For example, Lutz writes, "While the physiological aspects of emotional experience have not been considered in this work, it is important to stress again in conclusion that the biological basis of human experience, including that termed emotional, is not denied here." Similarly, Roy D'Andrade and Claudia Strauss state that "all humans have a built-in receptiveness to the form human cultures take, and all human cultures probably share some bedrock commonalities because of these coevolved features of human neurophysiology and morphology." But after paying such lip service, these anthropologists go on to conduct their analyses as if biology were irrelevant. What would we think of a biomedical scientist who said, *Well, yes, chemistry is undeniable*, but then proceeded to ignore it in all analyses? It is the mark of maturity of biomedical science that it now rests firmly on a foundation of chemistry and physics. Anthropology will mature when it similarly rests on and uses biology and psychology.

Meanwhile, ethologist Irenaus Eibl-Eibesfeldt has done an end run around the obstacle of language. Over four decades Eibl-Eibesfeldt has accumulated millions of feet of naturalistic film of people in exotic cultures, in every corner of the world. The Yanomamo of highland Venezuela, the Trobriand Islanders of Melanesia, the !Kung Bushmen of Botswana, the Ovambo of Namibia, the Samoans, the Balinese, and many other cultures have had their behavior archived. These films clearly and permanently show that human interactions and the facial expressions and postures involved in them are remarkably constant throughout our species. Mother-infant interaction, smiling and eyebrow raising in social greeting, playful pushing, shoving, and wrestling in male toddlers, alternately looking at and looking down and away in flirtation—all these interaction sequences and more are proven, species-wide features of human social and emotional behavior. It requires a great stretch of postmodern imagination to suppose that the emotions underlying such remarkably constant behaviors are secretly very different.

Mayr's two kinds of reductionism, physiological and evolutionary, are joined in neuroethology. Most of this work, devoted to tracing out the neural basis of fixed action patterns and innate releasing mechanisms, has been done in invertebrates and lower vertebrates, where behavior and brain are simpler. But groundbreaking work has been done on the squirrel monkey, and the collaboration that produced it is almost as interesting as the discoveries. The inspiration came from Paul MacLean, an American psychiatrist and neuroanatomist. In the late 1940s and early 1950s, MacLean was the major heir to the Papez tradition begun a decade earlier. In a 1949 paper, "Psychosomatic Disease and the Visceral Brain," he rescued the Papez circuit from relative obscurity, extended it, and named the resulting circuitry "the limbic system." It has by now won wide acceptance as the central brain network of the emotions, with demonstrated clinical, anatomical, physiological, and even immunological coherence.

MacLean began to attribute separate emotional functions to different parts of the limbic system and advanced it as the major circuit for psychosomatic diseases, such as stomach ulcers caused by social stress. In 1952 he proposed that the frontal cortex, long known as the seat of some of the highest human faculties— foresight, for example, and concern for the consequences and meaning of events—owe these functions to its intimacy with the limbic system. Thus, the newest, "highest" part of the human brain does not serve its functions by remaining aloof from older, "lower" parts of the brain but, on the contrary, works precisely *through* its relationship to the old emotional circuitry. This bold hypothesis about the anatomical relations of the frontal lobes to the limbic system was confirmed in the 1970s, when the great neuroanatomist Walle Nauta reconceived the frontal lobes as "the neocortex of the limbic system," suggesting that their main function lay in this relationship. This was prophetic. Just as other parts of the cortex had been identified as the highest report-and-control centers for vision, hearing, body sensation, and movement, so the frontal lobes have emerged as the highest center for the emotions.

Over five decades MacLean elaborated his theory of overall brain structure and evolution, known by his popular phrase "the triune brain." Some brain scientists reject this as a caricature, but they miss two points. First, without judicious simplification many thoughtful people outside of brain science will never have the vaguest idea what happens in our most important organ. Second, MacLean himself, as can be seen by a lifetime of careful experiment and scientific writing, especially his magisterial summary, *The Triune Brain in Evolution*, has never been misled by his own simplification. It remains valid and useful.

Briefly, it holds that the brains of humans, monkeys, and other advanced mammals contain within them three important evolutionary levels, which are to some extent functionally separable but have evolved many interconnections and usually act in concert. Thus, they are referred to as "triune" rather than, say, "tripartite." As neuroanatomists point out, the higher structures are not simply an invention of more developed mammals, since primordial versions of them are already present early in vertebrate evolution, certainly in amphibians. Also, the older, more primitive structures evolve and change; they don't just sit there as newly elaborate higher structures pile up around them. Nevertheless, the three levels helpfully sketch three major adaptations.

The first is the "reptilian brain," which in lizards and other reptiles is the dominant circuitry, and which most resembles the basal ganglia, thalamus, hypothalamus, amygdala, and related structures in the human brain. Think of it as a fistlike collection of structures, with your wrist as the brain stem going toward the spinal cord—or perhaps a child's fist thrust between your thumbs into the center of the cupped-hands model. This circuit figured prominently in Papez's "stream of movement." MacLean showed that these structures, whether in reptiles or mammals, are concerned not with movement alone but also with instinctive behavior—the fixed action patterns and innate releasing mechanisms of the ethologists. It helps explain why reptiles and birds, in which the brain is mainly built from the basal ganglia, have so many stereotyped actions and responses: a lizard turning sideways and dropping a bright red dewlap as a threat, for instance, or a finch repeating again and again the same territorial song. It isn't that mammals have no such behaviors but rather that amphibians and reptiles, such as newts and lizards, have so little else.

Field studies of the Komodo dragon, a fearsome eight-foot giant of a lizard, convey some sense of what this level of brain structure has to do. Since the part of the hypothalamus that regulates body temperature hasn't evolved, the Komodo moves in and out of the sunlight to keep from overheating or chilling out. It hunts large prey like water buffalo, killing them indirectly and unpleasantly: after a surprise attack and a bite wound, they hang around the dying buffalo to defend it from all comers, as it slowly dies of infection from the Komodo's mouth bacteria. A Komodo must engage in brutal wrestling bouts with other dragons, which may result in fatal injury. Fighting produces a simple linear dominance hierarchy in a neighborhood, based on nonfatal fights and threats—for example, taking a deep breath, straightening up, even rearing on the hind legs, all of which make you look bigger. Such fights are more common in males than females except in one classic situation: egg guarding.

Komodo eggs are fertilized internally through copulation—a pastime perfected by reptiles, who helpfully invented penises—although some salamanders have copulation without that nice appendage. In either species, a male and a female have to attract each other (by sight and especially by smell), approach each other, suppress their (especially her) fear, and at last gingerly touch. Hormones in each respond to signals from the other, an interleaving of arousing signs and moves that escalate the excitement. Just to ensure the female's cooperation, though, the male may invoke some moves that look a lot like those wrestling matches—in this case with a much weaker opponent. It's not in his interest to hurt her, however; she has to be in condition to gestate, guard the eggs, and fight off other females who want her good nesting site.

That's about the extent of Komodo talents. No parental care after hatching, no elaborate courtship, no coalitions or alliances, no cooperative hunting, no facial expressions, no tool use, and no play. Dinosaurs may have evolved some of these more complex things, which we used to think were just for their close cousins, the birds. Crocodiles carry their newly hatched young in their mouths and may stay near them for some weeks. But most of the reptiles that survived the Big Crunch—the asteroid impact near Cancún, Mexico, 65,000,000 years ago—come with the social and emotional equipment of the Komodo dragon: not much. This roughly reflects the early, generalized predinosaur reptiles of 300,000,000 years ago.

The second set of circuits is the paleomammalian, or "old mammalian," brain, on the theory that it evolved with the earliest mammals. This, in effect, is the limbic system, resembling Papez's "stream of feeling." In primitive mammals, such as rodents and rabbits, it occupies much more of the brain than it does in late-model mammals like monkeys. Parts of this system—the hippocampus and amygdala, for example—exist in simpler form in reptiles; it was their expansion and elaboration, particularly the cingulate cortex and what has been called the extended amygdala, that were crucial to this evolutionary step. This expanded limbic circuit, without replacing the instinctive functions of the reptilian brain, gives those functions emotional coloring, a feeling tone that facilitates expressive signaling, and—through a kind of signaling to the self, an internal highlighting of external experience—learning.

In thinking about what the early mammals accomplished, we generally talk in terms of advancing hip and shoulder structure, moving some bones of the jaw angle to the middle ear, and other things we can see in the fossil record. But from comparative evidence, including studies of primitive egg-laying mammals like the platypus and echidna, we also know that early mammals invented warm-bloodedness (birds and some dinosaurs invented it separately), evolved hair, and cared in substantial ways for their young, making milk, crouching over them, and retrieving them. From other comparative evidence MacLean proposed three key features of social and emotional behavior that the early mammals introduced: family, play, and the separation call.

If the basic, no-frills family is a mother with her young, no living reptiles really have it. Crocodiles do hear their young crackling out through their egg shells and gently move them from nest to stream couched in their awesome jaws. Impressive, and a glimpse of what was to come when the dinosaurs, birds, and mammals invented families. But living reptiles are socially more primitive than dinosaurs and don't have families. All mammals do. Likewise, only with mammals do we have young that show lots of useless, energetic behavior they seem to enjoy. Finally, if you are going to have significant care of your young, including retrieval, you'll need a separation call—even an ultrasonic one like rats have—so that mothers can locate straying pups. It has to be a sound that sears its unpleasant way into the core of the mammal mother, causes a deep uneasiness, and yet makes her deliver care instead of a lethal bite. These are all limbic system functions and, fittingly, they embrace a new kind of emotional warmth to go with the thermal control. Warm blood, one is tempted to say, warm heart.

The third and last set of circuits is the "new mammalian" brain, corresponding to the Papez "stream of thought" and so far achieving its evolutionary culmination—not *destiny*, since it wasn't a foregone conclusion—in the human brain. This circuitry became important in the mammalian population explosion that followed the implosive demise of the dinosaurs. It added much more of the new, specialized, six-layered cortex ("neocortex"), in small, columnar, computational modules, each holding millions of cells. As the sheet of six-layered columns grew, folded, and grew again, advanced mammals—wolves, elephants, dolphins, chimps—became capable of higher and higher levels of sensory information processing, and of the feelings evoked by the stimulation of the senses. In short, they could think, not just feel.

Michael Murphy and Sue Hamilton demonstrated in an elegant study the roles of the three levels. They discovered, rather amazingly, that removing a hamster's entire neocortex at birth leaves it with little or no impairment of instinctual behavior patterns. These are handled by the paleomammalian and reptilian parts of the brain. The surgically altered hamsters are none too bright, so you have to coddle them a bit, but they grow up to have sex, care for their young, fight, and (while growing) play. Neurobiologist Jaak Panksepp repeated this experiment in rats, with similar results for play. Further removal of the limbic structures damages maternal behavior and play behavior, these evidently being functions of the old mammalian brain. But the remaining unlearned species-typical behaviors— sex, aggression, food getting, and so on—can apparently be conducted in hamsters that have only some rough equivalent of the reptilian brain. This finding, astonishing to some who are too impressed with the cortex, has helped give credence to the triune-brain model.

Another of MacLean's collaborators was Detlev Ploog, an ethologically trained German psychiatrist. MacLean's knowledge of neuroanatomy and brain evolution combined with Ploog's ethological and psychiatric background were ideal for exploring the physiology of the emotions. Ploog went on to a distinguished career as a research psychiatrist, developing what he has called human neuroethology. But while working in MacLean's lab at the National Institute of Mental Health, he helped to show that advanced mammals do indeed keep within them a sort of archaeological record of their phylogenetic past. MacLean and Ploog's chosen species, the squirrel monkey, was a laboratory subject with complex emotional behavior controlled by *all three* brain levels.

Consider the unintentionally droll penile display. Our simian friend lifts a leg and spreads a thigh, exposing the genitals, while making a distinctive sound. It does this as a threat in aggressive encounters that may or may not lead to violence, and dominant animals are more likely to display to submissive ones—even to the point of, yes, sticking it in their faces. Males do it much more often than females in such encounters and also as a prelude to sexual intercourse. It is present in both sexes within a few days of birth, but the context is easily shaped by experience; one dominant female did it often, just as males do.

This display is a species-typical fixed action pattern. Lesion studies—experimental injury to the brain to trace functional connections—showed that it is controlled by a circuit passing through the globus pallidus, a part of MacLean's reptilian brain. Damage to the limbic system itself does not affect the display, but the globus pallidus has intimate relations with the limbic system. Perhaps the lesion interrupts some crucial interplay between the emotion of aggression or sex and the movements of the display. In any case, in normal monkeys the display goes with emotion, and that must come from the limbic system—the old mammalian brain. Finally, since context influences the display more than would be possible for a more primitive mammal, the new mammalian brain or neocortex must also be involved. Thus, the three levels all participate in the control of this instinctual yet flexibly deployed behavior.

One might think that brain science would find in such studies a reason to suppose that some similar throwbacks, and some similar integration, might be found in the ancient depths of the human brain. Not so. At least, not until the 1994 publication of Antonio Damasio's *Descartes' Error: Emotion, Reason, and the Human Brain*. This major advance in our thinking about the brain, followed in 1999 by *The Feeling of What Happens*, Damasio's emotion-based theory of consciousness, has made it impossible to speak of mind without also speaking of feelings. As might be expected from a mind that for eons felt before it could really think, the human mind thinks *with* feelings, not in spite of them. In that respect it is more like the mind Freud imagined than like that of Descartes and other philosophers. It is neither an airy spirit nor an exquisite computing device but a creaky old calculator sunk in a sticky swamp of feelings.

Most modern theorists of mental life seem to find emotion . . . well, not scientific enough. There are many grand, complex theories of consciousness and mental life that barely mention emotion and are thought all the better for it. Damasio, with Hanna Damasio and other investigators, has changed all that drastically in just a few years. The Damasios and their team studied people with brain injuries whose tragic trauma destroyed the underbelly of the frontal lobes, severing them from the limbic system. In MacLean's terms, a major part of the neomammalian brain had been cut off from older structures. Whatever was left of the frontal cortex—the neocortex of the limbic system—was now on its own.

The first patient known to have this problem was Phineas Gage, a railroad-crew supervisor who in 1848, at the age of twenty-five, had a freakish accident. An explosion drove a three-foot-long, inch-wide spike like a bullet through his brain; it passed under his left cheek (destroying that eye) and out through the top of his skull. Yet he lived to tell the tale, and became famous because his accident was so improbable and his recovery, with one crucial exception, so complete. Doctors carefully recorded every aspect of his medical and mental condition. After his death—Gage lived for another thirteen years and probably died from epilepsy—his body was exhumed and his skull saved in a museum at the Harvard Medical School. There, in the early 1990s, a team led by Hanna Damasio measured it, modeled it on their computers, and were able to figure out where the brain injury almost certainly was.

Gage's damage followed the pattern of the patients in Damasio's clinic. And, like them, his recovery was not quite complete: the injury changed his personality. Prior to the accident, Gage had been a model crew boss, a favorite of both his employers and the men working under him. But afterward he became a capricious, unreliable, vacillating, foul-mouthed boor who was permanently unemployable except, sadly, at the circus. He had lost no measurable intellectual skills. But as one observer said, "Gage was no longer Gage." Dr. John Harlow, who took care of him and wrote about his case, summarized it by saying that Gage had lost the "equilibrium or balance . . . between his intellectual faculty and animal propensities."

What made Gage's case so interesting to Antonio Damasio was that a patient of his, a man named Elliot, had very similar damage, although not from a freakish accident. Elliot passed every conventional cognitive test with flying colors, showed a superior IQ, and even came out normal on a standard personality test. Yet something—evidently something unmeasurable—was very wrong with him. He couldn't hold a job, entered ill-fated business ventures with disreputable partners, had two divorces in quick succession, failed to maintain his other relationships, and made one bad judgment after another until his life was a shambles. Elliot was "in some respects a new Phineas Gage," but without the profanity and other kinds of intensity. In fact, in many interviews he appeared to *lack* intensity, to the point of being nonchalant about his own suffering. His tragedy was "to know but not to feel." More tests showed that he was adequate or superior in every type of social knowledge. Even on tests of moral development he scored high; he could reason with full, mature capacity through all kinds of ethical dilemmas. But after one such challenging item of moral reasoning, he said significantly, "And after all this, I still wouldn't know what to do!" He could think through the problems but not make the choices.

Finally, some ingenious tests were devised that made his deficiency measurable. One was an investment-type card game in which normal players quickly learned to shift their strategy from high-risk, high-gain, ultimate loss to moderate-risk, moderate-gain, ultimate win. Elliot simply couldn't learn to win. Physiological studies confirmed that he wasn't feeling his losses "where he lived," so to speak. His defect was an inability to experience emotionally the consequences of his acts; he could reason perfectly except that where something he should care about was at stake, he *couldn't* care, and therefore he couldn't make decisions. Similar human cases, as well as research on monkeys, support the syllogism: damage the underbelly of the frontal lobe—I call it the Gage cortex—or disconnect that cortex from the limbic system, and you disrupt judgment, because the victim cannot *feel* the meaning of that judgment.

So we have an exquisite irony that grows year by year in scientific strength: crucial to the working of the human mental faculty—that noble reasoning apparatus of which we are justly proud—is that it is strapped to the engine of emotion and seated subtly in the body. The human mind can do what it uniquely does only because of how it fits in an apparatus of sense and feeling that predates it by hundreds of millions of years. Not only is Descartes's dualism not true, it is quite the opposite of the truth: far from being a unique and separate "stuff," mind is actually nothing but the best and most interesting function of the body. And as for that philosopher's cardinal principle, "I think therefore I am," it is a foolish saying; in truth, we think because we feel what we are.

Although these examples can give us some clues about what is involved, Damasio's idea of how the human brain works is far more complex than can be shown by Gage and Elliot. He accepts the now widely held theory that thought, like consciousness, is no function of one brain region but "a trick of timing," by which simultaneous activation of different brain regions gives us our sense of mental coherence. Thus, the self is not "a little person inside our brain perceiving and thinking about the images the brain forms. It is, rather, a perpetually recreated neurobiological state." This state includes the core self, a nonverbal ongoing consciousness of how we feel, and the autobiographical self, an emergent integration of whatever is *right now* with a dynamic delving into a library of conscious and unconscious memories. But what Damasio adds to the time-binding model is the immediate involvement of emotions and bodily sensations—the idea of the body as a theater for the emotions, but with an audience that enhances and inspires them.

The underbelly of the frontal lobes and the adjacent limbic cortex—the cingulate gyrus—are pivotal integrators of emotion and thought. If we revert to the cupped-hands model, there is an empty space staring up at us behind our fingertips that corresponds to this crucial region, the same one blown out of Gage's brain. Working memory, however, governed in part by the outer frontal cortex—the fingers themselves—tells us what has been happening over the past few seconds, which is vital to our sense of continuity. Still another part of the frontal lobe, back near the central motor-control strip—the knuckles in the cupped-hands model—integrates information to guide speech and other voluntary action. The temporal lobes, the thumbs in the model and just inside the tops of the ears in real life, are a convergence zone as well, but for *sensory* information, which gathers emotional tone and registers in memory as it goes through the deeper limbic structures.

A hidden part of the cortex, the insula, is also active; it reports concurrently on the internal state of the body—heart rate, stomach contractions, and the like. The precise quality of thought and feeling depends on the activation of what Damasio calls dispositional representations—working circuits disposed to fire in certain conditions because of innate structure or learning. These are what psychologists call templates or schemas. "Instinct" and "drive" are examples of innate knowledge, but many maps and representations in the brain are dynamic and constantly changing. Still, learned schemas can also have stability.

For Damasio, as for James and Lange, the body's state is central to emotional experience: "The mind is embodied . . . not just embrained." Whether this revival of an old idea will prove valid remains to be seen. It has to explain the obvious ability of paraplegics to have emotions, as well as the fact that different emotions sometimes have very similar physiological accompaniments. Damasio addressed some of these issues in his second book, arguing that emotion is indeed blunted in quadriplegics, and that the emotion they do have comes from a combination of memory and the sensations they have from their mouths, throats, and facial expressions. Above the neck, they still have working bodies. Also, they may have visceral sensation from the vagus nerve, which can do an end run around spinal-cord injuries. Research on emotion in quadriplegics and patients with severed vagus nerves should help to resolve these issues.

But many scientists continue to see emotion as independent of bodily functions, a process occurring more inside the brain than between brain and body. Even Damasio would agree that there are innate circuits, with limbic system components, corresponding to specific emotional states. A huge set of learned dispositions is added by experience, and developmental psychologists such as Carolyn Saarni have shown in exquisite detail how children acquire the many skills needed to negotiate their feelings and those of others. But genetic coding of what Damasio calls "nearly precise structure . . . seems reasonably certain for brain stem, hypothalamus, and basal forebrain, and quite likely for the amygdala and cingulate region." These are "the evolutionarily old sectors of the human brain. . . . We share the essence of these brain sectors with individuals in numerous other species." The use of the word *individual* here dignifies appropriately the emotional lives of other animals. We cannot share so much of basic brain structure without also having partly shared inner lives.

Even for skeptics, these findings have strong implications for how we think about the brain. Artificial-intelligence experts trying to model the brain as a general-purpose computer should in the early twenty-first century be feeling very uncomfortable, since there seems to be little general-purpose *anything* in the brain. Contrary to their claims, the brain cannot be a massively parallel processor with interchangeable computational elements. Neither is it a collection of "firmware"—generalized homogeneous learning networks that change in a simple way with experience. The brain is a makeshift, inelegant evolutionary pastiche. It is surely a parallel processor, but its structure of side-by-side or nested elements preserves ancient and outmoded chunks of circuitry forced to work in tandem with shiny, superfast new ones.

If you insist on the computer model, think, then, of a late-1970s Tandy desktop computer somehow hopelessly yoked to a Cray supercomputer, an early Macintosh, a Pentium multimedia-based system, an old mechanical calculator, an abacus, and a vast array of Webservers and software. DOS, MacOS, Windows, Linux, Java, Bluetooth, and various WAP enablers have no choice but to coexist, not just on a network but in every single machine. Parts of this system would do well to bypass the others, but they cannot. Aging, rusted, rudimentary, even broken components stay and play their roles. Now add all the things machines don't have to contend with: ongoing responsiveness to temperature, humidity, time of day, the ebb and flow of various chemicals, and, alas, parasites that constantly change the responsiveness of some parts of the circuitry. Let the system learn, but not just through programming; experience modifies software and hardware alike. Finally, make the system especially impressionable for certain kinds of experience, corresponding to love, lust, grief, fear, rage, disgust, resentment, jealousy, and pain. This is the spit-and-chewing-gum evolutionary mess that is the human brain, not a well-made computer solving problems by iterating neat little algorithms. And this mess is uniquely human. Future computers may well think, but until they feel like bodies they will not think like brains.

Brains think with and through emotions. But what are emotions, really? New Zealand philosopher Paul Griffiths posed the question in just these words, and answered that they are the subjective experience—the insider's view—of affect programs. And affect programs? They are the brain's internal pathways of rage, fear, joy, lust, love, and grief—among other possible candidates, combinations, and more subtle transformations. Glimmers of rage, fear, and lust probably arose with the reptiles. Most of the rest emerged with the early mammals, and with the advanced mammals, especially monkeys and apes, the common affect programs are undeniable. Ever since Darwin we've known that the details of facial expression are very similar across our lineage, and today we have growing evidence that the brain functions involved are also widely shared. That is why we can speak freely of rage and fear in monkeys, even though we must hesitate for beetles, which also fight and flee. This common evolutionary origin gives us our best basis for defining and assessing emotion, a scientific anchor in the turbulent sea of philosophic musings.

In the next chapter we will see what the human brain adds, and why *we* may have a *quadr*iune brain. For now, consider smiling in greeting. Irenaus Eibl-Eibesfeldt showed that it is, as Darwin claimed, a relatively fixed human action pattern, universal to all cultures despite some variation in context. Paul Ekman and his colleagues went on to show that the smile is both expressed and interpreted in remarkably similar ways in widely different cultures. It appears early in infancy as a result of events in brain development, not learning, and it can be triggered automatically thereafter. It relies in part on limbic circuits, but its main control may lie in the oldest, reptilian part of the brain—much like the squirrel monkey's genital display.

Finally, it can be faked or done on command, at least initially without the feeling of pleasure. *Smile though your heart is aching, smile even though it's breaking,* say the words of the old song, and such words, impossible in any species but ours, cause the ancient facial musculature to assume the classic expression. And we now know that such on-command facial expressions can produce some autonomic nervous system signs of the corresponding emotions. So we have to think in terms of mutual feedback between emotion and expression in daily life. Nevertheless, a faked smile can be detected by sophisticated observers, so it is not perfectly deceptive. Yet it begins without emotion, or even with hostility—a man can smile and smile and be a villain. As Walter B. Cannon knew—although he was a bit off in his anatomy—this ability is regulated from the newest, neocortical part of the brain, while authentic smiles are due to lower, older circuits. The distinction underscores the brain's status as a pastiche of ancient and modern structures.

Thus, the human smile—a gesture of joy, greeting, submission, deception, occasionally even contempt—can involve all three of MacLean's brain levels, and more besides. The sudden smile in greeting a casual acquaintance may be similar in neural control to the friendly or aggressive displays of lizards and robins. The slowly spreading smile on the face of a parent watching a loved child depends on the limbic system, that invention of the early mammals. And the less authentic, endlessly repeated flight attendant's smile could not be flashed without the command-and-control circuits of the more advanced, deliberately deceptive neocortex.

But most interesting is the smile that comes over us as we are sitting, aware of nothing at all, in solitary reverie. Somehow the flow of feeling catches a favorable wind, releases neural energy in certain ancient circuits, and without a conscious thought we find ourselves smiling. If the physiological sources of feeling were a well, the suggestion is that our thoughtful, conscious, higher, and more noble selves could dip into it and come up coated with warmth or joy, anger or sadness. This is something like what happens when we respond to the complexities of music, theater, or any of the arts. The well metaphor, however much it strains anatomical fact, seems if anything more appropriate than it did in Papez's day: the limbic structures contain neurotransmitters and neuromodulators, chemicals that traverse gaps or bathe neurons in subtly powerful fluids. And the ongoing electrical activity of the subcortical circuits, with their intrinsic, pulsing rhythms, is in itself a kind of imaginary fluid, a source of action, a well of feeling.

But the wording of the analogy will mislead us badly, if we see the emotional brain as a placid cistern. Far from being a static, tepid well, it is more like an artesian spring, upwelling forcefully at times of its own accord. And even this analogy may lull us into complacency, since it suggests only a bubbling of sweet water. In reality this particular spring, benign as it often is, can roil malignantly, and is equally likely to spew forth poison.

CHAPTER 8

Logos

... the consciousness of the human organism is carried in its grammar. Or the unconsciousness of the human organism.

—JOAN DIDION, *A Book of Common Prayer*

"In the beginning was the Word, and the Word was with God, and the Word was God." So begins the Gospel according to John. In the book of Genesis, Adam's first task is to name all other creatures, to gain dominion over them. And the first great cooperative human effort—the Tower of Babel—is foiled by a failure of speech. Such claims for the primacy of language are echoed today by intellectuals far removed from religious faith. In past ages, men and women had no difficulty specifying the source of the separation between themselves and animals: people had souls, animals did not. But with the fading of the soul—at least in conversations among scientists—language was about the best thing an expert could find to prove human uniqueness. Anthropologists have been great enthusiasts of this view, and to this day every student learns a catechism about symbolic speech as the moat around humankind that keeps all lesser animals at bay.

But the late twentieth century saw a frontal assault on this position. At least four different teams in as many universities used widely varying methods, including pictorial icons and hand signs, to teach chimpanzees some elements of language. Another group using sign language claimed to have taught a female gorilla four hundred words. Two chimps, Sherman and Austin, "talked" with each other in lexigrams. The graceful African vervet monkey was shown, in experiments in the wild, to have a few natural calls with something resembling symbolic meaning—warning calls that specify different predators. And the not-so-graceful rhesus monkey was shown to have aggressive and distress calls that, at least fuzzily, distinguish social categories—kin and nonkin, dominant and subordinate. Farther afield, an African gray parrot named Alex was shown to understand a lot of the phrases researchers thought he was just mimicking. And most impressively, an enculturated bonobo, Kanzi, proved under rigorous experimental conditions that he understood English sentences as well as a two-and-a-half-year-old child.

The unease produced by these claims depends on our definition of language. "Language is a purely human and noninstinctive method of communicating ideas, emotions, and desires by means of a system of voluntarily produced symbols." To the anthropologist Edward Sapir this seemed a "serviceable" definition of language. However, because it requires us immediately to say what we mean by *instinct, voluntary,* and *symbol*—tasks that are probably worse than the one we started with—it is for our purpose not so serviceable. Worse, since it includes the phrase "purely human," it excludes by fiat the very claims that interest us.

Virtually every definition of language suffers from flaws like these, so our sense that we know what language is is mostly implicit. This was not the case for Charles Hockett, a linguist at Cornell University, who advanced a definition that, if flawed, remains the best. It includes thirteen features, and for Hockett a true language must include all of them:

- the use of the vocal-auditory channel;
- broadcast transmission and directional reception;
- rapid fading of the signal;
- interchangeability of speakers;
- feedback, or comprehension of the signal by the signaler;
- specialization of the signal for communication alone;
- semanticity, or meaning;
- arbitrariness of the relationship between signal and meaning;
- discreteness of the elements, change in which can signal a change in meaning;
- displacement, or reference to things removed in space and time;
- productivity of novel signals;
- transmission by teaching and learning;
- and duality of patterning, or shuffling the same elements to make different signals ("team" and "meat").

By this definition only human speech qualifies. While each of the thirteen characteristics may be found in one or another animal communication system (and some in many of them), only human speech has every one. The definition is precise and well operationalized—that is, it provides a straight-out test of animal-language claims. Still, it has serious problems. For one thing, it excludes some things that clearly are language. Sign languages of the deaf do not use the vocal-auditory channel yet are otherwise complete equivalents of vocal-auditory human speech. To exclude them is to hold a definition of language that contains something inessential. Worse, the list, as thorough as it is, omits something closer to the essence of language than most of the list: syntax, or the organization of communication above the level of the single symbol. While Hockett was working on his definition and evaluating animal communication, linguist Noam Chomsky was designing his theory of syntax, which would give centrality to grammar, making language seem more remote from the apes than ever.

Chomsky composed the sentence "Colorless green ideas sleep furiously" to illustrate the higher level of organization. Although largely meaningless, it is easily recognizable as a grammatical sentence, and it even provokes us to try to find meaning in it. "Furiously sleep ideas green colorless," on the other hand, leaves us quite cold. Our ability to choose the grammatical sentence and engage with it shows that we know rules for stringing words together—rules that are to some extent independent of meaning. Syntax, like phonetics, has its own duality of patterning—"Jim hits Bill" or "Bill hits Jim"—and productivity—the invention of never-before-heard forms. Indeed, the infinite capacity of syntactic rules for generating new forms may be the crucial feature of human language. So these features must be added to Hockett's list.

Chomsky argued, as anthropologists had before him, that all human languages perform the same functions, reflecting basic capacities of the human mind and brain. For every sentence produced or understood there is a deep structure, and the deep structures of sentences that function the same way in different languages are isomorphic. The very possibility of translation suggests that this is so. The rules for the deep structure constitute a universal grammar, shared by all people of all cultures, and these rules arise from brain function. The key to any *particular* language is its transformational grammar—the rules for getting from the universal deep structure to the special surface structure of that language. Deep structure, along with knowledge of some special transformational grammar, adds up to something known as competence—the speaker's *abstract* knowledge. Performance, on the other hand, is what we say or hear—the fully realized, practical speech act—and it depends on much besides competence.

When syntax is added to Hockett's features of meaning, productivity, displacement, and arbitrariness, we begin to approach the essence of language. These definitions set the stage for Eric Lenneberg's *Biological Foundations of Language*, published in 1967, which summarized the then-existing evidence for the claim that language *capacity* is fundamental in human brain structure and that its emergence in early life reflects a genetically coded plan. This evidence has by now reached overwhelming proportions, summarized best in Steven Pinker's boldly titled *The Language Instinct*. Pinker is the modern heir to the Chomsky–Lenneberg model, but he has added a convincing evolutionary perspective. Most recently, neuroscientist William Calvin and linguist Derek Bickerton wrote a graceful and insightful book, *Lingua ex Machina*, that attempts to explain how language emerges from the machinery of the brain. As suggested by their subtitle, *Reconciling Darwin and Chomsky with the Human Brain*, their model is both anatomical and evolutionary.

Although we need not look far to find challenges to the biological view—the immense diversity of human languages and the obvious role of learning, for example—we can just as readily counter with the concept of competence, implicit in Hockett's definition and explicit in Chomsky's. Genetically designed brain circuits would explain the "deep structure" of languages and our mental capacities for meaning, productivity of novel signals, and displacement in space and time from the thing talked about. The learning environment would merely fill in the gaps or, more specifically, would set switches in brain circuits. These circuits, part of the brain's language module, were designed by evolution to expect certain kinds of information from the environment—whether to distinguish between "l" and "r," say, or whether to put tense markers before or after a verb root. This makes the environment crucial but far from all-powerful. Its impact on language, superficially enormous, would be to take advantage of these evolved brain circuits—a significant but small role in nature's linguistic plan.

Today the evidence for language universals is overwhelming and does not depend on Chomsky's or anyone else's theory, but it can be shown from wide study of many of the roughly 4,000 known languages. Furthermore, even generalities far short of the universal point to biological constraints. Joseph Greenberg of Stanford University, probably the world's most respected anthropological linguist, has often said that any time we find a nonrandom distribution of features of sound, symbol, or syntax as we look around at the world's languages, we have a generalized trait that requires explanation.

While we all feel that we understand language, and can compare them a bit if we speak more than one, comparative linguistics is a highly technical field almost as complex as comparative genomics. In both data sets there is a lot of noise and redundancy, yet it is possible with the right techniques to construct a phylogenetic tree that validly shows how things are related and what features they share. Linguistic variety is, of course, not genetic; only the constraints on it are, just as the constraints on genomes are the laws of physics and chemistry. In both cases, the constraints do not completely prevent all unpredictable things; they cause statistical biases. But these biases demonstrate their underlying power. To take one example, Greenberg's classic analysis of word order shows that subject, verb, and object—the way we usually order them in English, as in "I love you"—can actually occur in six different possible sequences. While there is at least one language using each of the six orders, the worldwide distribution of independent languages is heavily skewed in favor of subject-verb-object (English being only an example) and subject-object-verb (typical of Turkish, and often used in German). Verb-subject-object, which occurs in Welsh, is a distant third, and the other three solutions are rare. The word-order analysis can then be extended to include the little connecting words that, in English, are prepositions, but that in many other languages follow the noun; the order of the possessing and possessed nouns ("the dog's ears," reversed in Hebrew and many other languages); and the order of adjective and noun (opposite in English and French).

Greenberg calculated, as we can quickly confirm, that *even after eliminating the three least common of the six ways of ordering subject, object, and verb* there are twenty-four possible combinations of word order (meaningful-element order, to be exact) for the features just described: $3 \times 2 \times 2 \times 2$. Yet only fifteen of the twenty-four have been found among the world's thousands of languages and of these, four types overwhelmingly predominate. Furthermore, subjects precede objects in all four, so if we omit subjects from the analysis for a moment, there are two broad types of human languages: those that have verb before object, prepositions, possessed before possessor, and noun before adjective; and those that have object before verb, post positions, possessor first, and adjective first.

So what has happened to the vast and unpredictable variety of human linguistic creativity? As far as word order is concerned, most of it falls into two types. But aren't the other types important? Of course they are, and equally valid. Still, the fact is that what seemed completely unpredictable turns out to follow lawful patterns, and these are probably due to the circuits in the brain's language module, wired up under the influence of the genes. Whatever those circuits and switches are, if they are going to learn a rule that puts verbs before objects, then they prefer to put nouns before adjectives as well, while the rule that objects must precede verbs makes it easier for them to put adjectives first.

No one knows what makes the circuits favor these two patterns, but they must, or we would not see so many languages line up so neatly when so much more variation is possible. These are the kinds of patterns that psychologists and ethologists call prepared learning, and that comparative linguists describe as universals or universal tendencies. It is worth noting that authorities like Greenberg and Bernard Comrie have spent a lifetime immersed in and classifying the details of many languages. They have little or no commitment to the kinds of abstractions that Chomsky and his followers consider the underlying mental structure of language. On the contrary, they are interested in surface appearances—what the Chomskyans call performance. Still, after attending to and classifying these surface patterns, they have arrived at a similar conclusion.

Steven Pinker goes so far as to speak of a language *instinct*. Constraints on learning, against a background of strong and specific motivation, are precisely what we mean by instinct. But we don't have to accept the term in order to see the facts. As Chomsky put it in 1993, "By now, enough is known to indicate that the differences among languages may not be very impressive compared with the overwhelming commonality." Something about the structure of brain and mind makes languages converge from limitless variety to limiting constraint. As William Calvin and Derek Bickerton put it recently, "The evidence that language is an innate, species-specific, biological attribute that must possess a specialized neural infrastructure is . . . overwhelming." Fittingly, the statement is in a chapter called "Darwin and Chomsky Together at Last."

Consider the human skeleton. This exquisite structure has a genetically coded design that unfolds in a quite rigid maturational plan. Without a supply of nutrients, the plan is useless and the unfolding stops. Placed in an environment low in calcium, phosphorus, or vitamin D, the plan will go badly awry—the resulting structure may be markedly deformed. Just varying calories alters the size and growth rate of the skeleton and differences in diet and exercise account, in part, for subtle variations observed in normal skeletal structure. An aging person's unique skeleton owes its shape not just to genes but to diet, exercise, workload, type of employment, and posture. Even psychological stresses can change the structure.

Nevertheless, it would not occur to anyone to doubt that the plan of the skeleton is produced under the rule of genes over a largely fixed course of development. In Chomsky's words, "The embryo may fail to grow arms and legs properly, but no change in the environment will lead it to grow wings." On the contrary, in the normal expectable environment of a developing human being anywhere in the world, the skeletal structure special to our species will emerge. Similarly, we may now hold an analogous conviction with respect to language. But giving credence to this claim demands some details of the maturational plan and the linguistic anatomy of the brain.

Perhaps the first evidence for language localization was provided by anatomist and anthropologist Paul Broca in 1861. Broca noticed that patients who could not speak frequently had a brain lesion in the area just forward of that which controls the face and throat—roughly halfway between the top of the ear and the edge of the eyebrow. A few years later Broca published a further observation: in the vast majority of these cases of aphasia, or language loss, the damage in question was on the left side. Brain lateralization had been born. "Broca's aphasics," as they are generally called, can with great effort make a very telegraphic version of speech: "New York . . . bus" might stand for a whole idea about a trip to New York that would normally be expressed in a complex sentence. Most difficult for these patients are small transitional words. However, they often give evidence that they understand what is said to them—by carrying out instructions, for example.

In 1874, just a little more than a decade after Broca described his aphasics, an unknown twenty-six-year-old neurologist, Carl Wernicke, described a new pattern. Wernicke's patients were capable of fluent and rapid but largely meaningless speech, with some of the content words mistaken or missing and with loss of comprehension despite normal hearing. These patients often had lesions in a quite different region, now called Wernicke's area, behind the spot of the cortex involved in the first-level interpretation of sound—and coincidentally, just above the ear. Wernicke theorized that his area analyzed sound patterns for language, while Broca's transformed thought into speech. For a patient to repeat a sentence spoken by the physician, both Wernicke's and Broca's areas would have to be intact, with a preserved connection between them. As Wernicke guessed, this connection indeed exists—it is the arcuate fasciculus, an arc-shaped cable tunneling between the two areas under the surface of the brain. Much later a third form of aphasia, called conduction aphasia, was recognized. It is known by an inability to repeat what has been heard, but with a sparing of comprehension and fluency. The lesion is where we expect it, in the arc-shaped bundle, disconnecting comprehension from speech.

We are far from the 1870s, and armies of brain imagers fly the banner of Broca and Wernicke, but this outlook has held up rather well. Current accounts typically begin with a critique of the nineteenth-century view but, after reviewing much new evidence, come up with a scheme that preserves key elements of the old one. Certainly many refinements have been made. Two patients quoted in a recent account give a sense of what circuits these areas engage in. A Broca's aphasic said, "Little words, no," showing that he had a good grasp of meaning combined with a missing component of grammar. In contrast, a Wernicke's aphasic said, "The small words are too big for me," a perfectly formed sentence that poignantly contradicts its own meaning—for that sentence. Broca's area in the very limited anatomical sense is probably related to speech articulation, while larger lesions, possibly also damaging a part of the cortex called the insula (hidden behind the classic Broca's area), might result in the full syndrome. Wernicke's aphasia requires a larger arena of injury to the rear of the temporal lobe, including the wiring under it. And specific damage to the arcuate fasciculus makes it hard not just to repeat sentences but to send any forms of coherent meaning forward to the centers of speech production. These important refinements show us that premodern anatomists could not foresee everything, as we might expect. But they scarcely negate the attempt to localize language function in specialized neural circuits, as some have argued. Indeed, they make the case for localization *in circuits* more persuasive than ever. We can't expect the language module to contain structures that look like features of language. On the contrary, they will look like chains and loops of neurons generating the functions of language functions in ways we now find difficult to imagine.

Even classical neurology had an inkling that Broca and Wernicke had oversimplified things and that other circuits must be involved. For example, the loss of the ability to read and write, without a hearing or speech deficit, often follows a lesion just behind Wernicke's area—a transitional zone between language and visual patterning. Also, Broca's aphasics can often form good sentences when singing, or swear better than they speak. This suggests that presently unknown or partly known brain functions can make an "assist" when Broca's area is damaged; for singing it may be the right cerebral hemisphere, for swearing, the limbic system. And, consistent with all we know about the way the brain works, we can modify the Broca–Wernicke notion of functional centers in favor of a more complex notion of circuits. But none of these refinements make Broca and Wernicke wrong, nor do they undermine the idea that language is somehow hardwired in the brain.

Calvin and Bickerton's new theory in *Lingua ex Machina* brings that possibility to life. Based on years of stimulation experiments during brain surgery and a thorough knowledge of language and the brain, they argued that many of the language circuits also serve other, highly specific purposes. For example, categories and concepts are well-known functions of the temporal lobe and they can be very precisely localized—proper names at the tip of the lobe, color categories, and what are called "tool concepts" farther back. Speaking these names or nouns, though, requires a link to the frontal lobe through the massive arcuate fasciculus, second only to the corpus callosum as a bundle of cables. But the temporal lobe also classifies faces and facial expressions, and recognizing people involves remembering faces and names. The movement-control part of the frontal lobe can also set up an accurate throw, which involves planning and triggering an automatic rapid sequence, just as speaking a phrase does.

These processes of recognizing, naming, linking words together, setting up a simple phrase, and producing it are features of protolanguage, as found in emerging pidgins, early-childhood language, and ape language. They are not unique to language, yet they are not just general properties of an all-purpose brain, either. Adding more complex features of syntax—for example, small function words such as prepositions and auxiliary verbs—may require more specialized language circuitry. As sentences become longer, the setup of the "throw" becomes more complex. My seventh-grade daughter, reluctantly diagramming sentences, reminded me that their structure is hierarchical, or treelike. As our brains set them up, we probably form the subcomponents in parallel, then assemble them in the linear order of speech.

At first these seem to be language-specific functions. But syntax may have drawn on other brain capacities—say, for social reciprocity, or tool making. It is a paradox of the cortex and its evolution that nearly identical components, such as the half-millimeter cortical barrels, have multiplied and then become more specialized. Nor does this general feature contradict the idea of modularity. Language functions had to come from something nonlinguistic, since apes don't naturally speak. But subsequent genetic evolution strongly biased certain brain regions to serve their new, linguistic master.

One major form of language localization is not denied by anyone. In the late twentieth century the notion of two brains became so well known that it entered the popular lexicon and grew to be the stuff of New Age legend. A 1981 Nobel Prize went to Roger Sperry for his work on this problem; Michael Gazzaniga, Kathleen Baynes, and many others have continued the research. The "two brains" are, of course, the cerebral hemispheres, mirror images at first glance, but not in function. Each side of the brain controls the opposite side of the body. In right-handed people the left brain is dominant and it is also where language is localized; musical ability, spatial perception, and some emotional functions, on the other hand, appear to be managed better in the right brain. In left-handed people the situation is less clear: it varies from the reverse of that for right-handers to a mixed or middling state. But even two out of three left-handers have left-hemisphere dominance for language.

Some of the most impressive proofs are anatomical. In normal adults there are asymmetries near Wernicke's area. On the average, the left side is a third larger than the right; one groove in particular is a full centimeter longer—a vast difference in neurological terms. And this difference between the two sides is real—visible to the naked eye and confirmed by detailed microscopic analysis, defining the region according to its distinctive cell structure. What's more, it exists in the newborn, with no experience of language, and even in thirty-one-week-old fetuses. Evidently, it is coded at conception. Subtle experiments show a functional difference even in the youngest infants. Language capacity unfolds within us gradually, something like a slow metamorphosis.

Eric Lenneberg would welcome these discoveries. "Why do children normally begin to speak between their eighteenth and twenty-eighth month?" he asked in the late 1960s. "Surely it is not because all mothers on earth initiate language training at that time." Language *training*, if it exists, is difficult to identify in any culture. Anthropologists Bambi Schieffelin, who studied child language among the Kaluli of New Guinea, and Elinor Ochs, who did similar studies among Samoan children, failed to find adult responses that helpfully rephrase the child's speech or expand it to bring out more meaning. Such recasts and expansions of child speech pepper the parent-child conversations of the American middle class but are uncommon elsewhere. Still, with infants, mothers everywhere do speak "motherese"—psychological jargon for what anthropologists have long studied as baby talk. Mothers speak simply with exaggerated tonal variation. They repeat near-words, shaping them toward meaning, and they respond to a baby's pointing hand—often accompanied by a high-tone "Hnn?"—by naming something. If this is training, then it certainly exists for infants and young toddlers, in the San of Botswana as in American families. But it is no coincidence that mothers (and other people) act this way toward babies. Rather, it is the universal and natural responsiveness of those tiny linguophiles to certain patterns that *makes* adults speak motherese.

A more useful model of emerging infant language than the impossible dream of mother-determined development is that of John Locke (a modern psychologist, not the enlightenment philosopher) in *The Child's Path to Spoken Language*. Locke begins with the child's innate capacity to be social, and traces in detail how the earliest interactions and relationships might help form the basis for language learning. This model faces some of the same difficulties as other experience-biased theories, but it has the virtue of giving emotion and attachment a role in language—a great improvement over the dry, cognitive approach of the psycholinguists. As we will see, the role of emotion in language is experiencing a revival.

But first, consider the cross-cultural evidence on the resilience of child language. Among the !Kung San of Botswana, among native peasants of Guatemala, among the Luo of Kenya, and in Samoa, Korea, Japan, Russia, Israel, and virtually every Western European country, we now have proof of the rapid onset of true linguistic capacity during the ages specified by Lenneberg. For many languages, the evidence is good enough to show that not only overall timing but even the sequence of language-function acquisition is similar in widely disparate languages. Dan Slobin, who studied the acquisition of Russian, and Elizabeth Bates, who studied the acquisition of Italian, both concluded that some important aspects of children's language capacity are universal and must be genetically encoded in the growing brain. Although they see fewer universals than Chomsky does, both accept a major role for innate mental abilities.

More impressive, language does not even have to use the vocal-auditory channel to reveal such universals. Deaf children babble vocally in the second half of the first year of life, without ever having heard a sound, but these efforts lead nowhere and fade away. Yet they are also babbling with their hands, and just as the babbling of hearing infants later segues into words, hand babbling segues into signs. Later, acquiring American Sign Language, or ASL, deaf children show the classic rapid increase in phrase and sentence length during the sensitive second and third years. Laura Petitto, Amy Lederberg, and others showed that deaf children, although delayed, use the same strategies in acquiring sign language as hearing children do in the vocal channel. The brain's deep structure for language will out in the hands as well as on the lips. Lenneberg had an inkling of this from informal observations of deaf children. He knew that normal hearing children with deaf parents show little difference from those with normal parents, except that they quickly learn to use "deafisms" with their parents and normal speech with others. He studied Down's syndrome children and showed that language develops, albeit more slowly, in much the same way as in normal children, and that an IQ of 50 at age twelve or of 30 at age twenty does not rule out grammatical mastery of English. He also reviewed a tragic condition called nanocephalic dwarfism and found that these individuals, grossly subnormal in body weight, brain weight, brain-to-body-weight ratio, and IQ, usually attain verbal skills comparable to those of a normal five-year-old.

In the three decades since Lenneberg's untimely death, research on children with atypical brains has proven beyond doubt that some of the brain's circuitry is dedicated to language acquisition. In William's syndrome, for instance, there is a split between mostly normal language and social interaction and global retardation in other abilities. Refinements in the study of William's syndrome children, along with the usual controversy, have left this basic generalization intact. Critics challenge the idea that language has its own special circuitry in the growing brain by rightly pointing to hemispherectomy. If an infant has her left cerebral hemisphere removed, she will develop circuitry in her remaining right hemisphere that will handle language functions. While this tells us that there is great redundancy in the two halves of the brain—in some ways, in early life, they *are* mirror images—it does not mean that any old neural tissue could create the basis of language. It means that complementary right-sided circuitry can sometimes take over in the event of early, severe, left-hemisphere damage.

For a behavioral pattern to unfold in the face of such obstacles it had to be very deeply engrained in human biology. Language is as much a species-specific human behavior as is bipedal walking, and not even gross distortions of input routinely derail its development. Further, as Lenneberg knew, language cannot be explained by general properties of the brain, any more than erect posture is due to the overall size or hardness of the skeleton. Language capacity is specific—a dedicated set of circuits, a module. Lenneberg sought the characteristics of brain maturation that might explain language emergence—what psycholinguists were then calling the language acquisition device, or LAD. He didn't find it, but he did make some inroads and many more have been made since.

In the 1990s important theoretical work challenged but did not really change these conclusions. In books such as *Beyond Modularity* and *Rethinking Innate-*

ness, innovative psychologists—Annette Karmiloff-Smith, Jeffrey Elman, Elizabeth Bates, Kim Plunkett, and others—used a new approach to artificial intelligence, called connectionism, to help explain language acquisition. Old artificial intelligence (AI) programmed computers to solve problems that brains can solve—an algebra problem or a chess move, for example. But they did not build in any learning capacity, nor did they even try to make the computer program's algorithm parallel the brain's. Merely getting a computer to solve a puzzle the brain can solve would, they believed, tell you something about how the brain goes about solving that puzzle.

Connectionist models, called networks, are quite different. They are ordered webs of (somewhat) neuronlike elements that change the strength of their connections with experience. It turns out that they are capable of learning, including some language learning, and this has led some to conclude, quite implausibly, that no innate abilities are necessary. The *Rethinking Innateness* group is more moderate. They accept the idea that the brain has mildly specific preparation for language learning but they give learning a far larger role than genetic preparedness. Even these more reasonable thinkers, however, are misled by the AI error: *Because a certain machine can do something that the brain can, the brain must do it more or less the same way.* This error was more obvious with old AI theories than it is with connectionist programs because old AI was not even trying to simulate neural function, merely to solve problems brains can solve.

But the fact that connectionism tries to simulate neurons just makes the AI error more seductive. These software elements are not *really* like neurons, nor have connectionist circuits learned anything remotely resembling language. But even if they did, no amount of elegant machine learning can prove that the brain, that grotesque pastiche slapped together by evolution from old spare parts, learns language in any way similar to that of the machine networks. The central questions of language development must be answered by what the brain actually does, not by what machines can be made to do.

Nor even, for that matter, by what *brains* can do. The idea that if a brain *can* do something, then that must be how it *ordinarily does* it might be called Skinner's error. The great learning theorist B. F. Skinner, squaring off with Chomsky in the 1950s, argued that, because reinforcement produces verbal learning in the laboratory, inborn elements must play no role in the child's acquisition of language. But the syllogism was flawed and remains so. For example, Jenny Saffran showed in 1998 that eight-month-old babies can remember something resembling a word from a two-minute stream of meaningless syllables, based on the frequency of occurrence of the string that becomes the "word." This is interesting, but it does not in the least prove that babies naturally learn that way. In fact, as we will see, it is easily shown that they don't. On the contrary, in the chaotic environments where they grow they rely much more on innate preparedness than they ever need to in controlled lab settings.

Still, there is plenty of learning even amid the chaos. Cross-cultural research shows that children are bathed in an environment that is inherently and pervasively cultural. Such observations, the grist of anthropology's mill for generations, seem new to some psychologists, who attribute the idea to a Soviet predecessor, Lev Vygotsky. Swiss psychologist Jean Piaget had claimed that mental development and interacting with objects guided the growth of the child's thinking about the world. Vygotsky, influenced by Marxist-Leninist philosophy, made a more radical claim: that all thinking derives ultimately from relationships. In this view, the child's mind is inexorably embedded in—and molded by—society and culture, the link between mind and world being none other than speech.

One leading psychologist who began to think this way in the 1970s was Jerome Bruner. In Oxford, England, he studied interactions between mothers and their infants or toddlers prior to language learning. Certain preverbal interaction patterns, such as jointly pointing at or playing with an object, often preceded the first words. These seemed to provide "scaffolding" to assist the child in developing language. As the stone carver uses a physical scaffold to create a monumental sculpture, so the toddler's mind climbs the social scaffold to ever more impressive cultural heights.

There is some evidence to support this idea, but almost all of it has been gathered in middle-class, college-educated, English-speaking families. Such families, anthropologists know, hold more conversations with children—provide far more scaffolding—than most other cultures do. Don't the rest of the world's children learn to talk? The cultural narrowness of these studies makes their conclusions dubious. Second, most studies merely correlate language improvement with scaffolding, without proving a causal relationship. The parent could be responding to a verbally precocious child, or the verbal skill of both could be the result of social status or shared genes. Also, these interpretations lack any reference to the actual facts of brain development—even though what they are often supposedly disputing is how much of language learning can be explained by brain changes. Finally, just as psychologists have not faced up to the many questions about other cultures, they have given little attention to the classic cases of deviant language learning, such as occurs in blind children, deaf children, or hearing children of deaf parents.

A few studies support a role for interactional learning, but most are more consistent with the maturational theory. It was found, for example, that in middle-class, college-educated families, a number of young children's errors are followed by adult corrections, a part of what is called "negative evidence" that the child may use to learn. Such corrections often stimulate "recasts"—attempts by the children to get it right. However, numerous studies have shown that children have insufficient negative evidence and respond little to it. And of course, such studies almost never control for genes, despite evidence for genetic influence in many aspects of language development.

For example, in early childhood identical twins are more similar than nonidentical twins in their mispronunciation errors, vocabulary comprehension, and speech and language disorders. One adoption study showed that one-year-olds' verbal performance can be predicted better from the general intelligence of their biological parents than by the adoptive home environment. Ironically, this study discovered two features of maternal behavior in the adoptive home—vocal responding tied to the infant's sounds and imitation of the infant—that *did* significantly predict infant language. As often happens, the rare study that *controlled* for genes produced more convincing evidence for an environmental mechanism than the countless studies that do not use this control method. And increasingly, research on language deficits supports a role for genes. Although the genetic effects are not as precise as they were thought to be—they don't, for example, uniquely knock out plurals or any other feature of grammar—the familial patterns in specific language impairment are classically Mendelian, and increasingly shown to have clear brain correlates.

As with the learning-lab experiments of the 1950s, studies in the 1980s and 1990s did prove the power of some form of scaffolding in experimental conditions. These go beyond the concept that the child always learns by simply hearing adult speech. So now we know not only that middle-class American environments include such scaffolding but also that in certain experimental conditions these tactics enhance language learning. How *necessary* they are—the real question posed by Chomsky, Lenneberg, and Pinker—remains a completely different question, and one the environmental determinists seem determined to ignore.

Kathy Hirsh-Pasek and Roberta Golinkoff, in their book on the origins of grammar, attempt to split the difference, decrying the "hyperbolic dichotomies" and extreme statements on both sides. Theirs is a good synthesis of internal and external factors, but in fact the hyperbole is all on *one* side. Environmental determinists reject completely any role for biological preparedness and insist that language is only software programmed into a general-purpose computer, the brain. All the other side has tried to do has been to establish *some* role for biological preparedness in the form of *partly* dedicated neural circuitry for language. That is all we mean by "instinct"—evolutionary adaptations guiding the child through clouds of quite chaotic cultural input. Ours is a relatively weak claim, itself born in compromise, replete with concessions to culture from the outset. Yet even this limited role for biology is rejected out of hand by the cultural determinists.

This problem is in part one of ethnocentrism. One obvious solution is to go outside the Western middle class, but few psychologists have done this. When they do, they cast doubt on the role of specific cultural scaffolding. Roger Bakeman and Lauren Adamson, analyzing !Kung infant data, found that infants there have very little joint or shared attention to objects, common in the American middle class. We've seen that anthropologists like Schieffelin and Ochs did not find adult corrections and expansions of child speech in New Guinea or Samoa. As Ochs has said, "Expanding children's utterances, using leading questions, announcing activities/events for a child, and using a simplified lexicon and grammar to do so are cross-culturally variable." These and other tactics of social scaffolding are considered essential by many psychologists, yet !Kung, Kaluli, and Samoan children do without them. Not only do they acquire their respective languages but oratory is admired and is a route to success in all three cultures. And of course, children's speech shows strong cross-cultural consistencies despite differ-ences in input.

A fascinating cross-cultural case is the creation of creoles, languages that arise as blends between two others. This often happens in the culture clash of colo-nization, when a subject people must speak another language, usually a language of oppression. The first idiom to be formed is not the creole itself, however, but a pidgin, a substandard version of the language of subjugation with elements of one or more native languages. Derek Bickerton, the great modern student of this process, has noted that pidgins commonly arise in slavery or indentured servitude. When people are transported thousands of miles from home and thrown together in a polyglot world, they must somehow communicate with their masters and one another. To avoid the Tower of Babel effect, in which multiple languages make work and life impossible, they create a code from the linguistic material at hand. These first passes at a common tongue are crude and disorderly, meeting some of the semantic criteria for language but none of the syntactic ones. Of course, they are merely second "languages" or idioms used pragmatically by people whose first languages meet all known criteria.

Now here is the wondrous thing: when a new generation is born and grows up listening to the crude pidgin pastiche, they transform it as they acquire it. It becomes a true language, but one that resembles the language of children in many particulars. Amazingly, creoles—the second-generation blends—share those same resemblances regardless of where in the world they occur or what languages they blend. The Hawaiian creole, created from Japanese, English, and Hawaiian, converges with the creole of Mauritius, which draws mainly on French. Children throughout the world have used their inborn powers to fashion convergent creoles from disparate linguistic parts. This would seem to be the final proof of what Lila Gleitman and Elissa Newport have called "the invention of language by chil-dren." But there is more.

Special cases like deaf children or hearing children with deaf parents also need more study, but what we know hardly suggests that scaffolding is crucial. Signing parents of deaf children do exaggerate the size of their signs when addressing young children—manual motherese—and may make signs in contact with the child's body. But does such a pattern show that sign language depends on teaching? Those who study deaf children's communication consider this very unlikely. In fact, what deaf babies do in sign, with very different input, is amaz-ingly similar to what hearing babies do in sound, complete with insistence on their error-producing rules, in the face of negative evidence that they studiously ignore.

And what if there is no signed input at all? Susan Goldin-Meadow and her colleagues at the University of Chicago have for a quarter century studied chil-dren in an unfortunate but common situation: they are deaf, but their parents are not, and the parents do not sign. Many such children—"islands in the main stream"—are raised to speak orally and to lip-read, unsatisfactory but typical solu-tions. Some children, however, are initially exposed neither to sign language *nor* to systematic oral teaching. For them language is neither heard nor seen. Yet, amazingly, they invent a language of their own, "home sign," quite distinct from standard ASL. Years of research on this homegrown language show that these children structure their invented signs in ways that resemble true natural languages, whether signed or spoken.

The structure is evident both at the level of early one-word signs and at the level of phrases and sentences. "For example," Goldin-Meadow writes, "one child pointed at a tower, produced the HIT sign (fist swatting in air) and then the FALL sign (flat palm flops over in air) to comment on the fact that he had hit the tower and that the tower had fallen." These child-invented signs are conventionalized just as words are, and appear as two-word phrases at the same age at which hearing children, exposed to the full power of adult speech, achieve the same milestone. That children so deprived do not achieve the full subtlety of a natural language is hardly surprising; but what they do achieve without any relevant input is truly remarkable. Most remarkable, perhaps, is the evidence that a form of home sign has transformed itself into a true signed language, arising in much the way creoles emerge from pidgins.

This linguistic alchemy occurred in Nicaragua, when the Sandinista regime established the first schools for the deaf. Children completely isolated in a hearing world came together with others like them for the first time in their lives. Like the polyglot oppressed under slavery, they formed a crude code by which they could understand one another. Thus, they avoided the Tower of Babel that their varied home-sign systems could have produced. These children, mostly ten and older when they met, could not get past this pidginlike stage of signed communication. But as younger children joined them, some of the youngest began to transform their established code into a true language, one obeying the rules of creolization already seen by Bickerton and others in many parts of the world. Thus, a linguistic code with few rules at all became in one generation a true language with a childlike but valid syntactic order.

It appears that some aspects of language development are strongly guided by internal factors. What are these factors? Overall brain maturation is astonishingly rapid during the first two years of life, after which it slows markedly. This pattern of brain growth is one of the most distinctive features of our species. Apes and monkeys have a brief period of rapid brain growth followed, almost suddenly, by a long phase of slow growth. But there is this crucial difference: even in apes, the time of sudden slowing roughly coincides with birth. In humans, it is postponed to at least a year after birth, and this change in maturational plan may explain a great deal about our species—including, perhaps, our capacity for language acquisition. But overall increase in brain size, however distinctive and however rapid, cannot go very far toward explaining the sequence of changes we observe universally in human infants. For this we must turn elsewhere—to the differential growth of specific brain systems.

In a part of the frontal lobe—roughly, Broca's speech area—there are massive increases in the density of the web of fiber connections during the first two years, including the period from fifteen to twenty-four months. Although new neurons are generated throughout life and may even function, they represent only a very small fraction of the brain's neurons and probably are fewer than those lost through neuron death. Still, there is ample maturation in other ways. New dendrites, the listening branches of neurons, form new synaptic connections with the axons of other cells. Synapses broaden, and nerve cells grow larger. Most interesting—because its purpose is puzzling—is the population explosion of *nonneural* cells in the brain. This makes up most of the doubling of brain weight in the first year and has the paradoxical effect of thinning out the density of nerve cells in the cortex—a well-known fact with unknown meaning.

Part of the nonneural expansion lies in the growth of myelin—a sheath of fatty white matter surrounding many nerve fibers. Nerves can function without myelin—many normally do—and even among sheathed nerves, function begins *before* myelination. Still, myelin alters function dramatically: it increases conduction speed, raises the maximum possible rate of firing, lets the neuron work longer without fatigue, and protects it from irrelevant stimuli. For these reasons, neurologists have long been interested in the possibility that myelination sequences in the brain would help to explain the growth of function.

In the early twentieth century neuroanatomists described the pacing and sequence of myelination in the cortex. Primary sensory areas—such as for vision, touch, and hearing—and primary motor areas myelinate first, followed by nearby association areas, such as Broca's and Wernicke's. Basic descriptions, which were purely anatomical and did not concern themselves with theories about language, suggest that Wernicke's area develops somewhat earlier than Broca's. This is consistent with the fact that comprehension precedes fluency. It also helps to explain why a one-and-a-half-year-old seems a bit like a Broca's aphasic: understanding a good deal, but talking only in halting telegraphic utterances. (Even the ability to sing seems as true of the toddler as of the Broca's aphasic, although swearing, at this age, is not in the picture.)

Neurologist André Roch-Lecours, drawing on work done with Paul Yakovlev, found clues to the anatomical language acquisition device. He confirmed earlier observations on the growth of myelin in the cortex but went beyond them to trace the myelination of lower parts of the brain. Subcortical pathways for both vision and hearing are myelinated shortly after birth, consistent with the fact that in simple aspects of vision and hearing, six-month-olds resemble adults. But the two senses differ drastically in the myelination of their higher pathways—those projecting up to the cortex. In the visual system, myelination was fast and early, consistent with the mature vision of the infant; but in hearing, the complete myelination of the cable to the cortex was spread over several years, consistent with the slow growth of language comprehension.

Couldn't these myelination sequences result from, rather than cause, the child's changing behavior? No. Myelination is somewhat shaped by experience, but this is minor compared with the 80 to 90 percent of the process that goes on regardless of input, under genetic control. Many other aspects of brain development in infancy and childhood are also heavily influenced by genes. Arnold Scheibel identified left-right differences in the shape of neurons in the language-related cortex, and no one has yet shown that such distinctive cell shapes could arise from experience. Nor could experience alone produce the distinctive features of the brains of some dyslexic individuals, in whom Albert Galaburda has found defects in the ordering of cortical cells between the visual and language areas. Some of these transformations may be due to faulty neuronal migration coded, in an as yet unknown manner, by the genes. Ultimately, work like Scheibel's and Galaburda's may help to account for universals of normal infant and child language.

But this is an uphill battle against firm ideological commitments. Some still insist, against growing evidence, that language is entirely cultural. One tactic used by critics is to announce the demise of the biological approach every time a neural or genetic aspect of language turns out to be more complicated than originally thought. As we have seen, this happened with Broca's and Wernicke's aphasics, with William's syndrome, and with Specific Language Impairment. *Aha*, they say. *This behavior is not linked to the brain or the gene in the simple way you thought it was. Therefore, biology is not a major force.* This is disingenuous. They are attacking a straw man, and their syllogism fails. Scientists taking biological approaches are exploring an ignored frontier and over the years have found important clues as to how genes code language in the brain. True, they have sometimes been overconfident, announcing with fanfare a specific mechanism that later turned out to be more complex.

But as of the year 2001, lesions in Broca's area still cause language deficits that are distinct from those caused by lesions in Wernicke's area. Children with William's syndrome still have certain capacities, including some aspects of language, that are relatively spared against the background of pervasive mental deficits. And Specific Language Impairment still has a Mendelian pattern of inheritance in which a gene causes a problem in neural processing that has among its consequences certain errors in grammar. Did critics think that the biological theorists expected to find a neuron for nouns, or a gene for plural-forming S's? On the contrary, these discoveries were always seen as clues to a complex and indirect mapping of language on brain functions, and of brain functions on genes. New research that reveals complexities in the mapping just leads to a more plausible account of biological influence. And of course, however weak or strong that influence is, it has evolved.

For the biologist, the phrase "origin of language" has a very different meaning than it does for the child psychologist, and in this evolutionary sense the question is even more speculative. It is said that in the 1860s the Linguistic Society of Paris established a standing rule barring papers on the evolutionary origins of language. Such a moratorium may not be a bad idea. One influential approach combined anatomical evidence with computer simulations to try to show that Neanderthals lacked language because they could not make certain vowel sounds—"ay," "ee," and "oo." Aside from the fact that the anatomical judgments themselves were subject to great error, there was the clever letter to a scientific journal criticizing the hypothesis. It read in part: "Et seems emprebeble thet ther speech was enedeqwete bekes ef the lek ef the three vewels. . . . The kemplexete ef speech depends en the kensenents, net en the vewels." That letter laid to rest for some of us the notion that pharynx shape was the Darwinian key to language. At the same time, research on casts of the Neanderthal brain by Marjorie LeMay showed that these beefy, thick-boned hominids had some of the same specializations we have in brain lateralization. Language is not in the throat but in the brain.

Still, the Linguistic Society ban has been challenged many times recently, with varying degrees of success. Some theories propose slow and gradual evolution, some stagewise cognitive change, and some a sudden "catastrophic leap" from protolanguage to language. Some are anatomically informed, others are based on analogies with child development, and still others on processes such as the emergence of creoles. Some suggest that language emerged 2,000,000 years ago with the australopithecines, others claim that it only appeared with the last Paleolithic migration out of Africa some 50,000 years ago. The fact that scientists can have such strongly divergent views, based on completely different methods of inference, in itself shows that the data do not offer much guidance.

Direct research on ancestral brains is impossible, since brains do not fossilize. But endocranial casts—either natural or laboratory-made impressions of the inside of the skull—reflect to some extent the impression made by the brain during life. But in human ancestors, the skull is flexible enough during growth that the brain unfortunately does not make a very strong impression on it. Ralph Holloway studied the endocranial casts of australopithecines and found some evidence for lateralization in language regions at this very primitive stage. But the meaning of this is uncertain, especially since left-right asymmetry in language-related regions has now been found in chimpanzees. A group using brain scans made by James Rilling of the Yerkes Primate Research Center found a clear left-right difference in the part of the temporal lobe that in humans would help process language. Another group, led by Patrick Gannon and including Holloway, found the same hemispheric difference using different methods and another sample of chimps. If chimps share this difference with us, it may serve some mental adaptation unrelated to language. Or it could mean that a cognitive ability *tending* toward language was present five or six million years ago.

Oddly, one of the cleverest insights into the language of protohumans came from the spinal cord of the Turkana boy, the strong, tall, sturdy eleven-year-old who changed the world's view of *Homo erectus*. Studying every inch of him, Alan Walker and Richard Leakey's team published what is probably the most complete, and surely the most elegant, account of any protohuman fossil. Ann MacLarnon studied his spinal canal in cross section, and discovered a fascinating fact: in the chest area, his spinal cord was slim compared with ours. Primates in general, like other vertebrates, have bulges in the spinal cord corresponding to the nerves that control the limbs. But our own species has an *added* bulge in the chest region, possibly because humans need more subtle control of breathing in order to speak.

More control means more nerves, hence a thicker cord and canal. The Turkana boy's narrow canal was typical for apes and monkeys, which could mean that he lacked speech. Language, in that case, would have come late in human evolution, arriving with early *Homo sapiens*. This conclusion is at the opposite extreme from the suggestion that the common ancestor of humans and chimps may already have taken a step toward speech.

Timing the origin of language may remain one of the unsolvable questions. But if speculations about when language began are sometimes amusing, those about the original *functions* of language are often more so. It is reasonable to suppose that language in early hominids helped in planning hunts, teaching the young, mutual teaching of adults, and the modulation of emotional arousal. It certainly functions those ways in modern hunter-gatherers. Possibly, some adaptive advantage came through sexual selection acting on male courtship—that the boy who talked the best line got the girl. Or, language capacity may have begun to evolve first in females, who then produced it in males through female choice— you can't explain the relationship if the slope-headed lug doesn't get the point. Hunting, too, may have been a selection pressure for language, but storage and transfer of information about the location of plant foods was likely another. To this day men tend to find their way with maps, women by recognizing local markers— and, of course, by asking directions, which even our *female* ancestors couldn't do until they could talk.

Or the key might lie in the mother-infant relationship. Peter Marler of Rockefeller University made a film of chimpanzee vocalizations in their natural context. There is very little in natural chimp communication that resembles human language; excited hoots and staccato yelps are the main ways the reigning silence is broken. But in Marler's film a mother-infant pair lounging on the forest floor cooing at each other made sounds relaxed enough, soft enough, and continuous enough that if you closed your eyes, you might think it was human speech. Perhaps somewhere in our ancestry just such cooing mother-infant pairs—"talking," in a way, about love—were the crucible of language evolution. Recall the work of John Locke on the role of attachment in language emergence, a viewpoint that, uniquely in this field, recognizes the immense emotional power of words in relationships. If joint attention between mother and infant is indeed important in language acquisition, it is likely to be focused on the love that they share, not on some inanimate object.

Along these lines, Locke has fielded a hypothesis about language evolution based on mother-infant interaction. He suggested that our ancestors' first attempts at speech "were probably much like the infant's early forms because the hominid mother paid particular attention to what phonetic material 'went over' best with the infant in her audience." Finally, focusing on a very different intimate bond, Mária Ujhelyi sees the long calls of apes as prefiguring speech, because they involve strings of calls with meaningful changes of order. The duets of mated gibbon pairs are the most elaborate of these calls, putting individuality and attachment at the heart of communication. So we can be forgiven the temptation to find once again, as with the chimp mother-infant pairs, a possible prelude to speech in a natural language of love.

But gibbon calls are not replete with symbols and we, as Terence Deacon has said, are the symbolic species. Our direct adaptations for speech, like the structure of the throat, vocal cords, mouth, and tongue, are results, not causes, of language evolution. Of the deeper, mental functions, Deacon views symbols as central and unique to us. This differs from the earlier view—presented best by Steven Pinker, in *The Language Instinct*—that the crucial advance was grammatical. The centrality of syntax, also espoused by Chomsky, is quite plausible. But there is little doubt that the increasing need for symbols drove and was enabled by a tremendous expansion and transformation of the frontal lobes. As shown by James Rilling, using magnetic resonance images of monkey, ape, and human brains, the frontal lobe has more convolutions than expected even for a highly evolved higher primate, which means that the lobe expanded preferentially in our own line. In this view, the transformed frontal lobe matters as much as any changes in conventional language centers farther back in the cortex.

But Pinker and Chomsky are not so easily dismissed. For one thing, the human brain is *also* more convoluted in a region that includes Wernicke's area, suggesting that the dual power of meaning and grammar converged in effective synergy during our evolution. Furthermore, the concept of instinct deserves just the sort of revival, within limits, that Pinker tries to give it. Both he and Chomsky have been taken to task for ignoring brain function, for missing the difference between what languages do and what brain circuits do, and for advocating simplistic "language modules" or "grammatical organs" localized in over-reified centers. But they don't do any of these things. They appreciate the distinction between linguistic and neural functions and they know that language features cannot be mapped to the brain in any simple way. They may have overemphasized grammar as the unique human contribution to language, but semantics and syntax are ultimately inseparable. So in thinking about language and the brain, do we choose Calvin and Bickerton's hierarchical set-up of motor patterns as we "throw" sequential phrases? Pinker's switch-setting devices for the specifics of growing grammar, probably in Wernicke's area? Or Deacon's echoes of symbolic meaning around and away from the frontal lobe? The best answer is, *All of the above*.

Which brings us back to the talking apes and the question of how good they are at any of these things. If you'd been following them since the sixties, it would be hard not to be impressed. They have done any number of things that almost any linguist would once have insisted was impossible. Early on, they learned at least 150 different words, perhaps many more, and they use them fairly accurately and effectively. They generalize well and meaningfully among objects not obviously in the same class, and even from objects to pictures. For instance, Washoe, the first chimpanzee to learn American Sign Language, spontaneously used the word *hat* for a nightcap that didn't look anything like the hats she knew. She named it not by appearance but by function. They also clearly show displacement, the ability to discuss things removed in space and time, and they have mastered the elements of languages that have all of Hockett's characteristics except the use of the vocal-auditory channel, which is not a real concern.

They also have the rudiments of grammar. They have produced, spontaneously, combinations of previously learned individual words, and these combinations, often highly inventive, follow simple grammatical rules. They use the visual equivalent of a questioning intonation, represented by a distinct symbol in each of their languages, and they can formulate complex questions. They use language in everyday life to get what they want from the experimenters; this is not surprising, but they also use it in other contexts. For instance, Washoe was seen talking to herself in ASL—or at least practicing signs—on a number of occasions when she couldn't have known anyone was watching. Two chimpanzees, Sherman and Austin, using the visual-symbolic language Yerkish, talked to each other and communicated successfully their wants and needs. And Washoe, fully grown, communicated with her infant in ASL. Such findings raise the possibility that Lucy or even an earlier hominid form—*Kenyanthropus platyops* or perhaps *Australopithecus amanensis*—had already taken a halting step toward language.

Chimps have even been creative. The first time she saw a duck, Washoe invented the name "water bird," and she spontaneously called a Brazil nut "rock berry." Lana, who was learning Yerkish and knew the symbols for apple and the color orange, produced, after a few false starts, the question "Tim give apple which is orange?" to get an orange. And Lucy—not the famous hominid but a chimp by the same name—while learning ASL called radishes by the generic term "food" for three days, after which she spontaneously started calling them "cry hurt food." We may be forgiven, I think, a slight inclination to call these phrases poetry.

It was not surprising, then—although it was a major coup for ape-language studies—that as early as 1973 Roger Brown, a leading student of language development and a skeptic of chimpanzee language capacity, waved the white handkerchief. He conceded that Washoe, the only ape at that time that had learned a natural language, had attained a level comparable to that of a two-year-old child. The question since then has been, *How much farther can this ape talk go?* Even skeptics granted that the last word on the subject could not be spoken by a few pioneering studies, in which apes were at a great disadvantage compared with children. It was even possible that the wrong species was chosen—claims of a much larger vocabulary attained by the gorilla Koko, and of exceptional lateralization in the brains of orangutans, suggested that chimps might not be the best linguists. Bonobos, a separate species once known as pygmy chimpanzees, had not yet been studied.

Yet some concluded that these efforts had gone about as far as they could go. Herbert Terrace, leader of the ape-language group at Columbia University, backed away from strong claims. The group's subject was named Nim Chimpsky in wry homage to Noam Chomsky, for the hope was that he would be as bright a linguistic light in his own species as his namesake is in ours. Nim progressed to a phrase length of two signs, but no further. At age three and a half, when a human child (even a signing one) is racing toward sentences, Nim was still hovering between one- and two-word phrases, where he had stalled for two years. Also, he never produced utterances longer than two words that were grammatically rule-bound. Imitations of trainers' speech remained dominant. He rarely expanded on his caretakers' utterances, which children do frequently, and he never learned turn taking, the basis of conversation. In the end, Terrace expressed deep skepticism of all ape-language claims.

But no one was prepared for the advent of Kanzi, who was Noam Chomsky's counterpart if any ape was. Being a bonobo meant that Kanzi could be completely different from chimpanzees. Sue Savage-Rumbaugh, a scientist with an astounding sensitivity to ape behavior and consciousness, immediately understood that difference. Kanzi was more keyed in to human interaction and speech than any chimp had been and there was an utterly new glow of understanding in his eyes. By the early nineties, superb studies had demonstrated that this was no subjective illusion, nor was Kanzi a fluke—an individual ape genius unrepresentative of his species. Kanzi's Yerkish progressed faster and farther than any chimp's had. Studies with Patricia Greenfield and others showed that not only did he have a large vocabulary, but his utterances were rule-bound; he had about as much grammar as a two-year-old. His sister, Panbanisha, was not far behind him in competence.

One of Savage-Rumbaugh's key insights was that comprehension could be tested at least as easily as production and that such comprehension could be in English, not Yerkish or sign language. She knew that Kanzi understood much of what she said to him, in addition to his ability in the realm of iconography. In a rigorous series of tests, accepted as valid by the most stringent critics of ape-language research, she showed that Kanzi understood English exactly as well as a two-and-a-half-year-old child. If you were going to doubt Kanzi's comprehension, you had to apply the same doubts to the standard capacity of the child between two and three. Few were prepared to doubt that child's ability to understand English. Critics could still point out that the child would soon outstrip the ape, but they could no longer reasonably deny Kanzi his accomplishments—nor Savage-Rumbaugh the resources to keep inviting Kanzi into human culture.

Yet extreme skeptics keep raising the bar. Whenever an ape in the hands of a dedicated experimenter makes an inroad, critics find a way to diminish the accomplishment. "That isn't really what we meant by language," they say in one way or another, conveniently shifting the definition when apes are gaining on us. But if Kanzi doesn't understand, if he doesn't have a linguistic mind, neither does a two-and-a-half-year-old child. Is the child going somewhere that Kanzi will never go? Of course she is. That's evolution for you. It produced a tripling of brain size. No one makes the ludicrous claim that there is no difference, so that is not the point. The point is to find the glimmers of capability that somehow, over millions of years, led to the human mind. This is one of the many realms of science where we must tolerate ambiguity.

Glimmers of ape capability are found in vervet and macaque monkeys and glimmers of human capability are found in apes. Dorothy Cheney and Robert Seyfarth found that snakes and hawks elicit different, meaningful calls that send other vervets scrambling up or down trees to get away. Harold and Sarah Gouzoules found that rhesus monkeys modify their anger and distress calls to let others know whether they are under threat from kin or nonkin, dominant or less dominant opponents. Apes have cognitive abilities that enable them to learn important elements of language under favorable conditions. These abilities probably have some natural function, but apes did not invent language and they do not use it in the wild. They do not do what every human child does in the first few years of life, driven by brain processes so ineluctable that they virtually invent language even when they are only lightly exposed to it.

Still, there is a component in human language that has not yet been addressed in ape-language studies: the central role of emotion in thought and of face-to-face relationships in human mental life. This is not to resurrect the old error of deriving language from grunts and squeals, which have different circuits in the brain—although the meaningful calls of vervet and rhesus monkeys may hold some neurological surprises. But embedded in our brains, too, is circuitry for nonverbal communication and the emotions underlying it. On average, nonhuman primate species have some fifteen to forty different sounds or calls. But we have grunts and calls of our own, innate vocalizations such as laughing, crying, screaming with fright, crying out in pain, groaning, and sighing. Some theorists find this a dwarfish repertoire; we (they say) do it with words, so the calls wither away, just like our pesky appendixes or tailbones.

But what about cooing, grunting, whimpering, growling, purring, mewling, hissing, keening, coughing, yawning? What about the sounds transliterated in comic books as "phew," "oof," and "ugh," or such phrases as "a torrent of wordless fury," "she caught her breath," "the crowd roared approval," or "his cry at the peak of pleasure"? These may be bad prose, but they help to convey feelings from a deeper, less rational, more human and humane level, clumsily transferred to *less* informative words. Recall that some kinds of aphasics can curse, sing, or describe emotionally charged events much better than they can otherwise speak. Freud was impressed with these facts by 1890, and he invented a therapy that, like all the later therapies it spawned, was based on the power of words to transform feelings by expressing them. Yet we still attempt to model language in an abstract, computational, emotionless frame.

We can do better now. The centrality of symbols in human life lends itself to a natural synthesis of Calvin, Bickerton, Pinker, and Deacon with Damasio—ultimately a synthesis of language and feeling in the depths of the great frontal lobe. It would be a fool's errand to try to understand language and especially to give a central role to symbolic meaning, except in the light of emotion, because the *meaning* of meaning is inherently emotional. A symbol does not merely ramify in a dozen directions through the mind and brain, setting off echoes of verbal and visual imagery; it also makes you shiver or swoon or palpitate or feel a thrill at the back of your neck. A significant symbol stirs the emotions. The frontal lobe, so expanded in human evolution, is largely responsible for this symbolic capacity. It must also hold some of the specialized circuits that acquire and generate grammar—a whole new conveyer of meaning beyond single words. And it is the main organ of reason, which we now know depends upon emotion. So the distinctively human mind is inconceivable without feelings, and these in turn are transformed by talk.

Of course, apes in the wilds of Africa get along well without language, while we in the wilds of New York and Paris, Rio and Nairobi, clearly need it. Yet theories as to why are generally uninspiring. And of all the off-base conventional wisdom about language, the most useless, the most stubborn, and the most misleading is the conviction that it makes us morally superior. This is not an idle notion but is based on a sort of logic, which goes something like this: language enables us to keep our emotions, our baser instincts, in check. We have animal motives, of course, but we think about them, are conscious of them—a consciousness carried in language. Language thus enables us to master, divert, and resist them, and that is what we do every day. False, totally false.

I don't know what, if anything, dampens our enthusiasm for the pursuit of motives arising from baser instincts. Perhaps it is nobler instincts, perhaps learning, perhaps the desire to gain the regard of others—but it isn't language. As Steven Pinker straightforwardly said, language is one of the things we use to "acquire, share, and apply knowledge of how the world works, to outsmart plants, animals, and each other." Deception may even be one of its major functions— masking base motives or making them look pure. This may be one of the major functions of animal communication in general, carried in us to its most exotic realization, a signal system so complex and flexible that it raises to utter perfection the self-serving enterprise of lying.

This is the vision of an ape-turned-human in Franz Kafka's fable "Report to an Academy." Having recounted his forcible training away from apehood and showing his utter contempt for the moral aspect of the change, he says, "I repeat: there was no attraction for me in imitating human beings; I imitated them because I needed a way out, and for no other reason." He means this literally—a way out of his cage. This particular declaration comes just after his description of how he learned his first word—*Hallo!*—in a drunken stupor, while learning to drink schnapps. And this is the point of the epigraph from Didion: language is merely our way of advancing the motives we share with every other animal.

As behavioral biologists, we have to agree. What we do with language is very much like what other creatures do without it; we do it more intricately, more gradually perhaps, certainly more beautifully and on a grander scale. But we do it all the same and nevertheless. If language doesn't give us the mastery of emotion, then what does it give us?

Strether sat there and, though hungry, felt at peace; the confidence that had so gathered for him deepened with the lap of the water, the ripple of the

surface, the rustle of the reeds on the opposite bank, the faint diffused cool-
ness and the slight rock of a couple of small boats attached to a rough landing place hard-by. The valley on the further side was all copper-green level
and glazed pearly sky, a sky hatched across with screens of trimmed trees,
which looked flat, like espaliers; and though the rest of the village straggled
away in the near quarter the view had an emptiness that made one of the
boats suggestive.

In this, the most important, or at least the most pregnant, passage in Henry
James's novel *The Ambassadors*, we glimpse an unusually conscious creature,
slightly hungry, resting, at home in the surround, and about to discover an illicit
sexual bond between two other creatures he cares about, a bond that has so far
been concealed from him. His situation, and his impending shock, would be
familiar to any number of nonhuman animals.

But the words they are cast in are something else again. If words elevate us
above those other creatures, it is not for moral, but for aesthetic reasons. The
almost musical syllables, the echoes of metaphor and meaning, the painterly
phrases, the homage paid to a great tradition of language all combine to produce
something unique in the animal world; unique not because of its moral superiority but because of its beauty. And while there is nothing morally superior about
beauty, it is immensely valuable.

Dorothy Hammond, a professor of anthropology at Brooklyn College, used to
wave away the question of the origin of language by asking, "What would we talk
about, sitting around the fire at night, if we didn't have language?" It was meant as
at least a half joke, but we could take it more than half seriously. Sitting around
the fire at night among !Kung San hunter-gatherers, one sometimes hears passages as beautiful in their way as the one quoted above. The music is different, the
echoes of meaning are different, the visual surface is not the same, and the tradition is a world of its own. But the aesthetic achievement is similar. What would
we talk about, sitting around the fire at night, if we didn't have language? It is the
sheer useless joy of it, the entertainment, the touching with sound, the suspense,
the sadness, the humor, the elegance, the grandeur.

Of course, language has functions, too, or else it wouldn't be here; but
these may be the least interesting things about it. Many are achieved by other
creatures in other ways. Social cohesion is a function, but monkeys and apes
achieve that through mutual grooming, a behavior with many functional parallels to talking. Gossip itself has a bad name and it certainly does harm, especially in its mass media versions, but gossip compels us because it contains
information that, in our environment of evolutionary adaptedness, we needed
in order to survive and reproduce. Communication of emotion and information
critical to survival are also carried in speech, but other creatures accomplish
this in many other ways, from the dance of the bees to the song of the lark to the
wail of the howler monkey. Traditions are borne in language, but other creatures carry them without it.

No doubt, we do more. All of these functions are greatly expanded by human speech. Much greater amounts of information can be stored and transferred, concerning everything from the location of food sources to the behavior of predators to the movements of migratory game. Not only tales but also stores of knowledge are exchanged around the fire among the San; and even the dramatizations carry knowledge critical to survival. We will see in chapter 10 how a group of San men and women, wary of nearby lions, kept one another awake and stoking the fire by exchanging lion lore and natural history. A difficult way of life would be impossible without such knowledge. As for cultural traditions, those of other animals are so rudimentary compared with human culture that the difference seems categorical. This difference lies mainly in our capacity for symbols, both verbal and nonverbal, and to the access symbols give us to information and emotion alike. It is this information transfer, not only horizontally but vertically in time, that made possible the uniquely wide range of ecological adaptations we see even in hunting and gathering communities.

But to list such functions and stop there would be like talking about the functions of sex without mentioning pleasure. We would know all about why sex evolved but not why people do it, especially not why we go so much out of our way for it. As surely as we are naturally endowed with the capacity to feel pleasure in sex, so we are naturally endowed with a capacity to feel it in language, not only in the sounds but in the echoes, at level upon level of meaning, imagination, suggestion, challenge, appeal, innuendo. A sentence creates in the mind of the speaker as well as the hearer not just a picture but a realm of intricate mental events encompassing all five senses. Say what you will about nonhuman creatures—their admirable capacities, their beauty, their complex behavior, their consciousness—there is not a thing like it in the whole of the animal world.

Adrienne Rich asks in a love poem, "What kind of beast would turn its life into words?" We are that beast, and the words Rich has turned her life into include perhaps the most beautiful romantic and erotic verses written in English in the late twentieth century. That they are lesbian love poems does not prove the power of language over sex; nonhuman species also have homoerotic relationships. But it does show the power of language to communicate emotion, to transcend human differences and make the speaker and the hearer share the experience of wonder.

In an engaging study of the conversations of three-year-olds, in the context of a longer chat between a boy and a girl, this occured: "Hello, Mr. Dinosaur." "Hello, Mr. Skeleton." These three-word sentences probably contain enough complexity, enough levels of meaning, and enough imagination to ensure that comparable things will never be said by even the most brilliant talking ape. But even simpler utterances may be distinctively human. Consider a ten-month-old pointing at a butterfly: "Dat!" she says, emphatically. She has gotten into the habit, lately, of pointing at things in just that way, saying, "Dat, dat!" either questioningly or indicatively or emphatically or thoughtfully. She sometimes gives the impression that she is trying to drag a name out of an adult she is riding on and sometimes, when it is offered, she is satisfied. At other times, when she points and speaks, she seems to be announcing the existence of a relationship between a piece of the great world and her own presumptive mind, a relationship that has evidently surprised her.

This utterance, produced by a creature then the clear mental inferior of the various talking apes, was already in some respects distinctively human. Though very common in human infants, such behavior is rarely observed in apes. But beyond that, of course, we know what is going to happen next, and that makes it seem different. We are after all looking at a creature who is, in effect, going to suck a language out of the air during the next two or three years. The child has that potential, and much more; she might even eventually make paragraphs as beautiful as those of Henry James. Yet the distinction might go deeper still. I suspect that we are seeing the most rudimentary form of the key to being human: a sort of wonderment at the spectacle of the world, and its apprehensibility by the mind; a focusing, for the sheer purpose of elevation; an intelligent waking dream. In that capacity, perhaps, we find our greatest distinction, and it may be our salvation.

PART TWO

Of Human Frailty

We live in an old chaos of the sun,
Or old dependency of day and night,
Or island solitude, unsponsored, free,
Of that wide water, inescapable.
Deer walk upon our mountains, and the quail
Whistle about us their spontaneous cries;
Sweet berries ripen in the wilderness;
And, in the isolation of the sky,
At evening, casual flocks of pigeons make
Ambiguous undulations as they sink,
Downward to darkness, on extended wings.

—**WALLACE STEVENS**, "Sunday Morning"

CHAPTER 9

Rage

Many of our intellectuals rush to quell our fears by telling us that theoretically none of this has to happen, that violence is not part of human nature, that it occurs only because of evil intentions and circumstances that we can eradicate. They are the Christian Scientists of sociology; and they have not as yet solved the paradox: if we are not by nature violent creatures, why do we seem inevitably to create situations that lead to violence?

— LIONEL TIGER AND ROBIN FOX, *The Imperial Animal*

In the mid-1990s, popular attention was riveted by the criminal and civil trials of a famous athlete and culture hero accused of brutally murdering his ex-wife and her male friend. The criminal jury found him not guilty, citing reasonable doubt, but the civil jury found the weight of evidence against him. Still, his punishment seemed light to many and there was a widespread sense that justice had not been done. In any case, there was ample time for the world to think about the causes of such brutal violence. What underlies crimes like this, or for that matter the corresponding sense of justice? Consider two cases, from widely differing cultures, in which justice was convergent and swift.

One July evening in White Plains, New York, Richard James Herrin, a twenty-three-year-old Yale senior, went to the bedroom of Bonnie Jean Garland, a classmate and sometime girlfriend, and, with a claw hammer, bludgeoned her to death in her sleep. He fled, driving upstate to Coxsackie, where he surrendered himself to a priest and confessed his crime. He told the arresting officer that he had planned to kill the young woman and then commit suicide. The precipitating cause was romantic rejection; Garland had broken up with him.

At the trial Herrin pleaded not guilty by reason of temporary mental defect or disease. Evidence concerning his relative poverty as a child was introduced by his mother as a mitigating factor. Herrin testified that he did not know why he committed the slaying. Two psychiatrists testified that he had been psychotic at the time; two others testified that he had not. It became evident during the trial that Herrin was a well-brought-up, well-behaved, even religious young man who had never done anything to suggest in the slightest way that he was capable of homicide. There was support for him in the Yale community, but on the third day of deliberation the jury convicted him of manslaughter and he was sentenced to the maximum term of eight and a third to twenty-five years.

The jurors had been nearly deadlocked—three of the twelve had been holding out for murder two. They asked to go to mass on Sunday in the hope of "some divine inspiration," but they reached their verdict without that guidance. They took into account Herrin's premeditation but also the depth of his and Garland's love for each other, seen in their letters, and the emotional disturbance affecting Herrin during his crime. His two-year intimacy with Garland was to have ended in marriage, but she decided that she wanted to date other men and broaden her social life, plunging him into despair and the belief that he "could not live without her." Delivering sentence, the judge said, "Even under the stress of extreme emotional disturbance, the act of killing another is inexcusable."

On a fall day just over a year later, while Richard James Herrin awaited trial, Wang Yungtai—a twenty-four-year-old warehouse worker at the Materials Recuperation Company in Beijing—sought out Hu Huichin, a fellow worker he had wanted as his girlfriend. Near their lockers at the factory, Wang struck Hu seven or eight times in the head with a hammer. She survived this assault—just barely, after months of intensive care and with permanent brain damage. Wang Yungtai left the scene of his crime on foot, after swallowing a substantial amount of mercury, which he had prepared for his suicide. He became temporarily ill, but the next day confessed to his father. His father suggested he go to the police, but withdrew this recommendation when he realized that if the victim died his son would be sentenced to death. Wang was arrested several months later and subsequently confessed. The precipitating cause was romantic rejection; Hu Huichin had refused to become his girlfriend, after he had asked her several times, with all respect, in writing.

At the trial the judge asked Wang what he had been thinking while he hit Hu with the hammer. "I was thinking that she made me lose face by telling everything to someone or everyone," he said. "I was very angry and I wanted revenge, I wanted to teach her a lesson, to let her suffer. I didn't think of the consequences." The defense attorney pointed out that the defendant showed every sign of contrition, including contributing a substantial sum of money to aid in the victim's recovery, and that prior to his crime he had been a model worker guilty of no other criminal acts. In his summary statement Wang said, "I would like to repeat that when I struck her I did not intend to kill her. I only wanted to vent my anger. I had no other thoughts. The cause of my crime is my low political consciousness. I didn't study very much; I knew nothing about the rights of citizens and of the law and I have very bourgeois thoughts." Pressed by the judge, he added that he had come under the influence of Lin Biao and the Gang of Four during the Cultural Revolution. "I don't know what the law is and I have bourgeois ideas. Because I could not achieve my personal aims, I did not consider the interests of the state or of other people, so I threw everything to the wind and did what I wanted." With the judge's final statement, "In order to implement the law, preserve revolutionary order, protect the safety of all citizens, ensure the smooth progress of socialist modernization and strike at criminal activity, in order to strengthen the proletarian dictatorship," Wang was sentenced to life in prison.

The juxtaposition is striking. Two contemporary societies acting on completely different beliefs, with different systems of child training, education, work, and justice, confront almost the same crime and react in the same way, but with radically different explanations. In one, the defendant gives himself up to a father figure, a priest, and says, "I could not live without her." Those sitting in judgment hear medical testimony on whether the defendant could control himself and tell right from wrong. They are impressed by the romantic bond he shared with his victim, consider his impoverished childhood, and attempt to seek divine inspiration. In the other, the defendant says, "She made me lose face," and confesses his crime to the head of his family, his father. He attempts to explain his crime by his ideological inadequacy. The judge presses him regarding his "bourgeois ideas" and suggests the influence of the Gang of Four, a viewpoint with which the defendant concurs.

Now consider the similarities. In each case a young man is in love with a woman, is rejected, "loses control" of his emotions, and, after at least some premeditation, brutally bludgeons her with a hammer, striking many hard blows. One contemplates suicide, the other attempts it; each relates his crime to a culturally appropriate confessor; each is extremely contrite; neither has ever committed a crime or lost control before. Despite evident preparations both say they acted in extreme agitation. Both introduce culturally appropriate explanations and both receive the maximum allowable sentence. Both face judges who react first and foremost to the brutality and injustice of the assault, deeming other factors minor. Both go to prison, not because anyone thinks they will commit such crimes again but because of the two cultures' common sense of justice and as a warning to others to respect and fear the law.

Fortunately these are not everyday events. But in societies throughout the world, some young women meet death at the hands of men who supposedly love them. Frequently, there is a motive of rejection or jealousy, there has been no other criminal behavior, there is suicide or attempted suicide, and there is contrition. Take a step back and you find the widest range of emotions. The young man experiences lust; he has had or wants to have sexual intimacy with the woman. But the lust mingles with a much more respectful feeling, which, without the easy wisdom of hindsight, we would call love: a longing to be close to, stay with, and care for the one desired; a longing, we suppose, to possess. There is often a deep joy in the hope of a shared life.

But when the woman's affections do not mirror the man's, he experiences frustration, of course, and fear—of loss, loneliness, humiliation—that sometimes may be close to terror. From this frustration and fear comes rage, an impulse to take revenge, to destroy the obstacle and punish the object of the fear. In extreme cases the anger produces homicidal violence. And mixed with the fear and rage, supplanting them in the end, is grief, a mourning for the losses—of love, companionship, pride, sexual release, hope, and, after the crime, the loss of the beloved. Because of the range of emotions, such cases compel our interest more than most homicides. Conflict is more intriguing than unalloyed evil, and we sympathize because we have, however slightly, shared such conflict; its echoes remain in us, and they touch every chord in the human spirit.

Some try psychologizing, but except in the rare cases of psychotic derangement or real mental incompetence, such arguments seem out of place in a court of law. The law is not for explanation, but for justice, protection, redress of grievances, punishment. Social scientists may scrutinize these functions—judges, lawyers, and jurors are after all only human—but if the overall competence of the law is granted in a given instance, then what counts is the evidence and the moral and legal sense of judge and juror, tempered by simple human decency. Experts may rule out psychosis or mental retardation, but then they should withdraw, so the defendant can be judged by peers and members of the court, according to the evidence and the law. Yet attempts to explain acts of violence are not the same as condoning them. The law must judge, but the scientist can still ask questions. To explain is not to explain away, to understand is to get a firmer grasp on things, not necessarily to forgive.

The "unconscious" described by Freud is a marvelous metaphoric organ. With its ebb and flow of feeling, its stored record of experience, its concourse with the body, and its well-blazed trails of emotion and expression, it can generate dreams, slips, beliefs, symptoms, lifelong patterns of word and deed, while its mechanisms remain all but unknown to us. Freud did not invent the notion of the unconscious, but he and his disciples added greatly to our knowledge of it. Yet they did not go far enough, for the unconscious processes of the mind are by no means limited to the powerful, submerged currents of feeling proposed by psychoanalysis.

Among the things we might do without real awareness are driving sixty miles an hour along a highway and suddenly acting to avoid a collision, decoding the meaning and grammar of a complex sentence even as we become aware that someone is speaking to us, ticking through a bedtime ritual, walking or talking during sleep, reassuring a nervous child, or shouting a grave insult at someone we love—one of countless acts we regret as soon as they reach conscious awareness. One American president said of another that he could not walk and chew gum at the same time. Most people, though, can walk, chew gum, carry a package, scratch themselves, and daydream about sex all at the same time, and still avoid walking into a lamppost—mostly without real conscious awareness.

There is an enormous range of such actions. Some, like not urinating during sleep, are simple reflexes with little relation to mind. Others, like returning a fast serve in tennis while we watch ourselves, as if from the outside, are complex muscle-contraction patterns, concatenated marvelously but without much emotional content. Still others, such as raising the arms to protect the face from a blow, are self-protective. Some are designed to hurt others. Perhaps we should drop "the uncon-

scious" as a noun and simply keep it as an adjective. Replacing the general functions once called *the* unconscious are various circuits whose functioning may be
unconscious. Bladder control during sleep involves control of the autonomic nervous system from the brain stem, which happens to regulate much of the rest of
sleep as well. Returning the tennis serve engages the cerebellum, the corpus striatum, and probably the cerebral cortex as well, which inhibits irrelevant movements. Decoding a sentence involves Wernicke's area, a substantial portion of
(usually) the left cerebral hemisphere. Self-protective flinching could bypass the
cortex, using other connections between the visual system and motor-coordination
circuits; it might or might not involve limbic fear circuits, but these would come
behind the reflex, not in front of it. Shouting an insult could start in limbic rage
circuits, and would probably involve both cortical pathways controlling speech
and striatal ones for facial and bodily gestures.

Notice that these circuits are not confined to any one level of brain structure.
In terms of evolutionary antiquity, stage of brain development, or complexity of
connections, all brain levels can be involved in unconscious actions. Also, the
processes at issue, whatever their level, can be learned, unlearned, or both.
Sequences of action, thought, and feeling that are the products of complex, pro-
tracted learning—sometimes called overlearning—can be carried out as unconsciously as the simplest of unlearned reflexes. Third, unconscious processes may
be sanctioned or rejected by society, considered normal or abnormal by physicians, and viewed as desirable or undesirable by the conscious mind—though
they are no less a process of nature, regardless of these judgments.

What, then, and where, is consciousness? *Webster's Twentieth Century Unabridged
Dictionary* defines *conscious* as "aware of oneself as a thinking being; knowing what
one is doing and why." This is something like the philosopher's consciousness: self-
awareness. Gilbert Ryle, John Searle, Daniel Dennett, and Paul Churchland are
among the philosophers who fielded theories of that sort of consciousness in the late
twentieth century. During the same period, psychologists found that three-year-old
children cannot understand that other people might have a different view of an
object than they do, but four-year-olds can—an ability confusingly overinterpreted
as the child's "theory of mind." Still, the phenomenon is real, and provides the foundation for metacognition—the later ability to think about our own thoughts. It is the
most human form of consciousness.

But there is a simpler, more tractable meaning of consciousness, the sort that
a playful cocker spaniel has much more of than a sleeping philosopher has. Doctors call the loss of it "stupor" or "coma." It is not *self*-awareness but simply *awareness*, a global kind of integrated attention. It depends on working memory, the
short-term "keeping in mind" of things that gives awareness its ongoing quality. It
is not irrelevant to the big philosophical questions but it is not nearly as difficult.
And if it is unwise to suggest a center for it—like all brain functions it plays out in
circuits—we can at least say what sort of brain damage tends to change it.

This was a major concern of neurosurgeon Wilder Penfield of the Montreal
Neurological Institute, in a line of research carried on by the University of

Washington's George Ojemann. They and others showed that large parts of the cerebral cortex can be removed without disturbing consciousness in the simpler sense. Major damage to the peripheral nervous system, whether involving action or sensation, also leaves consciousness unharmed, as do blows to the cerebellum or corpus striatum, both involved in controlling movement. Deleting the lower brain stem makes life functions cease, so its role in consciousness is moot. That much was known by early-twentieth-century brain scientists. The one brain region that, if damaged, typically produces loss of consciousness without loss of life is the upper portion of the brain stem, the diencephalon, or "between-brain," which links the ancient lower brain with the new cerebral cortex. This includes the thalamus—the "bridal chamber" of incoming sensation just below the cortical cathedral—and the hypothalamus, the hub of the limbic, emotional brain and a major switching center for brain-body relations. These structures, or parts of them, head up the "reticular formation" of the brain stem—the central core of short, multiconnected, slow-acting cells that regulate the sleep cycle. In concert, they light up the brain with awareness, a first step toward mind. But awareness of what?

Nobel laureate Francis Crick of DNA fame, neurobiologist William Calvin, and others have fielded theories of consciousness based on time binding, or synchronized firing of arrays of neurons, which makes an object coherent. These models integrate general awareness—just being awake—with ways of understanding the fact that our awareness is almost always specific. They agree with Dennett that there is no "Cartesian theater" in the brain, no one key spot where "the real me" sits on a throne and watches events in the rest of the brain go by. According to this older view, all the events of brain function occur as on a stage, for the entertainment and edification of the audience—the real us.

There is no such theater and no royal audience. On the contrary, consciousness (in the sense of awareness) is only synchronized firing of nearby or distant neurons acting as reverberating circuits. Rhythms have been found in electrical brain activity that could reflect this synchrony, with pulses entrained to one another at a rate of about forty per second. Crick described his model in several papers with Christof Koch and in a 1994 book called *The Astonishing Hypothesis*—which refers to the not-so-astonishing claim that the mind is embodied in the brain. Following Penfield, Crick identifies not the cortex but the thalamus—the major hub between the senses and the cortex—as critical. Reverberations, possibly synchronized ones, from the thalamus to the cortex and back again, would be the essential electrical basis of consciousness. This kind of synchronized reverberation is at the heart of many recent models of brain function. But the thalamus is no "I," neither audience nor director; it is merely one brain region that has to be activated for awareness to emerge. Another is the reticular activating system in the brain stem, set to some mode other than sleep or stupor. Yet another is some specific circuitry of, say, vision or touch that could constitute and construct brain representations of the friendly face or hug that is the subject of our awareness.

In *The Cerebral Code* William Calvin proposed that small arrays of cortical cells, activated by sensations, can spread their activity in waves and either reinforce or counteract one another. We become conscious of sensations, thoughts, or feelings if and as their activations form a pattern. That pattern's coherence outcompetes other, more fragmented representations of weaker sensations or thoughts, "So the center of consciousness shifts about, from one cortical area to others, as the train of thought progresses." This helps explain why local damage to the cortex has little effect on consciousness. Although we know now that cortical activation reverberates constantly through the basal ganglia and thalamus, that doesn't invalidate the model. Either way, consciousness is dynamic and evanescent, a stream of electric waves that can carry almost any kind of experience.

The trouble with all these models is that they leave out a key component of consciousness: emotion. All are in the realm of what I call dry cognition—brain models that act as if all is rational and mechanical, as if the brain were encased in plastic and made of silicon. Not surprisingly, it was Antonio Damasio who met the challenge of including emotion in consciousness, in his 1999 book, *The Feeling of What Happens*. The title says it best. Recall that Damasio revised Descartes's claim "I think therefore I am" to what I paraphrase as "I think because I feel what I am." In *The Feeling of What Happens* Damasio made this almost explicit, showing how consciousness, first and foremost, is an ongoing awareness of the body. The positive and negative feelings that echo through the body color the brain's perception of reality, and the limbic system, including the Gage cortex at the bottom of the frontal lobes, makes consciousness inherently emotional. This accords well with Freud's insights a century earlier: conscious or unconscious, we are always feeling creatures.

But the vast realms of behavior, thought, and feeling that are often outside of awareness are not necessarily outside of control. Consider the most orderly aggressive behavior, predation. Prey catching in cats is as close to instinct as mammals get. It includes lying in wait, crouching, stalking, pouncing, seizing the prey between the paws, and clamping a killing bite at the nape of the neck, tearing the brain stem. A cat with no experience fumbles at first; but with a few repetitions, especially in moods of playful excitement, the sequence "clicks"—in a small fraction of the time cats need to learn equally complex sequences that do not rest on evolutionary preparation. The process bears no resemblance to the laborious conditioning that, after immense patience and effort by a talented trainer, leads a lion to jump through a burning hoop.

The gulf between this predatory behavior and the human acts described at the beginning of the chapter is vast, the differences crucial. First, prey catching is normal for all wild cats, while human homicide is rare and abnormal. Second, the cat behavior necessarily involves one action sequence, whereas the specific action sequence in the human cases was probably incidental. Third, the cat behavior serves the obvious adaptive purpose of getting food, whereas the human behavior, if it is functional, is much more obscurely so, and in any event has nothing to do with food. Finally, the cat sequence is carried out in playful excitement, whereas the violent young men were in moods of jealous rage.

Still, we view both as aggressive because both inflict damage. Aggressive behavior includes serious fighting, which might cause injury; play fighting, or "rough-and-tumble" play, which is generally harmless; dominance hierarchies, a settling out of winners and losers into temporarily stable patterns; threats, which can begin fights or prevent them; and, in some ways, predation. Threat, attack, and fighting serve a strange array of adaptive functions. These include competition for mates, food, and other scarce resources; play and exercise; forced sexual intercourse; defense against such forcing; protecting the young; killing the young, either one's own or those of others; competition between groups for territory and other resources; prey killing; and self-defense. While it is right to think of aggression as predominantly male, at least in most mammals, females have the full complement of aggressive equipment and actions. They show it fiercely in the unique situation of maternal aggression, but they also use it routinely in competition, dominance interactions, and other situations, not least of which is self-defense against abusive males. To these variations, add functionless aggression, the inevitable misfiring of so complex a system of hurtful acts.

Two things stand out from the list. Play fighting or rough-and-tumble play is universal in young mammals, and also occurs in adults. It is not violent, usually causes no damage, and involves different actions from those of real fighting. Still, it grades into real fighting, provides exercise for it, and leads to the stable dominance that regulates aggression. Prey killing targets other species, is done playfully or in a mood of skilled challenge, and arises from hunger, not anger. Yet it inflicts mortal damage using at least some of the same actions that the predator may use against its own species. To further complicate matters (or perhaps to help explain them), rage and fighting can be teased apart with specific brain damage. Cats may show real rage as a prelude to attack, with expressions under sympathetic nervous system control—widening of the eyes, growling and hissing, arching the back, and erecting the fur. But after certain brain lesions they will have only "sham rage"— the same expressive signs *without* attack. On the other hand, cats may kill prey with all those signs after stimulation of one part of the midbrain, but stimulation of another midbrain location produces only quiet-biting attack, the unemotional prey killing seen in the wild.

It is often useful to separate offensive from defensive aggression. Certainly these involve different motives—defensive aggression has a strong fearful component—and to some extent different behavior and physiology. Finally, aggressive behavior combines unlearned and learned components; every act in every species has a degree of both. But in some, genetically fixed action patterns and releasing mechanisms play a powerful role; in others, little or none, with the innate factors reduced to some aspects of motive and mood. Adapting a scheme put forward by the great ethologist Niko Tinbergen, we can say that to ask what causes aggression, or indeed any behavior, is really to ask a series of questions corresponding to various things we mean by the word *cause*.

1. What events in the environment immediately triggered the behavior, releasing stimuli that may be learned or unlearned?

2. What are the immediate physiological causes, the neural circuits and neurotransmitters, that produced the behavioral output?

3. How have slower-acting physiological events, such as hormone levels or disease processes, set the tone of the neural circuits?

4. What routine outside events, such as reinforcement, modeling, or stress, though not the immediate precipitating factors, may have altered the organism's response tendencies?

5. Were remote environmental causes at play, such as the special effects of experience, nutrition, or insults during sensitive periods in early life, including life before birth?

6. What events of embryonic development and their postnatal equivalents have shaped the relevant circuits and their hormonal context?

7. What genes directed the wiring-up of the circuits and coded the precursors, enzymes, and receptors for the needed hormones and neurotransmitters?

8. What adaptive function does the behavior serve? Or, what process of natural selection favored it in the natural environment? In effect, what caused the gene code?

9. What is the animal's broad heritage? The wings of flies come from thorax; of birds, from forelimbs; of bats, from fingers; and of human beings, from airplane factories. Each species solves this problem differently as phylogenetic history constrains the response to the same adaptive challenge.

These kinds of causes can be grouped as immediate or short-term (1 through 3), intermediate (4 through 6), and remote or ultimate (7 through 9). They also fall into organismal versus external groupings: 1, 4, 5, and 8 are environmental causes, and environmental effects are bound up in cause number 6. Indeed, it can be said that all behavior consists of responses to the environment at various levels of causality, beginning with natural selection.

Only in this framework can we give a more than partial account of what causes aggression or any other behavior. It would be misleading to suggest that behavioral biology can explain the homicides described at the beginning of this chapter. Although it might do better than the average court psychiatrist, such a claim would be of limited use to the court, which in the end must decide on a legal and moral basis. Research enhances our understanding of violence in general, which helps with prediction and prevention. But we must first go through a more detailed kind of analysis—one that takes us from the physiological laboratory through the field setting of the natural historian to the annals of human history. We begin in the brain.

By the 1950s we knew that damaging parts of the hypothalamus—the base-of-the-brain hub of the limbic system—could make rats violent, while other hypothalamic cuts could reduce violence. Likewise, stimulating different hypothalamic areas with electrodes could either raise or lower aggression, further evidence of the pivotal role of the hypothalamus in the limbic system and the emotions. But by then it had long been clear that other limbic structures mattered. In the late 1930s Heinrich Klüver and Paul Bucy had done monkey experiments in which they removed the end of each temporal lobe, a part of the brain behind the temples. This damaged several structures, including two limbic centers, the amygdala and the hippocampus. Changes in the behavior of the monkeys included tameness, rare in rhesus monkeys. This was not because of general debilitation (they were very active) or fear (they were bolder after surgery) but was a *specific* reduction in aggression. Later studies showed that tameness results from removal of the amygdala alone and that stimulation of the lower amygdala using the neurotransmitter glutamate induces aggression in cats.

Meanwhile, experiments damaging the septal area had the opposite effect: they *caused* rage, although in an undirected, inefficient way. This, combined with the known effects of amygdala damage, led to a hypothalamic model of aggression, with the hypothalamus regulated by higher limbic structures. The amygdala could enhance rage and aggression by exciting parts of the hypothalamus, and the septal area (or other limbic areas) could reduce them through other hypothalamic regions that have a dampening effect. Some specifics are controversial and refinements have been added. For example, the central amygdala can inhibit aggression even as the lower amygdala enhances it; this inhibition seems to use enkephalin, the brain's own morphine, to calm the lower circuits. And in male mice, as David Edwards showed, fighting depends on signals sent to the hypothalamus from the olfactory bulbs, as well as on input from the amygdala. But the broader idea is accepted: rage—along with many other aspects of emotion, motive, and mood—hinges on the hypothalamus, which integrates messages from other parts of the limbic system, producing a balance either for or against violence.

To trigger muscle action and arouse the circulatory system, the hypothalamus must send its summary of limbic activity down to the spinal cord and out to the periphery. It does so by way of the midbrain, specifically the central gray area. This was shown by John Flynn and his colleagues at the Yale School of Medicine, who discovered the distinction already mentioned between "affective" and "quiet-biting" attack in the brain. This distinction has held up well in research by Jaak Panksepp, Allen Siegel, and others, although some have preferred the terms "defensive" and "predatory" aggression. Siegel and his colleagues have found differences between the way the system is set up in cats and in rats, and have worked out much of the chemistry and anatomy.

In essence, exciting the middle part of the hypothalamus causes affective, emotional attack, complete with growls, hisses, arching of the back, and standing the fur on end. But exciting the lateral hypothalamus causes quiet-biting, a cool, predatory attack. Lesions and stimulation of the midbrain—the brain stem region between the hypothalamus and spinal cord—show that the hypothalamus controls attack, with or without rage, through circuits traversing different parts of the midbrain, particularly the cell-rich central gray area. The middle hypothalamus communicates the impulse toward rage through neurons ending in the *upper* part of the central gray. In contrast, the *sidestream* neurons of the hypothalamus, which process the less emotional prey-catching attacks, project to the *lower* central gray. These circuits in turn control parts of the brain stem and spinal cord that produce the attack itself and the sympathetic nervous system that expresses rage.

In a fascinating sidelight on the sidestream neurons, Panksepp showed that rats will stimulate themselves if they have an electrode in that region, proving that quiet-biting attack is pleasurable. This is not surprising, since the same stream is part of the river of brain reward. In contrast, rats quickly learned to turn *off* stimulation to the *middle* hypothalamus, tending to cause affective attack, showing that emotional aggression is unpleasant, something to avoid. The rewarding effect of stimulation for quiet-biting fits well with the fact that well-fed cats *worked* for a chance to attack a rat, even neglecting a full bowl of food to do so. Whatever motivates prey catching, it is not just hunger. Ethologists compare cat predation to play and trace its origins in kittens' play sequences, consistent with laboratory studies linking predation to pleasure. As I watch our plump cat pounce and dispatch chipmunks, voles, and little perching birds that she doesn't deign to eat but proudly lays on our doormat, it's hard to deny the pleasure in prey killing.

Still, rats and cats are one thing, people another, and most human violence is probably not pleasurable. People with tumors damaging the middle hypothalamus or the septal area have trouble controlling aggression, especially if provoked by a real or imagined insult. These effects support the idea that the septal area inhibits rage and the amygdala stimulates it, both perhaps by regulating the middle hypothalamus. In rare cases, a slowly growing tumor in the limbic system causes increasing irrational aggression over a number of years, whereas removing the tumor reduces aggression. Charles Whitman, a young Texan who killed his mother and wife, then climbed a university tower and shot thirty-eight people, was found at autopsy to have a brain tumor that may have chronically irritated his amygdala.

Although epileptics are very rarely violent, a handful with seizures in the amygdala have outbursts of aggression; furthermore, people with records of criminal aggression have more EEG abnormalities than people in general, and more than other kinds of criminals. Finally, a brain basis for human aggression is supported by studies of Vietnam veterans, followed for twenty or more years after head wounds in combat. Men with lower frontal lobe damage are more likely than men with other kinds of brain damage to have outbursts of rage at family members, friends, and colleagues. These outbursts are fortunately more often verbal than physical, but they are severe, reminiscent of Phineas Gage's uninhibited rages after the railway-spike injury. This finding is consistent with the idea that the underside of the frontal lobe is the cortex of the limbic system, monitoring and regulating emotional activity.

Brain-imaging studies of violent individuals suggest that weak activity in the left frontal and temporal lobes reduces inhibition, leading to outbursts of physical rage. Such studies are increasing rapidly and, with the growing precision of brain imaging, will soon show us just how the brain generates violence. For example, in an evaluation of thirty-one murderers, psychiatrist Jonathan Pincus found that frontal lobe damage often contributed. But combined with two other factors—a psychiatric disorder with paranoid symptoms and a history of childhood abuse—the chance of violence became very high. Thus, the role of frontal lobe dysfunction in violence must be seen in the light of other mental illness and experience.

But the most controversial insights have come from psychosurgery, a treatment with a long, unhappy history. Frontal lobotomies damaged many minds before doctors' enthusiasm for them waned; in cases of violence, they seemed little more than medical tyranny—as in the use of it against Patrick McMurphy in Ken Kesey's *One Flew Over the Cuckoo's Nest*. But even critics acknowledge the old medical saying that desperate maladies may require desperate remedies and so there is a growing sense that some surgical interventions may be justified in severe mental illness. Today there are far more subtle forms of brain surgery, and one type has been found helpful in the treatment of a very rare, violent form of epilepsy. The vast majority of severe epileptics have simple seizures: massive, uncoordinated muscle contractions, dangerous only to themselves. But in very few cases, the seizure is directed outward. A handful of patients in the United States, where opposition to psychosurgery is strong, and more in Japan, Europe, and Latin America have received surgical treatment for this disorder. These cases do not come to surgery, at least in the United States, unless the disorder is very severe and other treatments have failed. Still, the ethical problems are daunting and the results have not been consistent.

With these reservations, psychosurgery has occasionally been used to treat this rare syndrome, focusing on parts of the circuitry we have traced. One approach used in Japan and Argentina in treating extreme and frequent violent fits is destruction of an area three to five millimeters in diameter, in the rear of the middle hypothalamus. As might be expected from Flynn's cat studies, this procedure often reduces violent rage, though not without side effects. Another approach, used in Japan, India, and the United States, has been to destroy portions of the amygdala and this, too, is reported to have a calming effect. It is right to view such claims skeptically. But, as neuroanatomist Walle Nauta, beset by noisy protests, said in his Presidential Symposium at the 1973 neuroscience convention, critics of psychosurgery do not always have first-hand knowledge of the kind of suffering that leads patients, families, and doctors to consider such drastic treatment. These procedures could not be ethically contemplated in homicides like those discussed earlier, but only in cases of relentless, repetitive violent outbursts resistant to other forms of treatment. The danger of misuse in criminal or even political cases is great.

Another approach to the brain's violent circuitry is chemical. Research has been done in which lab animals are given drugs that influence neurons or neurotransmitters in the junctions between them. For example, if you provoke a fight between male mice and then examine their brains, you will find more norepinephrine absorption than in the brains of males that have exercised *non*violently.

Also, mice kept in isolation for several weeks are more likely to fight and they have changed levels or turnover of several neurotransmitters. In fact, genetic studies, including those involving gene manipulation, must often use isolation to bring out the added aggressiveness, further proof of the power of this experience. Moreover, drugs directly affecting those neurotransmitters can increase or decrease isolation-induced fighting, showing that chemistry and experience are to some extent interchangeable. Yet another paradigm involves mouse killing, a normal predatory activity for rats. As we have seen, this behavior is quite different from other forms of aggression, but it, too, can be influenced by psychoactive drugs that act through specific neurotransmitters.

As for emotional aggression, strong evidence supports a role for serotonin. In many animals reduced brain-serotonin activity is linked to fighting, due to a lowered threshold for aggressive reactions to frustration. In humans as in other mammals, decreased serotonin processing is reflected in lower levels of the byproduct known as 5-HIAA (pronounce the letters). Measuring 5-HIAA in the cerebrospinal fluid, the liquid that bathes and cushions the brain and spinal cord, shows that those who are impulsively violent and antisocial have low levels, although indirect measures may prove even more useful. This relationship is seen in children as well as adults. Interestingly, such low 5-HIAA levels predict suicide and suicide attempts as well as outwardly-directed violence. Since the old association between aggression and a *high* rate of norepinephrine activity has also held up in newer studies, serotonin may balance norepinephrine in controlling violent tendencies.

Targeting neurotransmitters to alleviate aggression is a new strategy. Drugs are less drastic and more reversible than brain lesions. They can also perhaps control aggression in the normal range—a notion with ethical problems of its own. Monkeys can be studied, though, and Michael McGuire, Michael Raleigh, and others at UCLA find clear relationships between serotonin and dominance. In elegant, controlled experiments they assembled groups of four vervet monkeys and allowed stable dominance to develop over time. Then they removed the top male in each group of four and formed another group consisting only of these "leaders of the pack"—a bit like high school football heroes being plunged into a college scrimmage with other high school stars. Bested monkey males—those *not* on top in the new group—had substantial *declines* of serotonin. Conversely, in other experiments, giving drugs that raise serotonin levels will increase a male's chance of becoming dominant.

This seems at first to contradict the studies of impulsive aggression, which show that lower levels of 5-HIAA, the serotonin byproduct, predict violence. But *impulsive* aggression does *not* lead to a stable dominant role, so there is no contradiction. Males must win fights to become dominant, but they must also control their own rage. The same pattern was found among females in two different species of macaque monkeys—rhesus and pigtails. Females with low 5-HIAA levels showed more evidence of high-intensity aggression, escalated aggression, and fight wounds requiring medical treatment, yet they also drifted toward the lower end of the hierarchy.

Similar reasoning applies to slower-acting chemicals, such as hormones. All the classical stress hormones—adrenaline from the core of the adrenal gland, cortisol from its outer portion (secreted in response to pituitary hormone ACTH), and others—are released into the blood in a "fight-or-flight" situation. This classic setting produces a critical adaptive demand to mobilize energy for muscle action. Accordingly, the stress hormones release energy from storage to free forms like glucose and constrict and expand blood vessels to make less blood flow to the gut and more to the muscles, supplying the energy for contraction.

These hormones also affect the emotional brain, but more slowly than neurotransmitters. The fastest hormonal event would be adrenaline release, because it is controlled by a neural signal. The signal may start in the higher limbic centers and pass through the hypothalamus, midbrain, and spinal cord to the sympathetic nervous system. This chain of neurons and way stations, holdovers from some wormlike ancestor, sends nerves out to our blood vessels and internal organs, including the core adrenal gland cells that release adrenaline. New manufacture of more adrenaline would be a different, slower effect, promoted by the outer adrenal gland, whose products seep into the core. One of the secretions of that part of the gland is cortisol, released by ACTH from the pituitary. ACTH, in turn, is regulated by a chemical signal from the hypothalamus, which integrates information from all over the limbic system. These events, all requiring bloodborne agents, are slower than neural signals but are needed for the sustained muscle action of fight or flight.

Sex hormones, too, especially testosterone, affect aggression in animals. This is more interesting than the effects of adrenaline and cortisol, which are more general in their action and are called up in all sorts of stress. As we have seen, testosterone promotes aggression, certainly in males and possibly in females, in a much more specific way. Indeed, generalized stress is likely to *decrease* testosterone. Yet in several species testosterone injections increase aggressiveness in certain situations and male castration decreases it. Human studies are more complex, but there is ample evidence that normal testosterone levels make aggression possible.

Although not consistent, some studies also suggest that steroid treatment, whether of androgen-deficient men, normal athletes, or ordinary volunteers, can enhance aggressive tendencies. Conversely, aggression can be reduced by antiandrogen treatment or by a drug that blocks gonadotropin-releasing hormone, the ultimate regulator of testosterone. Recall that in James Dabbs and Robin Morris's study of 4,000 army veterans, their natural testosterone level predicted their antisocial behavior. Interestingly, it does so more strongly among poorer veterans, suggesting that in a worse environment biological differences matter more. In another criminal population, high testosterone was linked with more violent and aggressive crimes during adolescence. Also, a study by Susan Chance in Dabbs's group showed that testosterone level helps predict aggressive behavior in five- to eleven-year-old boys, especially those of lower cognitive ability. Finally, a large, long-term, highly regarded study of Norwegian school bullies by Dan Olweus found testosterone, along with several social and psychological variables, to be a significant predictor of bullying.

Researchers have also done some limited cross-cultural testing of the testosterone-aggression hypothesis. A study that Carol Worthman and I did among !Kung hunter-gatherers showed that hunting changes testosterone levels in a manner suggesting exercise rather than aggression. But a study by Kerrin Christiansen and Eike Meinrad Winkler found that more violent !Kung men, many of whom had scars from fights, showed androgen levels correlated with their frequency of fighting. These intriguing findings suggest that in human hunter-gatherers, as in other predators, the biology of prey killing is quite different from that of defensive aggression.

Natural changes in testosterone accompany fighting, and winning and losing affect that level. When two groups of rhesus monkeys were made to fight, the losers showed a large decrease in testosterone after the fight—actually in two stages, the second perhaps corresponding to final acceptance of the loss. In a study of the Harvard wrestling team, all men competing showed a rise of testosterone during the match, but the winners had a significantly larger rise than did the losers and those who fought to a draw had levels exactly in between. Sociologist Allan Mazur has studied the impact of victory and loss on men's testosterone in situations from tennis matches to video games, from chess competitions to medical-school graduations, and found similar changes in most of them. Mazur's work strongly suggests that testosterone level is raised by victory and lowered by loss. The hormone also dips during the stress of basic training in officers' candidate school.

One paradox is that testosterone facilitates aggression but stress suppresses testosterone, so the stress of fighting should be counterproductive. The resolution is suggested by studies showing that dominant monkeys exhibit less physiological stress *before* fighting than do those at the bottom of the pyramid. Robert Sapolsky of Stanford University has for many years studied the hormonal changes in wild baboons in Kenya, in relation to their dominance rank and aggressive interactions. Dominant baboons have lower levels of cortisol until their rank is challenged, a stressful experience whether they win or lose. A very similar pattern has been found in studies of dominant versus subordinate rats, observed in the stressful context of a visible burrow system. Coming tantalizingly close to our own species, Sally Seraphin has begun related studies of chimpanzees in the Budongo Forest of Uganda and has found a variety of correlations between dominance rank, stress hormones, and sex hormones.

Ned Kalin, who draws a clear distinction between defensive and offensive aggression in his research on rhesus monkeys, has results that dovetail neatly with these findings. Defensive aggression is partly motivated by fear and is associated with high right frontal lobe activity and high baseline cortisol level. Offensive aggression is more impulsive, however, and is tied to low serotonin activity, high testosterone, and *lower* baseline cortisol. Still, all forms of aggression in Kalin's monkeys are influenced by experience, and disruptions in the mother-infant bond play a major role.

The pattern is consistent with the studies of serotonin, in which edgy, impulsive individuals get into fights that they might not be able to finish, whereas calmer animals give their all, but only in a fight they cannot avoid. Another clue to this process comes from human studies that do not involve aggressiveness directly: performance on skilled tasks is enhanced by some, but not too much, subjectively felt stress. Imagine a highly stress-reactive individual beginning a fight with a less reactive one. To the extent that fighting is a skill, the one who is calmer (though not apathetic) should do better and should also mobilize stress hormones more efficiently during the fight. For the lower-ranking animal, testosterone may even have the unfortunate effect of getting him into fights that he can't finish.

So the balance between stress and aggression in a fight is not simple. Adrenaline and cortisol mobilize energy and affect the brain. Testosterone acts on the neural circuitry of aggression and rage. In monkeys it acts in many parts of the limbic system and in rats it lowers the firing threshold of the stria terminalis, a major cable connecting the amygdala and hypothalamus. In male birds, it dramatically affects the nerves controlling territorial singing—basically a type of macho bragging—even causing the brain's song centers to make new neurons. Under certain conditions, testosterone implants will even make female finches attack other females. Finally, human studies have begun to show effects of androgens, in high doses at least, on aggression. A randomized, placebo-controlled, double-blind study led by psychiatrist Harrison Pope showed that in fifty-six normal adult men, testosterone injections increased aggression by several measures. Intriguingly, it also intensified their exuberance, known to psychiatrists as hypomania.

So there is little doubt now that the male hormone works to promote adult aggression, aside from its long-range effect in the womb. But why did evolution bring in these slow-acting hormonal effects when neural circuits are so much more efficient? One reason is that hormone levels have an adaptive life of their own, an ebb and flow tied to daily, monthly, annual, or other cycles, or even a set plan through the life course. The brain changes and grows, but hormones constantly fine-tune it to adaptive advantage. We are not equally likely to be stressed at all hours, so stress hormones vary at different times of day and night. Many animals breed only at favored times of the year and a seasonal rise in testosterone at those times serves two functions: it makes the male more likely to court and copulate, and it makes him more formidable in any courtship conflict, whether with another male or with the female herself.

Richard Michael and Doris Zumpe showed that the seasonal peak of breeding in rhesus monkeys coincided with the monkeys' peak in testosterone level. In the spring the songbird's brain changes under testosterone's influence so he can loudly proclaim aegis over a territory—celebration of spring, to be sure, although not quite the one Wordsworth or Shelley had in mind. But the actions of hormones and the brain are mutual. The hypothalamus and pituitary produce hormones to service behaviors they "decide" to engage in. The hormones, to the extent that they can cross the blood-brain barrier, will bathe the nervous system and make firing more or less likely in some circuits, as well as supply energy for nerve and muscle action.

Much earlier in life, hormones played a formative role. Such remote physiological causes include the events of "fetal androgenization" of the mammalian brain and body considered in chapter 6. The tendency toward aggression in adulthood is influenced by the amount of testosterone circulating in early life (before birth in primates, just after birth in rats), and this effect is almost certainly due to long-lasting changes in the brain. These are known as *organizational* effects of androgens, in contrast to the *activational* ones of post–adolescent life. In a classic study demonstrating both, David Edwards showed that giving testosterone to female mice on their first day of life makes them as aggressive as normal males in adulthood, provided they are given testosterone as adults, too, although events on day one only begin the process. Thus, both organizational and activational effects are required to make females fight like males. What Robert Sapolsky aptly calls "the trouble with testosterone" owes as much to what happened to males in the womb as to what affects them currently, in a fight or at the wheel, because those very early events wire the male brain a little differently.

But before fetal androgenization, and even just within one sex, the genes themselves have a strong effect on aggression. Gene technologists have already honed in on fifteen genes on two chromosomes that affect male aggression in mice, and additional ones for female aggression as well. This growing collection of "mean genes" includes one that codes for a serotonin receptor and another that makes an enzyme that removes norepinephrine and other neurotransmitters. There are others and the lab shelf of DNA bottles labeled "aggression" runs the gamut of mechanisms for how genes shape behavior.

To take one example, the X chromosome holds a gene for an androgen receptor. In wild mice the receptor combines with androgens and the resulting molecule turns on certain other genes in certain brain cells. But engineered mutations block the combination of androgen and receptor, producing peaceful male mice, even after social isolation. Chromosome 10 carries an estrogen receptor that works similarly, but in this case the mutant females are *more* aggressive than the wild ones—the opposite of the impact of the same mutation in males. Another way to increase aggression is by knocking out or inserting genes for the neurotransmitter enzyme monoamine oxidase A and the 1B subtype of the serotonin receptor. Knocking out one of the histamine receptors, in contrast, decreases aggression. And an enzyme that makes nitric oxide, a neurotransmitter involved in penile erection and other blood vessel functions, can have its gene knocked out as well. The result is a mouse that attacks not only more often but more lethally, by directing its bites more precisely at the opponent's neck instead of occasionally drifting down his back. If you are getting the impression that there is a rather complex genetic code for aggression in mammals but that this code is being broken with amazing rapidity, you're on the right track.

But if these new techniques of gene manipulation are dazzling in their wizardry, they were preceded by decades of classic experiments. Charles Southwick of Johns Hopkins University took fourteen purebred mouse strains and, after socially isolating them for weeks, brought four males together from each strain and counted the instances of chase, attack, and fight. Scores ranged from less than 10 to 80, an almost tenfold difference. Strain blending showed that aggressive genes are dominant, with the young resembling the feistier parent. In some crosses, unexpected, synergistic effects occurred, producing offspring much more violent than either parent. Taking either the father or mother from a given strain revealed some parenting or intrauterine effects and cross-fostering infants of one strain to parents from another supported both possibilities. It was possible for the foster mother to influence the offspring, but some important strain differences in chase, attack, and fight were due to genes themselves.

By the mid-1990s direct genetic effects had been shown for mice and men. In a large extended family in the Netherlands, a new form of mild mental retardation was found to be X-linked, making it far more common in males. It is also associated with attempted murder, rape, arson, and other acts of impulsive aggression that were not attributable to low intelligence alone. The syndrome was traced to a flawed enzyme—a type of monoamine oxidase, which helps remove the neurotransmitters serotonin, norepinephrine, and dopamine. The mutation put a stop signal in the middle of the gene, and the half-formed enzyme was useless. In an independent study, knockout mice were created with a flawed gene for the same enzyme. Their brains had up to 9 times the normal serotonin level and twice the normal norepinephrine. Chemical manipulations showed that serotonin made the difference. Adult males with the defect fought more with one another and tended to force their attentions on unwilling females—two symptoms shown by men in the Dutch kindred. And in another study of genetically engineered mice with this enzyme defect, drugs antagonistic to serotonin abolished their exaggerated aggressiveness.

Unfortunately for simplicity, these findings are hard to reconcile with studies already discussed, where *lower* brain serotonin *increases* aggression in various species. But the trouble must be subtler than a simple rise or fall—changes in rate of production, say, or in the number or sensitivity of receptors. At least serotonin is implicated in both lines of research. In addition, an entirely separate study showed that mice lacking the serotonin 1B receptor are very aggressive. This receptor is abundant in the central gray of the midbrain, the region that processes aggressive signals from the hypothalamus.

But it is vital to understand that the gene change is only the first step in a developmental process. The Dutch men were mildly retarded and isolative, with occasional outbursts of very serious aggression. The defective enzyme had months of opportunities to affect the brain in the womb. After birth, their self-imposed isolation may have gradually increased their aggressive tendencies, as it does in males of many other species. To say that there is a gene known to affect human aggression does not mean we ignore these environmental complexities. All genetics is developmental genetics, and therefore assumes environmental influences in the growing fetus, infant, and child. But these qualifications do not make it meaningless to assert genetic influence. In normal human beings, traditional studies leave little doubt as to the power of genes, almost certainly many different ones. Some of these affect general traits that may influence the growth of aggression, like pain sensitivity, impulse control, sensation seeking, and frustration tolerance. Such traits in a toddler could interact with environmental stress or cultural shaping to produce a variety of violent patterns, even without any dedicated brain circuitry for violence.

Consider the studies of juvenile delinquents in North America, Japan, and England. If one identical twin is delinquent, there is about a 91 percent chance that the other one is, while the corresponding number for nonidentical same-sex twins is 73 percent. That means that if you are a social worker looking at a convicted juvenile and you know that boy has a twin, you are justified in being more concerned for the unconvicted twin when he is identical. However, since the rate for fraternal twins is also very high, much higher in fact than their proportion of shared genes, large environmental factors must also be operating and we should not presume that the very high correlation between identical twins means that genes tell the whole story. But international data for adult felons who have a twin tell a slightly different story. In seven studies representing North America, Norway, Denmark, Germany, and Japan, the identical-twin concordance rate is just over 50 percent, while the fraternal-twin rate is 23 percent. These facts, without any other calculations, show that greater genetic similarity leads to greater similarity in lawbreaking, a similarity that for unknown reasons is greater for nonviolent than violent crime. The facts also leave room for—indeed they demand—environmental factors of even greater importance. *Fifty percent of the identical twins of adult felons are not criminals.* Something other than genes creates this immense difference.

Notice two contrasts between the juvenile and adult studies. First, concordance rates are higher in the younger offenders, probably because of high base rates for crime in the neighborhoods they came from. Second, genetic effects are clearer in the older criminals. This means that, contrary to common belief, the longer the environmental exposure, the more obvious the genetic effects. This kind of finding has come up in other studies. Various traits of toddlers look more heritable than do similar traits in very young infants, for instance, and heritability for scores on mental tests in children also increases with age. Common sense seems to tell us that the youngest infants will show the strongest impact of genes, because they have had so little experience. But for some traits genetic influence increases with age as behavior settles into "canalized" pathways, after the transforming disruptions of early development. Both Huntington's and Alzheimer's diseases show that decades after birth genes can command and control parts of the brain, without any prior effect or any necessary contribution of the environment.

Yet the evidence for environmental factors is overwhelming. We know that for both animals and humans, pain, irritation, frustration, and fear increase aggression in many situations. Of course, fear and pain, appropriately introduced, can also *decrease* aggression in a given time and place. But in many experiments the opposite occurs. For example, when two male rats are placed on an electrified grid and shocked, their classic response is to attack each other. In other experiments pain is not directly involved, but an animal is motivated, aroused, and then deliberately frustrated. This situation, too, can increase aggression. These simple paradigms serve as models for the high levels of aggression in impoverished inner cities, where pain is pervasive, tragically pitting its victims against one another. In animals, the well-studied phenomenon of redirected aggression occurs both in natural and experimental situations in a wide variety of vertebrates. We saw an example of this in chapter 3, from Robert Sapolsky's baboon troop, in which Nick ruled over Adam, who pushed around Scratch, who in turn dominated Absalom and Limp. Humans may do this sort of thing openly in a neighborhood bar or more subtly, in the overall ordering of society, but we do it either way, just as baboons do.

We have seen that in many experiments on rat or mouse fighting, the animals are isolated first. It is not clear what social isolation means to a mouse, but we know that a few weeks of it greatly increases the chance of a fight when the males are finally paired. In rats, the effects are more ambiguous but similar. Perhaps isolation results in multiple frustrations or in a general loss of conciliatory skills. But this and many other conditions of rearing affect the likelihood of violence and must be carefully considered in genetic studies.

Fighting can also be trained. German shepherd attack dogs are shaped through operant conditioning—the reward of naturally occurring behaviors progressively more similar to the desired ones. Pets can also be trained to inhibit chase, attack, and fighting, using punishment, reinforcement, and the animal's capacity for association and generalization. And of course, attack dogs can be brought to attack some people but not others through discrimination learning. In nature, the same learning processes help form dominance ranks. The human versions of violence training are very well understood, not least by Colonel David Grossman, whose book *On Killing* details the psychological evidence for such training against the background of his own experience raising the kill rates attained by American soldiers in battle. Grossman is convinced that the same desensitization process he took American soldiers through—repeatedly exposing them to violence until they no longer reacted with normal repulsion—is being carried out less systematically with all of our children, through violent television and especially video games. In his view, clicking thousands of times during childhood to produce a "kill" on a screen must make it easier for a young person to kill for real.

We do know that cultures take different approaches to encouraging violence or nonviolence in the young, and these help to produce widely different rates of violence, varying by as much as a thousandfold. This occurs partly through pragmatic conditions that promote war and peace, but it also relies on cultural differences in modeling, imitation, and identification—an emotional focusing of imitation on one model or class of models. There have been many studies of modeling and imitation in laboratory animals, and field studies have indicated their importance in the wild.

However, ethologists distinguish between imitation and social facilitation. This is an echoing of behavior too automatic to be imitative learning and seems to be more like a shared tendency that is released by its prior occurrence in others. A wave of coughing in a theater would be a human example, but the violence of a lynch mob might be in the same category. Psychologists have called this "emotional contagion." Subcortical mechanisms of mimicry activate the autonomic nervous system as well as facial and postural movements. The image of someone else's actions somehow becomes a template in our nervous systems, seeking a match among our own possible actions. The role of emotional contagion in aggression has been recognized by historians and social scientists for generations. But we are only now seeing serious efforts to bring these phenomena under scientific study.

True modeling and imitation involve higher brain functions. Some of the classic studies of modeling and imitation in aggression in human children were done by psychologist Albert Bandura. When a child watches an adult committing aggression, either live or on film, the chance of the child's acting similarly shortly afterward increases. This finding has become the basis of a vigorous research enterprise, much of which is focused on the role of media in modeling violent images—data that helped bring David Grossman to his sobering conclusion about learning to kill. But we must be very careful here. As Jeffrey Goldstein and his colleagues have shown in *Why We Watch: The Attractions of Violent Entertainment,* such images attract us because we already have aggressive tendencies and fantasies and feel a need to work them through. Like the Bible, Homer's epics, Shakespeare's plays, and many ancient texts outside the Western tradition, media are violent because life is. They may make life more violent, but they are not a fundamental cause of life's violence.

There is also a more complex, less understood category of environmental effects, more remote effects such as rearing conditions or other events of early life. Such effects have been shown in many species and, while their mechanisms are still puzzling, they contribute to the later level of violence. We know that in males from fish to mice a few weeks of social isolation will greatly increase fighting, but in monkeys there is another, more interesting effect. Early rearing in social isolation—for six months to a year, in the case of rhesus monkeys—makes them socially hyperreactive throughout life. Males threaten and attack very frequently, often inappropriately and without success. Some of the same symptoms are found in rhesus monkeys raised for a year with their mothers but without peer contact. Early research suggested that as little as twenty minutes a day of group play prevents the development of hyperaggressiveness in these monkeys. But more recent studies show that isolation from adults, even with peer contact, produces a monkey that is not only hyperaggressive but has a low level of 5-HIAA in the spinal fluid. This suggests not only that low 5-HIAA is a marker for violence, as in humans, but that this aggression-prone physiology can result from adverse early experiences.

Not surprisingly, dominance rank in monkeys is also affected by early experience. In free-ranging rhesus monkeys, high-ranking females have infants that grow up to be high-ranking themselves, and not just for genetic reasons. Infants frequently imitate their mothers' threat behavior, even chasing much larger adult animals. Obviously they couldn't defeat an adult in individual combat, but the infants make their moves in the mother's shadow, even if she is not right there—a phenomenon called "protected threat." In this context the young monkey has innumerable learning experiences and opportunities for imitation and social facilitation that lead to effective dominance behavior. This process also applies to female baboons on the African savanna, suggesting that the social transmission of dominance rank may be a widespread feature of natural hierarchies in higher primates. As we have seen in other studies, there is also evidence that the low 5-HIAA levels associated with violence do not result in higher social dominance and may even be lethal for the aggressor. In both males and females the behavior that leads to dominance is more sober and less impulsive, echoing Sapolsky's finding that dominant baboons are less stressed by daily life but mobilize all hormonal resources for a fight. This also fits with Raleigh and McGuire's studies of vervet monkeys, a very different species. The implications for human status and power hierarchies are intriguing.

Even predatory aggression is shaped by early experience. Rats reared with mice are much less likely to prey on mice, while bird dogs become better pointers and retrievers with experience. Big cats like lions and leopards gradually learn to chase and attack prey, and cheetahs may even teach their young, bringing stunned and injured prey back for the cubs to kill. Human children in hunting-and-gathering societies learn in similar and more complex ways. !Kung children are not discouraged from cruelty to animals. In fact, they have fun chasing, torturing, and killing small animals while adults watch approvingly—something that in our society is generally considered a warning sign for future delinquency. However, since !Kung behavior toward infants and small children is among the most generous and tender in any human society, encouraging cruelty to animals represents no general tendency to cruelty, merely a specific experience that helps in learning to hunt.

Interestingly, in Southwick's mouse-breeding experiment, maternal behavior altered the pups' later aggression in some strains. A mother that spent more time off the litter—giving the pups shorter feeding sessions in which they had to compete for a nipple and also lowering their body temperature, all of which increases stress and stress hormones—increased the aggressiveness of the young when grown. And while cross-fostering a nonaggressive strain onto an aggressive one made the genetically tame pups 80 percent more aggressive, the reverse procedure did not tame the genetically aggressive pups. This is consistent with many studies: training programs designed to increase aggression (in the military and combat sports, for example) are more successful than those designed to decrease it. Worse still, once acquired, such behavior is difficult to extinguish.

So there are powerful genetic effects on aggression and powerful environmental effects as well. It would trivialize the violent acts of Richard James Herrin and Wang Yungtai—and even more, the tragedy of their victims—to presume to explain what they did. Still, we must try to discern some influences. Both were men and such crimes are overwhelmingly committed by men—partly, at least, for genetic and hormonal reasons. Both were well bred in the moral sense, but both also grew up in societies that traditionally glorified violence, including male violence against women. Both were exposed to one of ordinary life's more stressful frustrations, that of romantic rejection, and both were young enough to be unfamiliar with this stress. Of course, all of these factors are at times present in millions of individuals who do not commit homicide and so do not help very much in explaining these cases. Such factors are not excuses, but they make these acts more comprehensible than the random shooting of motorists on a highway.

What of the remaining whys of adaptation and phylogeny? Psychiatrist David Hamburg pioneered the study of aggression in chimpanzees, finding many similarities to ourselves. Aggressive behavior is common, emerges early, shows sex differences, and plays a critical role in social hierarchies. Imitation and other forms of learning play a role but not an overwhelming one. Still, the use of tools in aggression by wild chimpanzees improves with practice and this suggests some unpleasant hypotheses about the role of violence in the evolution of human intelligence.

Hamburg's work foreshadowed much that was to come. The last decades of the twentieth century saw a sea change in the natural history of aggression, a shift from classical ethology to a more Darwinian sociobiology. According to the old view, championed by Konrad Lorenz, the function of aggressive behavior is to distribute members of a group over a territory. In theory, the group as a whole benefits and individuals stay out of one another's way. Threats and other aggressive displays reduce actual violence by spacing individuals and arranging them in a hierarchy. Studies of wild animals seemed to support this view and humans were said to be almost unique among animals in that we kill our own kind. One explanation for this distinction was that weapons such as projectiles give us a distance from our victims that weakens our natural tendency to limit the damage we do.

These claims are largely false. As E. O. Wilson pointed out many years ago, if a troop of baboons killed their own kind at the same rate as people in New York City do, you would have to watch the baboons for hundreds of years before you would see a killing. Not surprisingly, as hundreds of person-years of field observations were logged by naturalists, it became clear that many species do kill their own kind. In other words, "natural" mechanisms for the limitation of violence do not work much better in nonhuman animals than they do in humans. Animals, like us, might kill when they have more to gain than to lose by doing so and in all likelihood the absence of war in nonhuman mammals has more to do with their lower level of social organization than with any lack of aggressive tendencies.

Killing takes many forms in many species, but the grimmest is competitive infanticide, first observed by Sarah Hrdy in a sacred Indian monkey, the Hanuman langur. Langur troops consist of a hierarchy of female relatives with their young and a small number of males attached to the group, often for a year or more. From time to time new males appear, drive the old males out of the troop, and take over as resident males. Within a few days they kill all infants under six months of age and reimpregnate the infants' mothers soon thereafter. This and other patterns of competitive infanticide appear in chimpanzees, lions, wild dogs, and many other species.

Attempts to reduce this phenomenon to a manifestation of stress are of no fundamental interest. Evolution is stressful and the existence of such patterns is a key test for sociobiology. A scientific theory must not just explain what we already know but must make a verifiable prediction. Einstein's general theory of relativity predicted that light would bend around the sun and when it was found to do so in the eclipse of 1919, that theory was much more widely accepted. So sociobiological theory, with its emphasis on individual selection and competition, would have to predict the existence of something like competitive infanticide. Its discovery was a major setback for group selection, at least in higher animals. Here is a pattern that has serious disadvantages for the group—indeed it grotesquely disrupts the group—while giving a strong evolutionary advantage to a few individuals.

A similar imbalance exists in many human groups. Group selection—the proposed process by which one population as a whole wins out over another—has not been shown to occur in complex animals. Appealing as it is in certain ways, it simply does not produce as logical a picture of how evolution has taken place as does individual selection. The burden of proof is now on proponents of group selection like Elliot Sober and David Sloan Wilson, who have continued to defend it, more as a perspective on evolution rather than a necessary process. Most evolutionists remain skeptical, awaiting proof. In the meantime, we can provisionally accept the position of George C. Williams and other proponents of the simpler view that each individual is in competition with every other individual in its environment, and that group selection is a minor factor.

This is not to say that cooperation is not adaptive, merely that it is also fragmentary and transient. Aggression evolved to serve the interests of individuals and that is why some use it better than others. It garners resources for the victor and in the case of males, it means access to female reproductive capacity. As Irven DeVore showed in the 1960s, dominant male baboons have privileged access to females when they are ovulating. Glenn Hausfater later confirmed this in baboons and Kim Wallen found a similar process in rhesus monkeys. Jeanne Altmann and Susan Alberts showed that this access affects genetic paternity. It is likely that male predominance in all physical forms of aggression is mainly due to this evolutionary background of competition for fertile females.

As for violence in human phylogeny, there is little directly relevant fossil evidence until the emergence of our species itself. It was long thought that the emergence of hunting had important implications for human aggression—the "killer ape" hypothesis. This is not likely because of the great differences, physiological and behavioral, between predatory aggression and fighting one's own kind. All over the animal world are examples of vegetarian species whose members fight violently among themselves and who inflict great damage with natural weapons like beaks, hoofs, and antlers. Even our close relatives the chimpanzees, for whom meat makes up only a small part of the diet, have intraspecies aggression of the severest sort. This includes violence between groups at territorial boundaries, violent attacks on females by much larger males, and infanticide by females against the infants of other females.

Jane Goodall, the great naturalist and observer of wild chimpanzees, chronicled this violence and eventually changed her earlier view of these animals as harmonious and peaceful in all their relationships. One mother-daughter pair made a habit of killing the infants of other females and they were not the only perpetrators of infanticide. More important, Goodall and her colleagues observed systematic attacks on adult chimps by groups of males from adjacent communities. The victims were temporarily separated from their own communities and easily set upon by a gang that beat, stamped, dragged, and bit them to death. Victims could be of either sex, but females of reproductive age were often spared and incorporated into the other group—an outcome that incidentally demonstrates why individual selection is a better theory even in this situation. In two cases whole groups were eliminated through a combination of one-by-one ambush killings and female transfer. Interestingly, tribal warriors in New Guinea have nightmares about being separated from their comrades in combat, surrounded, and slaughtered by merciless enemies.

Bonobos, equally close relatives of ours—and as we have seen, better at language than chimpanzees—are very different. Males have fights, but much less injurious ones, and they never attack or abuse females. Takayoshi Kano, who has studied them for decades, believes that female power makes the difference. Although females change groups at adolescence just as chimps do, bonobo females forge intense relationships and effective coalitions. These relationships are deepened by food sharing and face-to-face sex—rubbing together of their large clitorises to what appears to be orgasm. Males cannot dominate females as they do in chimpanzees because females band together too effectively. Ethologist Frans de Waal sees peaceful bonobos as a better model than chimps for what our ancestors were really like. But, as Richard Wrangham and Dale Peterson argue in *Demonic Males*, male bonobos seem prepared to behave as badly as chimp males should the lid of female power be removed—which is basically what happened in human evolution. And in any case, to the extent that species selection is a factor, chimps have done far better than bonobos, who are very close to extinction.

But of course—ancestor-models notwithstanding—the best way to find out what humans are like is to study human behavior. This has been done by Martin Daly and Margo Wilson, who use Darwinian theory to generate novel hypotheses. For example, deadly fights may occur between men in inner cities because of an imagined slur or even the wrong kind of eye contact. How could so much be risked for so little? Daly and Wilson show that it is not little at all; it is something like honor, or respect, or status, and it is worth a great deal. Using computer simulations, they assigned various values to winning and losing in terms of reproductive success (in part a consequence of status), as well as to the risks of fighting (being killed, for instance). Within a wide range of reasonable assumptions, the tendency to have such fights "over nothing" are favored by simulated natural selection.

More dramatically, using real crime statistics from the United States, Canada, and England, Daly and Wilson tested the hypothesis that children in a household with their mother and an unrelated male would be at greater risk for abuse than those living with their mother and genetic father. As we saw earlier, not only was this hypothesis confirmed, but the difference was vast. Recall that a child living with a stepparent is about 10 to 100 times more likely to be killed than one living with genetic parents only. This exceptionally strong finding applies to all three countries and is independent of socioeconomic status. It is one of the most illuminating findings in the history of child-abuse research, and it came directly and exclusively from evolutionary psychology. In essence, it is the human version of competitive infanticide.

Male violence against females also has a long and complex evolutionary history, often in the service of sexual coercion, as shown in the 1990s by Barbara Smuts, a leading authority on monkeys and apes in the wild. Far from justifying male violence, she has prepared the groundwork for understanding human parallels. For example, Daly and Wilson found that women are at highest risk for being killed by their husbands or boyfriends when they are trying to leave these men, but the reverse is not true. The motivation seems to be, *If I can't have her, no one can*, and it is consistent with sociobiological claims about the proprietary nature of male sexual jealousy. The death of Nicole Simpson and at least one of the other two homicides discussed at the start of the chapter appear to be in this category. Male jealousy is also a main pretext for homicide in a wide range of hunting-and-gathering societies, including the !Kung, Eskimo, Mbuti, Hadza, and others.

What about the wider cross-cultural evidence? The easy way to suggest innate aggressive tendencies in humans is to describe the most violent of human societies: the Yanomamo of highland Venezuela, the Dani or Enga of highland New Guinea, the equestrian Plains Indians of the United States, the Aztecs, the Mongols, the Zulu of South Africa, the Germans of the Third Reich. Among the traditional Enga, studied by Mervyn Meggitt, 25 percent of adult male deaths were due to violence and social life was largely organized around it. The traditional Yanomamo, called "the fierce people" by Napoleon Chagnon as well as themselves, were comparably violent. Forty percent of Yanomamo men had killed at least one other man, and those who had killed had higher reproductive success than those who had not.

Such descriptions of the most violent societies can be multiplied at length, and easily give the impression that we humans are a malevolent species, composed of dysfunctional or "sick societies," as Robert Edgerton has called them. Irenäus Eibl-Eibesfeldt, in *The Biology of Peace and War*, reviewed many ethnographic accounts of warfare in primitive societies, including some long alleged to be nonviolent. Archaeological evidence assembled by Lawrence Keeley demolished "the myth of the peaceful savage." The tenacity of this myth is a study in itself, since it required an astonishing degree of blindness to hard evidence, in accounts that Keeley has aptly called "interpretive pacifications." The archaeological record, equivocal for prehuman species, leaves no doubt that homicidal violence was part of life from the time our own species emerged. With the Neolithic revolution and the spread of agriculture, evidence of warfare becomes decisive. Indeed, the whole of human history since the hunting-gathering era can be largely understood as a process of relentless, expansionist tribal warfare.

But cataloguing violent societies is easy. The greater challenge is to look at the *least* violent ones. Differences in the degree of violence among cultures are real and large, and understanding those differences should help us reduce violence. But what should we make of the claim that there are truly nonviolent societies? The !Kung San of Botswana are cited as a common example of the least violent end of the human cultural spectrum. Like most hunter-gatherers, they have not had war or other organized group conflicts in recent times. Nevertheless, as we saw at the beginning of this book, that is far from nonviolence—their homicide rate matches or exceeds that for American cities, and there are many nonlethal acts of violence as well. Moreover, their explicit contempt for non–San people, and even for !Kung in other villages who are not their relatives, makes it clear that if they had the technological opportunity and the ecological impetus to make war, they would be capable of the requisite emotions. Historical data and rock paintings show that they have made war in the past.

The traditional Semai, slash-and-burn gardeners of Malaysia studied by anthropologist Robert Knox Dentan, were a society almost as technologically basic as the !Kung. Violence was said to be virtually nonexistent and abhorrent to them. As Dentan wrote in 1968:

> Since a census of the Semai was first taken in 1956, not one instance of murder, attempted murder, or maiming has come to the attention of either government or hospital authorities. People do not often hit their children and almost never administer the kind of beating that is routine in some sectors of Euro-American society. A person should never hit a child because, people say, "How would you feel if it died?" . . . Similarly, one adult should never hit another because, they say, "Suppose he hit you back?" . . .
>
> It should be clear at this point that the Semai are not great warriors. As long as they have been known to the outside world, they have consistently fled rather than [fought], or even than run the risk of fighting. They had never participated in a war or raid until the Communist insurgency of the early 1950s, when the British raised troops among the Semai, mainly in the west. . . . Many did not realize that soldiers kill people. When I suggested to one Semai recruit that killing was a soldier's job, he laughed at my ignorance and explained, "No, we don't kill people, brother, we just tend weeds and cut grass."

But when the British engaged the Semai in counter-insurgency against Communist rebels in the mid-1950s, they gave evidence enough of violent capability:

> Many people who knew the Semai insisted that such an unwarlike people could never make good soldiers. Interestingly enough, they were wrong. Communist terrorists had killed the kinsmen of some of the Semai counter-insurgency troops. Taken out of their nonviolent society and ordered to kill, they seem to have been swept up in a sort of insanity which they call "blood drunkenness." A typical veteran's story runs like this. "We killed, killed, killed. The Malays would stop and go through people's pockets and take their watches and money. We did not think of watches or money. We only thought of killing. Wah, truly we were drunk with blood." One man even told how he had drunk the blood of a man he had killed.

Strong as this description is, it is less remarkable than what followed:

> Talking about these experiences, the Semai seem, not displeased that they were such good soldiers, but unable to account for their behavior. It is almost as if they had shut the experience in a separate compartment, away from the even routine of their lives. Back in Semai society they seem as gentle and afraid of violence as anyone else. To them their one burst of violence appears to be as remote as something that happened to someone else, in another country. The nonviolent image remains intact.

Despite this bleak reversal of a nonviolent cultural tradition, we know that cultural contexts *can* reduce violence. In a wide-ranging, cross-cultural study, anthropologists John and Beatrice Whiting discovered that when husband-wife intimacy is high, organized group conflicts occur less. Cultures where husbands and wives eat together, sleep together, and share the child care are among the least violent, while those that have organized themselves around constant or at least intermittent warfare tend to segregate men away from women and children. They have separate men's houses for eating and sleeping, and men's societies in which even young boys are severely stressed and actively trained for warfare. The Whitings' study indirectly supported a hypothesis put forward in a different form by Lionel Tiger in his book *Men in Groups*. Something happens when males aggregate; it is not well understood, but it is natural and it is often not quite nice. The Whitings' research gives substance to the old slogan "Make love, not war" and actually improves it, giving it new and interesting depth. It also recalls Kano's description of bonobos, in which the strong influence of females helps suppress most male violence.

But whatever cultural conditioning we do, human beings conditioned to be nonviolent retain the capacity for violence. We are animals, and the tendency to do grievous harm to others is part of our evolutionary legacy. As constrained as that capacity may be in certain contexts, it can come out in others. It is subdued, reduced, dormant, even forgotten, but it is never abolished. It is always there. Think of it as you would a virus or cancer. They are not accidental consequences of failed human tinkering, although they take advantage of such failures. On the contrary, they are ubiquitous natural forces. Recognizing their power is the first step in prevention and treatment. Complacency is deadly. The continued pretense by some social scientists and philosophers that human beings are basically peaceable has so far prevented little of human violence. Perhaps we could let it go for a while and see if another, truer assumption gives us a better understanding; if it does, it will give us a better chance at control.

CHAPTER 10

Fear

Of the world as it exists, one cannot be enough afraid.

— THEODOR ADORNO

Fear is what quickens me . . .

— JAMES WRIGHT

We have dropped into the bowels of the beast, where the snarl curls, poised to provocation. Lust thwarted, love rejected, thirst unslaked, hunger lingering, the frustration that chokes the gorge at practically any motive blocked, the example of creatures one admires, pain at the hands of those one hates, or simply being brutalized through a long, slow course of growth—any of these motivating states can trigger the violence that turns a good life into a nightmare. But there is one cause above all: fear.

Real fear bites deeply into our essence and it lasts. Consider this description by a Vietnam War veteran still disturbed by his memories many years later:

> I can't get the memories out of my mind! The images come flooding back in vivid detail, triggered by the most inconsequential things, like a door slamming or the smell of stir-fried pork. Last night, I went to bed, was having a good sleep for a change. Then in the early morning a storm-front passed through and there was a bolt of crackling thunder. I awoke instantly, frozen in fear. I am right back in Viet Nam, in the middle of the monsoon season at my guard post. I am sure I'll get hit in the next volley and convinced I will die. My hands are freezing, yet sweat pours from my entire body. I feel each hair on the back of my neck standing on end. I can't catch my breath and my heart is pounding. I smell a damp sulfur smell. Suddenly I see what's left of my buddy Troy, his head on a bamboo platter, sent back to our camp by the Viet Cong. Propaganda messages are stuffed between his clenched teeth. The next bolt of lightning and clap of thunder makes me jump so much that I fall to the floor. . . .

The responses are autonomic, biological, enduring, and in evolution their root is escape from pain. Nociception—the sense of pain—is the first function of nervous systems and the flinch from a noxious stimulus is within the ken of creatures with one nerve cell. This reflex is not for eating, drinking, reproducing, or finding prey. Rather, it leads the organism to withdraw, to flee, or in some situations to flinch and freeze, evading or baffling an enemy. It is said that every creature is in competition for survival with every other creature in its environment; if this is so, this creature has much to fear from those others—from each of them, however close, however similar, however allied they may be. Creatures that threaten run the gamut from the leopard at your hut door to the virus in your blood, from the man in your bed to the infant at your breast. And the threats do not end with living creatures. Light, darkness, heat, cold, heights, depths, rock falls, water, storms, earthquakes, wind, mere time spent without food or drink, mere distance from the things we know—any of these may spell an abrupt end to life or to reproductive success, which, for natural selection, amounts to the same thing. We are therefore amply equipped to detect threats and rebuff them or withdraw. Pain, stress, and fear, though often combined, are subjectively quite distinct perceptions and all serve notice of a possible threat to survival or reproduction. Pain and fear suggest an immediate threat, stress a slower-acting one.

Yet paradoxically, all three signals also have to be distrusted. Threats are so ubiquitous that if every detection produced a full-blown instinctive response, animals would spend their lives in flight or in hiding, unable to eat, drink, play, sleep, mate, or care for their young—any of which may involve pain, stress, and fear. Thus, we may watch a herd of caribou feeding calmly on a wolf-dotted stretch of tundra; a zebra nuzzling a foal just beyond striking distance of a pride of lions; a human mother cooing at a contagious, dying child. Fear may hold sway, but it can be overridden by a dozen other motives.

What are we to suppose, then, that the caribou is feeling? A clench of fear that persists but is transcended? A transient fear when the wolf appears on the horizon, which subsides and is followed by calm? A continuous mild fear below the surface of action, quickly intensified by certain of the wolf's movements? And what tells the human hunter to persist in pursuit of the caribou, in spite of the pain in his legs and chest? What persuades the young woman whose hymen is breaking, prefiguring a later, greater pain, to stay instead of flee? What makes the human animal (and other, less prescient ones) bear and transcend some pains but run from others?

To understand these balances struck among pain, fear, stress, and other motives and feelings demands an account of the actions that may ultimately resolve them: escape and flight. These we can measure; and they form an outward expression of the subjective feelings in question. Following the ethologists we can characterize the sequences of escape and flight in animals. Of course, caribou escape by running, trumpeter swans by flying, porpoises by swimming—not very interesting. However, other details do engage our attention: for example, certain stimuli reliably produce escape and flight, and many of these are species-specific.

In the caribou, escape awaits the moment when its flight distance—the distance from which it can still be confident of escaping should pursuit begin—is violated by the wolf. But the distance is not fixed. Consider an ill, aged, or disabled caribou grazing with its herd near wolves that are outside the flight distance of healthy herd members but well within its own. It can stay with the herd or detach itself and move farther away, but either way it is vulnerable.

Despite the varied behavior, its anatomical basis—from the recognition of patterns through simple or complex neural networks to the release of hormones or actions—is largely set by genetic plan, resulting from thousands of generations of selection. The behavior *and* the chance to change it through conditioning is shared in some form not only by all vertebrates—trout, newt, gator, pigeon, mouse, and human—but even by fruit flies and sea slugs. This means the instinct is over 500,000,000 years old. The action itself—surprise followed by flight or holding still—is a coordinated, delicately timed firing sequence of billions of nerve and muscle cells. An array of hormones that energize and regulate those cells is brought into play in a matter of seconds. What stretch of the imagination could lead us to suppose that such a system could be built up through experience in the course of a single lifetime? On the contrary, such sequences—fixed action patterns, with their companion releasing mechanisms—are laid down long in advance. True, the word *fixed* makes them seem more rigid than they are. The original German word for these behaviors was *erbkoordination*—"legacy coordination." The term implies a phylogenetic descent and a strong genetic component, but not fixed or rigid behavior. Thus, the patterns in question—fleeing, freezing, fighting, and other instinctive behaviors—can have substantial amounts of preplanned leeway. But not only are their basic form and function preset, they have been so since at least the dawn of the species.

Although such classic responses as fight and flight are primary, there are many other, less dramatic responses. One contribution of the classical ethologists was to identify what they called *intention movements*—a form of communication. When members of different species confront each other, they may ignore each other, attack, or flee. Either way, often they don't want to signal their true intentions; deception is ubiquitous in nature. But when two members of the same species confront each other, they may have to maintain proximity—to court, stand together against an enemy, or share the resources of a territory. How can two creatures capable of inflicting grave damage on each other stay together and still avert disaster? Put starkly, if we are always in danger from one another, why do we do anything other than flee?

One answer lies in the evolution of appeasement signals—communications that let the other know (or make the other think) that no harm is intended. Often these signals are easily understood as simultaneous expressions of tendencies to attack and flee—the intention movements of the classical ethologists. However, during evolution, the movements may become emancipated from the motives that initially gave rise to them, at which point they are said to be *ritualized*. This is emphatically not the "ritual" of a human culture, which resonates with countless learned echoes of meaning. Rather it is a genetically coded, stereotyped sequence. Evolution, not culture, has built the ritual.

For example, the zigzag mating dance of the male, three-spined stickleback, a little fish that populates the ponds of Europe, is essentially an attack intention alternating with a flight intention. The displays of some ducks combine wing movements that initiate flight with head movements and body postures related to attack. And dogs, wolves, or foxes face each other, bare their teeth, and lay their ears back flat—combining a potential threat with a protective, evasive move. These combinations, whether alternating or simultaneous, seem to be saying, "I fear you, but not enough to run from you; if you insist on fighting, I am ready." Obviously, these signals influence motivation in the other animal, as indicated by the frequency of subsequent attack or flight. If both animals give such displays, the chance of damage is lessened, and continued proximity is possible—along with all that proximity enables.

However, there are complications. First, not merely attack and flight but approach, feeding, courtship, sex, and many other motives may be aroused at the same time and combined in displays. Purity of motive is a psychological fiction. Most things we do for more than one reason, even while resisting several reasons *against* doing them. Second, each combination is a continuum. For example, in the attack-approach-flight triad, different degrees of each tendency or motive produce different displays. Even though the animal stands still, a predominant attack tendency may evoke a threat display, an approach tendency, a courtship display, a flight tendency, or an appeasement or submission display.

But how do we know what a display means? In each case the display is named not through empathic inference ("Gee, that goose seems angry.") but by considering the behavior that precedes and follows it. When an ostrich puffs itself up to its full height, fluffs its feathers, raises its tail, opens its beak, and snorts loudly in two-syllable calls, we label it a threat because other ostriches back away from it, giving up hotly contested food items, mates, and other scarce resources. When the other ostrich responds to the threat by dropping its tail, closing its bill, and lowering its neck and head in a U-shape toward the ground, we call that submission or surrender, not just because it looks weak but because it tends to prevent attacks and is associated with ceding territory or other valued resources. Sometimes ambiguity is built in. In the squirrel-monkey genital display discussed in chapter 7, genital exposure combines with facial expressions related to attack and to calls that express ambiguous arousal. It is thus not surprising that this display can be either a sexual signal or a threat. Its function depends not only on the display elements but on the age, sex, and size of the animal.

Because of this continuum, two individuals, each with some tendency toward attack, approach, and flight, will often begin or end an encounter with different displays. If most encounters between the two end the same way, we speak of a dominance relationship: A is dominant over B. If B should later challenge A with an escalation of threats, there may be a fight that confirms or alters the direction of dominance. Contrary to past belief, such ritualized displays do not always prevent violence, injury, and death. But when they do, important resources get distributed efficiently according to established dominance relations: territory, food, and access to the opposite sex. A dominance relationship is stable *only* as long as it is in the interest of both parties to leave it as it is. A male at the peak of his form and a brash young challenger soon enough become, respectively, an aging has-been and a male at the peak of his form, whereupon the direction of dominance changes.

Because dominance strongly influences reproductive success, it posed a classic question for evolutionary theory. The losing males are not getting access to females, so how are appeasement and submissive behaviors preserved and transmitted by natural selection? The initial answer depended on group selection. In this view, not just kindreds but larger social groups can be units of selection, so that characteristics beneficial to a group are preserved despite the disadvantage to the individuals that have them. But group selection is a superfluous construct, at least in relation to dominance and submission. As pointed out by the English ecologist David Lack, the alpha male can never get *all* the sex. DNA paternity testing shows that in many primates the dominant male is not getting quite the exclusive access to females he is aiming for, since females choose beta males on the sly. So appeasing, submitting, and other fearful movements and displays can be transmitted to future generations by the beta. Moreover, even the most dominant animal will find use for appeasement at some time or other, especially early and late in the life cycle. So there is no theoretical need to invoke benefits to the group, such as efficient spacing of individuals or general reduction of conflict, because there is evident individual advantage to the less strong: appease, submit, and live to reproduce another day.

Fear has an unmistakable look. The mouth is half open, the corners strongly retracted, the teeth mostly covered, the ears drawn back, the eyes wide open, and the brows raised and perhaps furrowed. Ever since Darwin's *Expression of the Emotions*—later extended by Austrian ethologist Irenäus Eibl-Eibesfeldt and American psychologist Paul Ekman—we have known that the facial expressions of people shown in frightening situations are quite consistent cross-culturally. Furthermore, photographs of those same facial expressions are labeled as fearful by people in widely disparate cultures. The point-by-point similarity to the facial displays of fear in other mammals is also unmistakable and it is likely that the phylogeny, physiology, and genetics in the human case generally follow the pattern for mammals. Teeth baring and brow furrowing reflect the mixing in of attack tendency, as is the case with other mammals.

Harder to categorize, however, and at first glance anomalous, is the main human appeasement display: the smile. Yet, as shown by the elegant research of Jan van Hooff and Signe Preuschoft, there is a clear phylogenetic continuum from the monkey's submissive grin to the human smile flashed timidly at a superior. That we often smile without feeling fear does not invalidate the hypothesis of a common structure, function, and evolutionary origin. In many species, appeasement displays have become emancipated through evolution from the emotions that once caused them—they have become, in the ethological sense, *ritualized*.

Of course, our smiles and especially our laughter may stem in part from the primate play-face, an open-mouthed grin occurring in young monkeys and apes as they tease and chase one another. On the other hand, the nervous smile we reflexively offer when passing the boss in the hall is parallel in every respect, including physiology, to the grin of an average baboon when passing the alpha male in the savanna equivalent of the workplace corridor.

And of course, the most dramatic continuities with our evolutionary past occur when prey meets predator. We have already encountered the visible burrow system, in which Caroline and Robert Blanchard studied dominance interactions and stress hormones in rats. They have also used the system to watch what happens when mice meet rats, rats meet cats, and cats meet humans. In each case, the vulnerable animal shows a continuum of responses, depending on how close the threat is. When the predator is right upon it, the prey shows defensive attack; when it is very close, freezing; when it is fairly close, flight; and when the prey is merely in a place where the predator might be, it exhibits an alert, aroused wariness that might correspond to anxiety.

So it is clear that our outward displays of fear are on a continuum with those of our mammal ancestors. But what goes on *inside* an animal during fearful displays? What neural and hormonal activity explains the observable behavior and the subjectively felt emotion? At least since Walter B. Cannon's work in the early twentieth century, we have known that the sympathetic nervous system is aroused during fear and flight as well as during rage and attack. This nerve net, balanced by the braking power of the parasympathetic system, spurs the increase in heart rate, rise in blood pressure, increased flow of blood to the muscles, and decreased circulation to the viscera that accompany fear and flight in many animals. The shifting balance also causes the reflexive emptying of bladder and bowel that helps to prepare an animal for flight and may humiliate a man on the verge of a battle he cannot flee.

The subjective experience is vividly captured by psychologist Michael Davis, who suggests how you may feel if confronted by a stranger holding a gun to your face: "Your hands will sweat, your heart will pound, and your mouth will feel very dry. You will begin to tremble and feel like you cannot catch your breath. You may feel the hair standing up on the back of your neck and your mind will race, trying to decide whether to stand still, to run, or to try to take the gun out of the assailant's hand. Your senses of smell, sight, and hearing will increase and your pupils will dilate. . . . Very similar reactions can be seen in animals." These are the basic physiological responses of full-blown, specific, realistic fear.

A more subtle yet specific reaction recognized by Darwin as part of the fear spectrum was blushing, which he called "the most peculiar and most human of all expressions." It is a response to being shamed, or to being seen or found out in an embarrassing situation, but it may be caused by any unwanted attention from others. Recent experiments link the response to the eyes of the onlooker; if an audience wears dark glasses, the speaker is less likely to blush. Audience size also matters. Vertebrate brains have an ancient ability to become aware of being looked at and people can judge this at a distance with remarkable accuracy. The importance for sensing danger is obvious. Blushing is hard to justify as primarily a signal, since it occurs in dark- and light-skinned people alike. Indeed, we detect embarrassment in others from shifting gaze and other facial cues more quickly than the fifteen or twenty seconds it takes to blush. Its main function is probably a signal to ourselves, to change our behavior as we become aware of risk. This would fit Damasio's somatic marker hypothesis, in which sensation from the body is intimately engrained in the central circuits of emotion.

Today we understand blushing as an activation of the sympathetic nervous system dilating the arteries of the face, neck, and ears. Unprecedented new understanding of the biology has come from a few thousand cases of pathological blushing, which is no joke to the sufferers. One young woman, Christine Drury, was described in a moving case report by Atul Gawande. The rising career of an NBC news anchorwoman working out of Indianapolis was stopped dead in 1997 because of her severe and growing inability to avoid blushing on camera. She tried everything from makeup to desensitization, from autosuggestion to drug therapy. Then she found out about a surgical procedure done in Göteborg, Sweden, to help pathological blushers and others with related medical problems. The Göteborg group, led by Christer Drott, had discovered the effect while operating for other reasons, such as angina, rapid heart rate, and excessive sweating.

The range of effects discloses the anatomy: the surgeons, using a minimally invasive tool called a laparoscope, enter the chest through a small incision and cut the sympathetic nerve trunks on both sides, at the level of the second and third ribs. The result? An end to pathological blushing, with minimal complications in the great majority of cases. The day after Drury's surgery, a handsome nurse took her blood pressure. Before, she said, she would have blushed intensely at this strange man's touch. She didn't blush at all. She felt "as if a mask had been removed." This language, intriguingly, suggests that after a part of her sympathetic nervous system was cut away, she felt less disguised, more herself. She returned to the station and began delivering news reports with a degree of professional calm that had been impossible before. To one observer, "She was, in fact, more natural than she had ever been."

The red mask was gone; her career thrived. People noticed a change in her but couldn't put their finger on it—and she kept it a secret. But a few months later she told a close friend the truth and he was horrified. She felt false, became self-conscious, lost her bearing, did poorly on the air, became depressed, and resigned—all without blushing. Fortunately this was temporary. She later told many friends and acquaintances the truth and their common reaction was enthusiastic support. This helped her recover and her career got back on track, but the take-home lesson is twofold: sympathetic nervous system surgery cures blushing, but self-consciousness, embarrassment, and humiliation are different and are best cured by sympathetic friends.

So the body displays ancient signals and concomitants of emotion; yet these reactions are not the emotion itself and they do not account for the control of fear and flight, which must be sought in the brain. An early clue came from research stimulating the amygdala—to be exact, the lower and outer portions of that almond-shaped lens of cells, a key switching center of the emotional brain. In cats, such stimulation with a tiny electrical pulse caused alertness and orienting— a bright-eyed looking around and a posture of aware readiness. A stronger pulse of current in the same spot caused all the facial and postural features of fear.

This was fascinating for three reasons. First, it implicated the amygdala as a place to start looking for fear in the brain, a suggestion that has held up beautifully. Second, since the amygdala also figures in rage, it suggests that the link between fear and rage is reflected in brain structure. Finally, it points to a continuum of arousal from alertness to fear, a suggestion with powerful consequences for the study of attention and even intelligence. But the amygdala is no more than a part of a relevant circuit. Recall that in one view of violence the amygdala and the septal area compete for the control of the hypothalamus. When the amygdala rules, violence may be motivated, but when the septal circuit dominates, it may be inhibited.

British psychologist Jeffrey Gray proposed that the septal-hippocampal circuit is the main *general* behavioral inhibition system. That means this circuit can make a frightened animal or person stop activity, survey the environment, and prepare for further action. In Gray's view, we may need the septal-hippocampal system to inhibit ourselves initially and the amygdala to discharge flight or other fearful reactions to things we associate with danger. As we will see, the notion of opposition between these two systems was too simple, but it is a piece of the truth. Inhibition can mean fear in certain contexts, but it can also calm fear in others. The role of the amygdala, though, is more straightforward.

Neuroscientist Joseph LeDoux has called the amygdala "a hub in the wheel of fear," because his research strongly implicates it in the learning of these most basic responses. Michael Davis, who differs from Gray and LeDoux on some points, has said of the amygdala's role: "The whole field now believes that this is the central fear system," and even accords the amygdala a role in unlearned fear. But consider first LeDoux's learning paradigm. When a rat hears a tone followed by a shock, it quickly learns to freeze or flee, but with what brain pathway? The sound of the tone sets a signal in motion in the auditory nerve and as with most sensations the signal goes to the thalamus. From there it goes to the lateral or outer part of the amygdala, then to the central part. This central amygdala generates the main bodily events of fearful reactions. Sever any part of this pathway, and you will block fear learning, probably because the rat cannot experience the emotion. Merely block RNA synthesis in the amygdala and you block the genes' power to code proteins. Any new functions, such as building synapses or manufacturing enzymes and receptors, depend on RNA synthesis and if you turn it off in the amygdala, you get no new fear.

But what if the scary event is more complex than a tone? Then higher centers are needed. Suppose, for example, we return a rat to the scene of the fear—the box where the tone and shock were paired—but don't sound the tone. The normal rat will act frightened just because of memories of the setting. Yet if the hippocampus has been removed, the tone will do it but the context alone will not. We know that the hippocampus functions to integrate sensory information from different parts of the cortex. These signals converge toward the organ and, as they reach it, stimuli that belong together are paired in memory. Howard Eichenbaum and others have shown that this kind of pairing is a key function of the hippocampus in learning.

Other approaches confirm much, but not all, of this picture. Michael Davis and his colleagues use a paradigm called "fear-potentiated startle." Teach a rat to fear a light by pairing it with a shock and then sound a loud noise after the light comes on, leaving the shock out of it. You now have fear-potentiated startle and you can study the impact of lesions and drugs on it. It turns out that this measure of fear produces results a little different from the LeDoux approach of interfering with learned freezing or flight. The amygdala is still key, but the hippocampus plays a less important role. Still, Davis and his colleagues agree that it often plays a role in establishing the context of fear. As we will see, it can be difficult to decide what is an animal model for fear and what for anxiety, but for the moment we will simplify matters by sticking to fear.

So how might a zebra that has once been chased and almost eaten by a lion become afraid of the setting in which the sound of a lion's cough was heard? It needs its hippocampus to match up the sound with the smells at a certain water hole and the light emblematic of dusk. The next time the zebra is at the water hole in the evening, it will get an uneasy feeling that prepares it to run even if the lion can suppress the cough. Now think of an ape, with its high capacity for intersubjectivity; she may glance frequently at the faces of her cousins and perhaps read the cues that they have heard something she hasn't. And if the ape is human, she will be mulling in her mind that annoying lecture her dad always let loose with, about lions at the north-slope water hole. The facial expressions and verbal admonitions are processed by higher cortical centers, converge in the temporal lobe, and are matched with simpler cues as they pass through the hippocampus. Thus, the hippocampus lets the amygdala know there is reason to fear, even in the absence of a single, glaring, unambiguous sign.

Thanks to modern brain imaging, we can test the idea that the human brain works like the lab models. In one study in Jeffrey Gray's lab, functional magnetic resonance imaging—a technique that reflects different states of brain activity— revealed how normal human brains interpret cues for fear. Whether the cues were facial or vocal, the amygdala was activated (cues for disgust, another negative emotion, activated the insula, which monitors the viscera). The role of the amygdala was even more apparent in a study led by Ralph Adolphs, an associate of Albert Damasio. Three patients with complete amygdala damage were asked to judge whether certain facial expressions of unfamiliar people made those people seem approachable and trustworthy. All three patients were more trusting than normal subjects were. The faces people with intact brains rated as least approachable and trustworthy were just the ones patients without amygdalas were most wrong about. Interestingly, they had no trouble judging trustworthiness based on *verbal* descriptions of people. In other words, without an amygdala we are too trusting and bold in approaching people who look dangerous to others.

The amygdala works through the hypothalamus, but an equally important way station is the central gray region of the ancient midbrain. The basic bias of the central gray is to promote the actions (and perhaps the feelings) associated with fear or rage. Most of the time the central, lower part of the hypothalamus *prevents* the midbrain from following that inclination, whereas the amygdala, probably integrating signals from many parts of the brain, sometimes *stops* the hypothalamus from its preventing. (Such double negatives are common in brain circuitry.) From the central gray, signals through the vagus nerve to the heart can depress heart rate when an animal freezes in hyperalertness, or release the heart to beat much faster in fight or flight. As Stephen Porges and his colleagues have shown, mammals have added an adaptation, the "smart vagus." This allows high-level cortical control of the vagal brake on the heart, so that exploration, courtship, and other positive actions can be energized without mobilizing fight or flight.

Undoubtedly, other limbic circuits influence fear. The cingulum, a major limbic fiber bundle that links the cerebral cortex to the hippocampus, has been severed in surgery designed to reduce some people's crippling phobias. Laboratory experiments also link the cingulum to avoidance or escape reactions. The lower frontal cortex and the nearby cingulate cortex help to process fears and anxieties initiated by higher mental processes—a forthcoming exam, say, or an Edgar Allan Poe story. Of course, inhibition due to anxiety about the consequences of our actions is a vital part of our normal adaptation. The cases of Phineas Gage, the brain-damaged railway foreman, and his modern counterpart, Elliot, described in chapter 7, both show how a life can unravel if damage to these cortical areas removes inhibition.

Since both rage and fear activate the body's general stress response, the hormones released are similar. According to Gray's original model, amygdala and hippocampus vie for influence over the hypothalamus, which makes tiny amounts of releasing and inhibiting hormones. These ooze through a small, private group of blood vessels to the pituitary gland, which seems to dangle from the brain by those slim vessels. One of these is corticotropin releasing hormone, or CRH. The pituitary, in turn, makes corticotropin, which courses through the blood to the adrenal cortex and stimulates the stress hormone cortisol.

The other major stress hormone, adrenaline, is produced more directly, by nerve activation, controlled by the brain. In fact, the cells that make it are modified sympathetic neurons, with an extra enzyme that makes adrenaline out of the much less potent norepinephrine. Classic experiments by Stanley Schachter, a psychologist at Columbia University, cast light on the role of this hormone. Schachter gave adrenaline to volunteers in three settings—one calculated to produce anger, one to produce excited happiness, and one a neutral setting not emotionally loaded in any way. Compared to volunteers who got the placebo, subjects receiving adrenaline felt either angrier or more happily excited, depending on the setup. Those in the neutral situation felt only a generalized agitation or even nausea, not any recognizable emotion. This means adrenaline is a common servant of several different states of arousal. On this view, hormones mobilize the energy needed for action but neural circuits determine what that action will be—and what subjective feeling will accompany it. This is consistent with a skeptical attitude toward the old James–Lange theory, which located emotional specificity in the body more than in the brain.

There is no good candidate for a specific fear-*enhancing* hormone, but there is evidence for a more general process that may be mobilized in fear. CRH, produced by the hypothalamus under stimulation by the amygdala, goes both downward (to activate the pituitary and adrenal cortex) and upward, further into the brain. Neurons from the amygdala to the hypothalamus use CRH to stimulate fear; and experiments infusing CRH into the ventricles of the brain induce fear or anxiety in rats. So it becomes very significant that early experience with different forms of maternal care in rats changes the amount of CRH available. As we will see, such experiments have become the basis of a bold new theory of post–traumatic stress disorder. But early experience aside, in the usual cascade CRH stimulates the pituitary to release ACTH, the hormone that goes to the adrenal cortex to provoke the release of cortisol. However, it also turns out that microinjections of ACTH into the midbrain's central gray will trigger vigorous flight. And there is evidence that another stress hormone, adrenaline, enhances fear conditioning in rats, even when they are under anesthesia. Reasoning from the Schachter experiments, adrenaline may create a kind of brain arousal even during sleep, and the conditioning LeDoux demonstrated in the amygdala can go on unconsciously when a noise is paired with a shock.

On the opposite side of the balance, it is likely that enkephalins and endorphins—"the brain's own morphine"—sometimes counter fear by suppressing pain. Testosterone may also suppress fear while enabling aggression. Certainly the association between testosterone and risk taking in both men and women is now firmly established. Testosterone is in any case suppressed *by* fear, as well as by defeat and other forms of stress, so fear and the gonads have something of a competitive relationship. Because testosterone enables aggression in various species when two males are paired for a fight, it reduces fear almost by definition in that situation. But fear returns the favor, suppressing testosterone in males and estrogen in females if it goes on long enough to produce chronic stress. This is nature's way of telling you that you can forget about reproduction when your survival is immediately threatened—not to mention the trouble you may get into competing with the wrong opponent for reproductive opportunities.

A fascinating sidelight on the physiology of fear is the role of lactate in panic. As exercise buffs know, muscles release lactate during extended exercise. It's not a hormone, just a by-product of energy use. But it lingers for a time after intense exercise and psychiatrists noticed that this sometimes triggered panic attacks in susceptible patients. Many studies later showed that you could do the same by injecting lactate. During our evolution, lengthy running or fighting might have caused panic in this way, which could have helped or hurt, depending on the circumstances. Efforts to locate a brain effect of lactate have not succeeded and other evidence suggests that it triggers panic indirectly, by increasing the breathing rate. We would then have an instance of the James–Lange theory: I am breathing fast because I am running, and my own rapid breathing makes me afraid. Like other body events outside the brain, this could enhance rather than cause the basic experience of emotion.

But experience is, of course, not the only factor in fear. In mice, rats, dogs, monkeys, humans, and other mammals it is clear that the fearfulness in an individual, a strain, or a breed is in important ways a function of genes. In rats there is a remarkable genetic coherence among factors, including greater fear conditioning, faster escape learning, and larger adrenal glands. In humans, twin studies show that the tendency to fears and phobias is moderately heritable, with experience playing an important role.

We have already looked at the inheritance of timidity in dogs and at Jerome Kagan's path-breaking work on the same trait in children. Timidity is a highly stable feature of temperament in some children, associated with a larger change in heart rate and with other physiological differences. It can be predicted from high reactivity in a four-month-old, and it in turn predicts shyness and hesitancy in teens. Kagan, Nathan Fox, and others have logically viewed the amygdala as the brain center that might be structured or set differently in these children. It responds to higher brain centers and in turn affects hormonal and other bodily factors—including heart rate, blood pressure, cortisol, norepinephrine activity, pupil dilation, and facial skin temperature.

Robert Adamec's research on timid house cats supports this concept directly. About one in seven cats is naturally timid, as with Kagan's timid children. The timid cats have very excitable amygdalas but other brain areas are average. Also, the *growth* of timidity in the scaredy cats parallels amygdala development, just as it does in human children. As for higher-level control of fear, Nathan Fox, Richard Davidson, and colleagues showed that, in timid toddlers, the right frontal lobe, which signals negative emotion, is more dominant than the left. This fits the amygdala model, since the prefrontal cortex sends messages to the amygdala through several routes. And a recent study in humans showed a reciprocal association between glucose metabolism in the amygdala and in the prefrontal cortex. Summarizing these and other findings, Davidson reasons that "the prefrontal cortex plays an important role in modulating activity in the amygdala," with the right and left sides taking on different, balancing roles.

But while few timid children become bold as they grow older, the timid category does expand with age because, as we might expect, some kids are made timid by life. This fact is compatible with twin studies, which provide an ample field for experience to modify genetic tendencies. Such studies show that about half the variation in timidity in children and adults is genetically determined; both genes and environment are powerful. In other words, if you had a printout of the genomes of all your friends and acquaintances and knew a lot more than we know

now about which genes influence fear, you still would have only partly explained the timidity and boldness those friends display in everyday life.

Of course, we can't do experiments on children to determine what experiences do balance or counteract their genetic tendencies, but rodents have been studied with more stringent controls. Rats, for instance, are admitted into a situation known as the "open field"—simply a large, shallow, unfamiliar, brightly lit box—and allowed to behave as they will. Since rats love dark, secret places, this is not a happy occasion for them and many are scared to the point of physiological embarrassment. Fear in the open field is measured by observing exploratory activity (the less exploration, the more fear) and defecation during the test, a simple indicator of autonomic nervous system changes before flight.

Beginning with rats that have identical scores on these measures, then selectively breeding the ones that are either more or less reactive in each generation, investigators produced two distinct genetic strains, one of which was *10 times* more reactive than the other, in approximately ten generations. Relaxation of selection after fifteen generations leaves the strains stably far apart for at least the next five generations. Cross-fostering (switching mothers) at birth has no effect, showing that maternal behavior does not explain the differences. Mating reactive, fearful males with nonreactive females produces offspring with middling fearfulness, as does the reverse mating of fearful females with fearless males. Because the offspring from these two converse types of mating are genetically similar, strong intrauterine effects are ruled out. Although other investigators have since proven that maternal behavior can influence timidity, it was not a factor in these classic experiments, where genes alone could account for the results.

In mice, the selection experiments have gone farther, breeding a thirtyfold difference in open-field activity in thirty generations. Two lines separately selected for high activity (low fear) had almost superimposable graphs for activity over the thirty generations, as did two lines separately selected for low activity (freezing, or timidity). A third pair of strains were unselected and randomly mated within each strain to serve as controls. They did not increase or decrease activity but fluctuated around a middling level. Remarkably, by the seventh or eighth generation, there was *no overlap* between these two middling strains and any of the four extreme strains. Behavioral scientists are usually comfortable with messy data, where the experimental groups grade into each other, producing overlapping results. They go on to use sophisticated statistical tests to show that the results are reliable in spite of the overlap. But in this case the first eight generations of separate breeding eliminated the blur between the groups. They had simply become, through selection, completely distinct.

As with research on aggression, the study of fear was revolutionized in the 1990s by the advent of gene technology, applied to both animals and humans. Models of fear based on gene-knockout techniques don't show a clear pattern, but some of them are consistent with prior knowledge. For example, we know that the role of the hippocampus in learning depends on a mechanism called long-term potentiation (LTP), a tendency of hippocampal neurons to be more likely to fire for quite a while after a long pulse of stimulation. LTP is linked to some forms of learning. So it is quite impressive to find that knockout mice with missing enzymes and weak LTP also have impaired fear conditioning. Even neater, the knockout mouse is fine where cue learning is concerned ("That lion's cough means run for it!") but impaired in context learning ("There's something worrisome about this water hole in the evening."). This particular gene deletion blunts the impact of context because the weakened LTP is in the *hippocampus*, which stores and patterns numerous environmental features, not the amygdala, which conditions individual stimuli to internal states of fear.

Another gene makes a kind of enzyme known as a second messenger—a molecule inside the membrane of a cell that reacts to outside stimulation, setting in motion a cascade of internal changes. A first messenger opens the nerve-cell membrane to calcium flow, which in turn triggers the second messenger. The resulting internal cascade changes the cell in ways that promote fear learning. But if you change the second messenger gene so that it no longer needs calcium to trigger it—instead, it is always switched on—the mouse can't discriminate between safe and scary situations. The result is a fearless mouse. In all tests, whether the conditioning is cued or contextual (open-field exploration, say, or even the approach of a dangerous intruder after being attacked), the mouse with the calcium trigger already tripped is a mouse stripped of its fears—for better or for worse. Furthermore, the switched-on calcium messenger gene is expressed in the amygdala, just as it should be, since we know that amygdala damage impairs all forms of fear learning. Last and most elegant, suppressing the altered gene brings fear back into the mouse's heart. Thus, the messenger gene makes a reversible change in brain chemistry, not a permanent change in structure.

In addition to these genetic alterations of fear learning, both knockout and "knock-in" gene models have been used to change fear by changing CRH, the hypothalamic hormone that rules the pituitary-adrenal axis. Since CRH is a peptide, genes make it directly, without intervening steps. Insert a gene that makes CRH in overdrive—a knock-in mode—and you get increased fear. Take out the gene that makes the CRH receptor, so that the brain is deaf to the hormone, and you get a pretty fearless mouse. These elegant results are consistent with all we know about CRH in fear and stress. However, neither these nor any other manipulated genes mentioned are *specific* to fear. They affect it, but they affect other things as well. As gene technology goes forward, this kind of multiple–role playing turns out to be common.

These knock-in and knockout studies start by changing known genes and following their actions up through the system toward fear. But it is also possible to take a top-down approach, in which studies begin with mouse strains known to be genetically scared or brave and then try to pin down which gene or genes make the difference. We've seen how easy it is to select rat or mouse strains that score consistently high or low on fear measures within ten generations. Now gene technology helps us find out why, using complex statistics to analyze complex traits—traits governed by many genes. It turns out that small regions of three chromosomes contain the genes that explain most of the genetic variation in fear. One is near the end of chromosome 1, a second near the centromere (point of attachment) of chromosome 12, and a third in the middle of chromosome 15. The first and third of these have also come up in other, independent studies. Fascinatingly, none of these is the same as the genes identified in the bottom-up approach; for instance, the calcium messenger is on chromosome 18. Fear, like other complex traits, will probably have most of its inherited variation explained by a handful of genes but with a number of other genes that affect it in some cases.

Our own species has not been overlooked. Recall from chapter 5 that in one study, about 10 percent of the genetic variation in novelty seeking was due to the D4 dopamine receptor. Novelty seeking is linked to boldness and contrasts with timidity, so the D4 connection may help explain fear. Furthermore, some of us who suffer from high anxiety may have a shortened version of a particular gene promoter. This means we manufacture the serotonin transporter less efficiently, changing the dynamics of that brain transmitter. This discovery could explain the well-known effectiveness of serotonin-related drugs—Prozac, Zoloft, Paxil, and the like—in anxiety disorders. But researchers have yet to work out how the drugs and the genes, both working through the serotonin system, have their different effects on the inchoate fear we call anxiety.

In the more statistical sort of gene search, anxiety and depression are highly correlated and a general trait that combines them is highly heritable. Interestingly, this trait is often called "neuroticism," echoing Freud's concept of a constitutionally sensitive person prone to be anxious, melancholic, or both, depending on personal history. And the dual trait coheres in brain chemistry as well as genes, since serotonin-reuptake blockers with very specific action work against both symptoms. Whether neuroticism proves to be related to measures of timidity in childhood remains to be seen. On the other hand, specific studies of anxiety, panic, and phobic disorders do show some heritability, but with a large environmental contribution. And close relatives of people with generalized anxiety disorder have one chance in five of having the same problem, compared with the average person's one-in-twenty risk. Similarly, phobias are more likely in relatives of phobic patients and identical twins are more likely to share them than nonidentical twins are.

Nevertheless, it is equally well proved that experience strongly affects fear, in us and in other animals. In fact, several kinds of simple learning shape the emotion. For example, in habituation, an innate fear is reduced when the frightful thing occurs again and again with no adverse consequences. In classical conditioning, a fear-causing stimulus is paired with a neutral one and the neutral one eventually becomes frightening. Passive avoidance conditioning repeatedly signals that a painful event is on the way, until we learn to exit the scene early. And active avoidance conditioning, in which escape is blocked, teaches the subject to press a lever or perform some other action that turns the shock off, in advance, when the signal appears. The sensory part of this conditioning—learning what to be afraid of—is largely due to neurons in the amygdala.

These learning strategies generate or reduce fear by pairing it with pain, other fears, or other consequences. The withdrawal reflex in response to a pinprick on the foot occurs in the human fetus by fourteen weeks of gestational age and soon after that there is already habituation, the first glimmer of learning. More serious forms of fear learning are postponed until after birth, but learning plays a role in fear throughout the life course. In an infamous experiment, John B. Watson, the controversial founder of behaviorism, gave a toddler named Albert a secure place in science by making him insecure about furry objects. At the outset Albert was easily frightened by the noise made by striking a steel bar; this was the experiment's unconditioned stimulus—the one that worked without training. When the noise was paired repeatedly with a furry object, Albert formed a classical association, soon responding fearfully to the furry thing itself.

This was like Pavlov's famous experiment in which a bell was rung when food was presented and the dogs eventually salivated at the sound of the bell alone. But before Watson's dubious study it was not clear that an emotional response, much more complex than salivation, could be conditioned the same way. Albert also generalized the new fear to furry objects of all kinds, including several he had never seen and, alas, to furry animals. From this Watson argued that not only children's fears but fears and phobias at all ages could be attributed to classical conditioning processes. Yet we don't need to look hard to find problems with Watson's experiment. First, how many children have arrived at their fears through a process of classical conditioning even remotely resembling the one that made Albert frightened of animals? Second, what made Albert fear the clang of the steel bar in the first place?

This latter question is more interesting than it seems because it opens the whole realm of innate fears. In the late 1940s Canadian psychologist Donald Hebb showed that infant chimpanzees with no prior experience of the objects and with no conditioning would show extreme fear when presented with a snake, the death mask of an adult chimpanzee, the cadaver of a chimp fetus, or certain other objects. There was no reason to think that these fears resulted from the usual learning processes, and in a famous paper about them Hebb speculated on a mechanism for their development. He suggested that they were the result of experience—not with the objects themselves but, paradoxically, with everything else. Against the background of knowledge already accumulated by the infant chimps, the new objects were different; they aroused many perceptual schemas or patterns stored in the brain but fitted into none, causing arousal and then fear. The brain was somehow designed to generate fear as the result of such a cognitive mismatch.

Meanwhile, ethologists in Europe were doing their own studies of innate fear. They used an animal model that was both simpler and more natural—the reflexive, fearful crouch of many bird chicks in response to the sight of a hawk flying over the nest. Early ethologists saw it as a classic innate releasing mechanism, a reflexive response, both in action and emotion, to a genetically wired stimulus pattern in the brain. The full response occurred when only a silhouette of a vaguely hawklike shape was passed over a nest. But passing the same shadow over the nest the opposite way—tail first—did not get the response. Evidently the backward movement resembled a goose- or ducklike shape—long neck, short tail—to which the chicks were not preprogrammed to respond.

However, since these early studies were on wild birds, a role for experience could not be ruled out. Experiments by ethologist Wolfgang Schleidt showed that the hawklike movement caused fear not because of an innate template for the hawk's shape but because of discrepancy, as with Hebb's chimpanzees. Discrepancy referred to an image similar enough to past experience to draw the chick's attention, yet different enough to cause a mismatch in the brain. Thus, if the chicks saw the gooselike cutout pass repeatedly over the nest, they crouched in response to the "hawk," as in the wild; but if, from hatching, they were accustomed only to the hawklike movement, then they crouched in response to the goose and let the hawk pass without fear. It was not its "hawkness" that really mattered but rather the rarity of the pattern and its resulting discrepancy.

So even the bird brain had an operating principle like the one Hebb proposed for his infant chimps: violation of a previously established schema gives rise to cognitive mismatch, or discrepancy, causing fear. But there are three qualifications. First, it may be less mysterious and easier to believe that the brain is designed to generate fear from a cognitive mismatch than that it is wired for a specific response to a specific shape. But this design may be no less innate than that of more classic releasing mechanisms, merely less specific. Second, some fears *are* highly specific. A rapidly looming object (or a simulation of one) can cause startled crying in even the youngest human infants. And the edge of a cliff (or an optical illusion that looks like one) will cause fear not only in human infants but in a wide variety of vertebrate young, including those with no experience of heights. It is not at all difficult to imagine a wiring diagram in the brain, constructed by the genes in the early embryo, that would produce fear in response to these simple visual patterns.

But it is the third proviso that is most interesting. If the brain is designed to acquire certain fears with special strength or speed, or if the normal environment in which the infant almost always grows up gives rise to certain fears more or less inevitably, then an equally certain adaptive reaction results without neurological specificity. In other words, if you wire the brain of the turkey chick for reaction to mismatch, you will achieve, in the normal environment of that creature, a guaranteed fear of hawks. And you will have done so without having to specially wire in a hawk schema. Better yet, you will have wired in a general mechanism—reaction to mismatch—that can serve in many other ways. This was the path of least resistance for natural selection.

Both specific and more general wiring may help explain human infants' social fears—fear of separation from a trusted caregiver and fear of strangers, particularly in a strange environment. Systematic study of these fears owes much to infant psychologist Mary Ainsworth, who developed a simple test of infant fear. First, place the mother and infant in a nonthreatening but foreign environment. Second, introduce a friendly, unfamiliar woman who chats with the mother, attempts to play with the infant, and remains when the mother leaves. Finally, have the mother return to the setting after just three minutes of separation. There is subsequently a separation without the stranger, followed by another reunion.

The Strange Situation, as it came to be known, has been used in hundreds of studies with infants throughout the world. Its results vary systematically with age, culture, and mother-infant relationships, yet there are also important cross-cultural constants. As we will see in the chapter on love, some theorists have emphasized the importance of the infant or toddler's behavior at reunion with the mother, using classifications based on that behavior to diagnose relationship quality. Whether or not this is valid, there are simpler results, like the emergence of wariness of strangers after six or eight months of age, that have widespread significance for our understanding of the development of fear.

This test admittedly confounds the two responses by arousing the fear of strangers and the fear of separation simultaneously. But similar results are obtained by testing the two fears separately. And the Strange Situation test has proved itself not only in the study of early relationships but in the psychobiology of the emotions. For example, Megan Gunnar and her colleagues showed that when parents were asked to rate their infants' fearfulness, these ratings predicted the level of the stress hormone cortisol when the same infants were tested in the Strange Situation at eighteen months of age. This prediction was made even more accurate when the home-based fearfulness ratings were combined with assessments of the mother-child relationship. Finally, the Strange Situation is vital to understanding cross-cultural similarities and differences in infants' reactions to strangers and to strange situations, and one does not need to accept all of attachment theory to find this research valuable.

One consistent finding is that in every sample of infants, in widely different social classes and cultures around the world, the percentage who fret, cry, or show other signs of distress when the mother leaves or when a stranger appears rises markedly after six months of age. Some infants do not show fear at all and some even engage the stranger. But the growth change is universal in this sense: before four or five months of age, signs of fear are for practical purposes nonexistent, whereas after seven or eight months of age they are quite common, if not predominant. In professional and working-class families in the United States, in an Israeli kibbutz where mother-infant contact is quite limited, in rural and urban Guatemala, and among the !Kung San of Botswana, who had one of the closest, most intimate, and most indulgent mother-infant relationships ever systematically described, separation protest and stranger fear rose in parallel and at the same age. Cultures differ in the percentage of infants who cry at a given age, in the age at which fear peaks, and in how fast it is outgrown. But the rising part of the curve, between six and fifteen months, is remarkably constant.

It is likely that the change is a type of growth, like the rise of social smiling during the first few months of life. It is also likely that this is due primarily to brain changes, which are in principle no different from those that go on prenatally.

These brain changes are many and varied, but one easily measured indicator is the formation of myelin around nerve fibers, extensively studied by Paul Yakovlev and others. During the period represented by the rising part of the fear curve, human infants go through rapid myelination in all the major limbic system circuits. This transforms function in the fornix, connecting the hippocampus with the hypothalamus; in the mammilothalamic tract, connecting the hypothalamus with the anterior nucleus of the thalamus and running from there to the cerebral cortex; and in the cingulum bundle, connecting the cortex with the hippocampus.

These large cables—the fornix is as thick as the optic nerve, which brings one-third of all external sense impressions into the brain—are the key paths of the Papez circuit, the core of the limbic system and the functional seat of the emotions. The stream of feeling cannot flow smoothly until these circuits are coated in myelin. Also occurring at this age is the myelination of the striatum, which may form part of the basis of fixed action patterns. To the extent that crying at separation is a kind of innate pattern, striatal myelin may play a role in its onset.

But what of the mismatch theory? If such flexibility applies to chicks, surely humans must be at least that malleable. And since human strangers do not fit a perceptual mold as hawks do, the possibility of a specific prewired template seems moot in any case. The mechanism must be a general one and the concept of mismatch is a good candidate; the infant stores the faces of known people and any stranger activates these person schemas but matches none of them. The mismatch circuitry generates fear. As for maturation, what is maturing is the capacity to experience mismatch, and to generate actions and feelings from it. This, too, requires functional neural wiring that must also grow into place.

There is other evidence of the growth of the mismatch capacity. Long before his studies of timid children, Jerome Kagan and his colleagues showed that all human infants are attracted to patterns moderately different from ones they know—neither too similar nor too novel. Yet some forms of intermediate discrepancy—a mask, for example—can cause fear. Similarly, if we were to take Hebb's infant chimps (the ones that feared certain novel stimuli), we would no doubt find less intense or less specific forms of novelty that would elicit only attention, not fear. Chimp infants, like our own, are curious about the new. Although these experiments on human and chimp infants involve perceptual mismatch while the cat amygdala experiments described earlier involved direct brain stimulation, they share a key feature: low-level stimulation produces alertness, arousal, or interest; high-level stimulation produces fear or flight. If we view mismatch as a kind of brain stimulation, then, we begin to get a glimmer of understanding. Yet this correspondence means little unless we get specific about how a mismatch might affect brain function.

We know that the hippocampus—a larger, more ordered structure adjacent to the amygdala—is involved in the process of comparing new perceptions with those stored in memory. The report of a mismatch to the arousal mechanism of the hypothalamus would therefore almost have to involve the hippocampus and its major outgoing cable, the fornix. So the ability of human infants to respond to perceptual mismatch, known to increase as the brain grows during the first year, may in part depend on the myelination of the fornix. The approach of a stranger to the infant at twelve months of age might occasion a rapid "filing through" of faces stored in memory followed by the reporting of a mismatch. At four months of age, with no myelin in the fornix, such a process might be so slow as to be impotent in generating arousal to the point of fear.

Mismatch detection cannot be the whole story, however; we need an emotional output from the mismatch. If Papez and his modern disciples are at all correct, then the myelination of the major fiber tracts of the limbic system during the second half-year of life could result in a quantum advance in the capacity to feel, above and beyond the advances in cognition. The same cingulum bundle that has little or no myelin at four months of age and a great deal at twelve months is cut by some neurosurgeons attempting to treat intractable phobias. The equivalent "lesion" caused by the absence of myelin at four months helps explain that age group's relative fearlessness. More generally, the four-month-old's emotional capacity is small compared with that of the twelve-month-old, even though the latter has none of the subtleties of emotion and expression later made possible by language. In addition, at the same age there are changes in the part of the frontal lobes associated with the limbic system. These developmental changes are primarily maturational and under genetic control.

Yet there are cross-cultural differences: the percentage fearful at any given age, the age of the peak of fear, and the shape of the declining slope vary markedly. Although we do not know how, it is likely that large cultural differences in the infant's social context—including mother-infant closeness and separation, contact with other caregivers, and the infant's experience with strangers—play a role. Studies in rhesus monkeys by Susan Mineka and Michael Cook show that observations of others play a major role in fear learning. Those who watch other monkeys acting frightened of snakes will show similar fears, but a prior chance to watch other monkeys interact playfully with snakes will "immunize" against such fears. Amazingly, this immunization by proxy, from observation alone, works better in preventing fear than does direct play experience with snakes. So even in nonhuman primates there is a robust tendency to adopt the behavior and emotions shown by others—a deep evolutionary basis for the later power of culture.

It is not the only basis. Many studies of laboratory animals have shown that critical early experiences influence later fearfulness. The open-field situation shows strong early-experience effects in rats. Thanks to the life work of comparative psychologist Victor Denenberg and his colleagues, we have a good idea of these effects. A variety of experiences is given to different groups of rats in the first twenty-one days of their lives. These range from what might be called stimulation, such as petting by a handler or being placed for a few minutes in a tin can filled with sawdust, to what would certainly be stress, such as being shaken in the can, given electric shock, or placed on a block of ice. These treatments, as different as they seem to us, have similar effects on rats. All reliably produce faster growth, greater adult body weight and length, longer survival times in tough situations, and less fear in the open field—whether assessed by exploration, which the rats do more, or by defecation, which they do less.

While the physiological mechanism for these effects of early stimulation is not yet understood, we have some important clues. Early on it was shown that rats stimulated or stressed in infancy—even for just a few minutes a day—and placed in the open field as adults had less activity in the adrenal cortex and smaller amounts of corticosterone (the rat equivalent of our cortisol). Other evidence suggested that the whole activity of the pituitary-adrenal axis, the control system from the hypothalamus through the pituitary to the adrenal cortex, is permanently altered by these early interventions. There seemed to be a kind of hormonal exhaustion—during repeated stress or stimulation in infancy, the pituitary-adrenal axis became depleted or played out, making it less responsive in adulthood. But it seemed likely that there were neural as well as endocrine changes, in the hypothalamus or other parts of the brain.

In the 1990s Paul Plotsky, Michael Meaney, and others showed that, consistent with work by Robert Sapolsky in the 1980s and Seymour Levine much earlier, stimulation in infancy may have its effect indirectly, by changing the behavior of the rat's *mother*. When infant rats are handled or otherwise stimulated, their mothers behave differently afterward; they lick and groom their infants at twice the rate they did before the infants were stimulated. Furthermore, there are mothers that naturally lick and groom at this high rate (about a third of all lab-rat mothers); their infants normally grow up with the characteristics of handled infants, but *without* being handled. They are bolder. They explore more in the open field, indicating less fear, and they have more receptors in the amygdala for anxiety-reducing drugs—receptors that they probably use to calm themselves naturally.

In addition, they are physiologically less stress-responsive, releasing smaller amounts of corticosterone. Although this is due to decreased CRH in the hypothalamus, ultimately the cause may be more receptors for the stress hormone in the hippocampus. Fewer receptors make the brain more sensitive to low levels of the hormone in the blood and in turn signal the glands to keep those levels low. This may be the basic brain mechanism changed by early stimulation: mild early stresses to infants cause their mothers to lick and groom them more when they come home. Separation made these ordinary mothers more like the ones who started out special for genetic reasons. In any case, if you are a helpless young rat, having a more licky-and-groomy sort of mom is going to change your brain, regardless of just why she fusses over you. These experiments are leading toward nothing less than a basis in the brain for the effects of early experience, which Meaney has aptly called "early environmental programming of neural systems."

Up to a point the pattern may be adaptive. Rodents that must store food in dispersed hiding places need to explore more readily. Instead of coding this trait in the genes, rats may take advantage of the effect of early experience. Mothers in some environments might have to leave the young behind while finding food. Upon return, they theoretically might lick and groom their pups more, and those pups would grow up to be fearless explorers. However, while this looks like a *healthy* effect of early stress (it does reduce fearfulness), more extreme stresses have precisely the opposite effects, both behaviorally and physiologically. Prolonged maternal separation and physical trauma make the same system *more* stress-responsive. This reverse effect works by decreasing gene expression for the adrenal stress hormone receptors, especially in the hippocampus and the frontal lobes. Fewer receptors in turn mean less negative feedback to the hypothalamus. With the hippocampal brakes off, the hypothalamus proceeds to make more CRH and supercharges the stress system for life.

In fact, a growing body of research is finding biological parallels between these more damaging stress effects in the laboratory and the syndrome of post–traumatic stress disorder in the clinic. And early stresses in other species can have quite maladaptive effects as well. Monkeys reared in isolation, for example, develop a syndrome resembling autism, in which they withdraw from social contact and rock or bite themselves, making few normal forays into their environment. As we've seen, monkeys so deprived in early life have a wide array of physiological differences from mother-reared monkeys. They are more, not less, reactive to stimuli, and their hair-trigger violence may be an overreaction due to fear.

It is no surprise that the extreme deprivation of isolation rearing leads to damaged adults. But less expected was the demonstration, by Robert Hinde and his colleagues at Cambridge University, that in the same species, the rhesus monkey, even short separations have long-lasting effects on exploration and fear. Remove a rhesus infant from its mother twice during the first six months of life, for just six days each time, and it will be more afraid in a strange environment at age two. This was a surprisingly large and long-lasting effect of what seemed a fairly minor intervention. It was viewed as a model for what might happen to a human mother and infant separated for a hospitalization, and it helped to liberalize the parental rooming-in policies of children's hospitals.

In the Hinde study, an interesting distinction was made between removing the infants and removing the mothers. The effects were far greater for monkey infants whose mothers were removed *from them*, suggesting that there are many subtleties in this matter of mother-infant separation. Also, recent studies of rat pups show that a mere twenty-four-hour separation can have permanent effects, increasing the fearfulness and stress-responsiveness of those pups when they reach adulthood. Yet it is well established that short separations from the mother have little effect on the development of normal children, so further research is clearly needed before we understand what these laboratory separations in monkeys or rats imply about human development.

Meanwhile, there is even evidence of prenatal effects. Rat pups' fear in the open field depends on the amount of fear experienced by their mothers during pregnancy. In an ingenious old experiment by psychologist W. R. Thompson, prospective rat mothers, *before* they became pregnant, were trained to avoid an electric shock after a warning signal. During pregnancy, they were placed in the shock apparatus and given the warning signal but no shock. This must have produced fear in the mother that affected the fetus physiologically, because those pups when grown had increased fear in the open field and more rapid avoidance learning.

These animal experiments clearly support three common notions about early experience and fear. First, some stresses in early life, probably mild ones, can decrease later fearfulness; second, other, presumably greater stresses can increase it; third, these early experiences do not necessarily have to occur after birth. The findings offer indirect support for the folk belief, common to New England Yankees and Plains Indians, that letting an infant cry will toughen it up; for the claim of some psychologists that phobias or even neurotic timidity may be traceable to early trauma; and—although this may be stretching things—for the folk belief that a calm pregnancy will produce a calm child. There may be a continuum of effects, with mild to moderate stress having a positive effect and severe stress a detrimental one, but this is probably too simple. Positive stresses are not necessarily mild, nor are negative ones always severe. We need to know much more before we can generalize with confidence. Still, some folk beliefs are basically consistent with what happens under controlled conditions in nonhuman mammals.

Wherever it comes from, fear is normal in our experience, both biologically and psychologically. But given the variation in living things, it is no surprise that abnormal extremes appear. Psychological defense mechanisms shield us from the anxiety that attends inappropriate rage or lust, deadlines, overeating, conscious thoughts about mortality, and a thousand other daily musings that threaten to disrupt our adaptive focus on work, relationships, and play. From such defenses arise many of our rationalizations, jokes, accusations, deceptions, and creative acts. But so complex a system must also often leak badly, and so the anxiety of psychological conflict is far from completely suppressed. Indeed it is always with us.

In the 1990s, Darwinian models of psychiatric symptoms added a new dimension to the analysis of many conditions. Randolph Nesse, a psychiatrist at the University of Michigan, proposed an adaptationist model of our seemingly continual, generalized fear. In his view, dangers were so omnipresent in the environment of our ancestors that natural selection had to make us chronically afraid, poised to react to a rival or predator without notice or thought. The steady, free-floating anxiety of the modern age and even some debilitating phobias may be held over from a time when they were more adaptive. In one experiment guppies confronting a smallmouth bass—its mouth was big enough for the guppies—were classed as naturally timid, average, or bold, according to how frequently they hid, swam away, or eyed the big fish. When put in a tank with the bass, none of the bold group survived, but 15 percent of the average group and 40 percent of the timid group did. This was an abnormal situation, and we can't claim that guppies share the emotions of mammals, but the experiment certainly shows the value of timidity. It seems obvious, but the demonstration is powerful: normal conditions in the wild produce populations of animals all of whom are constantly afraid, and we of course are descended from just such timid creatures.

In their 1998 book, *Darwinian Psychiatry*, Michael McGuire and Alfonso Troisi build on Nesse's work in placing clinically significant fears in an evolutionary framework. They emphasize not raw fears like that of predation but anxiety over the possible loss of social support, status, and sexual opportunity for ourselves or our children—any of which can be almost as serious a threat to reproductive success as a looming predator. Anxiety is thus "a future-oriented emotion," and one that signals us that a change in behavior may be needed to reach a goal. It may have evolved as part of an internal device for redirecting our own reproductive strategies. Up to a point it also elicits sympathy. Such strategic roles for anxiety may suggest a new classification of anxiety disorders based on their proposed evolutionary function. This view complements Freud's theory of anxiety, in which it is also an internal signal but one that flags a conflict between different instinctual goals, as when sexual wishes endanger survival. There is no contradiction here, since in either case anxiety signals the need for a change in strategy, whether conscious or not.

But McGuire and Troisi also recognize the fundamental connection between anxiety or fear and withdrawal or defense. Quoting Susan Mineka, they suggest that we and other primates have "what could be called 'evolutionary memories' that play a role in determining which objects and situations are most likely to become the objects of fears and phobias. It seems likely that these evolutionary memories underlie selective associations in fear conditions. . . . Through natural selection fear may have come to be associated with conservative cognitive biases which, under ordinary circumstances, are more likely to promote the reinforcement, enhancement, or overgeneralization of fear, rather than the forgetting of fear." This view has at least a little in common with Carl Jung's concept of archetypes—wired-in, phylogenetic templates, intensified by individual experience and routinized or ritualized by custom and culture.

Since fear is ubiquitous in our experience, partially present in arousal and even alertness, it is not surprising that it intensifies in some psychiatric disorders. Panic attacks are a particularly unpleasant anxiety disorder affecting perhaps 3 percent of people, although the fact that different drugs work against panic and anxiety suggests that they are fundamentally different. Explicit paranoid fantasies appear in many schizophrenics, and social withdrawal occurs in most. Childhood autism frequently looks fearful. Depression entails protective social withdrawal and in the manic-depressive syndromes may come in part from fears generated in the rush of the manic phase—what University of Pennsylvania psychiatrist Aaron Beck has called "the 'fear reaction' as a check on overly expansive or careless patterns." Among the neuroses, fears give content to phobias, but they may also underlie the strange rituals of obsessive-compulsive disorder. As for anxiety, as much the neurosis of our time as hysteria was in Freud's, its major symptom is a kind of inchoate fear. "Start worrying," reads the telegram in the old joke. "Letter follows."

Fear even pervades our dreams. According to J. Allan Hobson, a leading investigator of dreaming and the brain, anxiety is the most common feeling in dreams, and both it and the strange thoughts accompanying it stem from amygdala activation. A panic-laden dream that plagues college graduates even decades later involves taking an exam without having a clue as to what the course is about. In others the dreamer is on the wrong plane, has lost keys, is unprepared for a speech, misses messages because of a dead beeper or cell phone, and so on—what Hobson calls the "incomplete arrangements" of modern life. These may not be lions at water holes, but a lot can be at stake, and they unsettle us enough to invade our dreams.

If we return briefly to the animal models, it is easy to see that finding anxiety in a rat is more of a challenge than finding fear. Not surprisingly, then, the lab experiments have taken several different directions. The Davis group has tried to find disturbances in rats that have vaguer, less specific cues than outright fear. These have responded to a different brain circuitry, focused on a part of the hypothalamus called the bed nucleus of the stria terminalis, closely related to the amygdala and one of the areas sensitive to CRH.

It is of interest here that both the hippocampus and the amygdala project into the bed nucleus, because Jeffrey Gray's model, developed most recently in collaboration with Neil McNaughton, sees a division of labor between the two structures. Like Davis, they accept LeDoux's account of the amygdala's role in fear. But they see anxiety as very different, depending instead on the septo-hippocampal system, or the hippocampus combined with the septal area. This in their view is an inhibition system, causing normal hesitation in possibly dangerous settings, but triggering pathological anxiety when overactive. Based on drug and lesion studies, they distinguish phobias, panic attacks, and anxiety disorders as three very separate clinical and physiological entities. Perhaps all the animal models do converge on the hippocampus in part as a device for making us uneasy in certain times and places, without knowing exactly why. Phobias have specific triggers and are very much like fears. Panic attacks are extreme, time-limited physiological reactions that feed on themselves as fear turns inward. But anxiety is like a murmur under the surface, telling us something is not quite right, making us anticipate dangers that aren't there.

The role of fear can be traced in many mental illnesses, but consider just two examples. One of Freud's famous cases was the analysis of Hans, a five-year-old with an extreme fear of horses (a serious handicap in *fin-de-siècle* Vienna). Freud analyzed and treated the boy without seeing him, through correspondence with the boy's father, a concerned physician with an interest in psychoanalysis. Though limited, it was the first analysis of a child, so the case was of special interest—it would provide direct access to processes only *believed* to have occurred during the childhoods of adult patients.

Analysis of Hans did not require much delving to turn up thoughts and feelings usually hidden in adults. The boy spontaneously talked about his keen interest in penises, women's lack of them, intimacy with his mother, fears of his father, jealousy of his baby sister over his mother's affection, even his wish to have his mother touch his penis. After his sister's birth he became increasingly anxious. He gradually developed an intense fear of horses that prevented him from going out on the street or thinking about much else. He talked explicitly about horses' large penises, beatings of horses he had seen, and various imagined similarities between his father and horses. Hans's father and Freud began to analyze the boy's behavior, talk, and dreams. The father brought the associations concerning horses, genitals, sex, and the strong emotions of love and rage that Hans felt toward his parents into conscious awareness in conversation. Gradually the boy's fears abated, and both Freud and the parents attributed his improvement to the analysis.

Years later Freud had a visit from Hans at nineteen. He seemed healthy and happy, declaring that he was perfectly well and that he suffered from no troubles or inhibitions. Not only had he come through his puberty undamaged but his emotional life had successfully weathered a significant stress when his parents divorced and each married again. He lived by himself but was on good terms with both his parents and only regretted that the breakup separated him from a younger sister he loved. Remarkable (although not surprising to Freud) was the young man's complete amnesia for the phobia and all the associated emotions and events of his early childhood. When Hans read the case history, he didn't recognize himself. Thus, if he had come to psychotherapy at nineteen, with a therapist unaware of the earlier events, the process would have been to retrieve those lost memories. If the adult Hans in this hypothetical situation had at last remembered the early phobia in accurate detail, it would have provoked the same doubts as do the memories of early childhood sexuality and fear in many other adults' psychotherapy.

In other words, Freud was right to think that a child analysis would be inherently more plausible than an adult one. But we can't be sure that the analysis was correct, or that it had anything to do with ending the phobia. Many childhood phobias are outgrown through maturation, knowledge, and an almost inevitable desensitization through experience. Still, one psychoanalytic hypothesis confirmed by Hans is that emotional preoccupations of our early childhood—even fears strong enough to be traumatic and disabling—can be relegated to some backroom storage space in the brain. There they may be inaccessible not only to ordinary awareness but even to an effortful remembering. In the nineties, debates raged over recovered memories of abuse, which seemed to be more common the more therapists looked for them. Studies show that it is possible to plant through suggestion false memories in many people; but equally clearly, real memories that provoke extreme fear can be repressed or at least forgotten.

Phobias are among the most common mental problems, affecting 5 to 10 percent of people. Animal phobias like Hans's start earliest, often in early childhood, while social phobias, claustrophobia, and others start in middle childhood or even in adolescence. Around three-fourths of children have significant worries and fears, but they are really trouble for less than one-fourth. No one knows what causes most phobias, but there appears to be a biological vulnerability followed by a stressful life event, combined with misinformation and direct or vicarious negative experiences, somehow focusing fear on an inappropriate target. The most successful treatments include desensitization, a behavioral therapy that gradually increases exposure to what is feared, although cognitive therapy and other psychological approaches often help.

There are also drug treatments. Monoamine oxidase inhibitors (MAOIs), an older class of antidepressants, improve two-thirds of phobic patients. Recall that monoamine oxidase is low in thrill seekers, so it stands to reason that inhibiting it might combat some fears. In thrill seeking, monoamine neurotransmitters are removed less efficiently by one form of the enzyme (MAO-A); in the treatment of phobias that same inefficiency is artificially imposed by the drug. Nerves outside the brain have also been targets of treatment. Performance phobias, often experienced by musicians, can be treated with blood pressure drugs called beta-blockers. These work not on the brain but on the sympathetic nervous system, combating a racing heart, sweaty palms, shaky hands, and other fight-or-flight symptoms. Their effectiveness shows that feedback from the body does matter in an emotion like fear, even if the physiology is not unique to that emotion. On the other side, cocaine injections make mice more afraid of rats and rats more afraid of cats, and this strengthening of fear may involve release of norepinephrine, the same neurotransmitter muted by beta-blockers.

So phobia is a disorder of fear intensification, subject to evolutionary analysis, developmental emergence, and behavioral and pharmacological treatment. Consider anxiety again, the great neurosis of our time. Freud's brief book about it, *The Problem of Anxiety*, is an elegant account of his theory of neurosis in which anxiety stems from an instinctual wish that seems incompatible with survival. This would make anxiety a constant fact of life, since appraising and laying aside such wishes is continual. Freud described infants playing peek-a-boo, toying with their fears to gain control of them. He restated his view of Hans's fear of horses but this time generalized it: a deeply felt, intensely frightening fear—that of castration, triggered by sexual wishes—masquerades as a strong but easier-to-accept one. So he saw a continuum between anxiety and phobia, and between phobia and fear: "There is no difference between this anxiety and the reality fear normally manifested by the ego in situations of danger, other than the fact that the content of the former remains unconscious and enters consciousness only in distorted form." There was a continuum from the fear of real external danger through the anxiety raised by potentially dangerous strong wishes to the irrational fear attached to external objects that are not really dangerous.

Current behavioral biology can accept a surprising amount of this. We might question the "irrationality" of fear of large animals, which were certainly something to fear during evolution, but we would still have to explain why Hans experienced more of it than other Viennese children. We balk at giving a central place to fear of castration when—even accepting Hans's fantasies—it is only one of many imaginable dire consequences of failing to control our instinctive wishes. We also now have evidence that fear and anxiety differ, with different physiological correlates. For example, fear decreases pain sensitivity while anxiety increases it. As we have seen, they may even involve different parts of the amygdala. But Freud's view is consistent with the notion that a steady, or at least a steadily recurrent, degree of fear, whether of external threats or of possible dangers from our desires, is a healthy adjunct to survival. And this persistent murmuring will sometimes lead us to recoil from or even attack objects that pose no real danger.

Most of us do not handle this murmuring very much better than Hans did. Drugs like Prozac and Paxil that affect the serotonin system dampen anxiety even while alleviating depression—not surprising given the strong genetic association between these two symptoms. So it is likely that the serotonin system plays a role in anxiety, but another kind of brain chemistry is also crucial. Among the most frequently prescribed drugs in the United States, both before and since the Prozac era, have been tranquilizers like Valium, an "anxiolytic" or anxiety-dissolving compound. Valium (like other benzodiazapines) works by enhancing the release of GABA (gamma-aminobutyric acid), the inhibitory neurotransmitter that controls about a third of all synapses in the cerebral cortex. GABA tends naturally to dampen brain activity, and Valium-like drugs urge it along. These drugs and their actions on the brain have been the subject of many experiments in the laboratories of Gray, Davis, and others. Furthermore, in rats, receptors for both GABA and Valium-like drugs are altered by early experience.

Still a third neurochemical system, the opioid receptors, figure in fear, and elegant research led by Ned Kalin and Steven Shelton neatly separated this third component of fear from the GABA component. In their experiment, human strangers approached infant monkeys and either stared at them—a universal fear provoker in mammals—or made no eye contact. Both conditions frightened the monkeys, but with staring the monkeys tended to coo loudly—a call for their mothers—and bark nervously, while with no eye contact they tended to freeze and crouch. Drugs that affect the opiate system in the brain affected cooing, the separation call. Specifically, morphine reduced cooing, and its antagonist, naloxone, increased it. In contrast, Valium specifically quieted the other fear behaviors— barking, freezing, and crouching—much more strongly. This finding is consistent with the belief that separation protest is a biologically distinct system from general fear.

The suggestion is that separation distress is due to depressed activity in brain opiate systems, but other, more general signs of fear are due to lowered GABA function—which can be treated with Valium-like drugs. Further research by Kalin's group showed that the duration of freezing is related to the stress hormone cortisol, both in infant monkeys and their mothers. Baseline cortisol predicted freezing duration and both are stable over time for a given individual, at least from ages one to three, which for rhesus monkeys would be the equivalent of most of human childhood. Also, the more infants a mother had had, the lower the next one's cortisol, suggesting a possible effect of maternal experience. Finally, both cortisol and fearfulness were correlated with excess activity in the right frontal lobe compared to the left, and this right-brain excess is also stable as the infant grows.

Together with what we already know, this gives us a remarkably broad grasp of what goes on in the brain in relation to fear. The hypothalamus and midbrain generate flight or freezing along with sympathetic nervous system arousal. The amygdala mediates learned cues, and the hippocampus provides context. Neuro-chemical interventions that calm fear include drugs that bind with opiate recep-tors and those that potentiate serotonin or GABA, so all three systems are probably involved in generating fear itself. Mobilizing cortisol may be an ancillary stress response, but cortisol's ultimate controller in the hypothalamus, CRH, may be more intrinsic to the emotion. Finally, in monkeys and humans, the frontal lobes play an important role in emotion, and an excess of right frontal activity is associ-ated with greater fear.

One of the side effects of Valium-like drugs is depressed alertness, not surpris-ing if alertness and fear form a continuum of arousal. Ethanol, the active agent of wine, beer, and spirits, has an enhancing effect on GABA synapses, similar to that of Valium. An estimated 7 to 10 percent of Americans have alcohol abuse prob-lems in a given year, but more than 40 percent are self-described current drinkers, and many of us use alcohol regularly to reduce anxiety or fear. Booze has been a battlefront staple since ancient times. Sadly, we have a burgeoning teenage drink-ing problem, suggesting that the tensions and tremors of the routine departure from childhood now require artificial enhancement of GABA.

Yet distressing as it is, recreational psychic alchemy is not the most threaten-ing consequence of ubiquitous anxiety, or the worst solution to universal fear. What we really should worry about is the strategy of little Hans writ large. At vari-ous times and places in human history, vast numbers of people have allowed their natural fears to be turned outward and focused upon others whom they then vic-timized. Christians, Jews, Blacks, Muslims, Hindus, Sikhs, Tutsis, Kosovars, Tibetans, immigrants, homosexuals, "witches," "Communists," "reactionaries," and "capital-ist roaders"—to name a tiny fraction of the categories of people who, at various moments, were singled out—have become, for millions in the majorities around them, the psychological equivalents of Hans's horses.

Except that Hans of course did not, indeed could not, hurt the horses. Take the same emotional process from Hans's innocent little spirit, isolated in its con-fusion, and paint it onto a canvas depicting multitudes, with the objects of their fears few and weak among them. It is not difficult to imagine a degeneracy of decency such as history has shown us many times. Discouragingly, these episodes are abetted by the most natural of human social fears. Xenophobia draws upon the natural fear of strangers; conformity—the fear of appearing strange—draws on the fear of separation; and obedience to authority, including illegitimate authority, draws on the fear of stepping out of one's place in the dominance hierarchy. Fears that served us adaptively during our evolution ultimately cause reprehensible acts.

Imagine now an alternative situation. Instead of placing the human targets of the irrational fear in a small minority living among the multitude, we place them in a separate but juxtaposed arena and make them a comparable multitude. Now, make the turning outward of fear mutual. Soon the fears will cease to be irra-tional; each multitude, precisely because of its fear, will pose a threat to the other. This is the classic setting of mimesis, which anthropologist René Girard has identi-fied as fundamental to any episode of agonistic violence. Each antagonist mimics the other, justifying his worst fears and provoking yet more threats to be mimicked. But the irrational component can always be depended on to distort the threat upward, thus the positive feedback cycle that leads to war. And war may itself be largely a riot of fear. Consider the following description of Nicholas Rostov in *War and Peace*, which Tolstoy based on memories of his own youthful experience:

> He looked at the approaching Frenchmen, and though but a moment before he had been galloping to get at them and hack them to pieces, their proximity now seemed so awful that he could not believe his eyes. "Who are they? Why are they running? Can they be coming at me? And why? To kill me? *Me* whom every one is so fond of?" He remembered his mother's love for him, and his family's, and his friends', and the enemy's intention to kill him seemed impossible. "But perhaps they may do it!" For more than ten seconds he stood not moving from the spot. . . . He seized his pistol, and instead of firing it, flung it at the Frenchman and ran with all his might toward the bushes. . . . One single sentiment, that of fear for his young and happy life, possessed his whole being. . . .

Two centuries after Napoleon's wars, two fearful and fearsome nations, India and Pakistan, were poised to deliver weapons of mass destruction to each other's vastly populous cities within five minutes of launch. Observers in both countries described a mood of despondent, fatalistic fear mixed with patriotic pride. Intelligence-gathering systems that demonstrably failed in the past will have less than five minutes to decide how to respond to any suspicion of a missile launch. A sound analysis of human behavior must grant that irrational fear jeopardizes human survival much more than irrational rage. Real fear bites deeply into our essence and shapes our history.

Nicholas, at the front, also shows us what is distinctly human in our processing of fear: he talks to himself as he turns and flees. Unlike other animals, when we are afraid we talk to ourselves and to one another, sometimes to calm ourselves but other times, as with Nicholas, to highlight the danger and the magnitude of the loss that looms. Using words, we formulate a new plan of action. Exchanges of words between enemies add a distinctively human dimension to the reconciliations common in higher primates. Whole peoples that live in fear of each other can, even without decisive victory and defeat, use words to ratchet the fear down. This happened between Israel and Egypt in the late twentieth century, and it may yet happen between Israel and the Palestinians and between the warring factions in Northern Ireland in the early twenty-first. Perhaps it can also calm the fears that threaten disaster between India and Pakistan.

Once children can understand, we use words, not just the physical reassurances of infancy, to calm their fears. This was evident in a poignant, climactic moment in the saga of Elian Gonzalez, the Cuban boy who became the focus of family and national rivalries as the millennium turned. Among soldiers guarding against a potentially hostile crowd, the six-year-old and an anonymous female agent were locked in an unwanted embrace. The action may have been needed, but the boy, snatched by armed strangers in the night, had a look of terror on his face. Yet all the while the agent recited from a prepared script a Spanish version of these soothing words: "This may seem very scary right now, but it will soon be better. We're taking you to see your papa. . . . You will not be going back to Cuba. You will not be on a boat. You are around people who care for you. We are going to take care of you." How effective the words were is open to debate, but they were carefully designed to minimize and counteract fear in a situation guaranteed to generate it. They were chosen by child psychiatrists, with the child's past in mind, and they were effective enough so that the boy could be photographed in his father's arms, smiling, a few hours later. Age six is easily old enough for language to transport the mind away from a frightening context to a safer, better place.

The use of words to highlight, interpret, or calm fear was surely one of the great advantages our remote ancestors found in the evolution of language. Marjorie Shostak's *Return to Nisa*, recounts a dramatic incident in 1989 when she and a group of !Kung were on a hunting-and-gathering trip in the Botswana bush. One morning after a night's sleep by the fire, women collecting berries realized with some shock that there were fresh lion tracks near the camp, "something like a dog's, but magnified tenfold." The next night the fear was palpable:

> Throughout the day, lions had dominated most conversations: the tracks found by the women had made us all uncomfortable, evidence that the predators had passed so close to our vulnerable selves, deep in sleep, in the night. In the fast-paced conversations that swirled around my ears, the distinctive sound *n'!hei*, lion, seemed to be on everyone's lips.
>
> Uneasiness was also evident in the fires blazing in camp. Whipped by a steady wind, they were bigger and brighter than those of the previous nights. Logs, heavy limbs, and dead trees were piled nearby, ready reinforcements. And a third fire had been lit, the biggest by far.

The situation dramatically illustrates the role of fire in dealing with dangerous predators. Such predators were larger and more numerous several hundred thousand years ago, when fire was first brought under prehuman control.

But those fast-paced conversations peppered with the word for lion—what use were they? In Shostak's description, the people sitting around the fires retold all the stories of lion encounters they could remember, rethinking with and through these narratives the evasive maneuvers of those who survived and the errors of those who did not. The drama of these true stories, and the emotions they evoked, were strong reminders of the danger they were in, but also of the tactics that might protect them. Yet there was more.

> Now, hours later, the talk went on—unusual for that time of night. Fear moved from group to group like wind in the treetops. . . . Surely, I told myself, they had a plan. Perhaps they would keep vigil all night, taking turns

staying awake, stoking the fires. They were not talking of retreat, so neither would I. My imagination quieted. Their words filled our clearing, strong like protective armor, soothing like a lullaby. I slept.

Hours later I awoke to the sound of one voice: only Toma was still sitting, still talking. . . . A man rose and added wood to one of [the fires], his spear conspicuously in hand. . . .

Eventually even Toma lay down, still talking in low tones. His story, almost a recitation, was about lions. Gau, a younger hunter lying nearby, grunted "Mmm" or "Eh" every few phrases, either from respect or from genuine interest. Then the spaces between Gau's responses lengthened until only his steady breathing was heard. Toma continued, either not noticing or not caring that he was now talking to himself. Then his voice also halted. A word, then another, and he also slept, leaving his story unfinished.

To understand the role of language in fear, look to the lions. The !Kung in this setting had no place to run or hide, only a place to sleep, wait, and talk. Quite possibly, one stage in the evolution of language, permitting at least a simple, soothing exchange of phrases such as a two-year-old might utter, emerged with the mastery of fire. Talk, stay awake, keep others awake, even talk to yourself if necessary, but keep the vigil, touch the spear, stoke the fire, send the human voices out into the night. Let the lions know you are no sleeping target. And above all, perhaps, create the calm that lets you do what you must do: stay put and sleep, however fitfully, through the night.

Is it any wonder that night still frightens us? In the dead of night or in plain day our modern experience of fear is far less explicable and palpable, but it is not less immediate. Consider Mrs. Dalloway, an invention of Virginia Woolf, musing on her comfortable and safe life in London just after the first of the last century's great wars: "She felt very young; at the same time unspeakably aged. She sliced like a knife through everything; at the same time was outside, looking on. She had a perpetual sense, as she watched the taxi cabs, of being out, out, far out to sea and alone; she always had the feeling that it was very, very dangerous to live even one day."

No amount of comfort makes the fear go away. Natural selection has poised us on a knife-edge of uncertainty, destined to tumble luckily into knowing or, often, unluckily, into trembling. We wake and cringe, stir and cringe, eat and cringe, strive and cringe, even cringe in the midst of loving. Unlike other animals, we must look into the face of death knowingly, to find it a maw, insatiable. Aside from that, we are—not metaphorically but precisely, biologically—like the doe nibbling moist grass in the predawn misty light; chewing, nuzzling a dewy fawn, breathing the foggy air, feeling at peace, and suddenly, for no reason, looking about wildly.

CHAPTER 11

Joy

> . . . that they are endowed by their creator with certain inalienable rights;
> that among these are Life, Liberty, and the pursuit of Happiness . . .
>
> —THOMAS JEFFERSON and colleagues, 1776

In the phrasing of the third inalienable right lies the measure of Jefferson's wisdom. He was prescient to guess that happiness might be a major concern for Americans. It had not been an explicit goal for most past societies, which, when not preoccupied with survival, had to muster under such banners as Purity, Destiny, Nobility, Divine Right, or Glory. But if the bounties of the American land permitted this unprecedented obsession, it still seemed wise to insist upon no more than the pursuit.

Pursuit certainly seems to be the word. Ask people who trust you, "Are you happy?" and you don't get a no, you usually get a "Yes, but . . ." The "but" may be conveyed in just a facial expression; it usually means "I'm not sure I know what happiness is," or "I certainly ought to be happy," or "I'm trying hard, I'm almost there," or "Who are you to suggest that I'm not happy?" Of course, there are always the ones who grab you by the shoulders and shout at you about how happy they are: maybe they've just gotten married, or have been born again in God, or they've bought a new set of driving clubs, or gotten very drunk. They don't enter much into our thinking about happiness because they give the impression that they're riding for a fall, a fall one does not want to be in the way of.

As evolved animals, we are endowed by natural selection with the tendency, if not the right, to *pursue* happiness. But how are we endowed with the capacity to know it when we feel it? The previous chapter presents a part of the difficulty. To the extent to which nature is the war of all against all, living creatures must be alert to danger. This state may sometimes be a source of martial elation, but for most of us it's not what we mean by joy. As for the threats from within, the ones posed by our own wants—for water, food, sleep, warmth, or sex—they evoke an unease that hardly brings happiness, at least until those wants are satisfied. Yet both kinds of threats can lead to pleasure. This paradox is not hard to resolve: we must know when the external threat has been eluded or destroyed, or when the internal threat has been curbed. In general, external events cause responses on a continuum from alertness to fear. Internal wants produce a state psychologists call "motivated," ethologists "appetitive," and psychoanalysts "unpleasure." Whether externally or internally generated, these varied internal states share *arousal*, which readily verges on agitation.

Agitated arousal, then. This is not necessarily punitive. Even the psychoanalysts' *unpleasure* does not mean literally that we must dislike the condition. We may like it and seek it in the course of normal life. However, when we enter into unpleasure, whether deliberately or not, our usual response is to try to get ourselves out of it. Life is to some extent a game in which we catch the waves of unpleasure and surf toward pleasure, only to seek or be thrust into the waves again. In terms of neural circuitry, different states of arousal have a good deal in common. That is, the arousal caused by hunger, by the slow approach of danger, or by the needs of one's young may involve very different circuits of input and output, but they overlap a lot in the depths of the brain. By the same token, the resolution of these different forms of arousal—pleasure, which sometimes, mysteriously, becomes joy—involves a common neural circuitry in the limbic system. But before we turn to that circuitry, consider arousal and satisfaction in terms of observable behavior.

Freud's 1920 book, *Beyond the Pleasure Principle*, began:

> In the theory of psychoanalysis we have no hesitation in assuming that the course taken by mental events . . . is invariably set in motion by an unpleasurable tension, and that it takes a direction such that its final outcome coincides with a lowering of that tension—that is, with an avoidance of unpleasure or a production of pleasure.

With characteristic aplomb Freud plunged forward from these words to a major revision of the theory, by adding the "death instinct" as an independent motive. He ranged over many then new discoveries of invertebrate reproductive biology, embryology, and animal behavior and wove them into a sweeping instinct theory.

No modern biologist can follow the leaps and spins of this now outmoded theory; they are intrinsically impossible. Freud perhaps felt that it was time for him to return to the fundamentals of biology after more than twenty years away; but instead of incorporating the intervening discoveries in neuroanatomy and neurophysiology, those beloved subjects of his youth, he reached for irrelevant generalities. The result was one of those confused, pseudobiological theories of the purpose of all life—they still crop up occasionally today—from the crawl of the amoeba to the diplomacy of a nation; a theory that, by explaining everything, explains nothing.

Yet Freud could not write even a book full of blunders without saying something interesting. This came in the analysis of a disorder that had recently claimed his attention and that he badly wanted to explain: the traumatic neurosis. It had long been known, but only obscurely, as a sequel to railway and other disasters. But, "the terrible war which has just ended gave rise to a great number of illnesses of this kind." This kind of war neurosis—today we have broadened it to post–traumatic stress disorder, or PTSD—had forced its way into the consciousness of psychiatrists. When brain damage had been ruled out, the syndrome became psychologically intriguing. The oddest symptom of traumatic neurosis was a tendency to replay the trauma, with all its frightful emotion, in memory and dreams.

From this Freud proceeded to an account of child's play. A baby he knew, just on the verge of acquiring language, made a habit of playing games of disappearance. The child would hide and recover objects, or hide himself before a mirror and reappear, over and over again, especially when his mother was out or had recently returned. Obvious signs of pleasure were associated with the return of the object or the mirror image, however many times the game was repeated. For Freud, childhood games like hide-and-seek were linked to traumatic neurosis through the concept of repetition compulsion. In both situations a prospect that generates great anxiety or fear is reawakened by the mind, deliberately, to permit mastery, or at a minimum to permit survival—in itself a kind of mastery. In hide-and-seek at least the evasion brings pleasure.

There were echoes here of Freud's earlier book, *Jokes and Their Relation to the Unconscious*. According to that account, the listener puts herself in the hands of the joke teller—she is in the mood to laugh—and the teller brings her into a condition of arousal by creating uncertain anticipation; the punch line makes it clear that the built-up tension is unnecessary and the tension discharges through laughter. An evolutionary approach might converge a bit with this view, noting the similarity between human laughter and the aggressive hoots made by some of our monkey and ape relatives when involved in conflicts. Many jokes are acts of verbal aggression, so the tension has components of fear or anger. The laughter at the end of the joke includes the release of this anger and the pleasure of fear dissolving.

We all know the elation of passing through danger unscathed. We feel it in sports—more purely, perhaps, in those sports that pit human beings against nature rather than each other. People who have survived accidents or illness feel it, as do many women in childbirth. It can be an ecstatic elation, and even aesthetic contemplation and awe share it. The emotion of joy that overcomes us while walking on a mountain ridge, sailing in a small boat on a great expanse of water, or resting in a garden at night listening to the silence must have, in some measure, the pleasure of transcending fear. As the great German poet Rainer Maria Rilke wrote:

Then beauty is nothing
But the start of a terror we're still just able to bear
And the reason we love it so is that it blithely
Disdains to destroy us.

There is even some physiological evidence. In the famous experiment by psychologist Stanley Schachter, either elation or anger could be produced by an injection of epinephrine, a hormone secreted under conditions of stress or fear. The subject's emotions hinged on whether the person sitting next to him was acting happy or angry. But hormones are probably not involved in milder kinds of enjoyment, where events inside the brain suffice. Indeed, these events may be basic to thought itself. Early perceptual psychologists identified aesthetic appreciation and even pleasure itself with how intense the stimulus is compared with expectations. Freud thought that pleasure was the crossing into consciousness of sensations approximating stability and unpleasure, departures from stability, while between the two was "a certain margin of aesthetic indifference." This view was superseded, but it opened a way of thinking about pleasant sensations, reflections of the discrepancy between input and expectation; in effect, a perceptual theory of pleasure.

In modern cognitive psychology the relationship between perceptual processing and pleasure is better studied and more interesting. Jean Piaget—the Swiss genius who helped invent that field—believed that a smile occurs when a problem is solved or a stimulus pattern recognized. This "smile of recognitory assimilation" has been widely studied, especially in infants and children, and it leads to a new view: cognitive unpleasure is caused by a stimulus that is not well represented in the memory store, and as the discrepancy resolves, we smile.

We have already encountered discrepancy in our discussion of fear. To produce attention, a stimulus must evoke something in memory, but the pattern must also fail to match the schema already stored. That is, it must be "moderately discrepant" from an established schema. If enough attention to the stimulus can reconcile the mismatch, by creating a new schema or stretching the old one, a smile occurs. For the older child or adult solving a problem, the process is parallel: If the problem is too unfamiliar, it will not evoke attention; if it is difficult but doable, it will evoke interest, attention, arousal, and then, when solved, pleasure, often signaled by a smile. If the problem is too easy, we are not aroused; if attention is forced, we are bored. The same is true for four-month-olds who see a stimulus pattern with which they are completely familiar. Recall that Donald Hebb's fear theory was also based on arousal caused by moderate discrepancy. So partly different patterns with which we have no prior experience produce alertness, attention, and arousal, until their assimilation gives pleasure. But if they prove resistant after mental effort, the same patterns produce fear.

Perhaps the smile of recognition is a pleasurable discharge of arousal comparable, on a small scale, to the laughter at a punch line. To revise Freud's language, pleasure is caused by the crossing into consciousness of sensations that depart from and then return to stability. It follows that tasks that are either too simple or too repetitive can rarely be interesting or pleasurable. For this rule of thumb we must, of course, add a caution relating to age, prior experience, and individual differences. A retarded person might find both challenging and pleasurable a task that would bore a more average person to tears. An orderly person might see in certain routines an enjoyable challenge that would cause discomfort for someone else. Some people may even reach a kind of Zen satori—a state said to be serene and pleasurable—while slicing carrots or photocopying. But for others, endlessly repetitive activity leads to estrangement or alienation from what we are doing, not to joy in it.

A great advance in our understanding of positive states is the concept of flow, put forward by psychologist Mihaly Czikszentmihalyi (m'-HI chick-sent-m'-HI). The term was suggested by statements from people in all walks of life who described some of their ideal moments in terms like "I was carried on by the flow." Flow is a kind of channel running between anxiety and boredom. It can occur in almost any activity—physical or mental, competitive or solitary, gainful or playful, instinctual or learned, appetitive or creative—that transcends any goal it may have and becomes its own justification, an end in itself. Painting, gardening, baseball, dancing, having sex, doing surgery, conversing, rock climbing, fighting, sailing, or home repair can all induce flow. Consider the ancient parable of Ting, the butcher and cook for the lord of Wei: "Ting was cutting up an ox. . . . At every touch of his hand, every heave of his shoulder, every move of his feet, every thrust of his knee—zip! zoop! He slithered the knife along with a zing, and all was in perfect rhythm, as though he were performing the dance of the Mulberry Grove or keeping time to the Ching-Shou music." Complimented by the lord of Wei on his skill, Ting says, "What I care about is the Way, which goes beyond skill. . . . Perception and understanding have come to a stop and spirit moves where it wants."

In hundreds of studies "optimal experiences were described in the same way by men and women, young and old, regardless of cultural differences. The flow experience . . . was reported in essentially the same words by adults in Thailand and India, teenagers in Tokyo, old women from Korea, Navajo shepherds, farmers in the Italian Alps, and assembly-line workers in Chicago." One factory worker was in a state of flow because, although he performed the same operation about 600 times a day, he was proud of his skill and constantly trying to best his average time of twenty-eight seconds. Another man working in the same factory was preoccupied for several days with the thought that his disabled car might lead to the loss of his job; this man was in a state of "psychic entropy" or disorder, the opposite of flow. Thus, flow can be disturbed by pain, grief, fear, rage, anxiety, or jealousy, yet it is more than just the absence of them. Yoga has elements of flow, yet is missing an important one: yoga aims to dissolve the self, while flow integrates, enhances, even builds it.

The process of building starts very early in life. In infancy, joy is as fragile as it is in the realm of adult work. Independently, researchers Daniel Stern, Carolyn Rovee-Collier, and John Watson explored a region of early life in which the joy of mental activity meets the joy of love.

That region is the three-month-old's experience of her primary social world. Stern, a psychiatrist then at the Cornell University School of Medicine, measured mother-infant interaction and brought the joy of it in line with the theory of smiling.

In an influential paper, "Mother and Infant at Play," he and his colleagues proposed a sort of ideal mother who in face-to-face "play" would make expressions and sounds designed to arouse, but within limits, by challenging the baby's expectations. Probably this play also engages innate mechanisms of arousal. Loudness and pitch of voice, sudden brow raising and widening of the eyes, and vigorous physical stimulation may all cause reflexive responses in the baby's brain.

The "ideal" mother or caregiver is keying in to the infant's arousal, jazzing her up just enough to discharge smiling or laughing. Variations on this theme as the optimal level of arousal changes make up "the game." When the baby gazes, smiles, laughs, and coos, the stimulation is in the right range; averting a gaze, fussing, or crying mean the play has lapsed into boredom or spilled over into distress. Stern drew a key distinction between satisfaction, a state caused by meeting a need, and joy, a more complex rhythm of satisfying calm alternating with a desired level of arousal. Satisfaction is best exemplified by the sated babe at the breast, cooing contentedly after milk has quenched distressing hunger. But joy is most vividly seen in face-to-face play. Overall, the idea of optimal arousal fits well with an idea called adaptation level theory. What we are used to gets boring; what is too different and too arousing becomes unsettling. In the zone between them we find pleasure, and the pleasure is enhanced by a certain degree of control.

Rovee-Collier, a Rutgers University psychologist, approached the game differently, with an elegant experiment. A three-month-old lies in a crib, gazing up at a mobile. One end of a ribbon is tied to her ankle, the other to the mobile. When she shakes her foot the mobile moves, and with this alone as a reward, she soon learns to control the spectacle deliberately. Smiling and cooing show that she likes this entertainment; but when an experimental trick disconnects the ribbon from the display, frets of distress show that, having gotten used to it—especially, to the control—she is not pleased.

In Berkeley psychologist John Watson's lab, infants gained control differently. Under the pillow on both sides were levers that moved the mobile whenever they turned their heads. Not surprisingly, after a few weeks of this the infants were more adept at learning similar control tasks than others the same age. Technically speaking, they got used to "response-contingent stimulation"—not just events that occur and pass, perhaps drawing attention to their complexity or novelty, but events that change in response to our own actions. Babies like it.

More disturbing, though, is the result for another experimental group—infants whose mobiles moved on their own. These infants were not only less adept at learning control than were those with controllable mobiles; they were even worse than infants who had no mobile experience at all. In other words, they had learned to view the world as outside their control. One obvious warning based on this research was against too much passive activity, such as television watching. Recall that enriching the environments of rats does not affect their brains unless they grapple actively with those environments. Joy in learning may require acting on the world.

The second warning is more intriguing. Converging on Stern's thinking, Watson theorized that an ideal mother would be one who offers a response-contingent stimulus pattern. In other words, like the controllable mobile, she would change in a way that both challenged and responded to the infant's own mind. So far, so good. But the caution was this: if the infant is set up biologically to delight in stimuli she can control, then some ingenious toy manufacturer might come up with a convenience device that would fool the infant's attachment system and take the place of a relationship. We do not know whether this has happened to infants, but the strong attachments formed nowadays between some young people and computers, especially computer games, may be similar. The more extreme among these dependencies are not in the least amusing.

As for learning beyond infancy, the last few decades have seen interesting changes. The open classroom movement and the attempt to nurture multiple intelligences instead of just traditional academic ones are among them. Although not conceived as such, they may do for the schoolchild what the "ideal mother" does for the infant: challenge the child's arousal system just enough to bring her along at her own best pace, based on individual responsiveness. These methods also reflect the theory of learning based on the work of Lev Vigotsky: teachers must enter the child's "zone of proximal development"—the mental space just beyond what the child knows. There the teacher erects a psychological scaffold where the learner builds new skills.

Unfortunately, this has suggested to some that learning must be easy. But the joy of learning is destroyed by an absence of challenge, not just by repetition and rigidity. Even in infancy, even in animals, joy in learning requires challenge, and challenge, up to a point, entails distress. This also applies to play, which many higher animals do. Naturalists define play as energy expenditure that looks both impractical and pleasurable, but the fact that it looks useless does not mean that it is. It serves the functions of physical exercise and learning about the environment, and it may supply or sharpen fundamental subsistence and social skills. In some mammals, a lack of play in early life impairs later social and reproductive skills.

Not surprisingly, the most intelligent mammals—primates, whales and porpoises, and carnivores, including aquatic carnivores like seals—are the most playful. Intelligence and play coevolved, each strengthening the other. Also, if an animal is very short-lived, there is too little time for the young to gain much from playing. It is mostly the young that play, finding ample opportunity for observational learning—particularly if, as in many species, the playmates differ in age and developmental level. In cats, monkeys, and other animals, observing a task while or before trying to do it speeds learning.

We have already seen playfulness in the prey catching of cats. House cats can enter a grisly kind of flow while toying with chipmunks or voles that are stunned and struggling—releasing and chasing them, batting them around, and playfully pouncing again and again. These features become the basis of an almost unique transfer of skills to the young that in some cases may be genuine teaching. Cheetah mothers bring back half-dead prey, which their young then kill and eat. Lions and leopards lead cubs and kittens on expeditions whose main purpose seems to be to acquaint the young with stalking. And they partially kill prey on the hunt, leaving the young to finish the job and intervening only if the prey is about to escape. This highly evolved system uses the playfulness of the young to shape vital survival skills.

Did play have a role in learning to hunt during human evolution? Hunter-gatherer subsistence learning resembles the open classroom and the learning of nonhuman mammals, not the traditional setting of Western education. Yet to nineteenth-century educators the rigors of the old-style schoolroom must have seemed very benign compared with the daily chores or full-time work of most poor children. Indeed, anthropological studies show that intermediate-level societies—those that have left hunting and gathering behind in favor of gardening, agriculture, or herding—brought much more work and drudgery into the lives of children, and this continued well into the industrial revolution. So the modern open school has in some ways restored the hunter-gatherer pattern of playful learning.

But play involves serious challenges. All mammalian young engage in rough-and-tumble play, which is vigorous, arousing, and contains some elements of aggression. Hatred itself can be a source of pleasure, sadistic acts enjoyable. But that is not the spirit of rough-and-tumble play, which uses very different emotions. In our species, this and other kinds of play last into adulthood, one of several reasons for the name *Homo ludens* (Playful Man), coined by Dutch historian Johan Huizinga. Rough-and-tumble play can certainly be rough. Many American youths are killed each year while playing football; even in baseball, soccer, and basketball, serious, occasionally deliberate, injury is common. Standing your ground near the path of a ball hard and fast enough to break your skull must be arousing.

But where inside that skull does the impulse to play live? Little is known about how the brain generates play, but it is not primarily in the cortex. In studies by Jaak Panksepp of Bowling Green State University, rat pups whose entire cortex was removed played normally, rough-and-tumbling with their peers just like litters of untouched pups. Clearly this is very basic behavior, and both its patterns and its pleasures are regulated from older, more primitive parts of the brain. This and other research led Panksepp to put rough-and-tumble play at the center of his theory of joy. As he sees it, rough-and-tumble evolved first, for the purpose of honing social skills, and other forms of positive emotion were derived from it. Panksepp's research confirmed Paul MacLean's insight that play is part of the evolutionary achievement of the early mammals—dependent on the limbic system but not on the advanced mammalian cortex. Turtles show some simple elements of play, such as pushing repeatedly on floating objects with no apparent purpose. Some marsupial mammals have full play capabilities, and as Darwin personally observed, even primitive egg-laying mammals like platypuses play to some degree. These findings take the behavior back to the none-too-brainy dawn of mammals.

Panksepp has made another surprising proposal, "the admittedly radical idea . . . that the 50-kHz chirping of rats may be functionally or evolutionarily related to human laughter." These are ultrasonic vocalizations, one among many sounds that flood the social and emotional world of rats but that our ears are completely deaf to. Laughter has many functions in humans, one being a response to tickling, especially in childhood. Tickling works well in juvenile monkeys and apes, too, so Panksepp and his colleagues used ultrasound detection to see whether it might work in rats. It clearly does. Young rats seek human tickling and find it rewarding in learning studies, as long as it is deftly done. They issue particular ultrasonic chirps during tickling, and these tickle-chirps also anticipate tickling in situations where it has occurred before. Rats housed in isolation are much more likely to chirp when tickled than those that are socially housed, as if the isolates are hungry for stimulation and play. But there are also more basic individual differences, and when you take rats that have high and low chirp rates when tickled and mate them separately, their tendencies breed true within four generations. The neuro-transmitters glutamate and dopamine both play a role in tickle-chirping, and as we will see these are two of the main vehicles of brain reward.

We have now traced pleasure, play, and possibly even laughter to the primitive common ancestor of rats and humans. Still, larger-brained mammals play more; evidently the basic limbic system can generate the behavior, but an advanced cortex can build and elaborate on the fundamental moves, moods, and motives. There is no doubt that higher primates, the smartest of mammals, are also among the most playful. Superb comparative studies by Signe Preuschoft and Jan van Hooff have shown that the monkey play face is a true precursor of human laughter and smiling. These facial and vocal expressions in the same behavioral context suggest that the emotions of social play, especially in the young, are very similar. Among marsupial mammals the correlation between play and intelligence repeats itself independently; as that separate branch extended in its own evolutionary experiment, play—especially rough-and-tumble play—became very prominent in larger-brained kangaroos, which famously box both in fun and in earnest. At the same time, smaller-brained marsupials like the koala led a largely playless existence. Some birds play—brainier types like ravens, parrots, and macaws—but bird brains are very different from those of mammals, and we know little about how they generate the behavior. Since reptiles and amphibians show only the simplest forms of play, birds must have evolved the fancier forms independently of mammals. But we should not rule out play in some dinosaurs, especially the branch that led to birds.

Play is observable. But how do we start to think about the evolution of joy, which we can't quite see but can only ask about or, worse, merely guess at? Most people say that they are mildly happy, that they feel slightly positive most of the time. Common sense tells us that the experiences we have, even our sheer day-to-day luck, must determine how happy we are, but common sense is largely wrong. Between 1965 and 1999, more than 600 studies surveyed hundreds of thousands of people in sixty-nine countries to find out how different life circumstances affect happiness. Most of the effects are very small, of a sort that only such large studies readily detect. Self-reported happiness increases *slightly* with age, education, social class, and income, at least in Western countries. People in the developed world are happier than those in the developing world, but they do not become happier as they grow even richer. Greater disparity between rich and poor is associated with slightly lower average happiness in a nation.

Other demographic factors vary in importance. Minority-group status produces *slightly* less happiness, after controlling for the other variables. Unemployment has a more substantial negative effect, and enjoyment of leisure activities, including volunteering as well as play, is significantly tied to happiness. Religious people are somewhat happier, but this effect becomes very small after controlling for economics, education, and social contacts—churchgoing entails relationships. Being married or cohabiting has a small but real positive effect for both sexes, especially in the honeymoon and empty-nest phases. Overall, only 10 to 15 percent of the variation in level of happiness can be reliably attributed to the total sum of such demographic factors. Still, if you manage to achieve material comforts, marry the right person, have sex frequently, mind your diet and exercise, belong to a church community, and believe in a God who cares for you, it adds up to an edge in the happiness sweepstakes.

Yet the things about ourselves that we can't control—height for men, beauty for women, health and intelligence for both, and especially an extroverted, non-neurotic personality, make more of a difference than any of these basic life circumstances. Psychologists David Myers and Ed Diener summarized the evidence in a landmark 1995 paper titled simply "Who Is Happy?":

> A flood of new studies explores people's subjective well-being. . . . These studies reveal that happiness and life satisfaction are similarly available to the young and the old, women and men, Blacks and Whites, and the rich and the working-class. Four inner traits appear to mark happy people: self-esteem, sense of personal control, optimism, and extraversion.

The generalization remains true, and it seems to suggest that anyone can do it. But we know very well that individual differences are vast. Scrooge harrumphs at every hint of pleasure, while the bright-eyed birthday boy in the old joke digs and digs through the pile of horse manure, since "there must be a pony in here somewhere." Overall, personality is more predictive of happiness than are demographic factors. For example, all major models of personality include a dimension corresponding to extroversion, which is highly correlated with happiness, and another, independent dimension resembling neuroticism, which is virtually indistinguishable from unhappiness. The fact that these two dimensions are independent means that it is quite possible to be a neurotic extrovert (high on both unhappiness and happiness), a nonneurotic introvert (low on happiness but also low on unhappiness), a neurotic introvert (worst off), or a nonneurotic extrovert (happiest of all).

Psychologists Ed Diener and Martin Seligman have done a remarkable study called simply "Very Happy People." They screened 222 college students on several standard measures, identified the happiest 10 percent, and compared them to average or very unhappy people. The results were simple and elegant:

The very happy people are highly social, with strong romantic and social relationships. . . . They are more extroverted, more agreeable, less neurotic, and lower on several . . . psychopathology scales. [They] did not exercise significantly more, participate in religious activities significantly more, or experience more objectively defined good events. No variable was sufficient for happiness, but good social relations were necessary. The happiest group experienced positive, but not ecstatic, feelings most of the time, and they did report occasional negative moods. This suggests that very happy people do have a functioning emotional system that can react appropriately to life events.

Evidence is strong and growing for a genetic basis of such differences between people, though there are certainly other influences. Wealth and physical attractiveness (as rated by others) were greater in the very happy 10 percent.

To paraphrase two genetic researchers, happiness is a matter of probabilities, and the right genes merely improve your chances. But which genes are they? Studies of the serotonin transporter showed that a repetitive stretch of DNA just upstream from it can promote the transporter's synthesis. If you have a longer promoter, with more repeats, you make more of the transporter. This recessive brand of the promoter occurs in a double dose in about a third of us; that third is more likely to resist feeling anxious, sad, hostile, pessimistic, and fatigued than the two-thirds who are heterozygous or have a double dose of the short promoter. This gene, if confirmed, is likely to be only one of perhaps ten to fifteen genes affecting happiness, and all of them together probably explain only about 40 percent of the normal variation, leaving the rest to experience. Other studies put the total genetic effect even lower, at around 25 percent. Still, studies of how experience works have shown that winning the lottery, getting a promotion, or marrying the boy or girl of your dreams have surprisingly little impact on happiness. So whatever collection of genes is involved, they must be part of a trait vital to our well-being.

And perhaps not only ours. Researchers at the University of Arizona in Tucson have found some ingenious ways to study happiness in chimpanzees. Psychologist James King sent questionnaires to volunteers who work with chimps in thirteen American and Australian zoos. One hundred thirty-five apes were rated on how often they were in a positive mood, how much they enjoyed social contact, how well they met their goals, and how happy the rater would be to become the particular chimp. These were subjective judgments, of course, but as anyone who has worked with chimps can tell you, they are still worth a good deal. Every chimp could be paired with every other one for the purpose of calculating relatedness, which could be as high as 75 percent because of inbreeding. Heredity explained about 40 percent of the variation in happiness, roughly the same as in us. The implication is that genetic individual differences in happiness go back to our common ancestor with the chimp, at least five to seven million years ago. The ancient idea of fate begins to take on an eerie new meaning. If we are born to be happy or unhappy, regardless of what we do or experience, we may start to feel as if we have turned up in a Greek tragedy, our destiny spun and woven by remote genetic gods. But of course the situation is neither so bleak nor so simple.

Joy, flow, play, happiness—they share much in common and grade into one another. But they are not the same things, nor do the concepts of pleasure and contentment fully overlap. Barbara Fredrickson, a social psychologist at the University of Michigan, has developed what she calls the "broaden and build model of positive emotions." She recognizes that it is far more difficult to be specific about the positive emotions than about the negative ones. We have clear distinctions among rage, fear, grief, and disgust, and we find in lust and love two states that can blend while retaining some conceptual clarity. But joy? Contentment? Happiness? Pleasure? Diffuse definition seems to be part of their essence.

Fredrickson took this notion seriously in a theory that has diffuseness at its core. Research shows that people in positive and playful moods are more open to varied experience, and that they learn better and in more varied ways. The suggestion is that positive emotions are an adaptation for expanding our experience and knowledge—thus the phrase "broaden and build." This proposal matches the theory advanced by Brian Sutton-Smith in *The Ambiguity of Play,* which states that "play variability is analogous to adaptive variability; that play potential is analogous to neural potential; that play's psychological characteristics of unrealistic optimism, egocentricity, and reactivity are analogous to the normal behavior of the very young; and finally that play's engineered predicaments model the struggle for survival." The claim here is that play actually shapes brain development, because natural selection has designed it to do so. Or as John Byers said of playfully boxing joeys, "Most likely they are directing their own brain assembly."

These two theories of joyful mood and behavior converge on an elegant adaptive model, but unfortunately they leave much out. Recall the joyful, almost playful mood of Ting as he slashingly made short work of the ox: focus was the watchword, economy of movement the goal. Openness and variability would have separated Ting from some of his fingers. Ting's joy—like that of a concert cellist or even a speed typist—lay in precision, not variation, and the closed circle of that kind of flow is no less joyful than the wildest, most unpredictable play.

Or consider contentment, a different kind of happiness. As Stern argued, the sated babe at the breast is in a pastel-colored mood, but the ebb and flow of face-to-face play bears the bright hues of joy. This joy is a sought-after state despite its tensions, belying the claim that happiness is in essence a low ebb of wanting. Nevertheless, contentment—the absence of tension and want—is surely an important kind of happiness and is neither playful nor open. PET scans of the brain show a condition of low general blood flow in healthy women experiencing transient happiness. In the Zen master's *satori,* wanting itself has been deliberately abolished, as have fear, anger, risk, and tension of all kinds. Ting's brand of flow is motivated and active, but its happy mental state is the product of near-perfect predictability. Even play doesn't always broaden and build. The champion pool player's climax run is as exquisitely focused as Ting's joyful work, and a world away from the seemingly aimless antics of tumbling wombats.

So we have, at a minimum, joy, flow, play, contentment, and satisfaction. We have gaiety and utter calm, ambiguity and exquisite focus, questing and completion, all in the realm of what may make us happy. But at their core is pleasure, which can be achieved through the serious business of drinking, eating, and sex, or through mere quiet contentment devoid of challenges. While we have begun exploring how the brain handles play, we know more about how pleasure itself is processed, owing especially to the genius of James Olds. In 1953 Olds and Peter Milner made an observation that changed psychology. A rat with an electrode implanted in its brain was roaming the open-field apparatus and was given a tiny electric shock. The goal of the experiment had nothing to do with pleasure. But the rat returned to the place where the slight surge of current had been given.

Intuiting the importance of this lucky event, Olds and Milner shifted their line of research to explore that rat's behavior. They discovered that rats would return repeatedly for stimulation of certain parts of the brain. Given a lever controlling the flow of current, they would stimulate themselves. They would work hard to get that opportunity, and they would neglect other aspects of their lives — including eating — to pursue it. Since the current surges could reinforce behavior just as food or water could, they were referred to as brain reward; but later investigators could not resist calling them pleasure centers of the brain. They have been seen in animals from fish to us, so the experience of brain reward or brain pleasure is far older in evolution and more widespread among animals than is play.

The cells involved range over different brain areas but center on the hypothalamus, as though on the hub of a wheel. The spokes — nerve paths not symmetrically arranged — radiate mainly to the systems of smell and emotion, including the olfactory bulbs and their cortex, the septal area, the amygdala, the forward tip of the thalamus, the cingulate cortex, the hippocampus, and pathways running down to the limbic midbrain and lower brain stem centers. The list reads like a verbal sketch of the limbic system and contains the whole Papez circuit of the emotions. You can get rewarding effects in other areas, but it is much more difficult. Even the main areas are not equally rewarding. Electrodes evoke the most self-stimulation in the lateral hypothalamus, stretching back to the midbrain. Rates in the central hypothalamus are lower, and in the hippocampus and cingulate cortex lowest of all. In fact, you don't need your higher brain at all for these experiences.

This research confirmed the notion, familiar from limbic system studies, that the hypothalamus is the pivotal organ of motivation. The lateral paths of the hypothalamus, along with their forward and backward extensions, form a stream of nerve cells known as the medial forebrain bundle (MFB). The MFB controls simple acts involved in pleasure — eating or drinking, for example — by connecting down through the brain stem and spinal cord. But it also projects up to the limbic system and basal ganglia, using such neurotransmitters as serotonin, norepinephrine, and dopamine. Through these it controls more complex, emotion-based behavior like courtship or parenting. These projections feed back to where they came from, maintaining the behavior, its memory traces, and the pleasurable feeling.

Activation of these pleasure regions explains why certain behaviors feel so good. Some are inborn and some are learned through repeated firing of reward circuits. The main upward-running neurons from the pleasure centers use dopamine as their transmitter, and hundreds of experiments support the idea that dopamine is crucial for pleasure.

These dopamine neurons in the pleasure regions influence complex species-specific behavior—like courtship or the squirrel monkey's penile display—through the basal ganglia, and the associated emotions are experienced through the limbic system. The circuits for movement and emotion overlap in the nucleus accumbens—in effect an extension of the amygdala into the basal ganglia. The accumbens is crucial for dopamine's actions and the experience of reward. It is as if the essence of reward were where emotion and action meet, neatly dissolving the old Skinnerian debate about the difference between reinforcement and pleasure. Skinner would have liked to abolish concepts like pleasure, restricting the language of psychology to sensation and action, stimulus and response. The accumbens stands astride both realms, challenging Skinner or anyone else to preserve the false distinction.

In learning studies, dopamine itself is rewarding if delivered to the accumbens, yet the pleasurable actions of cocaine, amphetamine, and other drugs that promote or mimic dopamine work through that nucleus. Bartley Hoebel and his colleagues, summarizing decades of work on the nucleus, say that "learning how to release dopamine is the essence of learning to work." Yet a crack addict can "activate the accumbens by imaging something pleasurable" and if you block the dopamine receptors in that area, "the good things in life have little effect." Work in the laboratory of psychologist Darryl Neill has suggested that the accumbens is involved in the rewarding effects of appetitive stimulation—the sight and smell of food, for instance—while another structure, the ventrolateral striatum, feels the reward of consummatory behavior, like eating itself. Both systems use glutamate as well as dopamine in neural transmission. Not surprisingly for a pivotal axis of learning, the accumbens is changed by experience. If rats are raised in social isolation, the sensitivity of this mechanism is enduringly increased.

Separately, the endorphins—the brain's opiumlike balms—play a key role in reward. Morphine and its antagonists enhance or block pleasure by acting at opiate receptors. Rats will cheerfully self-administer opiates into the accumbens, just as they will electrify it with self-stimulation. But the rewarding effect of those opiates is not the same as dopamine-related reward. So at least two different neurotransmitter systems are involved. The dopamine system may highlight the rewarding stimulus, whereas the endorphin system supplies the pleasure itself. *Learning* through reinforcement may require yet a third input, from the amygdala, using the transmitter glutamate. Normally, these functions are integrated, of course. In a paper entitled "Pleasure, Pain, Desire, and Dread," Kent Berridge, summarizing his experiments on brain reward, separates working for something, or *wanting*, from sheer enjoyment, or *liking:* "Most rewards that are liked are also wanted. Most pains are feared. . . . It has been adaptive for these components to work

closely to achieve life goals. But their identity as components allows them to dissociate under some conditions." Darryl Neill believes that activational and reward
circuits are anatomically distinct but that both use dopamine and glutamate, modified by an input from GABA. This kind of functional separation within the brain's
reward system requires more research, but it will ultimately lead to circuit-specific
medical treatments that enhance pleasure, treat addiction, and improve learning.

Chemistry aside, we know some things about the behavior. Rats made to work
for brain reward will tap the lever no more than twenty times for one brain stimulus. But if the rat is given a signal before the stimulus comes on, it will tap up to
200 times for one surge, suggesting that brain stimulation is much more rewarding if anticipated. Perhaps this explains how looking forward to a trip or a tryst can
enhance the pleasure we take in it, by highlighting the joy in advance. However,
other experiments suggest that the same stimulation that produces reward if the
rat can *control* the timing can also provoke escape—a sure sign that the pulse is
unpleasant—if delivered into the brain at random intervals. There are echoes
here of the human infant's joy in controlling stimulation, and of grown-ups' enjoyment of horror movies but not of real horror.

The relationship of pleasure to arousal is also ambiguous. Depending on the
electrode's placement, rewarding pulses can either arouse or quiet. A surge near
the septal area slows or halts normal behavior, with a parasympathetic pattern of
body functions—lower heart rate, blood pressure, and respiratory rate. But a stimulus in the lateral hypothalamus, also rewarding, causes sympathetic arousal, *raising* respiratory rate and blood pressure while making the animal more active. This
paradox reflects our ordinary experience: contrary to Freud's claim that pleasure
always reduces excitation, it may either arouse or quiet. And we know from studies
of thrill seeking that different people inherently enjoy different levels of stimulation, novelty, and arousal. There is an argument to be made for arraying our psychic
states on two dimensions, one of pleasantness and another of arousal. The resulting four endpoints—pleasantly aroused, pleasantly calm, unpleasantly aroused,
and unpleasantly calm—may correspond roughly to happy excitement, serene
joy, anxiety, and depression. This is crude, but it covers a lot of ground and seems
consistent with common sense.

But we don't have to settle for musing on what a rat feels when a pleasure
center is buzzed, because brain surgeons have studied such effects in humans.
Operating for reasons having nothing to do with psychology, they often have to
stimulate the brain to find their way. The anatomy they see on the brain's surface
is ambiguous, and the best way to map it—and avoid a cut in the wrong place—is
to tickle various spots with a mild current and record the patient's responses.
Since we feel no sensations from inside the brain, patients can be conscious during these procedures. Psychologist Elliot Valenstein, who also studied brain stimulation in rats, described several cases:

A patient who had just attempted to commit suicide by jumping off the roof
suddenly started to smile when the electrode in his septal area was acti-

vated. It was difficult for him to verbalize his experience more explicitly than, "I feel good. I don't know why. I just suddenly felt good." Upon further questioning . . . he said: "It's like I had something lined up for Saturday night . . . a girl." Given an opportunity to press a button controlling a portable stimulation unit worn around his belt one patient reported: "When I get mad if I push the button I feel better. . . . [T]hat's a real good button. . . . I would buy one if I could." A woman with intractable pain said during the stimulation: "I'm feeling fine . . . feel like I could clean up the whole hospital." In several instances it has been suspected that a patient reached an orgasm during stimulation.

Work by Itzhak Fried and others during the late nineties dealt with a different aspect of pleasure in a different brain region. A sixteen-year-old girl with severe epilepsy had her left frontal lobe stimulated with a tiny electrical surge in eighty-five different places. One spot, near the area for control of mouth movements, made her laugh quite naturally. "You guys are just so funny—standing around," she told her doctors. This is not the part of the frontal lobe that is intimately involved with emotion, so the emotional aspect of her laughter must have been activated secondarily. Since Paul Ekman's group discovered that deliberately putting on facial expressions can influence subjective emotional states, it is not surprising that unmotivated laughter would seem amusing. But that does not negate the role of subcortical structures like the corpus striatum in displays of emotion, or the primary role of the limbic system in subjective feelings. Indeed, human brain imaging confirms the role of the nucleus accumbens and amygdala in positive feelings. This is not surprising; pleasure, we know from brain stimulation, lies deep in the ancient parts of the brain.

Then again, the orbitofrontal cortex, which figures so centrally in Antonio Damasio's theory of what it means to be human, also responds uniquely to certain pleasures. A group at Oxford University used functional magnetic resonance imaging to compare brain responses depending on whether the hand is stroked with velvet or touched with wood. The difference is striking. The wood (a "neutral tactile stimulus") activates the somatosensory cortex for the hand. But the velvet touch (a "positively affective tactile stimulus") instead lights up the orbitofrontal or Gage cortex, the crossroads of emotion and thought, one place where the ancient limbic brain meets the great human computer of the neocortex. Equally important, the orbitofrontal cortex contains different regions that respond to taste and smell. So three kinds of primary reinforcers, in the separate realms of gentle touch, taste, and smell, are represented in this crucial zone of the frontal lobe, which "helps to provide a firm foundation for understanding the neural basis of emotions." And, we might add, of pleasure. The mother-infant bond and the bond between lovers both depend on the pleasures of sweet taste, fragrant smell, and soft, velvet touch, and these in turn depend on the underside of the frontal lobes.

But of course these are just a few of the many kinds of positive experience. Brain stimulation seems to have its most dramatic effect on patients who start out in physical or psychological pain. But for others, too, the pleasure can be compelling. In one patient who had not been in pain, the desire to self-stimulate with the implanted electrode was strong enough to embarrass her, and she tried to conceal it or distract herself by reading, listening to music, and helping other patients. The resemblance to habits like smoking or overeating is striking.

Brain self-stimulation also illuminates more specific motives. Animals stimulate themselves in some brain areas when hungry, others when full. Also, at the sites where hunger decreases self-stimulation, sex drive increases it, and vice versa. For a woman who resumed menstruating while an electrode was implanted, stimulation had sexual overtones after menstruation but not before, evidently a hormonal influence. Valenstein's own research on rats showed that some reward centers also triggered specific behavior. Stimulation of the lateral hypothalamus— highly effective as a reward—also causes a rat to eat, hoard food, or perform other activities. Electrodes in the same site could produce food hoarding or maternal retrieving, depending upon whether food pellets or rat pups were in the cage. To make matters more complicated, some sites seemed to quench a motive instead of drive it.

Yet the findings are consistent with our subjective sense that pleasure can be arousing or quieting, general or specific. They suggest that the creature receiving brain stimulation will much prefer to control it actively, consistent with the link between joy and mastery. They also show that the brain is designed to pursue pleasure, but defines it only vaguely; it is not ideally suited to distinguish various sources of pleasure and to choose the one the animal (or person) really needs. It may even mean that the brain is not good at finding satisfaction. Finally, there is a strange mutability between pleasure and dreams. Animals deprived of REM or dream sleep increase their feeding, sexual behavior, and brain self-stimulation. And REM-deprived animals that are allowed self-stimulation don't catch up on REM sleep when given the chance, as REM-deprived animals normally do. Could self-stimulation of the brain somehow substitute for dreaming? This apparent equivalence, if true, would support both the notion of brain stimulation as pleasure and that of dreams as wish fulfillment. Yet some of the best work on positive emotions comes not from natural states like orgasm, flow, love, or wishful dreams but from the brain's reactions to some quite unnatural substances.

> Woe to you, my Princess, when I come, I will kiss you quite red and feed you till you are plump. And if you are forward you shall see who is the stronger, a gentle little girl who doesn't eat enough or a big wild man with cocaine in his body.

So wrote Sigmund Freud, the author of that same notion of dreams, in a letter to his fiancée in June 1884, when he was about to find fame with a series of studies on the medicinal uses of the coca plant and its derivative, cocaine. He was still a decade from his first steps toward psychoanalysis; his interest in cocaine preceded even his work on the neurology of language. He read the research on the drug, then very limited, and soon began experimenting on himself and others. Like many a modern scientist, he requested a sample from the Merck Company and began to explore its properties.

Self-experiment with drugs was not unusual then, and Freud's first paper was enthusiastically received. Like many medicines, cocaine was discovered by a traditional culture—in this case the Inca of Peru, who habitually chewed the coca leaf, a "divine plant which satiates the hungry, strengthens the weak, and causes them to forget their misfortune." Whether among the Inca or among soldiers in Europe, it could make brutal work tolerable and blunt the effects of fatigue. Freud's summary of the animal studies is valid today: cocaine causes general sympathetic activation, with increased respiration, heart rate, and blood pressure; at higher doses it causes poisoning, with motor agitation, convulsions, and finally death. He cited evidence that this sympathetic activation is carried down from the brain to the spinal cord through the medulla in the brain stem. At low doses, "dogs show obvious signs of happy excitement and a maniacal compulsion to move." He gave an almost correct chemical formula, although the structure was then unknown, and he recommended further study of it as a stimulant, aphrodisiac, digestive, antiasthmatic, local anesthetic, and substitute for morphine during withdrawal. But most interesting were his own and his friends' experiences:

> The psychic effect . . . in doses of 0.05–0.10 g consists of exhilaration and lasting euphoria, which does not differ in any way from the normal euphoria of a healthy person. . . . One senses an increase in self-control and feels more vigorous and more capable of work. . . . One is simply normal, and soon finds it difficult to believe that one is under the influence of any drug at all. . . . Long-lasting, intensive mental or physical work can be performed without fatigue; it is as though the need for food and sleep . . . were completely banished.

One is simply normal. Freud was heavily criticized for promoting cocaine and eventually regretted having done so. Among other things, it helped cause the death of a friend who was a morphine addict. More important in his ultimate change of mind perhaps was his naïve but firm conviction that psychoactive chemicals impede real change.

Interest in cocaine in the laboratory as well as on the street surged in the last decades of the twentieth century as it had in the nineteenth. Some fears were confirmed. It is certainly toxic and addictive, and at high doses it causes a florid psychosis similar to some forms of schizophrenia. But medicinal uses are also evident, as of course are the psychic effects. We now know that cocaine has a three-ringed chemical structure resembling that of tricyclic antidepressants. Because of its side effects and addictive potential, it is not a good treatment for depression. Yet, like antidepressants, it does block reuptake of norepinephrine and serotonin by nerve endings. Since this sponging-up is the main way those transmitters are removed, cocaine makes more of them hang around to excite the next cell in the circuit. But its leading effect is to block the dopamine transporter, the complex molecule that brings dopamine back into certain nerve cells after release. In fact, knockout mice that lack the transporter gene show no neural or behavioral response to cocaine. It is not surprising, then, that cocaine's pleasurable effects depend most of all on an intact dopamine reward system, including the nucleus accumbens—the crossroads of emotion and instinct. And this mechanism may be crucial not only to cocaine dependency but to substance addictions in general.

Michael Kuhar and his colleagues have enriched the picture with another brain gene. The CART gene, for "cocaine and amphetamine related transcript," makes a peptide that is expressed right through the reward systems. The gene is turned on by cocaine or amphetamine and makes its mRNA transcript, which in turn makes a protein. The protein is cleaved and spliced to make a smaller peptide, which appears to be a neurotransmitter between the nucleus accumbens, the brain's chief reward region, and the part of the midbrain that makes dopamine. If you inject this CART peptide into the same midbrain area, it mimics some cocaine effects, including increased behavioral activity and a learned preference for the place the rat was in when the injection came. Elegantly enough, these CART effects require intact dopamine receptors. The reciprocal exchange of CART and dopamine between the accumbens and the midbrain is a core reward circuit.

These findings support the dopamine theory of pleasure—undoubtedly oversimplified but surely part of the truth. Even under controlled conditions, cocaine produces the high described on the street, in a dose-dependent manner that declines neatly as the drug clears from the blood. The items on a survey that subjects under the influence of cocaine most frequently checked off were: "I feel as if something pleasant had just happened to me," "I am in the mood to talk about the feeling I have," "A thrill has gone through me one or more times since I started the test," and "I feel like joking with someone." In a study of three monkeys given a continuous choice between food and cocaine, "the drug was almost exclusively chosen. Periods of low drug intake did not coincide with increased food intake. . . . This exclusive preference for cocaine persisted for 8 days. Concern for the health of the animals prohibited extending the testing period." Amphetamine is also a stimulant and probably works in part by inhibiting reabsorption of norepinephrine, but it typically produces an agitated kind of euphoria, whereas one of the most frequent spontaneous remarks of human subjects in cocaine studies is "I feel more relaxed." Perhaps this was what Freud meant when he said, "One is simply normal."

The monkey cocaine study recalls the studies in which rats neglected food to press the lever for brain rewards. Efforts to more specifically localize the brain's reward capacity are under way. But the link through dopamine neurons is now well established. Early research showed that a chemical that poisons dopamine and norepinephrine neurons throughout the brain largely abolishes a rat's capacity to experience reward and perhaps even pleasure. In another study, injection into the ventricles of the brain of the same dopamine- and norepinephrine-depleting substance, this time in free-ranging monkeys, produced a syndrome resembling human depression. This makes sense in light of the fact that cocaine, amphetamine, and some of the older, tricyclic antidepressants work in part by enhancing the effects of these two neurotransmitters.

Today we know that depression can be treated with drugs that augment serotonin, norepinephrine, or both. The differences lie in side effects and in individual mood changes. We will return to this subject in chapter 14. But to the extent that joy requires first of all an absence of depression, these transmitters must figure in positive states. In addition, as we have seen, the chemistry of mood has another major focus, away from small neurotransmitter molecules like norepinephrine and serotonin. Peptides, chains of a few amino acids that, if longer, would be proteins, revolutionized brain research in the 1970s, adding several new dimensions. Because they are 5 or 10 times the size of the previously known neurotransmitters, they add complexity. Also, they may resemble some hormones of the pituitary and hypothalamus—hormones that flow up into the brain, not just down into the body. And they can be the breakdown products of proteins, those long, complex molecules only one step from the genes. This means that peptides may be involved in a vast assembly line of hormones with interchangeable parts and great information-bearing power. The CART peptide already discussed is one example.

Endorphins, among the most potent peptides, are another. They moved to center stage in brain science during the late 1970s because of work on a completely different class of chemicals—the opiates. Candace Pert and Solomon Snyder of the Johns Hopkins University, among others, spent the early seventies locating the brain's opiate receptors, the then hypothetical sites where morphine and heroin could act. Investigators gradually showed that the endorphins and similar peptides could mimic the effects of opiates on pain, so they came to be called "opioid" peptides. Two separate laboratories showed that tiny doses of these opioids—less than 1/100 of those required for pain reduction—made rats rigid, immobile, and unresponsive. Yet the most surprising fact was not that such low doses worked but that the chemicals involved occurred naturally in the brain, a system for pain control from within—the brain's own morphine.

In his provocative book *Optimism: The Biology of Hope*, anthropologist Lionel Tiger brought ethnography and sociology together in a biological theory of how we view the future. Religious beliefs, faith in progress, the desire for children, the search for money and power, the quest for utopia, even gambling and shopping sprees all share this: they dull our present pain with a questionable but dedicated focus on the future. Evidence shows that in every realm we exaggerate how well things are likely to turn out. An animal—especially one aware of its own death—simply cannot slog through the drudgery, fear, and hurt of daily life without self-sedation. Tiger proposed that natural selection produced the opioid peptides so that animals could release them when the going gets too rough. The resulting mild sedation and blunted pain make the next hour or day possible. Hopeful human behaviors and beliefs may be products of such sedation, causes of it, or both, in an animal so perversely able to keep an eye on a final, dreary future. Psychologist Shelley Taylor extended Tiger's analysis in *Positive Illusions*, arguing that survival itself, especially during life-threatening illness, may depend on creative self-deception.

At times we lose our taste for the truth. Yet surely, these mental tricks on ourselves are at least as acceptable as molecular elixirs—booze, 'ludes, dope, coke, crack, smack, speed, roofies, Ecstasy, angel dust. Not to mention the medical versions—like Valium, Ritalin, Prozac, Wellbutrin, and many others—a whole pharmacopeia of synthesized happiness. Children are not only eager to try, they are making them a way of growing up. By adulthood we are experts, even within the law. Here a recharge of serotonin to put the blush back on the mood; there a little GABA bump to dampen the faceless fear; over there a little molecular trickery to kick in the brain's own morphine and dull the long pain of life. *Reality*, we say half-jokingly, *is a crutch for people who can't cope with drugs.* What ever happened to the idea that life's pain is woven right through its joy, or at least is a needed coarse thread in the silken cloth of gladness? That the embrace of and triumph over hardship is more exhilarating than denial? In the sixties, in California, that fabulous state and state of mind, this was called the "beat-your-head-against-the-wall-because-it-feels-so-good-when-you-stop" school, and Americans in general rejected it. But for some the triumph over adversity stills seems a good hypothesis of happiness.

There is even evidence, in the lives of ninety-four men who were in college in the early 1940s—a sample chosen to represent that promising generation of young American men. In his book *Adaptation to Life*, psychiatrist George Vaillant summarized the first thirty-five years of research on them. Many men were happy and successful, by self-report as well as objective criteria; others were unhappy or were failures. There were ample records of the men's childhoods, with more information gleaned throughout their lives. No such study can be perfectly objective, but this one met high standards.

Its conclusions are simple. First, a stable, loving family in early life is a great advantage. Men with bleak childhoods usually continued to be unhappy, despite worldly success. Some were well adapted to life's challenges but were unsusceptible to the special joy that comes with intimacy; at least one such man was aware of his plight but could do nothing about it. A propos of such men, Vaillant quoted Joseph Conrad's *Victory*: "Woe to the man who has not learned while young to hope, to love, to put his trust in life." Second, stress is not necessarily bad for you. One-third of the men in this study spent at least ten days in continuous combat during World War II. All suffered major personal griefs, setbacks, disappointments, and losses during adulthood. None of this, per se, predicted poor adjustment.

About a famous man not in the study, Vaillant asked, "How can I give a logical explanation for the growth of Roy Campanella, a great Brooklyn catcher who at thirty-six broke his neck, was paralyzed in all four limbs; yet at fifty the crippled Campanella seemed a greater man . . . than Campanella the baseball star had seemed at thirty." Others have made the same observation about Christopher Reeve, the actor who became quadriplegic after a fall from horseback; he went on to greater performances and became a spokesman for the disabled. We tremble at such tragedies, yet those they strike teach us transcendence.

Vaillant's is not an isolated finding. After another lifelong study of Americans, psychologist Jean MacFarlane wrote: "Many of the most outstanding mature adults in our entire group, many who are well integrated, highly competent and/or creative . . . are recruited from those who were confronted with very difficult situations and whose characteristic responses during childhood and adolescence seemed to us to compound their problems." In still another study, psychologist Emmy Werner followed 698 infants born in Kauai, Hawaii, in 1955. One-third of them suffered severe deprivation and loss in early childhood, yet in their thirties and forties one-third of *those*—about one-ninth of the total sample—developed into "competent, confident, and caring adults." Looking back, they cited a significant supportive person outside their dysfunctional families. But they also clearly started out special, with a strong tendency to make the best of life.

As Vaillant put it, echoing the great physiologist Hans Selye, "It is not stress that kills us. It is effective adaptation to stress that permits us to live." In the 1990s he extended the follow-up of the college men another fifteen years, for a total of half a century. He compared them with lifelong studies of two other groups: forty women who, as children in the 1920s, had been intellectually gifted; and some 300 men from poor families in Boston, followed from junior high school into their seventies. All three groups proved the power of individuality, the ability of many to transcend adversity, and the self-healing bias of human mental life. Eleven men, chosen from the poor sample because of extremely bad childhoods, had seemed at age twenty-five psychologically damaged beyond repair. Fifty years later, eight of them were doing well. "Man is born broken," Vaillant concluded. "He lives by mending."

Every parent learns this lesson. When my first child was twenty-one months old, just on the verge of discovering sentences, she was left for a couple of hours with doting grandparents while her mom and dad took some time together. She was accustomed to this, but on that particular evening it distressed her. She had been crying for some time when we returned; she fell into her mother's arms but was unable to stop sobbing miserably. Her mother sat and held her, and in a little while my daughter looked up and there came, out of her still heaving chest, through the tears and the dejection on her face, the two-word phrase "Zana happy." I felt as if I were witnessing the dawn of her human consciousness. Such behavior must give a biologist pause. Here was a typical mammal, juvenile version, responding predictably to maternal separation, complete with ancient, species-specific cries and facial displays of sadness; and in the midst of it all, contradicting every animal gesture, two linked human symbols, pressing forward the claim that she was happy.

"There is more happiness in one real tragedy than in all the comedies ever written." So said Eugene O'Neill, who could always move audiences with either.

One thinks first of *schadenfreude*—the joy of witnessing someone else's misery. But that isn't it, or at least it isn't all. What we are watching is a reflection of our own challenged lives—our losses, our dismally failed good intentions, our weakness, our inevitable dying. We see our own, universal, individual human tragedy, but greatly ennobled, elegant, and far above the mundane. It is the triumph of the spirit in adversity that moves us; and if we come out of the theater exhilarated, it is because we feel that we have lost, as we inevitably must, but won anyway; we feel we have, even in the face of death, come through.

In Yeats's great poem "Lapis Lazuli," he challenged the call for moroseness he saw in some of his friends' political musings. Instead he cited Shakespeare:

> . . . tragedy wrought to its uttermost.
> Though Hamlet rambles and Lear rages,
> And all the drop-scenes drop at once
> Upon a hundred thousand stages,
> It cannot grow by an inch or an ounce. . .

And yet,

> Hamlet and Lear are gay;
> Gaiety transfiguring all that dread.

The lapis lazuli of the title is a carving in that fine blue stone of three men in ancient China, on a mountain, with a long-legged bird, a longevity symbol, flying overhead. He imagines them, after a long climb, staring down from the mountaintop upon the whole "tragic scene" of awkward humanity. Do they weep? Hardly.

> One asks for mournful melodies;
> Accomplished fingers begin to play.
> Their eyes mid many wrinkles, their eyes,
> Their ancient, glittering eyes, are gay.

CHAPTER 12

Lust

. . . Anguishing hour!
Last night, we sat beside a pool of pink,
Clippered with lilies scudding the bright chromes,
Keen to the point of starlight, while a frog
Boomed from his very belly odious chords . . .

— **WALLACE STEVENS,** "Le Monocle de Mon Oncle"

Take two organisms—make them adult humans of opposite sex—and bring them together to interact under our watchful eyes, in the manner formerly known as flirtation, courtship, and marriage, latterly known as hitting on, hooking up, scamming, and getting it on.

In the manner of engineers, we *could* view this pair as interacting systems. Behavioral scientists have often borrowed the system metaphor in the hope of achieving the elegant, resilient, predictable, and even mathematically precise analyses that engineers achieve. Thus, at the start, we might say each system received many inputs, some of which brought them closer together—a toss of hair, say, or the odor of aftershave lotion mixed with male musk, or a lilting laugh in a murmuring crowd. But as the two systems approached each other, the output of each became the other's main input, complicating the engineering considerably.

What could the woman be regulating, in the sense that a room regulates temperature with a thermostat (a classic engineering system)? Emotions? Perceptions? A hormone? The level of general arousal? Happiness? Loneliness? Self-deception? Risk? And how would this emotional thermostat be set? Would it change its setting cyclically over the day and night, like some clock-controlled thermostats? Could the system cool actively as well as heat up and turn off, like a combined air-conditioning-and-heating system? Would the set point change with energy supply? Do men and women have different settings, because of genes, rules, training, or experience? Can they tamper with each other's basic settings? The complexity is daunting.

We could simplify the problem by treating the couple as one system, designed to attain the goal of mating and reproduction. This outlook served in an earlier era, when "survival of the species" was considered a valid way of thinking about mating. If species were designed by nature to perpetuate themselves, then of course each should have evolved a mating dance that is a system in itself, getting male and female into intimate juxtaposition for a higher goal outside their own hierarchy of needs. The courting pair could then be viewed as one smooth goal-corrected system. But however smooth they sometimes look, that is not what they are, because they were not designed by nature (read: *evolution*) in that way. The idea of survival of the species as a central goal is basically dead, and present-day students of evolution see the gene, the individual organism, and its close kin as the most useful units of analysis. This is equally true of the mating process, which was classically—and wrongly—viewed as independent of individual adaptive purpose. If the individual purpose is basically just a gene's way of making another gene, then there is no good reason to think of the two courting partners as sharing *any* simple goal.

In other words, they are not one system. Their respective purposes are exquisitely individual and are not happily wedded—pun intended—in the design. Neither, of course, are their goals at desperate odds in every way. But they are individual, complex, poorly matched, and, to the observer as well as the actors, guessed at but unknown. Even accepting the distinct adaptive goals, is the system model useful? Each person (or system) in the pair would be regulating some quantities (arousal?) around an intermediate or fluctuating set point, like the thermostat, but minimizing others (disappointment? fear?), cyclically varying others (lust?), and maximizing others (attractiveness? control?). Even if the long-term goal is perpetuating genes, what are the immediate ones? "Scoring?" Orgasm? Tenderness? A few flirtatious minutes of arousal? A couple of hours of sweaty intimacy? A year of companionship? A lifetime of loyalty? Love?

And consider what must be transcended for this intricate dance to lead to mating. Whenever two creatures meet for any reason, there is and should be fear. In courtship, risks include rape and other physical trauma, infectious disease, unwanted pregnancy, the humiliation of unrequited love, loss of status, seduction and desertion with the attendant pain and grief, and making the most colossal mistake of your life—entangling your destiny with that of the wrong person. Viewed as a system, the intricacies become unmanageable. Engineers can feel superior if they like, but they never attempt anything remotely this complex. Multiple moods and wants are regulated, uncertain goals are approached, and many fears are modulated and minimized (or at least balanced against the fear of loneliness) as the dance goes on. Even from what little we have said so far, how utterly trivial seem such concepts as need, drive, instinct, reflex, habit, imitation, schema, rule, game, script, narrative, ritual, and many others like them in the face of these complexities of subjective human experience. Unsurprisingly, intelligent people with some understanding of human experience sigh at the simple-minded world of behavioral science, where stick figures dance around on all-too-visible strings.

Some of the poignancy, even tragedy, in human affairs—particularly in the presence of desire—is inaccessible to behavioral and social scientists precisely because they believe in the systems model. That is, they think human interaction was designed to run smoothly. Thus the large gap between the human relationships in the classics of every language and those described by behavioral and social science. Literary artists of stature have tended to perceive and record the often nasty, sometimes deadly, illogic of relationships, but social scientists build models of human experience in which everything is okay—or if it isn't yet, it soon will be. This is as much a product of their penchant for systems logic as it is of a sunny disposition. Systems should work, whether they are individual people, interacting pairs, small groups, large populations, or even groups of populations. If they don't work, something is wrong with them; you fix them and then they will. If they haven't yet been fixed, keep working at it, and eventually they will be.

I call this the tinker theory of human behavior. Its practitioners, far from being itinerant, are well established in every field of behavioral science, in think tanks of the left and right. For psychotherapists and economists, marriage counselors and social revolutionaries, policy wonks and congressional aides, the tinker theory is daily bread and butter. They work along as if on an intricate jigsaw puzzle, fitting a piece here, a piece there, and they truly believe it is only a matter of time before the pretty picture becomes whole. Unfortunately, they are working on a puzzle in which the pieces do not fit. Or if they do, it is in another dimension, beyond the tinkerers' awareness. That dimension is evolutionary time. A jigsaw puzzle is logical and solvable because it is designed by a mind like that of the solver. Unless the designer was perverse, the pieces will all be there, they will fit, and the picture *will* become whole. But in evolutionary time—a stream in which not only Methuselah's life span but even the time since is barely a drop—there is no designer. There is only the blind action of natural selection, amoral and inherently disordered, sifting genes.

A person is only a gene's way of making another gene. Richard Dawkins may have said it first, but it is a thought that comes to the mind of anyone who understands evolution. Of course, it is oversimplified. The gene, even in a virus, does not act alone; it cooperates with other genes, at least temporarily. Together, they make a microbe, a lily pad, a pussycat, a person. That creature, to be sure, has subsidiary purposes: to be moist, take the sun, eat a meal, learn, embrace another creature, tear a mouse limb from limb, sniff mustard, play, feel safe from harm, get out of a cage, empty its bowels, bomb a city, see a sunset, get a reward in heaven, ejaculate, win the Nobel Prize.

All that is fine with the gene. It is not distracted by the existence of such purposes, by the creature's unawareness of its own teensy molecular purpose, nor even by an outright denial of that purpose. How could it be? It has no machinery of awareness. It doesn't know anything. It simply arranges the chemicals of life around itself, assembling protoplasm to provide one or another kind of container. If this container carries the gene around, protects it from disintegration, and facilitates, at crucial moments, the gene's effort to assemble some molecules around it into a mirror image of itself, then the gene will remain part of the stream of DNA—the conduit of life since life began. The stream meanders, has eddies, tributaries, deltas, bayous, branches—it is the central fact of evolutionary time. But it is constantly changing, even around the edges of long-static pools. The changes can be point mutations—alterations in the chemical structure of the gene itself. Or they can be rearrangements of the genes in relation to one another, also important for the shapes the gene carriers (the organisms) will take. But the elementary process is the change in gene frequency. And a fundamental contributor to that change, amid the environmental and statistical noise and chaos, is natural selection—a change in gene frequency due to the success of the gene's clumsy carrier.

For a human gene, that clumsy oaf is us. There was a time when all reproduction on the planet was asexual. It sounds boring, but the creatures that practiced it could not have cared less. They were invertebrates, elegant and intricate in their way and quite successful at reproducing without sex—a feat carried forward in many present-day organisms. In effect, back then there were only females. They were the only bearers of DNA and they all participated equally and directly in generating the young. The question, an unsolved one that occupies some of the best minds in biology, is, *Why in the world did they ever invent males?* Males are impractical. If you have a balanced sex ratio and males are about the size of females, you are wasting half your biomass on creatures so useless they can't even make eggs. One traditional explanation, that sexual reproduction increases variability—to speed up evolution, say, or to baffle microbes—seems unequal to the challenge; asexual species find other ways to solve those problems. Another theory, that males can help with the young and provide for the common defense, runs aground when we consider that in many sexual species the male doesn't hang around long enough even to wave goodbye. Any way you look at it, the first tentative males must have been females who saw an unexploited opportunity and pioneered intimations of masculinity.

We know that a sexual species may displace asexual competitors, as has happened with geckos in the Hawaiian and Fijian islands. Males of the sexual species are larger and more aggressive than the asexual females, who withdraw from the best insect-hunting grounds when the sexual males appear. One intriguing suggestion is that females created males to make this kind of thing possible. But in a sense, the whole history of sexuality has been "one vast breeding experiment" in which females were the breeders and males the selected breed. According to this view, the first sexually reproducing females may have gained a great deal by sequestering half of their DNA away from themselves and locating it in expendable new creatures—namely, males. These females reduced competition among themselves while intensifying it among the males. Many males could be discarded without reducing egg production, and a relatively elite group of chosen males with desirable traits could father the females' offspring. Meanwhile the females could focus on making more.

John Updike's 1997 novel, *Toward the End of Time,* claims hilariously, "If women were fastidious, the species would go extinct." But in fact the whole course of evolution is due largely to the fastidiousness of females, a motive so strong it often trumps all others. Female widowbirds prefer males with very long tails, even those who take their tails to otherwise maladaptive lengths. Peacock tails owe their glory to peahens' pickiness, and even the mutual bashing of male elk or elephant seals takes place against the background of choosy females patiently waiting to tap a winner. In contrast to the public displays of elk and peacocks, the process is often hidden; in many insects, genital shape is literally the key. In short, this sexual selection theory of Darwin's now rests on a vast body of evidence, and increasingly the neural and genetic mechanisms of female choice are being discovered. In fact it is so powerful that some naturalists now speak not just of female choice but of female *control*.

The origin of males will always be debatable, but we now know a great deal about how they evolved afterward. For example, some things about males are predictable from how much direct care they give their offspring. In about 8,000 species of pair-bonding birds and in some mammals—including the prairie vole, the small South American monkeys called marmosets, the gibbons or lesser apes, and a few carnivores such as coyotes and bat-eared foxes—the male tends to contribute substantially to the care of the young. Males and females tend to be about the same size and color, lack fancy or threatening appendages, and grow at about the same rate. They pair off for the long term and guard their mates, fighting off same-sex strangers. This does not mean they approach some romance-novel ideal. They may cheat on each other, flash deceptive signals, desert, fight, separate, and remate with others. All of this, in fact, was observed by Konrad Lorenz in his own famous group of devotedly pairing geese. But they are not promiscuous, and in principle they form mated pairs. If they don't quite forsake all others until death, at least they tend to stick together over the longer haul, and the death of a long-term partner leads to something that looks like grief.

Part of this pattern is the male's participation in care of the young. Male marmosets, for example, may carry the young (usually twins) two-thirds of the time, giving them to their mother only to nurse. Pair-bonding mice show much more paternal care of the young than do those from a closely related but multiple-mating mouse species. The same is true of vole species that do or don't pair, and in this case the paternal male has a distinct pattern of brain receptors for vasopressin, a hormonal key to fatherhood. In some species of pair-bonding birds—the ring dove, for example—males have elaborate physiological adaptations for feeding chicks that rival those of the female. In pair-bonding species—monogamous seabirds, for instance, or us—sexual selection tends to be mutual, with choosy males as well as choosy females.

But the other end of the spectrum tells a different tale. Here males take little interest in infants and juveniles. They differ markedly from females—they are larger or gaudier, possess more dangerous weapons, grow more slowly, or often some combination of these. Males may be triple the weight of the female, as with the elephant seal; may display gorgeous colors, as does the peacock; or have antlers for prancing or battering, as do many antelope. These are called tournament species because fierce competition for females may put males at odds in an annual breeding tournament, or lek, as in the Uganda kob or the elephant seal. One result is extreme variation in male reproductive success. This is achieved partly through differential access to females (winners take much more) and partly through higher mortality (losers lose all) than for females—a fact of life for most sexual species, including ours, but more so for tournament species.

Competition can be extreme. In one breeding season among the elephant seals off the California coast, studied by ecologist Burney LeBoeuf, 4 percent of the males accounted for over 85 percent of the copulations and these few males impregnated the great majority of females. The rest of the males—96 percent— were left to compete for the attention (and the wombs) of the last few females. Even over many seasons, only 9 percent of males ever reproduce. This is sexual selection with a vengeance, and males in such species do little or nothing for the young. In elephant seals, breeding fights among the enormous, dangerous bulls may result in injury and death to newborn pups. Theory says that only in pair-bonding species do males "know" where their genes are—meaning they have been selected throughout evolution to act as if they did. In pair-bonding species, the male has the female more or less corralled. She has *him* corralled as well, but for reproductive certainty that is beside the point. In species with internal fertilization, a female is rarely "in doubt" that the offspring she cares for carry her genes, whereas no male in any such species can ever "be sure" he has not been cuckolded.

In tournament species, top males career around, coupling where they may, with females almost as willing in one place as another. A male caring for *any* offspring would run the risk of spending energy on genes not his own. And he would be wasting precious time that he might have spent finding and fighting for sex. His own genes, which sparked that failing strategy, would be gone in future generations. Of course, none of this assumes a conscious recognition of the process; evolution has merely produced a design that makes animals act, however blindly, to enhance their reproductive success. We don't know how species get to be pair bonding or leklike in the first place, and the correlations are not perfect. But once they have diverged, pair-bonding species tend to have unspectacular males, less variation in male reproductive success, more male care of offspring, less promiscuity, and less competition among males. Tournament species tend to have large or florid males, high variability in male reproductive success, little or no direct care of offspring, high promiscuity or polygamous mating, and fierce male-male competition. Females are at a loss compared with dominant males, but subordinate males—sometimes the majority—lose everything.

Elegantly for theory, even the exceptions prove the rule: in species such as jacanas—sometimes known as Jesus-birds because, striding on lily pads, they seem to walk on water—*females* are larger and *males* care for the young, so males become a scarce resource that females compete for. A female may sequester several males, each of which sits on some of her eggs. Phalaropes show a similar pattern, and wild English moorhens tend in the same direction as well. They have a penchant for small fat males, who then incubate eggs better than their less plump competitors. Females who snare such well-rounded beaux actually bring more broods to hatching in a season.

The most dramatic reversal is probably that of the spotted hyena of Kenya, where feisty females rule the roost and meek males tag along for the ride. Females are fully masculinized in this species, not just behaviorally but anatomically, with an enlarged, virilized clitoris. Females are larger and much more aggressive than males, who typically migrate into a group that is thoroughly dominated by a hierarchical matriline, and basically beg for acceptance, sexual and otherwise. This unique type of female is created in the womb, where the placenta largely lacks an enzyme needed to convert androgens to estrogens, so that a backed-up inventory of androgens virilizes the female fetus. This is an intriguing example of extragenetic influence on a trait central to survival and reproduction. The likely adaptive cause is a very strong correlation between female rank and reproductive success in a setting of high population density, although exceptionally intense and aggressive competition between nursing littermates may also be a cause.

Incidentally, hyenas' reputation as scrounging scavengers is nothing but a bad rap. Ninety percent or more of their food comes from hunting they do themselves, females as well as males. All in all it is not surprising that young women studying biology have embraced the hyena with open arms. Hyenas, jacanas, phalaropes, and other species with partly reversed sex roles illuminate the evolutionary process and prove that gender as we know it is not irrevocably set in the genes. But important as they are for theory, they are exceptions. In most species of birds and mammals, males are at least somewhat larger, more conspicuous, more competitive, more variable in reproductive success, and less caring toward their offspring than females. In pair-bonding species (including most birds), the differences are small; in tournament species (including most mammals), they are large. Most species are somewhere in the middle.

Consider, for example, the humble bank swallow, known to its devotees as *Riparia riparia*. The sexes resemble each other in size, form, and behavior. Mated pairs form and persist in large colonies, each pair digging and nesting in one of hundreds of burrows in a sheer sandbank above a natural waterway. Both sexes live by foraging for insects, but in the first week after pair formation, the male follows the female every time she leaves the nest, chases her, and stays within a meter of her, which calls for some impressive aerobatics. He does this up to 100 times a day. It is always male-chase-female, never the reverse, and the female can never leave without being followed. The pair often attracts a third chaser; out of more than 100 such instances studied, the interloper was male every time. In fact, there may be up to five males at once. As observers Michael Beecher and Inger Mornestam Beecher put it, "All the males follow the intricate maneuvers of the female, giving the chase its spectacular appearance." The mated male "will loop back and attempt to fight off the chasers," bumping or attacking them. When the odds are against him he tries to chase the female back into their burrow, but interlopers at times complete copulations. Underscoring the behaviors' functions, both the chasing by interlopers and the guarding by mated males are restricted to the female's fertile period. And some of the interlopers may even be males mated to other females who currently are between their fertile periods.

Or look to the very different ring dove, *Streptopelia risoria*. This beautiful bird's courtship and reproduction became classic, thanks to psychologist Daniel Lehrman, who used to delight students by climbing on tabletops and mimicking bow-and-coo displays. Under natural conditions it is a strongly pair-bonded, tree-living African dove in which the male parents intensively—and is biologically built for it. But the most elegant studies have been in the laboratory. Pair formation requires an elaborate, prolonged courtship period in which the male bows and coos to the female. Watching this behavior—or even a film of it—directly stimulates the female's pituitary gland to secrete luteinizing hormone. This in turn stimulates the ovary to make progesterone and estrogen, the hormones that prepare her, physically and behaviorally, to breed. The pair build a nest together and copulate, and the female gestates and lays the eggs.

The male's initial gestures depend on his testosterone level, which favors the responsiveness of brain circuits controlling bowing and cooing. In fact, the weight of the female's oviduct can be plotted as a function of the male's testosterone level, a powerful anatomical change that depends on another individual's behavior. Both male and female brood the eggs; this stimulates each of them to develop crop milk, a specialized baby food produced in a side chamber of the esophagus. They regurgitate it when they see and hear the newly hatched young. Males are as adapted for feeding as females. And, as theory predicts, they guard their paternity carefully. Following up on Lehrman's work, Patricia Zanone, Eleanor Sims, and Carl Erickson showed that males can detect females that have already been courted by other males, and will reject them. Further, if the male of a mated pair is separated from the female and brought back to her, there is little effect, *unless* the female is in the interim exposed to another male; in that case her returning mate assaults her with a severe pecking.

Or consider the redwing blackbird, *Agelaius phoenecius*. Like many smallish perching birds, it is highly territorial. Males do not care for the young to any great extent, but they do establish territories, defend them, advertise them with a classic, recognizable spring song, and attract a female, whose young are fed from the territory's resources. Intruding males are attacked by the resident male, who while attacking undergoes rapid and large changes in levels of luteinizing hormone, which stimulates testosterone and other androgens, as well as changes in the androgens themselves. A male on a superior territory can attract and keep more than one female, while a male on barren territory has none. But here is the rub: in experiments in which mated males were vasectomized, the female still often managed to lay fertile eggs, demonstrating under natural conditions that the threat to a male's paternity from unseen strange males is real. Thus the selective pressure for territorial defense.

These three species exemplify mating in pair-bonding birds. But the continuum extends through to species where females take a malelike role. In addition to jacanas and English moorhens, there are three species of phalaropes—sandpiper-like shorebirds—in which the females are bigger, more bold in plumage, and more aggressive than the male. The female Wilson's phalarope arrives first on the breeding ground, where the pair will nest amid marshy vegetation. *She* picks a male from among the arrivals and jealously guards him as they swim together, feeding from the water surface. She threatens and drives away other females. Males may fight among themselves, but less aggressively than females. And in all three kinds of phalaropes, *all* care of the eggs and young is done by males. Thus, it is the males that invest most in offspring, and they are the resource that must be fought over.

These exceptions show that the patterns are not ultimately intrinsic to one sex or the other: evolution has reversed the roles in a few cases. But these are a handful among 8,000 species of pair-bonding birds—they have already taken up a proportion of this discussion many times larger than their role in nature. The bank swallows, ring doves, and redwing blackbirds are much more typical. Males are adapted for greater aggressiveness, whether toward other males or toward females, and the physiological changes leading to sex also increase their tendency to fight. Males use this to guard their mates and prevent cuckoldry, at the same time trying to steal sex outside the pair or to attract other females to their side. They act as if they want it both ways, and theory holds that they should, since sperm are much less costly than eggs and since males can never be sure that their main mate's offspring is really theirs.

Although eggs can be surreptitiously slipped into her nest—cuckoos do this for a living—the female has much less "doubt" that she is the mother and has little to gain, in offspring quantity, by ranging away from her territory for stolen copulations. She can increase male quality and buffer herself against desertion, particularly in species in which mates may pair off differently in successive seasons. These potential gains make her receptive to stolen copulations. Males have evolved mechanisms to protect themselves from such cuckoldry, including aggressive defense, corralling of the female, detection of cheating followed by abandonment, and the counterpolicy of casting their seed widely. Females, having less to lose by male cheating, have evolved less dramatic forms of prevention. Females cheat less often. But genetic studies have shown that some males in pair-bonding species cannot have begotten the offspring they are raising. Females are not so committed to bad-boy males that they just sit back and take it; they have their own strategies.

Again, these are examples of the better-behaved end of the mating continuum. Males in non–pair-forming species (few birds but most mammals) are typically more brutal toward one another, and more aggressive and neglectful toward females and young—although to be sure they are not all as bad as elephant seals. Perhaps a kind of carnal chaos is most basic. Consider the mating habits of the common domestic dog, not the sweet ones on short leashes but those that are "natural" in the urban wilds of India. Their mating is something of a free-for-all. Bitches in heat exercise choice but mate serially with several males, and some mate promiscuously. Preferred males mate more, but dominant males use force to get their way even when not chosen. Fights between males in mating spots outnumber fights between females, and successful males will mate wherever they can. Such promiscuous mating does not lead to the sexual dimorphism of elephant seals, but neither does it produce the commitments of pair bonding.

Even our closest relatives show great variety. Dominant male baboons are big, and they succeed genetically. Gibbons are more or less monogamists, and the sexes resemble each other. Gorillas tend toward harems, with males about twice the size of females. Chimps grade toward the free-for-all of dogs, with modest size differences and multiple partners for both sexes. Females attract males in a timely way with great pink vulval swellings while in heat, whereas males have colossal testicles, designed to enter intense sperm competition with, well, their best shot. Bonobos, in contrast, play it cool, with an easy, promiscuous but relaxed approach to sex, both lesbian and hetero, concealed ovulation, and no estrus drama. Still, all is not sweetness and light: females try to dislodge males from rival females right in midthrust.

Humans are not hyenas of course, but neither are we elephant seals. We seem to belong in the part of the continuum occupied by bank swallows, ring doves, and redwing blackbirds. We are clearly pair-bonding, and clearly imperfectly so. Males and females develop at different rates, with girls reaching maturity a year or two earlier. Males become larger than females and in most populations have conspicuous hair on the main display organ: the face. Females have breasts and hourglass shapes. But as mammals go, human males and females do not differ drastically. Mating habits settle out similarly. George Peter Murdock, one of anthropology's great systematizers, found that of 849 human societies in the ethnographic record, 708 of them (83 percent) practiced polygyny—one man married to two or more women. We are talking here about *official* marriages, not just sanctioned dalliance. These 708 were about equally divided between those with usual and those with occasional polygyny. Most hunting-gathering societies have the latter, and often it is the older men in such societies who have several younger wives, causing considerable frustration among young men. Among the traditional Australian aborigines, this gerontological polygyny pattern was common enough to prompt a formal question at a distinguished symposium: "Why do all those old men get all the girls?"

The answer seems to be, *Because they can.* But this is only a special case of a general rule: as in redwing blackbirds, males with the best resources garner extra females. In hunter-gatherer societies, only the aging process brings a man that kind of status and power. But in even the simplest stratified societies, wealth differentials give young men some advantages as well. Monique Borgerhoff Mulder studied marriage and reproductive success among the Kipsigis, herders and farmers of Kenya. As in many traditional societies, bride wealth must be delivered by the groom to the bride's family. But more bride wealth is given for plump, early-maturing women who will tend to have more children, and less for a woman who has a child by another man. Rich men take several wives and have up to eighty children. Yet this only hints at the possibilities.

Laura Betzig, giving a new interpretation to unquestioned historical evidence, decisively showed that one of the main consequences of despotism has been to garner females for leading males. Despots on all continents have kept harems, usually numbering in the scores or hundreds of wives and concubines. These women produced up to thousands of children for one such man. It is perhaps a mark of democracy that leaders, even those guilty of sexual indiscretions, do not have many children. Ronald Reagan was the first U.S. president who had divorced and remarried. President Clinton refused to have sexual intercourse with his White House intern, although the image of her on her knees servicing him in his office is not exactly egalitarian. But even in modern France, with its open contempt for puritanical America, nothing like despotic sexual favors is possible. President François Mitterand was limited, at least officially, to one wife and one mistress—both of whom, with their children, came to his funeral.

Monogamy rules in 137 (16 percent) of the societies in the ethnographic record, but in most of these an individual may have more than one mate in succession—life is short, divorce is allowed, or both. And because of the starkly different reproductive life spans of men and women, men are more likely than women to have more than one family. This is true even in a culture like that of Pitcairn Island in the South Seas, where centuries-long historical records show consistent, strong monogamy. Thus, in our species monogamy grades strongly toward serial polygyny. Polyandry, in which one woman marries more than one man, occurs in four of the societies on record (less than half of 1 percent), and in all there are special conditions. For example, a woman in polyandrous highland Tibet may be marrying three brothers so that they can avoid breaking up their father's farm. The pattern is not a mirror image of polygyny. So in summary, the human species can be said to be pair-bonding with a substantial polygynous (but minimal polyandrous) tendency.

Other cultural comparisons are relevant. Murdock reported the exchange of goods or services at marriage in a sample of 860 cultures. In 3 percent of them women were directly exchanged between the families of the two grooms. In bride-wealth cultures such as the Kipsigis of Kenya—64 percent of the worldwide sample—the groom or his family gave goods or services to the bride's family. In 30 percent there was either no exchange or an equal exchange of gifts. Dowry, in which the bride's family gives gifts to the groom's family, occurred in only 22 cultures, less than 1/20 of the number that had the reverse custom, bride wealth. And dowry, like polyandry, requires special conditions. As Mildred Dickemann and John Hartung showed, dowry occurs where monogamy prevents powerful men from accumulating women. These men's scarcity, together with their potential for raising equally valued sons, makes a link to them worth paying for. But in most human matchmaking, wombs are the scarce resource.

Social psychologists have long known that in choosing lifelong mates, men give looks and romantic love more weight than women do, while women weight status and wealth more heavily. Sexist distribution of status and wealth is often blamed for this discrepancy, but the great recent success of women in the workplace has not changed the pattern. Early in 2000 the sublime became ridiculous as hundreds of millions of television viewers watched *Who Wants to Marry a Multi-Millionaire?*, a show in which a modestly rich middle-aged man selected a bride from among hundreds of eager applicants. His choice, like the other finalists, was poised onstage in a wedding gown. She was young, blonde, beautiful, and had a fine career as a specialized nurse. They married on the spot, and the union, probably unconsummated, was over in days. Yet the age difference and the marriage of wealth to beauty were classic, even if the spectacle was grotesque. The choices were wrong on both sides, but they were nonetheless predictable, and, as William Irons has shown, they occur in most cultures.

To test the claim that cultural shaping is key, David Buss, a leading evolutionary social psychologist, studied human mate choice in thirty-seven different societies. In all thirty-seven, women were more concerned with status, men with beauty. Cultures show surprising agreement about who is good-looking, although beauty in women has a relatively narrow definition, while in men good looks are defined more broadly. Preferred women tend to look like Lara Croft, the cyber-creation in the popular Tomb Raider game—large eyes, small nose and jaw, ample breasts, and a waist-to-hip ratio of about three fourths. She trounces men at every turn, yet matches much of the classic ideal of beauty. Cultures do differ in weight preferences, with most cultures preferring greater plumpness than the modern Western ideal, yet the waist-to-hip ratio preference persists. In addition, in every one of Buss's thirty-seven cultures men prefer slightly younger women and women prefer slightly older men. All these patterns, outlined decades ago by anthropologist Donald Symons, fit well with Darwinian theory.

The "double standard," according to which women are more strongly punished for infidelities than men are, is very widespread in human societies. In many societies women adulterers are punished by death. According to the Kinsey Reports on sexual behavior among Americans during the 1940s, men in most age groups were at least twice as likely as women to have had extramarital intercourse. There is evidence that permissiveness has increased markedly since that time and that the sexual revolution is real; with that change, the double standard has weakened, but it still exists, at least in the United States. Men and women differ substantially in every measure of sexual attitude and behavior, with men consistently practicing and approving more and more varied sex with a larger number of partners. To take one comparison out of hundreds that all point the same way: of Americans self-identified as "conservative," 63 percent of men and 94 percent of women say that sex with a stranger appeals to them "not at all"; among Libertarians, however, the corresponding figures are 20 percent for men and 61 percent for women. Men are also more likely than women to predict that their marriages would end if their partners committed adultery.

Does the male strategy actually enhance reproductive success? Patricia Draper and Henry Harpending proposed a distinction between "dads" and "cads"—men who care for the young they father and those who instead devote their efforts to finding more sex. Evidence suggests that either strategy can work, which may explain why neither has been culled by natural selection from the human repertoire. Alexandra Brewis, an anthropologist at the University of Georgia, studied husbands of both sorts on Butaritari, a Micronesian atoll. Cads—the classification was based on interviews with their wives and others—have a higher number of live-born children and shorter birth spacing, but their children have a higher mortality rate. In the end, they have as many offspring surviving to early childhood as the dads do—but at a tragic cost.

Part of the reason for the double standard is that men in every society are more aggressive than women, and men account for the overwhelming majority of homicidal violence. Such violence, whether the victims are male or female, is often occasioned by sexual infidelity. Among the !Kung San, noted for gender egalitarianism, men have been known to commit homicide over adultery and, in one case, to kill the wife herself. Severe wife beating often follows the discovery of adultery; whereas women's response to their husbands' infidelity, while equally angry, is much less dangerous. Barbara Smuts of the University of Michigan has done the first systematic work on male violence against females in evolutionary perspective, and her work exemplifies how evolutionary analysis can be separated from ethical judgment. In some species coercion is part of normal sex, but that is not a common *human* adaptation. Indeed, coercion in humans appears to be pathological. For example, psychologists Neil Malamuth and Mario Heilmann have found that coercive male sexuality results from high hostility combined with strong promiscuity.

Still, there is unfortunately much evidence to suggest that rape has played a role in many societies. Among the Yanomamo, "the fierce people" of the Venezuelan highlands, groups of men traditionally attacked other villages to seize land and wives. If they succeeded, they killed the men in the village and took the new widows as wives. If these women had infants, the infants might be killed as well. For the Yanomamo and for the Shavante Indians of the Mato Grosso in Brazil, there is good genetic evidence of enormous variation in male reproductive success. One Shavante man had sired 23 children by various wives, while 16 men in the same group had only one or none—a disparity 3 times that for Shavante women. Napoleon Chagnon showed that Yanomamo men who have killed other men have more offspring than men who have not. And William Durham, surveying the ethnography of the Amazon basin and beyond, found a Yanomamo-like pattern of village raids to be widespread and predictable under certain environmental conditions.

But men exceed women at both ends of the continuum. In most human populations male mortality is higher than that of females at all ages, which means that males are more numerous at both the zero end of the scale of reproductive success and at the extreme high end. This skewing is exacerbated by sexual competition. Even in the United States there are far more men than women who have had ten or more partners, but there are also far more men who have had none.

Some cultures or subcultures have attempted to suppress sexual urges to the point of dispensing with sex. Despite the famous sins of some popes, Catholic priests have always served to challenge the idea of a relentless male sex drive. Even more strongly than they renounced sex, they renounced reproduction, which the church rightly recognized as their most impressive sacrifice. Still, for men to truly suppress their sex drive would create serious problems for biological theory. So it is of great interest that a former priest, psychotherapist Richard Sipe, did a sympathetic, multi-decade study of celibacy in the priesthood. Sipe concluded that of those who have taken a vow of celibacy, 2 percent have achieved it unquestionably at a given time. Eight percent have "consolidated" celibacy, with some backsliding. Forty percent subscribe to what Sipe calls celibate practice; they are celibate in principle and make a serious effort to avoid sex. The other half of all priests have incorporated heterosexual (28 percent), homosexual (10 percent), pedophilic (2 percent), or other active sexual behavior (sex with adolescents, "problematic" masturbation, or transvestitism) into their lives. While many honestly strive toward the ideal of celibacy, and the quest itself is a valid one, few truly succeed. "The Church's teaching on human sexuality," Sipe concludes from respectful, lifelong study, "is not credible." Perhaps success is greater with celibate women, but there is no comparable research.

One clear gender distinction is that almost all rapes are done by men. It seems obvious, but is it? Men often rape other men, especially in prisons, and certainly lesbians could attack women who attract them, but they do not—or they do it in such small numbers as to make little impression on the statistics. As for heterosexual rape that might result in conception, only men are capable of enforcing it. Intercourse, at least until Viagra, has always depended on male desire, a physiological legacy of eons of biased sexual selection. And it is not the only one. There is much about which to say *vive la différence*, but sex differences in arousal and orgasm are not necessarily cause for celebration. For most mammals there is no evidence that females even have orgasm, although females among our closest relatives, the higher primates, probably do. There can be no such question regarding males, since orgasm is for most purposes synonymous with seminal emission. The genes of males who do not decisively ejaculate are quickly culled from the stream of DNA.

Females who do not have orgasm are missing one of life's great experiences, and there must be evolutionary consequences. Perhaps they are less inclined to copulate and are therefore more choosy; perhaps orgasm even plays a small role in fertilization. But it is perfectly possible for females to achieve high reproductive success *without* orgasm. If they are reluctant, they can be urged; in some species, if they are very reluctant, they are forced. And in many species there is a moment—annually or semiannually in most birds and mammals, monthly in monkeys and apes—when the female is overcome with a rush of hormones that coax her not just into cooperation but into enthusiasm. This is estrus. The word is Latin for "frenzy," but the behavior is known to pet lovers simply as "heat."

Humans do not have estrus—one of the distinctive facts about our species. But the notion that human females are continuously receptive—popular in older anthropology textbooks—is just naïve male optimism. What they meant was, the times women feel sexy are distributed more or less randomly over the days of the month or year. The exception, in many cultures, is lowered receptivity during the days of menstrual flow, when sex is considered polluting or taboo. But the otherwise random receptivity is an impressive evolutionary change. In most monkeys and apes there is a clear monthly surge of female sexual posturing and signaling toward sexually active males. In some, such as the savanna baboon or the chimpanzee, there are distinctive physical signs as well—genital color change and swelling. Many monkeys and apes also have annual breeding patterns, and in some, like the South American squirrel monkey, whose males fatten up for sex, annual breeding entails physical changes in both sexes.

Such cycles are regulated by neuroendocrine clocks. Annual cycles in most vertebrates depend on the pineal gland, a pea-size organ between the cerebral hemispheres that Descartes thought was the seat of the soul. The pineal's activity is partly controlled by light, which in some species, like frogs and newts, strikes it directly through a paper-thin skull. In higher primates and humans it lies too deep for that but still responds to darkness and light. A sympathetic nerve path starts at the retina and stimulates the pineal, which in turn pours out melatonin, made with a few key chemical changes from the neurotransmitter serotonin. Through dark winter days melatonin and other pineal secretions suppress the gonads until light becomes long and strong enough to stop the pineal from stopping sex—and when the time comes it does so promptly, due to a sensitive molecular switch. The pineal may not tie the soul to the body, but it detects the growing light of the spring day and creates a new hormonal balance, whereupon newts and squirrels start to act . . . well, like the birds and the bees. In some species, though, it can be difficult to separate the effects of light from those of annual changes in behavior. For example, among sifakas, a polygamous prosimian primate of Madagascar, males have large seasonal fluctuations in testosterone. But Patricia Whitten and Diane Brockman have found that social factors like dominance challenges and group transfer are as important as climate in determining these seasonal changes.

The monthly rhythm of monkeys, apes, and humans is more complex. Only females have it and it depends on balancing influences among the hormones of the hypothalamus, the pituitary, and the ovaries. First the hypothalamus makes minuscule amounts of a small peptide hormone—usually called GnRH, for gonadotropin releasing hormone—into the tiny local circulation leading down to the pituitary. Intriguingly, for fertility it must be released in pulses, some ninety minutes apart. At the pituitary it activates two protein hormones—LH (luteinizing hormone) and FSH (follicle stimulating hormone). Both are named for their role in the ovary, but males have them, too, and in males the final gonadal hormones are androgens—especially testosterone and dihydrotestosterone, which together account for most of biological masculinity. In males this brain-to-gonad stream does *not* flow in cycles. Females, humans included, have a clock. The ovaries do not kick out random amounts of estrogen and progesterone, nor does the pituitary constantly ooze LH or FSH. Both pituitary hormones surge about fourteen days after menstruation starts. Before this, estrogen rose gradually, reaching its first monthly peak several days earlier. Progesterone rises slowly *after* the fourteenth day and peaks about a week later—a time when estrogen, which has dipped, makes a second peak. The pituitary hormones fall about as fast as they rise, so that their heights last just a day or two. Estrogen and progesterone both fade during week four.

Meanwhile the ovaries change, both cause and result of this fluctuant chemical cocktail. During the first two weeks, an egg-containing follicle grows and releases estrogen. In response to the pituitary LH surge, the follicle pops, releasing the egg, which may then be fertilized. The egg leaves behind a sort of discard organ, the corpus luteum—hence the name luteinizing hormone—a fatty blob that raises blood progesterone to its twenty-first-day summit. Progesterone builds the uterine wall in preparation for a budding embryo's soft, blood-rich landing. Without fertilization or implantation, progesterone and estrogen start their monthly slide, and their withdrawal causes the womb's new lining to slough off— menses, day 1 of the next cycle. We take menstrual cycles for granted, but in our ancestral condition they were fairly unusual. Boyd Eaton and his colleagues have estimated that because of pregnancy and lactation, hunter-gatherer women had only around 150 cycles in an average reproductive lifetime, compared with about 450 in women today. This means that both physiologically and psychologically women evolved under greater exposure to the hormones of pregnancy and lactation than to those of monthly cycles. The change has profound implications for the risk of women's cancers, but it also must have psychological and cultural consequences. For instance, if bleeding from the womb were rare except for infertile women, it is easier to understand why it may have seemed dangerous.

All these hormones affect behavior, especially sexual behavior—male and female, animal and human—and a genetic dissection of the mechanisms is now possible. For example, mutations in the gene for GnRH make a mouse with very low levels of it, as well as of LH and FSH. Reproductive behavior is absent. Yet, if you implant a few genetically normal cells—cells that *can* make GnRH—in the hypothalamus of the adult mutant female, you will restore her sexual life. This is because in addition to its effect on pituitary hormones, GnRH has its own direct effect on brain tissue, since some of the cells that produce it (and other hypothalamic releasing hormones) also send axons to other parts of the brain. There the releasing hormones may operate as neurotransmitters with varied effects on sex. For example, if appropriately delivered into the brain of a rat, GnRH activates female lordosis, in which the body is arched downward and the rump raised in response to the male.

This brain hormone also has contradictory yet powerful effects on the sexual behavior of higher primates, us included. Maryann Davis DaSilva and Kim Wallen showed that GnRH suppresses testicular function and sexual behavior in rhesus monkey males, and it also suppresses ovulation and sexual activity in females. In a human study, men with temporary suppression of sex hormones using a drug that blocks GnRH had far fewer erections and sexual fantasies, as well as other signs of lowered libido; giving testosterone at the same time compensated, keeping sexual interest (and organs) up. Clinically, GnRH has also been used (in the form of a nasal spray) as a treatment for male impotence, as well as for severe sexual deviance such as pedophilia; it looks as if it may improve both. The resolution of this paradox may have to do with the importance of secretion in packets, or pulses, under natural conditions. If given continuously, the hormone can have the reverse of its normal effect.

LH and FSH have direct effects on sexual behavior, partly because they target certain cells in the brain. Contrary to earlier theories about the function of the blood vessels connecting brain and pituitary, we now know that these vessels carry hormones both ways. More important, some of the hypothalamic cells that make releasing hormones send axons up into the brain. But perhaps the strongest sexual chemistry still lies in the gonads: testosterone and other androgens in males, estrogen and progesterone in females. All are steroid hormones, built from the basic four-ring structure of cholesterol. All are made in other organs in addition to the gonads, notably the outer adrenal gland, and all are present in both sexes. Recall the effects of these hormones. Testosterone removal by castration or hormonal suppression causes a decline, rapid or gradual, of sexuality in the males of many species, although extensive experience before surgery—a long and active sex life—combined with patient partners afterward slows the decline in cats and humans. Testosterone replacement reverses the trend and also enhances normal sexual behavior in uncastrated males of some species.

Similarly, estrogen facilitates female sexual activity in various species, and ovary removal typically reduces it, although as with male castration, experience plays an important role. Progesterone, on the other hand, has the paradoxical effect of enhancing sexual activity at some times or in low doses, while inhibiting it at other times or at higher doses. Its enhancing actions seem to depend on estrogen and are most evident at low estrogen levels, after a couple of days of estrogen exposure; the effect requires progesterone receptor gene expression. Finally, testosterone also enhances female sexual drive and may often be needed for it. Richard Michael and Doris Zumpe first showed this in rhesus monkeys, and growing clinical experience with testosterone replacement for menopausal women shows that it often revives a libido lying dormant even after estrogen replacement. And rising testosterone at puberty predicts the onset of sexual behavior in girls as it does in boys.

In fact, all of the steroid sex hormones rise markedly at sexual maturation, causing many of the physical and psychological events of puberty. Yet some of this may start much earlier. Growing evidence, summarized by Martha McClintock and Gilbert Herdt, points to the psychological importance of rising levels of adrenal androgens long before gonadal puberty. By about age ten, these hormones have increased tenfold. Significantly, a number of studies of both heterosexual and homosexual men and women show that recall of their first romantic attraction or fantasy centers around age ten. Some prepubertal physical changes, including the first pubic hair, oilier skin, and changing body odor, depend on these adrenal androgens. They also enter the brain and probably have effects there as well.

But these are just preliminary events heralding much greater ones: the maturation of the gonadal axis in puberty. At this time, the female cycle begins in all higher primates, including humans, unless they are pregnant or starving, and the cycle influences some mental processes. The cyclicity in turn depends on an intact hypothalamus and is a basic part of the *initial* genetic plan of *both* sexes. However, for males, testosterone circulating in early life permanently masculinizes the hypothalamus, abolishing its potential for cycling, while females among many monkeys and apes have obvious behavioral changes with the cycle. Some aggressively seek sex at ovulation, and males avidly pursue them. Irven DeVore showed, and Glen Hausfater and Jeanne Altman confirmed, that among savanna baboons dominant males tend to monopolize ovulating females. Kim Wallen and his colleagues found that sexual behavior of female rhesus monkeys increases greatly at mid-cycle, as long as they are living in groups, and that female competition increases near ovulation. Interestingly, the sexual behavior of low-ranking females shows stronger hormonal dependence than that of dominant females; evidently the subordinates need the cyclical boost.

Might human females also have such cyclical changes in behavior? As we have seen, higher primates vary surprisingly. Even our two closest relatives, the chimpanzees and bonobos, differ a good deal. Chimp females have behavioral estrus around the time of ovulation, along with a big pink swollen genital area very visible from behind. Bonobo females have nothing like this, and tend to be fairly sexually receptive any day of the month. So as on some other measures, our two closest relatives diverge on this issue of cyclicity, frustrating our attempt to find our essence in either of them.

What about us? Martha McClintock found menstrual synchrony in women who live together, and this synchrony has been shown to be due to pheromones — olfactory signals not consciously detected. This proves that important processes entrained to the cycle are outside of conscious awareness. But is there a direct effect on sexual mood or behavior? To summarize many studies, there is strong evidence of a weak effect. In cultures as widely different as Americans and the !Kung San, there is an increase of sexual interest or activity at mid-cycle, the time of the pituitary hormone surge and of ovulation. The effect is most apparent with large samples; one study asked 4,000 women to report the day they experienced "increased" sexual desire. There was a sharp peak at ovulation, and virtually all such reports occurred in the ten days before and after it. This curve is virtually superimposable on the curve from Wallen's study of group-living rhesus monkey females. The cross-cultural evidence makes it difficult to attribute this phenomenon to cultural factors, and it may be in part testosterone-dependent. As Sarah Blaffer Hrdy has warned, "Ovulation may be hazardous to your judgment."

However, the effect is subject to external cultural influences such as the day of the week and internal psychological influences such as false beliefs about what part of the cycle you are in. All else being equal, we might see an important effect of mid-cycle hormonal changes on an individual woman's sexual activity. But all else is never equal, and external influences such as cultural events, aesthetic suggestion, availability and behavior of partners, and even the weather, as well as internal influences such as fear of pregnancy, fatigue, illness, imagination, and mood, are powerful enough to swamp the effects of cyclic changes.

But if cyclical moods are minor, what is the essence of sex? To really understand human sexuality, we must first range far afield to see the mechanisms at play in distantly related sexual species. Throughout evolution, the most direct and dramatic triggers of sex have been signals from partners—the courtship and mating signs, displays, and rituals of countless animal species. They mainly follow the rules of sexual selection set forth earlier in the chapter, but we are only now beginning to understand the mechanisms by which evolution implements these rules. In the fruit fly, for example, even the developmental genetics of courtship are being unraveled. Genes for slow and fast mating are known, and another gene, called "fruitless," makes males bisexual in one form and sexless in another. Fruitless males lose the urge for pursuit, stop playing their courtship wing-song, and make no attempts at sex. The fruitless gene, expressed in nine small groups of nerve cells, apparently codes a transcription factor regulating other sexual genes.

Fireflies, those intermittent candles in the dark, are sending courting messages in syncopated beats of light. Responsive males sometimes find romance but sometimes, alas, find another destiny in the jaws of a predator mimicking the prey species' pickup line. Some snails shoot little darts into each other to stimulate arousal. And in the stickleback fish, the zigzag dance of the male—a classic approach-and-withdraw flirtation—instigates courting displays in the female, which cause the male to go to the nest, which invites the female to follow, and so on, in a manner so automatic and interlocked that at each step a model can replace the real fish and still keep things moving. But the system seems too strange, too automatic and exotic, to be really relevant to us until vertebrates climb out onto dry—or at least moist—land.

The newt, an amphibian resembling the first land animals, has some basic elements of our own courtship dance. From research by Frank Moore and his colleagues, we know that in the rough-skinned newt both sight and smell bring male and female toward each other, where touch and taste can start to play their roles. Initial signals elevate androgen in males and estrogen in females, and the hormones feed back to the behavior. But in the clasping that leads to copulation, other hormones and circuits interact. A brain peptide, vasotocin—which in mammals evolved into oxytocin and vasopressin, two hormones vital to sex, pair bonding, and maternal behavior—is primed by androgens and helps govern mating through direct brain effects. Genital stimulation can trigger clasping, through brain stem cells that also respond to vasotocin. Either stress or the stress hormone corticosterone promptly stops clasping and trips up the mating dance. The neurotransmitter GABA, we know, can inhibit fear and flight, but give a newt enough of it and it will inhibit sex, too, just as alcohol does in us.

David Crews and his colleagues have contributed much of what we know about the next, reptilian stage of sexual evolution. Although Crews stresses species divergence—garter snakes and geckos are quite different—some generalizations are still possible, and whiptail lizards provide good ways of testing them. Lizard species that have two sexes show brain differences, with the preoptic area of the hypothalamus being larger in males, just as in mammals. The lower, or ventro-medial, part of the hypothalamus is larger in females. Both findings reflect control of mounting from the preoptic area and receptive posturing from the ventrome-dial—again, true of both lizards and mammals. The mating dance is much like what we saw in Komodo dragons. The male approaches, explores the female with his forked, darting tongue, and if she is receptive enough to stand still for it, he mounts her. Gripping her back or foreleg with his jaws as he mounts, he then scratches her sides, presses her body down, and maneuvers his tail under hers. Here she must lift her tail, or his hopes are lost. As their genitals touch, he inserts one branch of his forked penis, then shifts his jaw grip to her pelvic area—a con-tortion Crews calls the doughnut, and one not even in the Kama Sutra. Five to ten minutes later he dismounts. The two have gone through something not fun-damentally changed for about 400,000,000 years.

But here is the really stunning thing: some whiptail species are unisex—any individual can play the male or female role. The hypothalamic areas are not spe-cialized, yet hormones have the same effects on those areas. Androgens in the pre-optic area cause mounting, but in the ventromedial area they do nothing; estro-gens in the latter area cause receptive posturing, but they have no effect in the former. Now, under natural conditions, ovarian cycles govern the behavior. Androgens don't cycle, but progesterone can mimic them, and it surges in tan-dem with estrogen. So a unisex lizard mounts like a male when its progesterone tide is high but switches to female receptive posturing when estrogen is ascen-dant. This elegant research shows that a species can use either hormones, brain circuits, or both to differentiate sexual acts; natural selection makes endless use of the physiological options. It even makes use of behavior; as in the ring dove, the behavior of partners shapes whiptail physiology, regulating the amount of mRNA expressed for hormone receptors in the hypothalamus.

The system has evolved, yet common threads run from newts to mammals. Donald Pfaff has done for rodents what Crews has done for lizards, and his book *Drive: Neurobiological and Molecular Mechanisms of Sexual Motivation* is a grand and elegant summary of three decades of work. As Pfaff has put it, "sexual motiva-tion is very fine, both to experience and to explain." In the late twentieth century his laboratory explained rat sexuality at every level from gene to behavior. In the rat, the male and female alternate in a much less rigid but still predictable dance. Every fourth or fifth night a female is in heat and approaches a male presenting her rump. If he is unresponsive, she may dart away, wait a bit, and return, wiggling her ears and presenting again; rat males love this kind of tease. If it works and he tries to mount her, she will also arch her back in a crucial move called lordosis, sweeping her tail out of the way to facilitate his entry. This is basically the lizard's tail-lift response, and it is one of the hallmarks of vertebrate female consent.

The sequence helps to distinguish three different components of female sex-ual readiness, which Frank Beach first did in a classic paper. An attractive female just looks and smells good to a competent male, while a receptive one lets him have his way with her. But a proceptive female actively seeks sex, striking out and away from a traditional passive role. These are not strict distinctions, and all three respond to estrogen. Also evident are two components of drive. The first is *arousal*, a general component that activates an animal and motivates an undirected seeking; it involves the brain stem's reticular activating system and the lateral hypothalamus. The second is particular to a biological need, in this case for sex. Arousal grades into a more directed *appetitive* behavior—not just activity, but flirty or courtly activity. Finally, specific motivation ideally ends in *consummatory* acts.

Much of the physiology is known. For instance, during estrus the surge of estrogen expands the sensitive region of the female's pelvic nerves. Males attracted by the vaginal odors sniff or touch around the female's pelvis, and the widened sensitivity there triggers a neural cascade that further arouses limbic sex circuits. The ventromedial or lower hypothalamus is crucial for triggering lordosis. Males will work to get to a female, then sniff her mouth and rump. In rats the key signals are not only visual and tactile but also aromatic; abolishing a rat's sense of smell seriously weakens its sexual responsiveness. If the female's posture permits, the male will mount her, insert his penis, clasp her body, and thrust repeatedly. Ejaculation is associated with arching his back and lifting his paws off her body; he may deposit a sperm plug that blocks the sperm of other males. After withdrawal, he croons an ultrasonic song while breathing rapidly and shallowly; he then becomes lethargic and drifts toward sleep.

In mammals, these behaviors use a fully evolved limbic system. For generations students have memorized the functions of the limbic system as "the four F's": feeding, fleeing, fighting, and . . . sex. The same regions and circuits come up over and over again in the experiments: the hypothalamus, the amygdala, the hippocampus, the limbic midbrain, the septal area, and the regions of cortex associated with these lower structures. All these and the fiber paths connecting them have for decades been implicated to some degree in sex, as in the other three F's. For example, stimulation of the lateral hypothalamus under the right conditions produced sexual behavior. But under the influence of *other* environmental conditions, the same stimulation produced feeding, fighting, fleeing, hoarding, or even maternal behavior, at least in rats. A more specific center for sexual behavior proved to be the medial preoptic area of the hypothalamus, the same area where sex differences are found. Also, damage to the forward part of the midbrain—a final common pathway of emotion—could decrease the time between one ejaculation and a subsequent erection and copulation in male rats, a change involving depletion of norepinephrine.

These were glimpses gained at an early stage of research, but we know far more today. For three decades psychologist David Edwards has also pursued various paths in the brain that help explain sexuality. Following Beach, Edwards distinguished the three aspects of female sexuality: attractivity, receptivity, and proceptivity. In addition, he measured sexual motivation, defined as the degree of preference shown toward a sexually active member of the other sex. All these aspects of sex, in males and in females, have somewhat different brain regulation. Male rats without olfactory bulbs will not have erections at a distance from active females, but they will still show preference for receptive females and have sex with them under the right conditions.

However, lesions of the forward part of the preoptic hypothalamus eliminate copulation, while lesions that decrease partner preference tend to be in the bed nucleus of the stria terminalis, the place in the hypothalamus where that amygdala cable lands. Most of the males that don't copulate after these lesions still show some preference for receptive females, provided their olfactory bulbs are intact. Using new techniques to measure the activity of *fos*, an immediate early gene that promptly reflects neuronal activity, Edwards and his colleagues have also shown that during sexual activity *fos* lights up in precisely those neurons that also show androgen-receptor activity simultaneously. This technique beautifully delineates the limbic, hypothalamic, and midbrain neurons that take part in male sexual behavior. Intriguingly, some forms of brain damage that hurt male performance can be counterbalanced by an especially active female.

Female receptivity, as measured by lordosis, has its own laws and logic in the brain. Damage to the central gray area of the midbrain eliminates lordosis, which depends on input to the central gray from the ventromedial hypothalamus. Midbrain lesions abolish lordosis, but they spare sexual motivation as measured by preference for intact males. And even males will show lordosis if they have lesions in the sexually dimorphic nucleus of the preoptic area, suggesting that the androgens affecting this region, which masculinize them at birth, also defeminize them. Pfaff has traced the lordosis circuit in full, and concludes that "the main source . . . must be ventromedial hypothalamic neurons." From there axons project to a crucial module in the midbrain's central gray area, which suppresses pain while it synergizes with brainstem neurons that control the deep back muscles. Higher brain centers integrate visual stimuli to influence the hypothalamic module even in rats, and of course in humans the integration includes dress, language, dreams, and fantasy. These higher centers are also sources of inhibition, responding to environmental stress and danger.

Many of the important neurons throughout this system contain GnRH, the brain's governing signal for sex hormones, and these neurons in turn originate from the olfactory system of the embryo. This undoubtedly recapitulates the evolutionary history of the system and gives it an elegant coherence that goes beyond the connections among the modules. Men with Kallmann's syndrome unfortunately lack sexual motivation, due to a failure of this same embryonic migration; they can, however, be treated with testosterone.

More evidence of these sexual connections can be seen in damage to the amygdala or its cable, the stria terminalis, which links it to the hypothalamus—both affect sex in male rats. Typically, such damage increases the time elapsed between the first intromission of the penis and ejaculation. The amygdala-hypothalamus connection plays an important role, partly through pathways from the hypothalamus to the midbrain, partly through hormonal influences of the hypothalamus on the pituitary and sex organs. In males, the preoptic area of the hypothalamus is crucial for sex, while in females the ventromedial hypothalamus plays the pivotal role. All these brain structures capture and concentrate sex hormones, as do some of the nerves outside the brain that are involved in sex. For example, the number of testosterone receptors in the preoptic area shows a telltale difference between studs and duds among male rats. And a growing body of evidence implicates oxytocin in the sexual acts of both males and females, as well as vasopressin, particularly in males. These two hormones, derived from a common hormone in our reptilian ancestors, may be activated by estrogens and androgens. So limbic system circuits can respond to and influence the hormones of reproduction, as well as—over faster, neural pathways—the sensations and acts of sex.

Not surprisingly, the estrogen receptor plays a key role in activating these circuits, as shown by Sonoko Ogawa and others in Pfaff's lab. Parallel findings come from Emilie Rissman and her colleagues, in a paper brilliantly titled "Sex with Knockout Models"—sure to disappoint curious male undergraduates but delight brain scientists. Female mice whose gene for that receptor is knocked out attack males when they should be receptive to them, and fail to show lordosis when their flanks are stimulated, even after estrogen injections. Because androgens are converted to estrogens in many cells in the male brain, the estrogen-receptor knockout male is also sexually deficient, although less so. In the normal mouse the estrogen receptor binds the hormone in the nucleus of a neuron, and the combined complex binds to certain genes. Many consequences follow. For example, in the ventromedial hypothalamus of the female rat, the genes for oxytocin and its receptor are stimulated to produce more of both, giving estrogen a powerful multiplier effect on oxytocin's functions. Injecting antisense DNA into the hypothalamus, which blocks the messenger RNA for the oxytocin receptor and thus prevents its synthesis, reduces lordosis. Estrogen also turns on the genes for enkephalin, "the brain's own morphine," and one of its receptors in the hypothalamus, suggesting an impact on the capacity for pleasure. As Pfaff has aptly said, sex has now been shown to involve a "symphony" of genes and transcription factors, and the circuits involved can now be understood electrically, neurochemically, hormonally, and genomically.

But for all the importance of these hypothalamic circuits and hormones, many lesions there do not abolish interest in the opposite sex, which may remain avid despite all lack of ability. This means that even for rats, higher limbic and cortical centers in the brain store more complex thoughts, emotions, and memories that motivate closeness even when sex itself is impossible. This is all the more true of the human animal, and our greatest insights into these higher aspects of sex come from clinical studies.

Some of the most intriguing epileptics have a seizure focus in the temporal lobe, in or near the amygdala or hippocampus. Temporal lobe epileptics often have diminished sexuality. In a classic study by Dietrich Blumer of 50 such patients, 29 described low levels of desire, imagery, and activity, reporting that they felt aroused less than once a month—20 of them less than once a year. They found it difficult or impossible to experience orgasm. An obvious objection is that general effects of epilepsy or the drugs used for it may have caused the dampened sexuality, but this is easily met: epileptics with a seizure focus in other brain regions are not hyposexual. Even more convincing, 24 of these patients were operated on in a standard procedure attempting to remove the epileptic focus, and for all but one of the 8 patients whose seizures stopped, there was also an increased sexuality. Finally, all but one of the 16 patients whose seizures continued *remained* hyposexual. Thus, the limbic structures in the temporal lobe seemed to play a key role in sexual behavior, a suggestion confirmed by later research. In one case, a thirty-seven-year-old woman had spontaneous orgasms associated with a sharp wave in her right temporal lobe, a problem that resolved promptly with antiepileptic medication. This does not contradict the general finding of hyposexuality; it merely suggests that an exquisitely sharp seizure focus can stimulate sexual feelings, while the typical, blunter type of temporal lobe seizure overwhelms and inhibits them. Alternatively, the hyposexuality may be characteristic of the long periods between seizures rather than of the seizures themselves.

Recall that norepinephrine was depleted in one experiment on increased sexual activity—a glimpse into the role of brain neurotransmitters in sex. Research of this kind is fraught with problems, but it can readily lead to new treatments. Drug molecules—at least those that get into the brain—have easy access to neural circuits by invading the synaptic cleft, the window of communication between each neuron and the next. For example, many patients suffering from Parkinson's disease experience increased sexual desire when they receive dopamine replacement therapy, in a pattern which suggests that more is involved than just generally feeling better. Neurotransmitter stimulants and antagonists may one day play an active role in the modulation of our sexual emotions—that is, we may find true aphrodisiacs.

Viagra, as the package insert clearly states, is not one of them. This means that it works not inside the brain, to make men want sex more, but at the nerve endings that control blood flow to the penis. It certainly does work. Neurochemically, it frees up the cascade set in motion by the smallest neurotransmitter, nitric oxide, or NO. (Among neuroscientists, "Just say NO" ironically invokes Viagra as a liberal counterpoint to abstinence.) Thus the official line: no libidinal arousal, just good, solid erections. Yet it is naïve to think that getting easy, reliable erections has no effect on desire. The effect is indirect, but it is, so to speak, big, and often increases desire in the partner of the man taking the drug. There can be a mechanical quality to Viagrafied sex, but for a man with a soft penis it is a godsend, resulting in "a near normalization of erectile function." Some women are taking it—off-label, without specific FDA approval—and there is reason to believe that it may work for them, too. Clitoral erections and a general blush of blood suffusing the vagina and vulva are among the sexy effects.

But are there true aphrodisiacs, ones that would work in the brain? Brain neurotransmitters play a role in sexual behavior, so it should ultimately be possible to increase desire by manipulating them. Yohimbine may work this way; it blocks certain norepinephrine receptors, affecting the sympathetic nervous system and enhancing erections, but it also enters the brain and affects mood. Most mood-enhancing drugs, however, act more like *anti*aphrodisiacs. The Prozac revolution has bettered millions of lives, and for the severely depressed, lifting mood is almost always a sexual boon. But for much larger numbers of mildly depressed, mildly anxious people—the walking wounded—blue moods have often been traded for decreased sexual drive. The most widely used twenty-first-century antidepressants so far work selectively in serotonin synapses, gumming up the serotonin transporter a bit, preventing the neuron from sponging it up again. This means serotonin hangs around in the synapse longer—good for mood but apparently bad for sex.

Drugs like Viagra countermand mechanical failure, and many people take both a Prozac-like and a Viagra-like drug. If each works for them, they can have a sort of executive-command sex life, making decisions in the cerebral cortex and effectively mobilizing lower parts, but perhaps without the strange, grand surge of lust that, when the system is working normally, fills us with desire. In that happy, helpless state, what poets call the heart—the limbic system, alas—drives both brain and groin. It is a mood avidly sought and intensely feared, and humble herbalists as well as giant pharmaceutical houses will keep trying to bring it on. But for now, no traditional or synthetic aphrodisiacs are proven. Some drugs, like alcohol or marijuana in small doses, help by enhancing GABA, counteracting fear and inhibition, but in large doses both sedate us and put sex to sleep. Cocaine can intensify the pleasure of sex as it does other pleasures, stimulating dopamine in the brain's reward system, but it does not stimulate sex when the impulse is not there.

Even food can affect desire, at least in the short run, as it should. Food is made up of molecules, some of which seep into the brain just as drugs do. As shown over three decades by Richard Wurtman and his colleagues at MIT, neurotransmitter levels in the brain change in specific ways when we eat foods containing their chemical precursors or otherwise playing a role in brain chemistry. Eat enough eggs, which contain choline, and the amount of acetylcholine in brain synapses rises; eat enough cereal, which contains tyrosine, and you increase brain dopamine, made from tyrosine. Eating a large noncereal carbohydrate meal increases the brain's level of the amino acid tryptophan, by shifting the balance of amino acid competition at the blood-brain barrier. Increasing tryptophan increases serotonin, the neurotransmitter made from it. Since serotonin inhibits sex in lab animals, this finding lends credence to some old kitchen folklore: if you want a man to perform, don't fill him up with a pot of starch. But this may only be because serotonin induces sleep.

Patricia Whitten's laboratory focuses on phytoestrogens, the estrogenlike compounds found in many plants, with a view toward understanding how some plant foods in our diets might change our internal hormonal climate. She and her colleagues Heather Patisaul and Larry Young have shown that certain phytoestrogens, including some contained in soy and other components of traditional Asian diets, act as anti-estrogens at two types of estrogen receptors in the brain. Furthermore, commercially available supplements containing phytoestrogens reduce lordosis behavior in female rats. Taken post–menopausally, such preparations or foods appear to have protective effects against heart disease, bone loss, and other menopausal symptoms related to those of estrogen. These studies open a very important avenue of research into the role of diet and increasingly common dietary supplements in sexual behavior and reproductive health.

Another interesting sidelight on human sexuality comes from research on the role of smell. We have seen that in lower vertebrates and even rodents, olfactory signaling is vital to sex, even if not always absolutely required. Parts of this pattern are also true of the rhesus monkey, a much closer relative of ours, as shown in impressive experiments by Richard Michael and his colleagues. Plug up the nose of the male of a courting pair, and you effectively extinguish his interest. Take out the ovaries of the female, leaving the male's nose unplugged, and he still shows no interest. But take a female long since ovariectomized—one who has not been able to stir a male for months or even years. Smear her rump with secretions taken from the vagina of an intact female at the time of ovulation, and an experienced, unplugged male will once again take an interest in her. The secretions in question have been analyzed, and the active principle is a cocktail of five simple, small, straight-chain fatty acids. When this cocktail is artificially synthesized, mixed in the right combination, and smeared on the rump of a female without ovaries, it has the same effect as the real thing: the male's ardor is roused.

Work by David Goldfoot and his colleagues cast some doubt on these findings, however, by showing that under the right conditions anosmic males will get aroused, and that it usually takes more than just smell to get things going. Other investigators, including Richard Michael's group, have long acknowledged that monkeys are too complex to be ruled by vaginal odors. Even rats use signals other than odors, and the idea of olfactory determinism in monkeys was always something of a straw man—a convenient target, but not a claim that anyone really believed in. Pheromones have always been viewed as having signaling and priming, not just releasing, functions; that is, they let the prospective mate know that you are around and ready and they initiate some hormonal changes. These are important functions even if there is no dramatic *release* of full-blown instinctive behavior. Martha McClintock has usefully added *modulating* functions to the list—changes in sensitivity and the integration of sensation and action.

Could some such effect work in humans? Three facts were suggestive. First, human vaginal secretions have the same five fatty acids, although in a different balance. Second, when men smelled (in a test tube) vaginal secretions taken from women at different times of the month, they rated midmonth secretions as "less unpleasant." Third, the five fatty acids rise and fall with the monthly cycle just as in monkeys, peaking around ovulation. Since the test-tube study was surely biased against male arousal, there may be something here resembling the rhesus monkey pattern. As McClintock has said, "We have clear evidence for the existence of human priming pheromones" in a different realm—that of synchronized cycles in women who live together. This means we must take seriously the hypothesis that courting couples stimulate one another with chemical signals, and that the ancient search of perfumers for an airborne aphrodisiac is not necessarily in vain.

But we should be cautious about this. In the stumptailed monkey, a near relative of the rhesus, odors play little or no role in courtship, and even rhesus monkeys are no olfactory determinists. More important, a major trend in human evolution has been to shrink the olfactory brain and expand the visual system. Odors may have their effect, but most of us respond more powerfully to visual stimuli. The pictures of gorgeous men and women in Calvin Klein ads arouse us well enough, but they do not give off odors. The onset of sexual attraction, remembered by many people to have occurred around age ten, is predominantly visual. For example, a young gay man remembered watching a *Star Trek* episode and feeling excited when Captain Kirk took off his shirt; this began a consistent pattern of same-sex arousal.

The rhesus monkey aroma effect may be one of the cases when an animal closely related to us is not necessarily the best mirror for our behavior. The ring dove, for instance, is barely a distant cousin yet shows important parallels to humans. It uses visual and auditory signals in courtship and finds them powerfully arousing; also, pair bonds last as the male stays to help care for the young. This bird, separated from us by perhaps 200,000,000 years of evolution, may have things to teach us about our own courtship and sexual behavior that we can't learn from our closer monkey kin.

Other studies of true romance and naked lust in the laboratory have ranged from "homosexual rape" in flatworms to "For whom does the female dove coo?" to the "ultrasonic post-ejaculatory song" of the male rat. But none have surpassed the classic studies of rhesus monkey sexual development carried out by psychologist Harry Harlow and his colleagues. What they add up to is that for animals like us, doing what comes naturally does not come naturally at all, if you have been brought up in abnormal conditions. This is true for both sexes but particularly for males. In a series of colorful papers—one was called "Lust, Latency and Love: Simian Secrets of Successful Sex"—the ineptitude of males who were motherless in early childhood, reared by surrogate mothers, or even reared by normal mothers but deprived of peer play, was mercilessly detailed. Unlike most adolescent rhesus males, who get the idea fairly quickly, socially deprived ones may try to copulate with the partner's side (Harlow called this "working at cross-purposes to reality") or even her face ("the head-start program"). Such sexual retardation might be remediable with an extremely patient, sophisticated female, but only with long practice. In anything resembling normal conditions in the wild, these males would see their reproductive success plunge to zero; behavioral failings like these—or much milder ones—would cull a male's genes from even the shallow end of the next generation's pool.

Indeed, everything we know about higher primates points to a crucial role for experience, and this is as true for sex as for other behavior. Social play, which includes play mounting, is universal in young monkeys and apes, and it evidently provides the "normal expectable environment." Genetically coded behaviors, even those vital to reproduction, may depend on it. In one study, chimps were raised by Roger Davenport and others at the Yerkes Regional Primate Center, under conditions like those imposed on Harlow's monkeys:

> Of the five males who reached sexual maturity and were given sufficient opportunity to engage in copulatory behavior, all but one have done so. For these animals, considerable learning appeared to be involved. Initial attempts at copulation were very poorly coordinated. For example, males with erections might mount the side or head end of the female and thrust against her, but with experience, particularly with the helpful tutelage of sexually proficient females who assisted with positioning and penetration, these animals have improved in frequency and style to the point that they are approaching normal species typical sexual behavior, except that they lack the usual signaling systems exhibited by wild born males. One (now fully adult) restricted male has neither attempted nor solicited copulation, and females rarely approach him. He masturbates frequently, sometimes to ejaculation, and occasionally uses a 55-gallon drum for thrusting.

There are lessons here for our own sexual growth. First, in order for normal sex to develop smoothly, certain early-life experiences are necessary. Second, deficits can often be fixed—at a cost. This is a tribute not so much to the flexibility of the program (the usual explanation) as to the fact that these behaviors are deeply *canalized*—engrained in the nervous system by the genes. Third, some individuals can't recover from early social restriction, at least in the complex realm of sex. Animal studies show that early experiences are crucial and that bad ones can sometimes leave incurable emotional and behavioral disability. But we need a new, more flexible, better-informed notion of "the normal expectable environment" for the social development of our own species.

Indeed, the middle-class nuclear family in turn-of-the-century Europe, which formed our notion of normal and expectable, was quite unrepresentative of our species for most of history. Subtly, deeply, persistently, it continues to serve as a yardstick for us, and that has to stop, especially in view of the dramatic changes under way in the structure of families. Compared with the extended family of many traditional societies, including hunter-gatherers, the nuclear family is a virtual pressure cooker of emotions; no wonder the first psychotherapy patients came from it. Isolated from the extended family and the wider social world, it was unusually severe in all aspects of child rearing, including weaning, toilet training, modesty training, and other early childhood tensions. Fathers were relatively distant from children and more authoritarian than in many societies. In a 1940 study that compared middle-class Chicago families to a wide range of other cultures, the only realm in which parents were rated as *more* lenient than the average non-industrial culture was children's aggression.

And of course, the Western family was sexually repressive. In the United States at mid-century as well as in Freud's Vienna, it allowed children much less experience with sex, whether playful or serious, than did the average nonindustrial culture. Little wonder, then, that the late twentieth century saw a sexual revolution, spearheaded by a bulging population of teenagers who finally found out what you do after you kiss. This revolution was real. It helped create an epidemic of sexually transmitted diseases and added a new, graver one, AIDS. There has been some modification of sexual practices among those at high risk, but it "is inadequate in magnitude to prevent the transmission of infections." Even AIDS couldn't scare us back into the nineteenth century, and the difficult, costly, only partly successful treatments for the disease have led to false complacency and more unsafe sex. AIDS aside, the epidemic of STDs is growing. For teens generally, there was evidence of a small increase in abstinence as the century ended, but sex in the modern era is unsuppressible. No warnings, religious or medical, can get this biological genie back into its cultural lamp.

Little wonder, too, that unprecedented numbers of adults began seeking professional help for complaints relating to sex, problems they were once very quiet about. Direct study of sex (as opposed to what people say about it) came late to our prudish culture. In the nineties, conservative members of Congress were still trying to stop some of our most important sex surveys—studies vital to disease control. But we have made up for lost time. Thanks to private funding, researchers at the University of Chicago and elsewhere completed the best survey since Kinsey's—in fact, better. Ironically, and contrary to conservative fears, they found that Americans are sexually conservative. For example, 80 percent of Americans had one or no sexual partner in the past year. But they also found enormous variation; despite the media culture, people differ to a stunning degree in their sexual lives.

Yet even more compelling than surveys is how sex works in living, breathing humans. The epochal and still the greatest work is William Masters's and Virginia Johnson's *Human Sexual Response*, a twentieth-century medical masterpiece that is one of the most courageous intellectual efforts of any era. It simply and brilliantly turned the searchlights of modern behavioral and medical science on the dark, old acts of sex. What it illuminated remains beautifully visible.

One focus of their work was orgasm, so it will help to develop the background to that particular wonder of nature. The word has origins in Greek (*orgasmos*—to grow ripe, swell, be lustful; and *orge*—impulse, anger) and in Sanskrit (*urj*, meaning nourishment, power, strength). Scientific descriptions are more detailed, realistic, and valid but fail to convey the thrill of an experience unique in human and animal life. Orgasm is known in all cultures, for women as well as men. As a !Kung woman interviewed by Marjorie Shostak put it, a woman "wants to finish too. She'll have sex with the man until she is also satisfied. Otherwise she could get sick." Margaret Mead discovered considerable cross-cultural variation. Among the Mundugomor cannibals of the Yuat River of New Guinea, lovemaking is "like the first round of a prizefight," but with biting and scratching. Yet "women are expected to derive the same satisfaction from sex that men do." In traditional

Samoa "a highly varied and diffuse type of foreplay . . . will effectively awaken almost all women, however differently they may be constituted." This includes preparing a woman's mind with songs and poetry and her body with skillful, gently playful hands. Yet even among the Arapesh of highland New Guinea, "In spite of most women's reporting no orgasm, and the phenomenon's being socially unnamed and unrecognized, a few women do feel very active sex desire that can be satisfied only by orgasm." Cultures may criticize women for having orgasms or for not having them, but the potential is always there.

A remarkable description, written in 1855 by the French physician Felix Ribaud, contains many elements we recognize a century and a half later:

> The pulse quickens, the eyes become dilated and unfocused. . . . With some the breath comes in gasps, others become breathless. . . . The nervous system, congested, is unable to provide the limbs with coherent messages: the powers of movement and feeling are thrown into disorder: the limbs, in the throes of convulsions and sometimes cramps, are either out of control or stretched and stiffened like bars of iron: with jaws clenched and teeth grinding together, some are so carried away by erotic frenzy that they forget the partner of their sexual ecstasy and bite the shoulder that is rashly exposed to them till they draw blood. This epileptic frenzy and delirium are usually rather brief, but they suffice to drain the body's strength.

But after Riboud, alas, Freud turned the study of orgasm onto a wrong path, hurting women for half a century. He reasoned, as many have before and since, that so impressive an event must serve reproduction. The male peak of pleasure at ejaculation seemed designed to occur in intercourse; by analogy—or rather by mistaken homology—the vagina should produce feminine ecstasy while milking the male organ that so relentlessly seeks a berth in it. This led him to postulate a mature female orgasm in the vagina, as opposed to an immature one centered on the clitoris. He wrote in 1920: "The transition to womanhood very much depends upon the early and complete relegation of this sensitivity from the clitoris over to the vaginal orifice. In those women who are sexually anesthetic, as it is called, the clitoris has stubbornly retained this sensitivity."

The colossal error here was first suggested in Alfred Kinsey's interview studies. Kinsey found that women's orgasms *typically* were not achieved through intercourse alone. Most women needed clitoral stimulation and, if they had to choose one route exclusively, would be more likely to achieve *la petite mort* through the clitoris's exquisite responsiveness than by any amount of pure penile thrusting. As for the alleged transfer of sensitivity from clitoris to vagina, Kinsey rightly called it "a biologic impossibility." But it was Masters and Johnson who—after studying how hundreds of women actually had sex—opened the trap door under Freud's armchair. Their inventive, meticulous studies took the field out of interview psychology (talk being always suspect in science) and into the realm of physiology.

Armed with electrodes for measuring heart rate and breathing, devices to chart the strength of muscle contractions, scoring systems for sexual flush, and, amazingly, video cameras inside the vagina, they set out to see how people really do it. Their initial studies, done at the Reproductive Biology Research Foundation in St. Louis between 1954 and 1965, involved 382 women and 312 men as active participants. They were interviewed and observed in sexual intercourse, in ordinary masturbation, and, for women, in directed vaginal masturbation using a plastic penis that doubled as a cold-light camera. Some were old, some pregnant, some homosexual, some prostitutes, but most were ordinary people in the prime of their lives, volunteering, getting adjusted to the program, and then, well— doing what comes naturally.

Debunking was something Masters and Johnson did a lot of. Claims that sex in pregnancy was harmful to the fetus and unwanted by the mother were alike proven wrong. Pundits who thought that aging removes both impulse and ability gave way to reports of eager septuagenarians with lubricating jelly. And the routine details of sexual physiology, which amazingly had never been properly studied, were often startling. In the "EPOR" model, they defined four stages of sex, affecting almost every part of the body: excitement, plateau, orgasm, and resolution. Excitement is the buildup—blood-vessel engorgement; increased muscle tone, heart rate, and respiration; erection of penis, clitoris, and nipples; vaginal lubrication; skin flushing, and other changes. Plateau, a variable phase that became controversial, refers to hanging fire at a high excitement level. Orgasm is the discharge of the built-up tension, muscle contractions in spasms in many areas of the body—particularly intense in the "orgasmic platform" muscles of the genital area—disgorgement of the collected blood, ejaculation in the male, and intense pleasure in both sexes. Resolution brings rest and lowered responsiveness—that phase of which Ovid said, "After sex all animals are sad." Not sad perhaps, but at least quiet and dreamy.

The penis and clitoris, Masters and Johnson found, function similarly—not surprising, given their indistinguishable embryonic origins—and many farther-flung reactions, from the rhythmic contractions of the anal sphincter at four-fifths-of-a-second intervals to the red sex flush over the chest, are the same in men and women. But the sexes differ in one crucial way: the resolution phase in men is always long, expanding with age from a few minutes during the late teens to a day or more in the elderly. But in women of all ages it is variable. In some it can be so short as to be virtually nonexistent, with multiple orgasms in rapid succession or even one seemingly endless ecstatic state. Compared with men, who are mere sprinters, women are sexual long-distance runners.

The Victorian myth of female unresponsiveness was dead, and a new liberation dawned. Men would forever after be called upon to see the full potential of women's sexuality and to help their partners reach it. But the most important casualty of the Masters and Johnson assault on ignorance was the so-called mature vaginal orgasm. Like Kinsey before them but with far more direct physiological evidence, they challenged traditional doctrine. The whole spectrum of body changes was the same whether the orgasm was clitoral or vaginal. They went so far as to suggest that the vaginal orgasm does not exist—that it works through indirect stimulation of the clitoris by means of labial movement, a sort of Rube Goldberg machine.

Others have carried on this work. For example, Marie Carmichael, Julian Davidson, and others confirmed some basic findings on orgasmic physiology and included hormonal measures. Oxytocin peaks during orgasm, and it may produce the rhythmic contractions of the uterus and other smooth muscle. Research led by Michael and Natalie Exton monitored physiological function during arousal and orgasm. In one study, ten healthy young women watched a pornographic film and then masturbated to orgasm. Their heart rate and blood pressure rose, as did blood levels of adrenaline and norepinephrine. Prolactin increased substantially after orgasm and was still up sixty minutes later, suggesting that it may be involved in sexual satiation. There were also small increases in LH and testosterone, and a parallel study in young men suggested similar mechanisms. Other research confirmed a role for oxytocin in women's orgasms.

Not surprisingly, some aspects of sex are independent of consciousness. Objective measurements of penile swelling during sleep show that it depends on blood flow first and foremost, and sleep erections can last for thirty minutes. Similar objective measures show that men with spinal cord injuries can respond substantially to a combination of direct and psychological stimulation, depending on how severe their injuries are. Women, like men, can have orgasms during sleep. And women with severe spinal cord injuries may still have lubrication and orgasm, possibly because the vagus nerve supplies signals that bypass the spinal cord.

As an account of subjective experience, too, Masters and Johnson left room for further advances. Even as they were calling the clitoris the emblem and core of female sexuality, others were digging in for a defense of vaginal orgasm. There is now evidence that at least a large minority of women can have orgasm with vaginal stimulation alone, without Rube Goldberg's help—especially, though not exclusively, through stimulation on the upper third of the forward wall. This has been called the "G spot," a nod to Ernst Gräfenberg, the first physician to describe it—although it has been known to writers in India since the seventh century, and to women since time immemorial. In fact there is no one spot, but there often is a region of heightened sensitivity.

Furthermore, women report subjective differences between orgasms achieved through clitoral as opposed to vaginal stimulation. If asked to choose, most women find it easier to achieve orgasm with clitoral rather than vaginal stimulation alone, but say they prefer a climax that blends both. One curious implication of the Masters and Johnson work was that orgasm seemed just a pulsing of the body. These amazing events were happening below your waist—or, if we count heart rate, breathing, nipple erection, and sexual flush, at least below your chin—while your brain observed them at a distance. It found them quite nice, to be sure, but it wasn't directly, causally involved. Their account amounted to a sort of

James–Lange theory of orgasm, and it seemed insufficient for an experience described by thoughtful people as the earth moving or as a little death.

But this part of the model was soon undermined. Robert Heath, a neurosurgeon, using electrodes deep in the human brain, found areas that, when stimulated, produced a feeling described as "pleasure" and "like sex." His unique paper, published in 1972 in the *Journal of Nervous and Mental Disease*, dealt with two epileptic patients—a twenty-four-year-old man and a thirty-four-year-old woman—in whom electrodes had been implanted in various parts of the limbic system, among other areas, for therapeutic purposes. Within the limbic system, the septal area—a crossroads between the limbic brain and the hypothalamus—seemed critical. Stimulation there produced a pleasurable sensation "like sex"; other regions produced nonsexual pleasure. More important, during intercourse and orgasm the septal area showed electrical patterns not found elsewhere: "[I]n both patients, the most striking and consistent changes were recorded from the septal region, with characteristic spike and large slow-wave with superimposed fast activity." Although these forms can reflect seizure activity, there were no seizures—except to the extent that normal orgasm *is* a seizure. Most interesting is the fact that these spike-and-wave changes *preceded* orgasm—they were not a response to muscle contractions but may have caused them.

So orgasm is in the brain, and is no mere twitch of the sexual flesh. Physiologist Julian Davidson proposed a "bipolar hypothesis" to explain the subjective experience. Orgasm has many of the features of an altered state of consciousness: it destabilizes normal consciousness, with changed perceptions of time, space, and motion; it is physiologically measurable and real; it requires a letting go of inhibitions; and descriptions of it resemble the experiences of mystics. Davidson proposed a hypothetical "organ of orgasm" to mediate between mind and loins. The organ—almost certainly including parts of the limbic system and quite possibly the septal area studied by Heath—generates the muscle contractions of the pelvis and also receives feedback from that region's sensations. Meanwhile it creates an altered state in the cortex, and that all-too-responsible part of the brain takes some time off. Still, right up to the brink of release, the cortex is sending feedback—sight, sound, thought, and fantasy—downstream to the limbic organ of orgasm.

Sex differences at this level are minimal. In one study, experts could not tell from reading written descriptions whether they were by men or women. In addition to vivid descriptions of the physical sensations, there were accounts like this: "Often loss of contact with reality. All senses acute. Sight becomes patterns of color, but often very difficult to explain because words were made to fit in the real world." For women, the "willingness to relinquish control"—as indicated by measures of hypnotizability, enjoyment of alcohol, and inability to control thoughts near the end of intercourse—predicts the consistency of orgasms. All this suggests something mental.

Yet even if orgasm *is* a mental state, it is not an exclusively human one. It's plausible to think that the male monkey who lets out a raucous whoop at semen ejection feels something like what a human male feels. But for a female, with no event that compares with ejaculation, the reasoning is far more conjectural. Still, there is circumstantial evidence.

Interestingly, some of the clearest cases involve females with females, rubbing their genitals against one another's rumps or, in the dramatic case of bonobos, making face-to-face genital rubbing a major part of female friendships. But similar reactions are seen in matings with males. Rhesus monkey females observed by Doris Zumpe and Richard Michael would, at the moment of the male's ejaculation, turn and reach back toward him with spasmodic arm movements—the clutch reflex. In a frame-by-frame film analysis, 97 percent of 389 copulations included it, as long as estrogen levels were normal. Since it was unlikely to be a mere expression of concern for the erupting male, some form of whole-body reflexive release in the female seemed plausible. Francis Burton, using artificial clitoral and vaginal stimulation in rhesus monkeys, found most of the physiological changes that Masters and Johnson saw in humans. And Suzanne Chevalier-Skolnikoff observed behavioral evidence of orgasm during both homosexual and heterosexual encounters in female stumptail monkeys. Another stumptail study, by David Goldfoot and his colleagues, supported this. At the very moment of what looked like a sexual climax on their films—a round-mouthed or ejaculatory face along with repeated vocalized expirations—they recorded a sudden increase in heart rate and a prolonged, intense contraction of the uterus (measured with force transducers outside the uterine wall). Human contractions are rhythmic, occurring once every 0.8 seconds, whereas the stumptails' are not; still, this was good evidence that nonhuman orgasm is physiologically real. Further work by Dutch researcher Koos Slob confirmed these results.

But why does it exist—in animals or humans? There are claims for the adaptive value of female orgasm, such as the idea that it promotes retention of sperm and thus fertilization (there is some evidence); or that it induces rest in a horizontal position with the same effect (more persuasive). Another claim is that it provokes uterine contractions that prepare the uterus for the contractions of childbirth (weakly supported); and, of course, it rewards the female for doing what comes somewhat naturally but can still get better with experience. Considering some of these explanations, evolutionist Stephen Jay Gould observed (as he often does) that not everything has to be specifically adaptive. Just the simple assumption that development is economical could leave the female with a vestige of male orgasm—something like the nipples on male breasts.

But an adaptationist can handle more than one explanation at a time. Female orgasm could have kept some women at rest while the sperm found their way, perhaps with a little help from the climax contractions themselves. In one study of thirty-four couples, an average of 35 percent of sperm were ejected by the women within thirty minutes of insemination. But if the woman had an orgasm any time from one minute before to forty-five minutes after ejaculation, significantly more sperm were retained. This suggests not only a general promotion of conception but an even more intriguing possibility: in climaxing, women may be "choosing" the sperm of a particular man—whether because of his skill, concern, patience, attractiveness, or loyalty. Such unconscious, physiological choosiness would be in line with patterns in other animals and would support the emerging evolutionary concept of female control.

And of course the reward function must also have made an impact. Patterns of reward in sex are different for men and women. One of the laws of learning is that consistent reinforcement produces the highest rate of performance of the behavior—thus, perhaps, the urgency of the male. But habits that are reinforced intermittently are the hardest to extinguish; this would explain the female's persistent willingness despite inconsistent reward. Still, the en face position is an evolutionary triumph of bonobo and human nature, allowing gazing or nuzzling during copulation. Davidson's model emphasizes the importance to orgasm of something as far from the genitals as a long, loving gaze. And nothing precludes an adaptive role for the pure *clitoral* orgasm: before intercourse it puts a woman at ease, and afterward it serves the classic reward function.

Adaptation or no, there is plenty of dysfunction in both sexes. A 1992 sample of more than 3,000 Americans representative of all adults showed sexual dysfunction in 43 percent of women and at least 31 percent of men. Behavioral treatments involve problem-oriented counseling as well as techniques of conditioning and desensitization used in bed. Men who ejaculate prematurely are patiently desensitized, while anorgasmic women are taught to stimulate themselves. Expert opinion once held that sexual inadequacy always requires psychotherapy, but now we know that sex problems can often be treated on their own.

Treatments are certainly needed. At least 20 percent of women report that they have orgasms only sometimes, rarely, or never; for 5 percent or more it is "never." About two-thirds of adult women say they have faked orgasm, and many have done so frequently. Men, for their part, sometimes find it expedient to pretend that they have been fooled. Large proportions of young men report problems with premature ejaculation, and older men have trouble getting or keeping erections. Viagra has transformed the field of male impotence and it may help women, too. But for women who want to become more orgasmic, masturbation has proved to be a useful part of treatment. Almost all college-age boys have masturbated, but even today a large minority of women never have. Studies in the nineties confirmed Kinsey's finding that educated people masturbate more, so knowledge and attitudes make a difference. And the power of mind over pelvis has been proven once again by Eileen Palace of Tulane University, who gave anorgasmic women make-believe biofeedback "showing" increased vaginal blood flow; women who got this fake feedback were more aroused by soft porn than those who didn't; thinking made it so. But whether we use drugs or desensitization, for many with sexual problems the future looks much brighter than the past.

Yet all is not sweetness and light. It has been said that a common response of women to sex is fear—not surprising, since one basic response of any creature to any other is fear, especially if the other is large, aggressive, and not yet an object of trust. The wiring of the autonomic nervous system supports the lessons of daily life: in sympathetic activation (such as fear promotes), ejaculation is triggered. But it is the calm parasympathetic branch that promotes erection of the penis or clitoris, relying in part on neurons that use nitrous oxide. Natural selection would have favored an ability to detect conditions that threaten either survival during the mating dance or the fate of the resulting young, and to suppress sex in such conditions. This mechanism may even be a kind of "density-stat"—a device to limit births temporarily in crowded locales.

But apart from these general mechanisms, women have special reason to feel afraid. In our species, as in most mammals and birds, males are better equipped to do damage, and women risk much more in any sex act. Beyond this, males may have evolved a brain in which aggressive and sexual tendencies are compatible if not mutually enhancing. In many bird and mammal species, male gestures of courtship and sexual invitation resemble those of aggressive threat and dominance, particularly in species that do not form pair bonds. In squirrel monkeys the genital display acts either as threat or sexual signal. A male baboon or macaque shows his dominance over a lower-ranking male by mounting him from behind as he would a female for sex. In orangutans, the female and young are usually alone, and the dominant style of the male's sexual advance during his brief, infrequent visits is rape. In humans, as we have seen, high male status is desirable to women in all cultures. Experimental psychological manipulations that increase dominance make men, but not women, more attractive. And some sociobiologists have proposed that "men have psychological traits that are designed for the specific purpose of rape."

Anthropologist Barbara Smuts of the University of Michigan has devoted decades to this problem and concludes that sexual coercion is widespread among mammals, including primates, and that it has shaped the evolution of both sexes. But coercion is not common in all species, and comparisons are illuminating. Baboons practice it, but the closely related macaques do not, and Smuts believes that female coalitions protect female macaques against males. As we have seen, such coalitions may also explain why bonobos are much less aggressive than chimps. Yet male coercion may have been more important in evolution than we would like to think. Female preference for the winning males in mating contests, for example, may have served to protect them from other males.

Given this evolutionary background, it is not surprising that some of the same conditions that elicit fighting can induce male sexuality. For example, painful electric shock or a tail pinch to a male rat will make him fight if he is in the presence of another male, but it will lead to sexual activity if he is in the presence of a female. Considering how often in nature males have to fight for sex, this association between pain and sex is not entirely surprising. In another experiment, a male and a female mouse spend a week together, after which the female is removed and a strange female is introduced; this causes a large immediate rise in a male's testosterone—which can facilitate either fighting or sex.

In humans, it is males who not only rape but who buy most pornography involving fantasies of violence or coercion; it is males who, overwhelmingly, pay hard cash for sex—including sex with humiliation, bondage, and sadomasochism. Such sexual practices are more common in male homosexual couples than in lesbian couples. There is growing evidence that men's and women's sexual fantasies are different, and that male sexual fantasies are more heavily laden with themes of violence, dominance, and submission.

Still, humans are predominantly pair-bonding, and in such species courtship often involves demonstrations by the male that he is capable of caring for young. He may build the female a bower, or bring her a morsel of food and feed her, or show a meek, submissive side, enabling her to judge the kind of father he will be. But in other sexual situations, even in pair-bonding species, swagger, bluff, threat, and force may edge out the romantic gestures of courtship. This unfortunately remains true in modern human societies. In the year 2000 the U.S. Army's highest-ranking woman, Lieutenant General Claudia Kennedy, the commander of two military-intelligence battalions, credibly accused a superior officer of groping her in her office. If she could not avoid harassment, no woman can. Such tendencies, however natural, are not admirable, desirable, or unchangeable; but intellectually dishonest claims that there are no significant biological bases for sex differences can only obscure the path to a more equal future. The facts may not be pleasant, but concealing them cannot help us.

Nevertheless, female choice is everywhere: at the time of marriage, before it, instead of it, between marriages, in stolen embraces during them, and even in the strangely discriminating responsiveness that enables a woman, through orgasm, to abet some sperm and reject others. So what if biology decrees some awkward sex differences? Being human, we don't just throw up our hands and accept a dismal evolutionary discordance. True, men can't reproduce without orgasm and women can. True, too, that to the extent that the vagina itself is orgasmic, "bestial" copulation—the front-to-back pattern characteristic of most nonhuman mammals—probably maximizes its chances, giving the penis leverage at the most sensitive part of the vagina (as opposed to the clitoris). But women in various studies (like our bonobo cousins) prefer face-to-face intercourse, and that, too, leads to interesting anatomical possibilities. Why settle for the mere results of organic evolution? Every culture improves on them, from the Kama Sutra to *Cosmopolitan*, from the ribald stories told by !Kung San hunter-gatherers to the pep talks of high-tech sexologists. Vaginal lubricants, contraception, hormone replacement therapy, Viagra—these are the first steps toward real human control of sexuality, and ironically they will give culture unprecedented power to deeply transform the most biological acts. Wisely used, they make sex more human.

"Sexual intercourse is dangerous . . ." is the sensible opening phrase of "Risky Business," Kim Wallen's analysis of primate sexual desire, and the quotation "Sexual passion can only exist outside normal life" stands as his epigraph. Yet sex is ineluctable, drawing us toward it as a vortex of water that may drown us or of wind that may uplift and carry us beautifully out of ourselves. In an insightful commentary on Nietzsche's *Birth of Tragedy*, Adam Konner writes of the demise of Dionysian inspiration in our too-orderly Appolonian lives:

How can "technological man" still connect to that universal sea of emotion? Does the wild soul of nature have any remaining inlet into his systematic life? In every modern culture and in every culture since the beginning of time, there is one influence that always brings us back to the turmoil of nature, one inescapable act that plugs us in to the roar of Dionysiac passion: Sex. . . . As far as science and technology may advance us, as disconnected as we may become from our natural surroundings, sex will always bring us back to what we are made of. It is the common human passion that will never cease to make us feel . . . the filthy, base, animal instinct that keeps us united with the inspiration of Dionysos. Even when all other connections to the Dionysian impulse have been torn away, sex will always remain.

For a sexually reproducing species, that base animal instinct is the main evolutionary stream, our living connection to past and future, the essence, biologically, of everything. But "we don't see a lot of art inspired directly from the act of sex. . . . Sex is a formless slice of the Dionysian impulse, but it engenders an encompassing form: Love." And love, of course, *does* inspire great art.

Camille Paglia's brilliant, irascible study of Western art and literature, *Sexual Personae*, reminds us that sex and gender have been grand, enduring themes for 5,000 years. Great spirits in this tradition have toyed relentlessly with definitions of male and female and with the playing out of sexual appetites, always in the shadow of both hatred and love. Say what you will about evolution, we are cultural creatures, even at the core of our sexual selves. Examples in myth and song, sign and symbol, story, admonition, gossip and iconography, not to mention stage, screen, and television shape, as they have always shaped, every romantic gesture, every sexual act. Sex itself, as an animal act, is perfectly transparent, but the color of human sex is culture. More than the sweetest technique, more than the most transforming medications, art is what makes sex human.

But improving on evolution requires openness both in science and in art. So lyric and gentle a novel as D. H. Lawrence's *Lady Chatterley's Lover* was banned in the United States just a few decades ago. In a sense, it threw down the gauntlet to both religion and evolution, rejecting the puritanical past but also the emerging cult of the natural. Here is how the lover, Mellors—"mellow," as well as "better," despite his low social origin—is said to have brought about fulfillment in the lady:

He had drawn her close and with infinite delicate pleasure was stroking the full, soft, voluptuous curve of her loins. She did not know which was his hand and which was her body, it was like a full bright flame, sheer loveliness. Everything in her fused down in passion, nothing but that.

We should recall, as we attempt to keep sex perfectly natural, endlessly seeking the grail of simultaneous orgasm, that this very superior, very masculine lover made the magic with his hand; that their joining was suffused with grace and tenderness; and, not incidentally, that he loved her.

CHAPTER 13

Love

The mothers were always there. Sitting on stools, they rested their upper torso and head on their child's bed and slept holding the small hands. . . . Fifty yards away in Emergency he had heard grown men scream for their mothers as they were dying. . . . This was when he stopped believing in man's rule on earth. He turned away from every person who stood up for a war. . . . He believed only in the mothers sleeping against their children, the great sexuality of spirit in them, the sexuality of care, so the children would be confident and safe during the night.

—MICHAEL ONDAATJE, *Anil's Ghost*

We had brought our firstborn daughter to the pediatrician for the six-week visit. My general impression as a scientist—that newborn babies all looked alike and were quite unappealing—was confirmed by fatherhood. Not only that, but this one didn't sleep. (Energetic babies are otherworldly spirits sent to certain new dads to punish us for our sins.) There we were, the three of us, in the bosom of medical wisdom, and I had a question. I held the baby up to the light, squinted at the physician out of one bloodshot eye, and spoke starkly: "Tell me, Doctor. You've been in this business a long time." I glanced meaningfully at the baby. "She's ruining my life. She's ruining my sleep, she's ruining my health, she's ruining my work, she's ruining my relationship with my wife, and . . . and she's ugly." Here the gentle reader may well imagine that my usual professorial reserve was doing battle with other forces on the strained, small field of the vocal cords. Swallowing hard, I managed to compose myself for my one simple question: "Why do I like her?"

The physician, a distinguished one in our town, wise, old, and virtuous, was unfazed. "You know," he shrugged, "parenting is an instinct, and the baby is the releaser."

"Doctor," I said, "that is one of the worst clichés from one of my own worst lectures." Suppressing a shudder over the fact that the language of so new an enterprise as ethology was already the common coin of the consulting room, we took our leave. I sank back into my misery of love: a desperation of affection for the tiny, whining monster mounting a relentless assault on my nerves.

The adoptive parents of another newborn we knew (they went through the same feelings in almost exactly the same way) found a wonderful way to describe it. It was, they said, like nothing so much as an adolescent crush. You sighed, gazed, mooned around, dreamed orgies of tenderness, saw, in your mind's eye, decades of future mutual love, dignified, courtly, publicly known. Meanwhile, you suffered every known variety of emotional abuse, neglect, rejection, anguish, and humiliation. If you managed to steel yourself for an hour, became convinced that you could stay on an even keel, you were thrown a scrap—here an appropriately timed belch, there a fleeting mutual gaze—and you tumbled back down into the well with glazed walls, stewing in your own affectional juices. This set you up nicely for the next diaper change, when, literally, you would have more offal dumped on your pitiful self.

The question the doctor evaded echoes with fascination. Here was a fairly complex creature—a grown-up male college professor in his calm, intelligent thirties, full of experience, including various loves as boy and man. He was not a recently parturient mother rocked with hormonal changes, filled breasts urging from within. He was not a tiny baby with a fairly simple brain, clinging for blind comfort and protection. He was not even a teenager surging with rough humors, growing too fast in too many directions. He was in short no easy mark for any of the usual push-pull, click-click explanations favored by behavioral biologists. Yet there was also little comfort here for confirmed social determinists. Would they seriously propose that these crazy emotions were products of cultural nudging— that our protagonist felt these things because someone told him he *should* be a good father? What sort of training during early life, what admonitions or rewards in young adulthood might build up, through conditioning, not just the acts but the love? None, at least not in the sense that pigeons are trained to play Ping-Pong; more, perhaps, in the sense that cocks are trained to fight. That is, quicker, with greater ease, drawing upon a deep well of ancient, stereotyped emotion, thought, and action—a well in the brain, with its source in the genes.

Since the newborn human is incapable of love, its mother and father must make up the shortfall; if they don't, the infant will lose its life, and the parents their reproductive success. Later in infancy the child's emotional assets grow, and the parent starts to get something out of the deal. But not all animals play this waiting game. In fast-maturing birds like ducks and chickens, the hatchlings leave the nest almost immediately; natural selection has made an infant that can hold up its end of the attachment from day one. The parent is designed to defend and protect it, and to want to, but she does not have to chase it all over the landscape, because it will shortly use its walking ability to do not much else but follow her. This does not result from genetically coded images of the mother. It comes, primarily, from imprinting: the one-day-old chick or duckling forms an indelible penchant for some prominent object in its environment. Normally this turns out to be the mother.

But not always. The man who made imprinting famous by studying goslings attained fame himself by becoming the imprinting object for some goslings studying *him*. Konrad Lorenz, who later won the Nobel Prize for his work in behavioral biology, described imprinting in a 150-page monograph published in 1935, "Der Kumpan in der Umwelt des Vogels." It is usually translated as "The Companion in the Bird's World," although *Kumpan* connotes "partner" or "buddy," and *Umwelt* in ethology means "subjective world." It was not only informative and convincing but sweeping, incisive, beautiful.

Lorenz introduced all bonds seen among birds. For each important partner as the bird might see it—the parent, the infant, the mate, the social partner, the sibling—he used examples from several species and discussed the expectations apparently wired into their nervous systems: what the partner may look like, what it will probably do, what you can count on it for, and what you do in return. Such relationships are real, ubiquitous, reliable (up to a point), strong, long-lasting in many cases, and crucial to survival or reproduction. Furthermore, they are highly patterned—stereotyped—and although experience guides their emergence, major features are independent of learning. Or to put it as Lorenz does, to say that they are innate is about as much of an exaggeration as to say that the Eiffel Tower is made of metal.

Take imprinting. The object the infant bird must slavishly follow is targeted partly by genes and partly by prehatching experience—the sound of the mother's calls, for instance—and that influence makes the hatchling tend to pick the mother bird, or at least a member of its own species if one is around. Yet, failing that, some doting goslings zeroed in on Lorenz. Some got hooked on inanimate objects, like a big orange ball. Some, in later experiments in other laboratories, even imprinted on the stripes painted on the wall of the box they were housed in. This is a powerful effect of experience, depending mainly on who or what is around just after hatching, and the choice can last a lifetime.

However, and this is crucial, the rest of the process is basically wired in. The chick or duckling pecks its way out of the egg, gets on its feet, and begins ambling around. For a few hours it will tend to approach any object easily discernible against the background. If an object is highly salient, especially if it has certain characteristics of the mother—a certain squawk, say, or a shape or style of movement—the hatchling will tend to approach and follow it. Head-and-neck shapes, for example, trigger an innate preference. The more it follows, the more it wants to, and after a certain point punishing it for following tends to increase rather than decrease following—exactly counter to the predictions of learning theory. Sight and sound converge to intensify the process. Meanwhile, the chick becomes less inclined to approach other objects and creatures, and ultimately it fears them.

All this takes a few days at most. The tendency to follow is greatest on the first day and declines exponentially over the first week. Attachment to the mother or other imprinting object and fear of others persist through much of growth. Many laboratories around the world have extended these observations, notably those of Eckhardt Hess at the University of Chicago, Patrick Bateson and Gabriel Horn at

Cambridge University, Katharina Braun at the Institute for Neurobiology in Magdeburg, and Johan Bolhuis at the University of Leiden. The phenomenon is more flexible than previously thought, but it is still quite quick and mysterious. The brain's chemistry and anatomy change during imprinting, and increasingly we understand how it has been prepared by its structure and pattern of growth to do just that.

For example, high in the chick's forebrain are two connected areas that correspond roughly to our association cortex, a crossroads of neuronal activity. In this circuit, but not in other forebrain areas, there is a massive pruning of the number of spines on the dendrites *during the course of the imprinting process.* Evidently, dendritic pruning in this region narrows the chick's focus as it learns, and this pruning depends on activation of certain glutamate receptors, which in mammals are implicated in fear. Blocking these receptors prevents pruning and extends the sensitive period for imprinting. As imprinting progresses, GABA neurons in the same circuits become genetically activated—imprinting switches their genes on. It is tempting to suggest that imprinting is a peculiarly emotional form of learning in which fear is activated through glutamate receptors and calmed by GABA.

Yes, but is it love? Probably not. We have difficulty enough in making proper use of the word to describe our own emotions and behavior, and it's a stretch to include apes, monkeys, and other mammals. This should give pause to any attempt to apply it to ducklings; yet imprinting may be *relevant* to love. The very difficulty gives us a certain freedom. All we have to go on, except in humans, is behavior; so if dogs love their masters, then ducklings, perhaps, love their mothers. It is not a trivial issue. Brain research on imprinting and on mammalian attachments will soon test the hypothesis of similarity. If it is borne out, then these behaviors may involve some of the oldest parts of the brain.

Still, how much weight can we give the chick's sudden devotion when there are at least four other kinds of companions in the bird's world alone? And how do we know that this first relationship has anything to do with the others? There are several arguments in favor of common processes. The first is an adaptive one: in the words of John Bowlby, the great twentieth-century theorist of attachment, all love is inextricably intertwined with fear, "that very archaic heritage that is placed at the center of the stage."

> A tendency to react with fear to each of these common situations—presence of strangers or animals, rapid approach, darkness, loud noises, and being alone—is regarded as developing as a result of genetically determined biases . . . present not only during childhood but throughout the whole span of life. Approached in this way, fear of being separated unwillingly from an attachment figure at any phase of the life-cycle ceases to be a puzzle and, instead, becomes classifiable as an instinctive response . . . to an increased risk of danger.

In this view, the imprinting of hatchling birds resembles not only our own infants' attachments but also many other forms of love.

Harry Harlow, who pioneered the classic studies of rhesus monkeys described in the last chapter, drew his own theory not mainly from evolution but from observed similarities in attachment behavior over the life course. Like Lorenz, he began by enumerating various forms of companionship—he called them "affectional systems"—linked by a thread of continuity:

> The first of the affectional systems is maternal love, the love of the mother for her child. The second is infant love, the love of the infant for the mother. . . . The third is peer, or age-mate, love, the love of child for child, preadolescent for preadolescent, and adolescent for adolescent. . . . The fourth love system, heterosexual love, is one in which age-mate passion is augmented by gonadal gain. . . . The fifth love system is that of paternal or father love.

Harlow didn't claim that each system is separate. Rather, "there is always an overlap . . . affectional motives are continuous," and "each love system prepares the individual for the one that follows." There are many parallels with Lorenz. Neither takes theory as a starting point, nor do they assume that one of these systems gives rise to the others. The two lists meld if we add the sibling system to Harlow's and the paternal to Lorenz's, for a total of six distinct relationships. But the six have a good deal in common, and various combined and intermediate forms occur. In a sense, they are six peaks in the primal landscape of affection. Brain evolution would have been most unparsimonious if it could not scramble from one peak to another while backpacking some of the same basic tools.

We know that such continuity appears in simple imprinting. When the goslings that imprinted on Lorenz reached the age of courtship and mating, some relentlessly courted *him*. Such erroneous eroticism can easily make adult birds court the wrong species and even inanimate objects. Some species—zebra finches, for instance—develop slowly and imprint only as half-grown fledglings. No matter. Imprinted on the wrong species when young, they will court their own kind if no other choice is available. But a zebra finch raised by a society finch female will, given a choice, choose a society finch rather than his own kind— even if he has already produced a clutch of eggs with a female like himself. He has already mated and fledged chicks with his own kind, yet somewhere in his brain he carries a torch for society ladies, the sort he was raised by. Early bonding has etched them indelibly in his brain, and the mechanism has much in common with the process in hatchling chicks: pruning dendritic spines for a focused, permanent preference. This conclusion is no surprise to the psychotherapist, who helps patients search for just such continuities, on the theory that they underlie romantic impairments. But it would be nice if we could support the conclusion with animals more like us than birds.

Harlow's work helps. He began in the 1950s, trying to pin down "the nature of love"—meaning something like, *What does the rhesus monkey really see in its mother?* and *How does the infant ever move on?* The first question initially had a dismally minimal answer: baby monkeys would become attached to a sloping cylinder of wire covered with terrycloth and warmed. If the wire column also gave milk through a nipple, the monkey's clinging and contact increased slightly. But given the choice between a milkless warm terrycloth icon and a wire-only one with a milk-dispensing nipple, monkey infants spent almost all their time on the cloth model, switching only to feed. When a scary object—a wind-up toy teddy bear banging a drum—was brought into the cage, infants always went to the milk-less cloth model. This put aside (at least for rhesus monkeys) the idea that oral gratification was the root of attachment in infancy. Leonard Rosenblum, then in Harlow's laboratory, showed further that equipping the cloth model with a device that gave the infant a periodic blast of cold air—a punishment, in conditioning terms—*increased* the infant's time in contact with the "mother." Paradoxically, it sought comfort from the source of the punishment, exactly as imprinted birds did—and also, unfortunately, resembling many abused children.

But what of the second question—the requirements for normal *future* development? Monkeys raised on cloth models grew up with far fewer abnormal behaviors—rocking, self-clasping, self-biting—than did those raised in isolation *without* such simple surrogates. But they had more of them than monkeys reared by normal mothers. In social situations, they tended to withdraw into "autistic" states. As they grew and became sexual, both sexes were inept. In social situations as adults, deprived males were more likely than normally reared ones to threaten or attack other monkeys. And some females, if they could be forcibly inseminated, were clumsy, negligent, and even brutal toward their infants.

There were many experiments by others—Rosenblum, William Mason, Gene Sackett, Gary Mitchell, Stephen Suomi, Gary Kraemer, Jeremy Coplan, and Maribeth Champoux, to name a few. They showed, for example, that opportunities for contact and play with peers are also essential, in some ways more important than normal mothering. They also demonstrated that even closely related species may differ in their response to deprivation, which must make us hesitant to generalize to humans; that with the right encouragement males can "mother"; that rocking the model reduces later deficits caused by cloth-surrogate rearing; and that infants raised to the age of six months or even a year old in total social isolation can be substantially rehabilitated by being placed for a few months with a younger infant monkey. So strong is the attachment system that even placing an isolated monkey infant with a long-haired dog will result in a lasting relationship: the infant clings to, rides, and shows affection to the canine surrogate mom.

All this suggested that strong neural and hormonal controls ensure the development of attachment even in very abnormal circumstances, and some of these controls are coming into view. In rhesus monkeys the effects of a year of total social isolation beginning at birth—essentially abolishing affectionate behavior— can be largely reversed, although by a difficult and costly method. Individual differences in temperament, studied by Stephen Suomi and his colleagues at the National Institute of Child Health and Human Development, have proved important. Some deprived females are adequate mothers the first time out, and others seem to have a particular nervous-system vulnerability—possibly analagous to the timid children in Jerome Kagan's studies. But some vulnerability is widespread; on average, early experience matters, and any relationship depends to some extent on prior ones. This has become increasingly clear as laboratory testing has widened to approximate natural conditions. For example, peer-reared rhesus monkeys are more successful than mother-reared ones in some restricted social tests, but when jockeying for dominance in groups the mother-reared do better.

During the 1990s Suomi and his colleagues also greatly advanced our understanding of the physiological impact of rearing conditions. In laboratory programs of wide-ranging power and significance, they described in detail the consequences of differential rearing for neurotransmitters, brain function, and gene regulation. Among other comparisons, they looked hard at the differences between mother-reared rhesus monkeys and those reared with peers for five months after an initial month of human nursery rearing. All infants in both groups were in similar social conditions after six months of age, in mixed groups. Recall that Harlow and others initially thought that peer rearing was as good as, and in some ways better than, mother-rearing.

Leaving the issue of "better" aside for the moment, the results of these rearing conditions are clearly divergent. Peer-reared monkeys are more reactive and impulsive throughout life. Males are more aggressive, a problem that worsens at puberty, and due to this lack of self-control they sink through the dominance ranks. Females grow up to be more abusive and neglectful toward their own first infants, and they cradle their subsequent infants abnormally. Peer-reared females also tend to groom their companions less. These enduring differences owe much to physiology. Peer-reared monkeys consistently show lower levels of 5-HIAA in cerebrospinal fluid. As we saw in chapter 9, this serotonin by-product is low in men who are impulsively violent or suicidal. It is also low in wild-reared rhesus monkeys that are impulsive for *genetic* reasons.

But the 5-HIAA difference is just the beginning. Peer-reared adult males are harder to sedate, requiring larger doses of the anesthetic, and they show more whole-brain glucose metabolism in PET imaging. They consume more alcohol ad lib and develop a greater tolerance for it, a difference in turn dependent on serotonin metabolism. This may be a model for one type of human alcoholism, in which male impulsiveness and aggression go along with excess consumption. If so, it could mean that inadequate mothering contributes to the disorder in some individuals, although it is almost certainly genetic in others. Other evidence links the monkey syndrome to a more active serotonin transporter, sponging serotonin from the synapse. Thus, high serotonin transporter binding and low levels of 5-HIAA in cerebrospinal fluid go with impulsiveness, aggressiveness, and excess alcohol intake.

Finally, the gene for the serotonin transporter provides other clues. Since some of the most effective twenty-first-century antidepressants work by blocking this transporter, it has been a focus of intense interest. The region the gene is in may be longer or shorter, and in both monkeys and humans the short version has lower gene expression. But it is only among peer-reared monkeys that this translates into lower 5-HIAA levels and vulnerability to symptoms. This fits with growing evidence that the effects of experience must be understood in relation to genetic background.

Cross-fostering effects also work this way. Genetically high-reactive monkey infants fostered to highly nurturing females are behaviorally precocious, rise through the dominance ranks, and become nurturing mothers like the ones who reared them. But infants with the same reactive genetic background are *worse* than average in all these dimensions if they are fostered to so-so control mothers. So the reactive genotype can be an asset or a detriment, depending on the adoptive mother's nurturing style. This is a transgenerational effect that works outside the genes, through maternal behavior, yet it is strongest against the background of one genetic predisposition. In the mother-infant bond, monkey or human, there are no simple explanations.

What about subtler effects than peer versus mother rearing? There is growing evidence that the marked deprivations of the classic studies are not required to produce lasting effects. Not long after Harlow began his research, Robert Hinde of Cambridge University studied less drastic interventions. Also working with rhesus monkeys, he used two separations from the mother of only six days each; as we saw in chapter 10, these separations resulted in increased fear at age two, with far greater effects when the mothers were removed than when the infants were, suggesting a subtler mechanism than mere separation.

Leonard Rosenblum, now at the Downstate Medical Center in New York, was also dissatisfied with the extreme deprivations of the classic studies, and he and his students designed ingenious variations in the environments of mother-infant pairs. His laboratory was situated in a bad, inner-city neighborhood in Brooklyn, yet he knew that even children raised there were not subjected to such severe and obvious stresses as those of the Harlow laboratory. His solution was to mount more meaningful models of stress on human mother-infant pairs in the real world.

Rosenblum and his students raised bonnet macaques in three different conditions of food availability. Mothers facing low foraging demand found food predictably and easily. Those under high foraging demand found food only with difficulty, after prolonged searching. But those under *variable* foraging demand had to face both conditions in unpredictable alternation. All of the mothers were adequately fed, but they had to behave differently to achieve adequate intake. Contrary to expectation, those with unpredictable foraging needs were the ones whose infants suffered the most severe long-term consequences, substantially worse than those whose mothers faced consistently tough foraging demands. When grown, these infants showed more anxiety, and their symptoms were linked to increases in corticotropin-releasing factor and somatostatin, two hormones of the hypothalamus, as well as of serotonin and dopamine.

This brings us a long way from what Harlow thought was sufficient for normal development. No isolation or severe deprivation was involved here, just unpredictable changes in how easy it was for the mother to get food. Rosenblum has suggested that the mother's role is threefold: to modulate the infant's arousal, to mitigate environmental uncertainty, and to mediate kinship and other relationships. His four decades of research on monkey development now point to what he has called "the maternal demand matrix" as a crucial determinant of the child's future coping skills.

There are many species differences, and it is difficult to develop a consistent model even for rhesus monkeys. In the early nineties Gary Kraemer, of the Wisconsin Primate Center, attempted to integrate the results up to that time. In his framework, the development of catecholamine neurons in the brain can be put off their normal course if deprivation occurs early in life. But with normal growth in an adequate social environment, the brain develops greater resilience, so more severe deprivations must occur to get development off track. Early insults can be followed by recovery but often leave latent long-term vulnerabilities. Today we understand much more about the physiology. Not surprisingly, given what we know about behavior, early-experience effects are found in many important brain systems.

This is as we would expect. Catecholamine neurons are implicated in fear and reward, serotonin in aggression and mood, corticotropin-releasing hormone in depression and stress. It stands to reason that early-experience effects on these domains of behavior and mind would work through underlying brain systems, and they do. But we have a long way to go before we have a theory of early deprivation's lasting damage. Why do bonnet macaque infants react more calmly to maternal removal than rhesus or pigtails do, yet sustain a profound and lasting impact of fluctuations in their mothers' ease of foraging? Why such a large difference for rhesus monkey infants depending on whether they are removed from their mothers or the mothers from them? We just don't know.

And what of *human* ties of affection? As Stephen Suomi sadly but wisely wrote in 1997: "The idea that early experiences disproportionately influence adult behaviour has long been a fixture of mainstream developmental theory . . . although unambiguous empirical support for this view at the human level is surprisingly sparse," compelling evidence from animal studies notwithstanding. The ambiguity of the empirical support for this widely believed idea has repeatedly allowed intelligent observers to declare that the influences have been greatly exaggerated. This situation is likely to persist until child psychology becomes more serious about scientific method, especially in the sense of ethically sound experimental or intervention studies. In the meantime, we can learn from experiments in animals, and observations of humans in different cultures, where child-rearing methods vary much more than they do in our own.

Given the thousands of known cultures, it might seem impossible to generalize, but they show both a surprising degree of uniformity and a lawful variation beyond it. Consider two cases, !Kung San hunter-gatherers and the United States.

!Kung infancy research taps into the cross-cultural range and may debunk false notions of the universality of Western wisdom. However, it adds a further, historical dimension: to some extent, we can guess by extrapolation from modern hunter-gatherers what adaptations in infant care and development may have characterized ancestral foragers.

From birth and through at least the first year, !Kung infants are carried in a sling at the mother's side, held vertically in continuous skin-to-skin contact. Reflexes such as crawling movements in the legs, the use of the arms to move and free the head, and grasping responses in the hands allow the infant to adjust to the mother's movements and avoid smothering in her skin and clothing. These movements also signal the infant's changes of state, teaching the mother to anticipate its waking, hunger, or defecation. The hip position lets infants see the mother's social world, the objects hung around her neck, any work in her hands, and the breast. Mutual gaze with the mother is easy, and when she is standing the infant's face is just at the eye level of keenly interested ten- to twelve-year-olds, who frequently initiate brief, intense, face-to-face interactions. When not in the sling, infants are passed from hand to hand around a fire for similar interactions with adults and children. They are kissed on their faces, bellies, and genitals, sung to, bounced, entertained, encouraged, and addressed at length in conversational tones long before they can understand words.

The mother indulges the infant's dependency completely in the first year and in the second year resists it only slightly. Nursing is continual, four times an hour throughout the day on average, triggered by any slightly fretful signs. Close contact for the first two years allows a much more fine-grained responsiveness by the mother than can be attained in a culture where mother and infant are often apart. During the first year the average time elapsed between the onset of an infant's fretting and the mother's nurturing response was about six seconds.

Was the !Kung pattern characteristic of our early human ancestors? Ethologist Nicholas Blurton Jones has studied infants and children among the Hadza, hunter-gatherers of Tanzania. Some features of their childhood are different from that of the !Kung—for example, older children are expected to forage for themselves a good deal—but the pattern of indulgence in infancy is similar. Among Central African Pygmies studied by Gilda Morelli, Edward Tronick, Barry Hewlett, and others, there are interesting variations, such as greater participation by fathers or by women other than the mother. Still, indulgent care in infancy is the rule. Hunter-gatherers outside of Africa—such as the Ache and Siriono of Amazonia, the Agta of the Philippines (where women hunt but are still indulgent mothers), the Paliyans of India, and even the Eskimo—are similar.

This may be our original, species-typical pattern. Extending work by Ben Shaul and others, Blurton Jones compared the infant care styles of a number of mammals. Those in constant proximity differ predictably from those that cache or nest their young. The "cachers" have young that feed infrequently, suck with a rapid rhythm during each feed, and have rich milk high in protein and fat. Carriers and followers feed more or less continually and have low sucking rates and thin, watery milk. Humans, along with monkeys and apes, have the milk composition and sucking rates of carrier species; this, along with frequent nursing by hunter-gatherers, suggests close contact during human evolution.

!Kung nursing bouts are not passive events but interactions, and they continue until weaning, usually after age three. Infants often play with the free breast, make languid arm and leg movements, coo, talk, have face-to-face interaction with the mother, and do various forms of self-touching, including occasional genital play. Separation is initiated by the infant very gradually, with little urging from the mother. She rarely leaves the infant's immediate vicinity until the later part of the second year, and then only occasionally until the birth of her next child, during the fourth year. But the infant begins to stray a bit with crawling and walking, using the mother as a base for exploration. Although getting lost in the bush would be fatal, the infant's consistent return and intense fear of strange situations prevent it. The infant passes slowly to a child group ranging in age from near-peers to adolescents, a familiar and safe context. While weaning is neither abrupt nor punitive, it is relatively firm and often results in weeks of depressed and fretful behavior. Still, there is the consolation of an accepting group of children, who become a major focus of the child's life.

How can we approach the cultural and species differences in infant attachment and its role in development? We can start with a theoretical framework that is reasoned, elegant, and testable—John Bowlby's, as presented in the 1970s in *Attachment and Loss* and extended during the decades since. It is not the last word in attachment theory, but it is durable, grounded in Darwin's worldview, and consistent with clinical psychodynamic theory. Debates go on about how to define and test Bowlby's claims. Daniel Stern has emphasized the role of parent-infant attunement in the development of the self, while Susan Goldberg and her colleagues believe that studies should emphasize maternal protection of the infant rather than responsiveness as the core concept. There have also been decades of research on infants' social relationships that are not based on the attachment concept at all. Current scientific approaches to relationships in early infancy are exemplified by the studies in Philippe Rochat's *Social Cognition: Understanding Others in the First Months of Life*. Because this is an age when attachment is minimal and other emotions are primitive, it lends itself to simpler approaches.

In addition, a new body of theory has grown up, grounded in late-twentieth-century evolutionary models. Starting from life-history theory, this approach assumes that different life plans are adaptive in different environments, and that patterns of early attachment are *designed* by natural selection to produce different kinds of adults. There have been significant contributions by Patricia Draper, James Chisholm, Jay Belsky, and others, culminating in Chisholm's insightful *Death, Hope, and Sex: Life History Theory and the Development of Reproductive Strategies* and Sarah Blaffer Hrdy's masterly *Mother Nature: A History of Mothers, Infants, and Natural Selection*.

Hrdy emphasizes that mothers must constantly assess the care they give their young in light of the realities of their life situation, including the needs of other present and future offspring. Draper, Belsky, and Chisholm emphasize that severe or unpredictable environments stress mother-offspring relations, but the resulting offspring may be better adapted to survive and reproduce as adults in those same environments. Overall, this approach seems to reassess as normative variations in childcare that have sometimes seemed pathological. But the clinical and evolutionary models are not in contradiction; in fact they must be reconciled if we are to understand our choices. At the same time, we must not abdicate those choices, which is what we would be doing if we tried to derive *ought* from *is*.

Here is what *is*. All species have life histories designed to maximize reproductive success, but many have more. They have flexible life histories that change direction, depending on environmental consistency and quality. They may reproduce earlier or later, try for a few offspring of high quality or a multitude of sketchier young, even change sex with temperature, food availability, or the stress of dominance challenges. Darwinian theory predicts that life histories will be adjustable so that reproduction can be maximized in different environments. The result may be trade-offs between growth and sexual maturity, between number and quality of offspring, between sex and parenting, between a short but busy and long but sedate life. Over time, a species may evolve a range or norm of reaction, a predictable relationship between environment and the life course. As we saw in chapter 2, flexible adaptation does not mean that "anything goes," only that certain things go in certain environments and not in others.

In the light of this body of theory, different patterns of childcare and attachment are not better or worse for infants, they are just strategies for maximizing reproductive success in different environments. For example, if a poor environment makes for a relatively rejecting mother, and the result is a child who becomes sexually promiscuous or delinquent, that may be nature's way of maximizing reproductive success in that poor environment. Abuse, neglect, extreme favoring of males over females, and certainly such milder variants as multiple caretaking and wet-nursing all may be interpreted as adaptations made by parents to demanding environmental conditions. Hrdy's approach in fact restores compassion for parents, especially mothers, to a field that has traditionally ignored their happiness in favor of children's at every turn. But apart from mothers' well-being, the human infant's brain may be designed to detect environmental conditions that require different reproductive strategies. Chisholm has suggested that the amygdala might be a device for setting life-history strategies in primates, a notion consistent with what we know about the lasting biological effects of early stress.

Still, there are two broad problems with this line of work. First, despite fascinating reinterpretations of data collected for other purposes, including historical and ethnographic descriptions, it has not been subjected to many specific tests of its own hypotheses. In fact, after decades of research we remain embarrassingly uncertain of the effects of early nurturing in human development. Second, it cannot solve the great clinical questions about what is good for children and families. It provides insights that will ultimately aid in solving them, and it can help temper the sense of superiority that upper-middle-class clinicians show toward families in

environmental circumstances very different from their own. But we all still yearn
for guidance about what is best for children and families. We might say that the
life-history theorists have provided a sound antithesis to Bowlby's thesis, which
was smug in its assurances about how evolution works. Now a synthesis is needed,
and it remains to be made. It awaits evidence about the lasting effects of early
experience that we simply do not yet have.

So much for the evolutionary theory of attachment, which is clearly in a
dynamic state. As for attachment *behavior,* Bowlby, Mary Ainsworth, and others
described it in classic writings, and their account holds up well. The relevant
behaviors include visual-postural orienting; turning to the breast and sucking; crying and stopping of crying; smiling, cooing, and talking; grasping and reaching;
separation protest; approach, following, greeting; climbing; exploring; burying the
face against the caregiver, using her as a base for exploration; fleeing to her; and
clinging. When these occur *preferentially* in relation to the mother or primary
caregiver, attachment has begun. They are never exclusively directed at one person, but there is a strong gradient of preference. !Kung and Western infants do
these things at similar ages, as do those in many other cultures.

So human infants, like those of many birds and mammals, are born predisposed to form attachments. These can be blocked by deprivation, but given the
normal, expectable environment of a newborn, they are inevitable. Bowlby,
being Western (English) and male, emphasized attachment at a bit of a distance,
through looking, smiling, crying, and cooing. The nipple-finding and sucking
reflexes of breast-feeding and the need for tactile stimulation and comfort
seemed less important to him. Later in the infant's first year, he realized, proximity-
maintaining mechanisms come into play. Infants grasp, cling, scramble, and
climb on the mother; later they follow her and use her as a base. Like Harlow,
Bowlby rejected secondary-drive theory, which stressed hunger reduction as a
reinforcer. He proposed a radically new ethological outlook: *Attachment unfolds
in accord with an unknown genetic program. Attachment is itself and not derived
from anything else.*

This behavior system "seeks" a target, in something like the way the eyes and
brain of a newly hatched duckling seek an object to follow. The fact that in primates the process takes much longer—up to eight months in human infants—
does not make imprinting irrelevant. The behaviors in question will fully emerge,
change, and function predictably only after an appropriate object is found, and
the baby—bird or human—will be uneasy until then. Because immature organisms, especially in slowly maturing species, have to stay close to caregivers or get
eaten, the underlying brain mechanisms and any genes behind them were under
powerful selection throughout evolution.

Although evolutionary, Bowlby's theory speaks to development: because early
attachment was vital for survival, it eventually became essential to later mental
health. A prolonged and close early relationship to one individual—a mother or
permanent substitute—produced a lifelong ability to love. Natural selection
relied on it and relaxed its genetic specification of adult behavior, depending

more on the child's environment. If this environment was anything like that of
!Kung infants, some interesting implications follow. Mary Main, a leading investigator of attachment, put it this way:

> Infant attachments and the emotions that accompany them in contempo-
> rary settings do not necessarily serve an obvious immediate survival func-
> tion, but the selection pressures on ground-dwelling infant primates and
> hunter-gatherers are believed to have been similar to each other, and the
> brains of hunter-gatherer infants are unlikely to have been significantly dif-
> ferent from ours. Therefore infants in contemporary settings often behave as
> though their survival were at stake in situations in which we believe them to
> be safe.

Thus, the emphasis on early attachment to a nurturing figure, common to
Bowlby, Freud, and others, seems appropriate, with the qualifications raised by
life-history theory. Despite variation in the environments of evolutionary adaptedness, many hunter-gatherer groups had a mother-infant relationship considerably
closer, more delicately responsive, and more nurturing than the Western pattern.

Compare the American context. In most editions of *Baby and Child Care*—
including the 1998 edition, the last he worked on before his death—the pioneering American pediatrician Benjamin Spock advised mothers consistently. They
should, he wrote, become suspicious of possible "spoiling" of babies by three
months, and exercise "a little hardening of the heart." If by five or six months the
baby still expects to be picked up every time he cries, the mother should follow a
program of "unspoiling," including pretending she is busy to "impress the baby"
that she just can't respond. Dr. Spock thus encouraged the cultural tendency of
American parents to start shaping self-reliance in infancy. His advice, faithfully
paraphrased to a !Kung mother in her own language, produced a mixture of surprise, amusement, and contempt: "Doesn't he realize it's only a baby? It has no
sense, that's why it cries. You pick it up. Later on, when it gets bigger, it will have
sense, and it won't cry so much." She considered Spock's method abusive and
unethical.

Granted, the !Kung situation is quite different. An American mother is not
surrounded by a network of relatives who help with the practical and emotional
burdens of baby care. Nor is her baby surrounded by children of all ages who provide an attractive alternative, taking pressure off the mother. Hrdy might fairly
argue that the mother's most adaptive choice is to keep her distance from the
infant, teach it self-reliance, and (if she can afford it) hire others to care for it most
of the time. Then, too, the dangers of "spoiling" might be greater in the American
context; life-history theorists, like traditional pediatricians and psychological
anthropologists, might argue that American parents have just the child care pattern they need to create adults well adapted to their culture. Many cultures differ
from the !Kung, to no apparent ill effect. Under different conditions, they provide
a variety of forms of multiple caregiving. As Margaret Mead pointed out in an

early critique of Bowlby, cross-cultural studies, from traditional polygamous cultures to the modern Israeli kibbutz, have failed to show ill effects of multiple caregiving as long as the two, three, or more caregivers offer a nurturing and stable environment.

Even among some hunter-gatherers, such as the Efe Pygmies of Zaire, mothers may be the main caregivers less than half the time, with the rest of the care given by other women and girls. Among the Kikiyu, farmers of Kenya's highlands, Herbert and Gloria Leiderman showed that young children form multiple attachments; although the mother is the main object of their affection, other caregivers contribute substantially to their emotional development. Robert LeVine and his group made similar observations among the Gusii of Nyansongo, Kenya, and the same is true of the Hausa of Nigeria, the Dogon of Mali, and many other cultures, including some for which tragedy has made maternal attachment impossible. In the Israeli kibbutz, as in many other Western settings characterized by multiple caregivers, attachments to figures other than the mother have been shown. Nevertheless, infants consistently choose one caregiver over the others available, and attachments to others are less strong and less secure by various measures.

Finally, in industrial cultures, where the risk of infant death is small, there is no clear biological advantage in leaving infant care to women. Mothers have a nine-month head start on fathers, and there are still certain advantages to breast-feeding, so infants and mothers will always be, on average, more disposed toward each other than infants and fathers. But there is no evidence that it is detrimental for a man and woman to participate equally in the care of an infant, or even for a man to be the exclusive caregiver. Single fathers headed at least a million families in the United States as the millennium turned. There is no particular reason to think that their children are growing up with inadequate opportunities for attachment. Still, this is an unprecedented experiment, and it bears close watching.

In traditional societies as in our own, a key influence on early relationships is the mother's workload. In many farming cultures, the organization of work results in several hours a day of mother-infant separation, which precludes a !Kung-like pattern of contact. The mother works in the fields, while her infant is with a girl or young woman (often an older sibling) back in the home compound. In some cultures women manage to garden with a baby on their backs, but often that is impractical. In any case, the notion that balancing work and motherhood is something new seems bizarre to anthropologists and to anyone who has spent time in the developing world.

However, there is little variety in one aspect of care: in all traditional cultures, mothers and infants sleep in the same room, usually in the same bed. Of 90 societies in a worldwide sample, mother and infant slept in the same bed in 41, in the same room with bed unspecified in 30, and in the same room in separate beds in 19. In *no* culture did they sleep in separate rooms. Also, in all higher primates and among hunter-gatherers mother and young sleep close, if not touching. This pattern was probably selected for early in higher-primate evolution; an infant sleeping alone would almost certainly be eaten by a predator. It has even been suggested that the contact and stimulation in co-sleeping could protect against sudden infant death syndrome; ongoing research by James McKenna of the University of Notre Dame is testing this possibility.

But the dominant tradition in the United States derives from farming cultures of northern Europe that swaddled infants and kept them in cradles. Compared to the more indulgent traditional societies, mother-infant contact is low in the United States, and this is clearest in sleeping arrangements. Infants often sleep in separate rooms, alone or with siblings too young to nurture them. The syndromes of bedtime protest and night waking that afflict many infants and toddlers are artifacts of our sleeping arrangements. Northern European and American cultures have devoted great energy to combating the natural tendency of infants and mothers to fall into one another's arms. This was not easy, and the results are not yet known.

Dr. Spock's reputation is not that of a hard-hearted fellow urging parents to stand back from their babies. He was no such villain, of course, and is properly known for liberalizing baby-care advice. An excerpt from John B. Watson's 1928 book, *Psychological Care of Infant and Child*, which was influential in America, gives the flavor of the pre-Spock approach:

> There is a sensible way of treating children. Treat them as though they were young adults. . . . Let your behavior always be objective and kindly firm. Never hug and kiss them, never let them sit in your lap. If you must, kiss them once on the forehead when they say good night. Shake hands with them in the morning. Give them a pat on the head if they have made an extraordinarily good job of a difficult task. Try it out. In a week's time you will find out how easy it is to be perfectly objective with your child and at the same time kindly. You will be utterly ashamed of the mawkish, sentimental way you have been handling it. . . .
>
> In conclusion won't you then remember when you are tempted to pet your child that mother love is a dangerous instrument? An instrument which may inflict a never healing wound, a wound which may make infancy unhappy, adolescence a nightmare, an instrument which may wreck your adult son or daughter's vocational future and their chances for marital happiness.

This came not from some extremist on the fringes but from the dean of American behaviorist psychology, the same man who had trained little Albert to fear furry animals and who later had a successful career in advertising. One wonders what he would have advised in a marriage manual.

Even today the standard pediatric advice in the United States concerning sleep problems is the same as Dr. Spock's: let them cry it out. For some infants and children this works easily, but for others it may take several nights of crying for half an hour or more—perhaps these are the more timid or reactive infants in Kagan's or Suomi's studies. Spock's counsel:

It's hard on the kindhearted parents while the crying lasts. They imagine the worst: that the baby's head is caught in the slats of the crib, or that she has vomited and is lying in a mess, that she is at least in a panic about being deserted. From the rapidity with which these sleep problems can be cured in the first year, and from the way babies immediately become much happier as soon as this is accomplished, I'm convinced that they are only crying from anger at this age. . . .

If the several nights of crying will wake other children or anger the neighbors, you can muffle the sound by putting a rug or blanket on the floor and a blanket over the window. Soft surfaces of this kind absorb a surprising amount of the sound . . .

Yet parents may have to be even more steadfast:

Some babies (and young children) vomit easily when enraged. The parent is apt to be upset and shows it by anxious looks, by rushing to clean up, by being more sympathetic afterward, by being quicker to come to the baby at the next scream. This lesson is not lost on children, and they are likely to vomit more deliberately the next time they're in a temper. . . . I think it is essential that parents harden their hearts to the vomiting if the baby is using it to bully them. If they are trying to get the baby over a refusal to go to bed, they should stick to their program and not go in. They can clean up later after the baby has gone to sleep.

This advice is in a child-care manual said by its publisher to be the best-selling new book since 1895, when best-seller lists began. Before the 1998 edition, it had sold 50 million copies, most of them carrying the same advice. Interestingly enough, the first edition—full disclosure: I was raised by it—was much milder in its handling of sleep and spoiling problems, no doubt helping to account for Spock's indulgent reputation. But by 1968, all these hard-hearted passages seemed set in stone.

Today the spectrum of sleep-problem advice runs from Richard Ferber's *Solve Your Child's Sleep Problems*, at the Spock extreme, to William Sears's *Nighttime Parenting*, which takes a strongly contrasting approach. Parental emotions run high, and viewpoints almost seem to be coming from armed camps. For example, customer reviews of Ferber's book on the World Wide Web include several who felt that Ferber's method is akin to selfish and cruel behavior, while others swore by it. One mother said it took two weeks for her frightened daughter to let her out of her sight after the Ferber method failed. Another said that she had two nights of hell, but soon after that her two-year-old slept through the night without a problem.

Most parents in traditional cultures would agree with the first assessment. There is no basis for the claim that the infant is crying from anger, and little for the claim that the procedure is harmless. These claims may be true for some babies, but for others they may not. Clearly this is an area where Hrdy's evolutionary perspective, with its heightened concern for mothers, might apply. But the experience of night waking most mothers outside of Western society have had, whether in the distant or the recent past, would have been something like the one described in Jill Hoffman's poem, "Rendezvous":

> Summoned from a dream of your summoning
> by your cry, I steal out of bed and leave
> my doting husband deaf to the world.
> We meet, couple, and cling, in the dim light—
> your soft mouth tugs and fills and empties me.
>
> We stay that way a long time it seems, till
> on your brimming face, where milky drops glide,
> I see my body's pleasure flood and yawn.
> We turn each other loose to sleep. Smiling
> your smile of innocence, I return
> to the bed of your begetting, and a man's warm side.

The prediction that products of such care will grow up tied to their mothers' apron strings is false. At age five, !Kung children strayed significantly farther from their mothers than did their London counterparts in a parklike setting; they also had more interactions with other children and were less often nurtured. At fourteen, a !Kung boy may go out walking alone or with a friend, drive lions from an antelope carcass with sticks, and carry the meat home to his parents. A !Kung woman in her early twenties may go out into the bush in labor, deliver her infant and afterbirth without assistance, cut the cord, and bring the baby back to the village. However inadvisable this may be, it hardly shows lack of independence. Indulgent nighttime parenting may not be the best thing for couples, but there is no basis for the claim that it is bad for babies. There is a large, hidden subculture of co-sleeping in the United States that cuts across ethnic and economic status. In fact, studies by Marjorie Elias of extremely indulgent mothering by core members of La Leche League International showed few later differences between their infants and a control group from middle-class Boston.

Independence in children indulged in infancy would not be surprising to Bowlby or to a theorist like Erik Erikson, who held that establishing "basic trust" is *necessary* for later independence. This proposed link between early indulgence and later reduced dependency runs so contrary to classical notions of learning— responding to distress signals should theoretically *increase* their frequency—that it is hard to believe. But it is supported by many studies. An early study by Silvia Bell and Mary Ainsworth showed that infants whose mothers respond to them more in the first three months cry less in the second three months than infants whose mothers were less responsive earlier. Another by Alan Sroufe and Everett Waters showed that toddlers who have had close mother-infant relations tend to respond more maturely to separations than do other toddlers.

Such findings have been challenged by learning psychologist Jacob Gewirtz, whose argument is related to the "spoiling" concept: if you respond to crying and other dependent behavior, you should strengthen that behavior, working against the growth of independence. It is reasonable to think that this is part of what happens, but ongoing research suggests that early secure attachment is predictive of later psychological adjustment. Such continuity is strong in some studies but modest or equivocal in others. The astounding claim that early nurturance has little or no long-term impact on psychological development is controversial but sustainable in twenty-first-century debate. As we have seen, a scientist as committed to the power of early experience as Stephen Suomi has labeled the support for its role in human development as "surprisingly sparse." Only intervention studies are ultimately persuasive, and few are being done.

The best, by Dymphna van den Boom in the Netherlands, demonstrated the enduring effectiveness of a skill-based program designed to enhance maternal sensitivity. Mothers of six- to nine-month-old infants who were objectively determined to be irritable were taught to respond to both positive and negative signals but not to become intrusive or detached. Fifty mother-infant pairs received the intervention, 50 did not. Eighty-two of the pairs—43 from the intervention group and 39 controls—were evaluated at a year and a half and two years, and 79 again at three and a half years. There were observations of free play, everyday interaction (including dinner with Mom and Dad), problem solving, and peer interaction, as well as laboratory tests of separation and mental development.

Most of the lasting behavioral changes in the second year had to do with attachment security, maternal sensitivity, and child cooperation. Strikingly, only 26 percent of the untreated infants were securely attached, whereas 76 percent of the intervention group were. In the fourth year there were enduring effects on parental responsiveness and child cooperation, and the intervention group had better interactions with peers. This is the gold standard, a randomized controlled experiment. Combined statistical analysis (meta-analysis) of a number of studies, especially intervention studies, together with the van den Boom results, enable us to say, as Jay Belsky did, "that Ainsworth's core theoretical proposition linking maternal sensitivity with attachment security has been empirically confirmed." This is an experiment, not a correlation. The hypothesis was jeopardized, and it was confirmed.

If attachment behaviors are part of normal biological functioning—instinctual, if you will—then it makes sense for them to be hard to extinguish. They are not randomly occurring Skinnerian operants. Some, at least, such as crying and contact seeking, stem from organic distress. It may be as inappropriate to ignore them as it would be to ignore cold-induced shivering. Ignoring attachment signals simply increases the distress. The basic-trust model is not completely convincing, but there is at least as much evidence in favor of it as there is against it. As for the spoiling theory, it clearly fails, at least for indulgence of infant dependency. (This argument, incidentally, has nothing to do with the sort of spoiling we are so good at in our culture, namely, the indulgence of the older child's desire to consume, disrupt, or destroy.)

Recall, too, the context of !Kung parent-infant relations. Social support for the mother and the lure of the child group have no real parallel in our own culture—or at least the parallels are much weaker. The multiage child group fosters separation from the mother and gradual independence, aiding the growth of emotional competence in a warm and psychologically rich environment. This includes competition, real fighting, and rough-and-tumble play. It also includes "gentle-and-tumble play"—mutual touching, tangling of legs, clinging and rolling while lying on the ground. It may take imaginative forms in which the older child takes the role of parent or a boy and girl play husband and wife. Adult models abound. While !Kung parents try to conceal sexual activity, they do not always succeed, and young children have a high awareness of sex. They do not play "doctor," they play sex, even simulating intercourse. Interview studies by ethnographer Marjorie Shostak confirm that such play occurs and remains vivid in adult memory. Even at puberty it may still occur, and the transition to real sex may be (although it is not always) gradual. The hormonal changes of puberty transform the heterosexual bond among the !Kung as among ourselves, and—as Harlow suggested—the playmate relationship underlying it.

!Kung marriage, however, is arranged, as it has been in most cultures throughout history. If the young people have strong inclinations, they may be respected, but most marriages are formed without romantic love, even though many !Kung experience it. Once in a lecture I said I was puzzled about the adaptive function of romantic love, since it has so rarely in history been the way pair bonds were formed. One of my more cynical students came up after class and explained (dismally but, I now suppose, correctly) that its function is to get people *out of* a pair bond—either a previously established one or one that is being arranged. The student's claim fit the theory later proposed by Robert Frank about emotion generally: that the irrationality of emotion has rational consequences. In his *Passions Within Reason: The Strategic Role of the Emotions,* Frank showed that "irrational" feelings appropriately (although not deliberately) deployed could produce gains for those having the feelings by showing others that a certain course of action may be more costly than it is worth. In plain words, tantrums can be adaptive. Romantic love, in this view, is a kind of tantrum, a passion just not worth opposing.

The !Kung, in any case, often end these early marriages, although rarely after a child is born. So they have something like the trial-marriage idea once proposed by Bertrand Russell (for which, among his other reckless ideas in those timid days, he was barred from teaching mathematics and logic at New York's City College). This instability leaves room for romance and sex to figure in pair formation. It also leaves room for adultery, which for the !Kung involves romance as much or more than sex. As Marjorie Shostak showed in her classic study, *Nisa,* and its sequel, *Return to Nisa,* !Kung men and women may be far more preoccupied with their extramarital lovers than the number of actual sexual encounters would suggest. A woman should work for her husband, as Nisa said, "but she should still have a few lovers. Because each one gives her something." Nisa was not just waxing poetic—she meant material, not metaphoric, gifts; yet she went on to describe passionate love:

> Her lover may have been away. But when he comes back and she sees him, her heart knows that he is around once again. She lives, waiting until she has a chance to be alone with him. When they meet, he says, "Perhaps you didn't think about me?" . . . She says, "What? I thought about you often. What could have stopped me? . . . Am I not a person?" Because when you are human, you think about each other. He says, "I thought maybe you had forgotten." She says, "No, I thought about you often and with strength." He says, "Mm, that's why I came to talk to you, to see what you were thinking." She says, "And how do you feel now that you have seen?" He says, "You . . . you really made me miserable! The month I left, my heart pained for you and wanted you very much." She says, "It's been the same with me. I also wanted you and my heart also pained for you."

"The appeal of affairs," as Shostak says, "is not merely sexual; secret glances, stolen kisses, and brief encounters make for a more complex enticement. Often described as thrilling adventures, these relationships are one of the subjects women spend much time discussing among themselves." Nisa's strong, romantic, including adulterous, ties demonstrate clearly the power of female choice in the most basic type of human society.

But how do we understand the role of romance when it can work so differently in different cultures? In the 1989 classic *Dreams of Love and Fateful Encounters: The Power of Romantic Passion*, psychiatrist and psychoanalyst Ethel Spector Person considered theories about romantic love. They have run the gamut from patronizing to denigrating—Y to Z. Most psychological theorists simply ignored it, while others have compared it to obsession, mania, psychosis, infantile dependency, childish wishes, and a host of other abnormal or immature mental states. It is seen as disruptive to productive work and destructive of human relations, producing unwanted consequences, from divorce to unwed parenthood to suicide, and at the very least causing some of life's most colossal disappointments. Dorothy Tennov, a respected authority, tried to change its name to an ugly and contemptuous bit of jargon, "limerence." One would never know from this official view that it also produces marriages lasting more than half a century, mundane loyalties, inspiring sacrifices, children, grandchildren, even dynasties that share this love as it grows and changes, and personal transformations that rival any in psychotherapy—not to mention transcendence, which may be transient but is still one of life's best experiences.

Of course, there is a downside. As we saw in chapter 9, some men turn violent when rejected by their lovers—an extreme adult form of separation protest—and for those women love may be fatal. But countless ordinary and extraordinary human beings have risked much for romantic passion, and they have not been simply deluded. Transience alone does not invalidate an experience; unhappy endings do not negate love's value. Passion outside of the usual rules—extramarital, homosexual, chaste, or across large age gaps—can be as lasting, fulfilling, and vital as more conventional love. Person's thesis is simple: "The capacity for romantic love is inherent in human nature." Hers is a work in which the prose itself is essential to the viewpoint: "Sex is a sacred rite in the religion of mutual love"; "Love is the only appetite for which an 'excess' is allowed"; and "Love serves to assuage the sorrows and wounds of some old developmental conundrums by binding the present to the past. It repairs the lingering humiliations of early life, melds the sensual to the tender, the body to the soul, and provides continuity at the same time it separates the lover from the past."

How differently Freud's precise but chilling jargon might have expressed this same concept. Person's work righted a long-standing wrong—the consistent slighting of romantic passion by psychologists, philosophers, and other "experts." She turned instead to the real experts, writers. Take Stendhal's concept of crystallization, a mutual idealization but one not easily achieved. In Person's words:

> The metaphor . . . is that of a branch or bough stripped of its leaves in winter and thrown into an abandoned salt mine. Months later the branch is pulled out and is covered with brilliant crystals. . . . Even the tiniest twigs are spangled over with sparkling, shimmering diamonds, and the bare bough is no longer recognizable.

Still, great writers characterize this passion as a decidedly mixed blessing, as in Carson McCullers:

> The most outlandish people can be the stimulus for love. A man may be a doddering great-grandfather and still love only a strange girl he saw in the streets of Cheehaw one afternoon two decades past. The preacher may love a fallen woman. The beloved may be treacherous, greasy headed, and given to evil habits. Yes, and the lover may see this as clearly as anyone else—but that does not affect the evolution of his love one whit. A most mediocre person can be the object of a love which is wild, extravagant, and beautiful as the poison lilies of the swamp. A good man may be the stimulus for a love both violent and debased, or a jabbering madman may bring about in the soul of someone a tender and simple idyll.

Nor is it lost on them that women are often oppressed by love, as in this quote from Virginia Woolf:

> Yet, she said to herself, from the dawn of time odes have been sung to love; wreaths heaped and roses; and if you asked nine people out of ten they would say they wanted nothing but this—love; while the women, judging from her own experience, would all the time be feeling, "This is not what

we want; there is nothing more tedious, puerile, and inhumane than this; yet it is also beautiful and necessary." Well then, well then?

But despite outlandishness and even oppression, "Love is an act of the imagination. For some of us, it will be the great creative triumph of our lives."

Romantic love, we now know, is universal, in the sense that *some* people in every culture experience it, despite the fact that only a few societies routinely base marriage on it. There is a widespread myth among historians that it is a product of the purely Western tradition of courtly love that began in the Middle Ages. But four *millennia* ago, an Egyptian poet wrote this:

> Last night made it seven my eyes missed my kitten —
> I'm a tottering outpost invaded by sickness,
> Love lethargy leaves my limbs logy,
> I stagger about with a low-grade fever,
> Sometimes I drift into reverie. . .
> sometimes don't even remember my name . . .

The translation may be liberal, but the feelings are remarkably familiar. For the author of the *Song of Songs* a millennium later,

> . . . love is strong as death;
> jealousy is cruel as the grave:
> the coals thereof are coals of fire,
> which hath a most vehement flame.

And in the 1980s, Bedouin women described in Lila Abu-Lughod's classic, *Veiled Sentiments*, used love poetry as "a discourse of defiance [and] of autonomy and freedom" in the context of a strong and confining patrilineal kin group.

Princess Yoza, a poet in Japan's seventh-century imperial court, wrote this *tanka*:

> Now the nights grow cold
> and cold winds return to howl.
> With you gone,
> my whole life is torn by winds.
> I wonder: Do you sleep alone?

Thirty-one Japanese syllables, yet enough to demolish the myth. What blend of arrogance and ignorance can have led any "expert" to claim such precious, universal human ground for one culture alone?

Some anthropologists have claimed that romantic love is absent from primitive cultures, but this, too, is prejudice. William Jankowiak and Edward Fischer found romance in 146 out of 166 cultures sampled from the anthropological record, and Jankowiak went on to compile a fine book full of examples.

On the Polynesian isle of Mangaia, where Helen Harris did her research, romantic love (*inangaro*) is common and valued, with seven elements recognized in the West: desire for physical and emotional union, idealization, exclusivity, intrusive thinking about the beloved, emotional dependency, changing life priorities, and a powerful sense of empathy. A woman in her forties said, "I didn't want any other men, and I wanted him to look only at me, not at other girls." A man in his sixties said, "It's both feelings and sex. It's when both things come together; that is falling in love." And an elderly man recalled, "We felt together—close. When you look at some ladies, you know they are bright and good and beautiful, but someone else keeps coming into your mind."

All three were talking about their spouses, but many cultures show a disjunction between love and the marital bond. This was the case in the courtly-love tradition of medieval Europe. Marriages, as in most of the world, were always "of convenience" but sometimes life was more pleasant for a lady if she were chastely courted by a knight not her husband. Saga and song testify to the power of such passion at a distance. Our image is of the forlorn knight mooning beneath a castle window, plucking his mandolin. But chastity belts, the brutal iron traps that still dangle from hooks in the museums in those castles, prove that courtly love was not always chaste. Passions sometimes sought evolutionary advantage even outside the bounds of reason.

Jankowiak invokes Abu-Lughod's "discourse of defiance" to explain real and fictional Romeo-and-Juliet cases in China. Irrational feelings empower the young to reject parental plans and in some cases create a nest of privacy protected from the state. Communist China shows profound continuity with ancient times: "While romance was seldom the basis for choosing a marriage partner, it existed well before the founding of the Han dynasty, and, in some cases, actually thrived in the face of powerful parental opposition." By the 1980s, urban Chinese had widely accepted the role of romance in marital choice. And something similar is going on in traditional India, where Elizabeth Ahearn has found that romantic courtship by mail is replacing arranged marriage. As in the West a century earlier, the age-old, romantic discourse of defiance became bound up with the ideals of individuality, democracy, and freedom.

But how did such passions intersect with sex or pair bonding in evolution? Most primates do not pair-bond, and those that do are imperfect at it, to say the least. Marmosets, the small and varied group of monogamous South American monkey species, move toward plural mating in some situations. Gibbons, too, those graceful acrobats of the Asian forests, have proved less strictly faithful than once thought. Yet even imperfect pairing among our relatives tells us that primate brains can form lasting attachments between adult males and adult females. What is tricky about humans is that, unlike in marmosets and gibbons, the pair belongs to a much larger group, a situation charged with other opportunities. But is there incipient pairing among other monkeys and apes, in settings where the couple must persist despite being buffeted by the winds of group life?

Barbara Smuts's classic study, *Sex and Friendship in Baboons,* is grounded in evolutionary theory but acknowledges the complexity of individual lives. In olive baboons, a large, ground-living, Old World monkey, sex is often inseparable from male-female alliances—nonexclusive sexual friendships. Foreign males join troops and must form friendships that may eventually become sexual, but this may take up to a year or more. One young male, Ian, never made the transition. He had a hard time forming relationships with females, often provoking alarm, and unlike most males his age, he did not seem to know how to calm a partner by sitting at a distance, making friendly sounds and gestures. Instead he chased and frightened them, even evoked screams, bringing on group action that drove him away. Eventually he gave up and disappeared. Another male his age, arriving around the same time, behaved appropriately toward females and was soon fully admitted. The consequences for reproductive success? For the young male olive baboon, patient courtship is vital.

Something similar is true of bonobos, creatures intensely yet tenderly, almost sweetly, sexual. Sex, apparently including orgasm, binds females to males and to each other; it is somehow both promiscuous and gentle. In contrast, Jane Goodall's chimps seem at first glance quite vulgar. Promiscuity is rife, and at the female's rear end, ripeness is all. When she swells, she's swell, or at least males think so, and they take turns spritzing their variously viscous, densely teeming blobs of semen into the vestibule of one ready womb. The ensuing microscopic race pits sperm against sperm until one speedster crash-lands in the egg's outer zone, opening one slim path to a genetic future. In such a world large testes matter far more than love.

Yet even chimps can form temporary couples: a young male and female often like each other enough to go off on "safari," a tryst in the bush away from the sexed-up crowd. Could they be an ape version of Romeo and Juliet? To answer that, we would have to know much more about what the chimps are feeling. But these consortships, studied by Goodall, Caroline Tutin, Patrick McGinnis, and others, lasted from days to weeks and produced numerous pregnancies. Fear evoked classic signs of attachment: "During sudden alarms (such as the calls of strangers or the sound of passing fishermen) the female often ran to the male and the two embraced and kissed." Yet both were far calmer than in other mating contexts; "there were many sessions of relaxed social grooming when the male almost always groomed for longer than his female." When girl met boy and kept him for a couple of days out of eye- and earshot of otherwise occupied, broad-shouldered, heavily hung males—when *her* fastidiousness chose helpless little *him*—may have been, among some distant ape ancestors of ours, the dawn of romantic love.

The physiology of love, formed in that evolutionary crucible, is poorly understood, but that is changing. We can start with what Sappho knew 2,500 years ago:

To me that man equals a god
as he sits before you and listens
closely to your sweet voice

and lovely laughter—which troubles
the heart in my ribs. For now
as I look at you my voice fails,

my tongue is broken and thin fire
runs like a thief through my body.
My eyes are dead to light, my ears

pound, and sweat pours down over me.
I shudder, I am paler than grass,
and am intimate with dying . . .

Barbaric as it must seem to say it, these are signs of autonomic nervous-system turmoil. Sappho explicitly implicates fear, probably fear of loss, which may cause some of the arousal. Her fear is associated with jealousy, since she loves the woman beside the man. But the core response is love itself. In Person's account:

Falling in love is often accompanied by physical sensations—loss of appetite, breathlessness, and sleeplessness. Lovers feel the growth of love in their pound-ing hearts and in less traditional (or less poetic) sites as well—their stomachs, arms, groins, and lungs. Love becomes a delirium and is spoken of as a fever. These are the physical counterparts of the excitement and the fear that accom-pany falling in love. And it is no wonder that we are frightened. To fall in love is to risk opening up, revealing one's true self, and then being rejected. . . . Falling in love is an agitation, a mixture of hope, anxiety, and excitement.

This recalls infant attachment, but the infant has no more mature self, look-ing on as if from a distance asking, *What is going on here?* Also, unlike in infancy, the provenance of the beloved in romantic attachment is baffling. Why him (or her)? And why so much arousal, so much fear, and (if love is lost) so much grief?

The autonomic nervous system can explain only part of it, but research by Stephen Porges during the nineties illuminated its role brilliantly. He proposed a three-stage model of the evolution of the vagus nerve, a key branch of the system. In the first stage, a primitive, unmyelinated, *visceral* vagus fosters digestion and responds to threat by depressing metabolic activity—the freezing response. In stage two, the *sympathetic* nervous system can increase metabolic rate and *sup-press* the visceral vagus to promote fight or flight. Stage three, in mammals, adds a fast-acting, myelinated vagus that can quickly regulate cardiac output. Because the *mammalian* vagus is linked to the nerves for facial and vocal expression, it helps regulate social engagement in courtship. We know that fear of the partner must be suppressed to become close, and that, once formed, a bond mitigates fear of other threats. So it makes sense that a system designed to process fear would evolve further, to mitigate fear in the midst of love.

But this "polyvagal" theory needs higher circuits in the limbic system, where we have already found tracks of other emotions. For example, oxytocin and vaso-pressin connections link the hypothalamus to brain stem centers that govern the vagus, and we know from the work of Sue Carter and others that those brain hor-mones are vital to social bonds. But there must be some still higher structures in the brain whose function will make attachment and affection understandable. Arthur Kling and Dieter Steklis of Rutgers University showed long ago that the affiliative behavior of monkeys can be reduced by removal of the underside of the frontal lobes (the Gage cortex) or the ends of the temporal lobes. This is a crude intervention and a very general effect, so it cannot tell us much. But studies that Kling did with Jane Lancaster and Leslie Brothers focused on the amygdala, the almond-shaped emotional region inside the tip of the temporal lobe. They found that African vervet monkeys in the wild reacted more mildly to removal of the amygdala than did those in the lab. This was important, since it showed that social context has a big effect even on something as dramatic as brain surgery.

Still, monkeys of several different species became socially isolated after amygdala damage. They experienced a fall in rank, a deterioration of maternal behavior, and a decline in social grooming, which is the basic glue of monkey relationships just as conversation is in ours. The amygdala—which, as we have already seen, plays a role in both fear and rage—is at the crossroads of a brain system that processes relation-ships, including the temporal lobe cortex overlying it, and the lower middle frontal lobe, which figures so prominently in Damasio's theory of emotion. Because the amygdala relays messages from those areas of the cortex to the hypothalamus and so to the body—partly through the mammalian vagus—the anatomy gives it the oppor-tunity to embody the "warmth" we feel in affectionate social ties. Another clue lies in the timing of development in the major cables of the limbic system, such as the fornix, the stria terminalis, and the mammilothalamic tract. Myelination of these massive connections, occurring during the second half of the first year of life, could help form a basis of the fear of separation, as well as other components of attach-ment. Study of those structures in animals would illuminate the physiology of love.

Meanwhile, research on the physiology of attachment in children, by Nathan Fox, Megan Gunnar, and others, has made initial inroads. Older infants and tod-dlers show physiological signs of stress in response to separation, but it is hard to separate infant temperament from the effects of different kinds of attachment. Gottfried Spangler and Karin Grossman, working with Klaus Grossman in Regensburg, Germany, have found stable physiological responses from birth to age six, suggesting temperament is the reason, but they argue that maternal con-tact mitigates stress responsiveness—a cortisol-coping hypothesis. Gunnar has a similar model, in which inhibited, highly reactive infants use secure attachment as a physiological buffer against stress. As we saw in chapter 5, research by Fox and Richard Davidson showed that right frontal activation is associated with negative emotion, whether caused by a chronic reactive temperament or a bad situation, while left frontal activity goes with positive emotions and relationships. This means that the frontal lobe must be asymmetrical in its exchange of information with the vagus nerve and the heart, as well as with subcortical limbic circuits that mediate affiliation and pleasure.

To make matters more complex, affection and attachment are different emotions, and we know little about what makes the one deepen into the other. As for romantic love, that is something else again. The pubertal blooming of the gonads and their control systems transforms affection in both sexes, enhancing as well as limiting it. But it is challenging enough to see how pubertal hormones promote sex and lust, much less the subtler, more mysterious emotions of romance. Still, the fact that crushes emerge around age ten suggests that *prepubertal* changes — whether the surge of adrenal hormones or some unknown process of limbic system growth — must be involved. And the later emergence of truly erotic affections and attachments depends on pubertal hormones to produce sexual feelings.

When we get past adolescent and youthful pairing to parenting, the third major part of the attachment triad, we are on firmer ground. Many studies have focused on the physiology of parenting in animals. In mammals, we know that the hypothalamus is involved, both as a hormonal regulator and in brain circuits for retrieval and protection of the young. But here the sexes must part company. The only higher vertebrates where males have physiological adaptations for parenthood rivaling those of the females are pigeons and doves. There is little evidence of bodily adaptation for male nurturance in other species. Still, the brain adaptations must be there, since the caregiving is there, and it isn't done by magic. Thanks to the work of Sue Carter, Thomas Insel, Larry Young, and others, we now have some idea what they are.

Carter has spent much of her life studying the physiology of pairing and parenting in voles, and has focused on oxytocin, which she believes is the fundamental hormone of all relationships. Insel, working at first with Carter, continued and expanded her work. Why voles? Laboratory rats and mice, the main subjects of physiological research, neither form pair bonds nor show male parental care. Voles vary. Mouselike as they seem to the casual observer, they range from species as promiscuous as lab mice to those that mate for life, like ring doves. Prairie voles pair-bond; think of a prim farm couple of the American plains, with paternal care comprising a full share of parenting. Montane voles (found in high meadows in the Rocky Mountains), on the other hand, make multiple matings; think, perhaps, of a mountain-man kind of miscreant male, leaving the young to maternal care only. Alliteratively speaking, prairie voles are pair-bonding and parental, including paternal; montane voles have multiple matings and are only maternal. Helpfully, two other alliterative species, the pine vole and the meadow vole, follow the same patterns, respectively.

Insel asked, *Do prairie and montane voles have specific brain adaptations that can explain their behavioral differences?* A decade of elegant experiments have shown that they do and that the differences center on not one but two hormones, oxytocin and vasopressin. These two peptides, now among the most important behavioral molecules, are made in the hypothalamus and sent either down to the rear of the pituitary gland or up to the wider brain. Remember sex in the rough-skinned newt? Like our amphibian ancestors, the newt has one hormone, vasotocin, which combines features of oxytocin and vasopressin. Vasotocin, in these primitive vertebrate ancestors, enhanced sexual responses culminating in mating. They probably had neither parenting nor pair bonding. But as they evolved into reptiles and then mammals, vasotocin gave rise to two derivatives, oxytocin and vasopressin. Each differed from vasotocin by only one out of nine amino acids, yet they changed mammalian destiny.

Oxytocin is involved in both milk ejection and the pulses of the uterus during labor; vasopressin helps the body retain water. But these physiological functions say nothing about their role in behavior. Insel and his colleagues showed how differently the brains of males and females achieve pair loyalty. Oxytocin is involved in female loyalty as well as maternal behavior. It is also, in all likelihood, part of the most basic apparatus of affiliation, and in this sense may serve the universal function proposed by Carter. Studies in oxytocin-knockout mice—male mice lacking the gene that codes the hormone—show how fundamental its functions are. Jennifer Ferguson, working with Insel and Larry Young, found that oxytocin-knockout males paired repeatedly with the same female show no memory of her; they explore her each time they meet as if she were someone new. Replacing their oxytocin cures this boorish social amnesia, while giving genetically *normal* mice an oxytocin blocker makes even them socially stupid.

There are no other memory deficits, so this study supports others suggesting that social memory involves a different set of brain circuits from other kinds of memory, probably including the amygdala. But alas, this is not loyalty, it is merely social oblivion. Genetically normal male mice, because they remember the female well, soon lose interest in her, responding strongly to a new female. It must be little comfort to the lady that her partner stays interested because he can't seem to remember who she is. But such is the way of the mouse—loyalty is just not part of the plan. In the pair-bonding prairie vole, though, it *is* part of the plan, and so we need an explanation that goes beyond oxytocin and social memory.

Male loyalty and paternal behavior in prairie voles depend on vasopressin, which simply does not function the same way in the multiple-mating montane vole—or for that matter in the ordinary laboratory mouse. Yet the variety of mating and parenting patterns owes little to the hormones themselves—these are similar across species—but much to the way the brain reacts to them. Genes for the two hormones and their receptors are distributed differently in the brains of different species and respond differently to the hormonal surges of puberty. So the pairing, paternal species and those whose males normally roam have similar hormones but different brains.

The findings are specific. Partner preference, with or without mating, depends on oxytocin in females and vasopressin in males. Injecting vasopressin into the brains of prairie-vole males enhances partner loyalty, and blocking vasopressin decreases it, but neither changes the sex act itself. These effects do not work with male montane voles because they don't have the needed brain receptors. As for female prairie voles, oxytocin does the trick for partner preference and again does not affect sex itself. But for the montane vole, even at high doses delivered into the brain, oxytocin doesn't change its fickle habits.

Where in the brain do these molecules work, and how? For one thing, oxytocin receptors in prairie voles lie in the brain's reward centers. Direct stimulation of reward circuits—through the D2 dopamine receptors in the nucleus accumbens, for example—promotes partner preference, so it is likely that oxytocin recruits dopamine to make closeness rewarding. Overall, the distribution of both oxytocin and vasopressin receptors confirms the role of limbic system circuits in attachment, but to go beyond that generalization will require much more anatomical work. In the meantime, Larry Young, who started out in Insel's lab, has brought the power of gene technology to bear on these problems.

Young showed that the flanking, promoter region of the gene for these receptors is highly subject to mutation, which makes it a perfect candidate for rapid evolutionary change. Small mutations here can produce large changes in gene expression in the brain—just the sort that would make attachment systems diverge in otherwise closely related species. Also, quite astoundingly, Young has managed to insert some vole genes into mice. Transgenic mice that get the prairie-vole vasopressin-receptor gene express the receptors in the brain in a most un-mouselike pattern resembling that of the prairie vole—including the cingulate cortex, the claustrum, and parts of the thalamus. They also show a marked rise in social affiliation when injected with the hormone, losing their species-specific detachment and replacing it with a volelike closeness. In one genetically altered line, the changes in the male brain were passed on consistently for four generations. Perhaps women will one day turn to gene therapy in their eternal quest for more dependable men.

But is this the secret of *romantic* passion? Probably not. First of all, species differences are great, even among pair-bonding rodents. Second, we have no idea what feelings are associated with pair bonding and mate guarding in nonhumans. Given our more recent common ancestry, primates make more likely pair-bonding models for us than voles do. For this reason, Insel and Zuoxin Wang are studying the two brain peptides in common marmosets. In these small, pair-bonding South American monkeys, vasopressin receptors are found mainly in the limbic system, including the nucleus accumbens, diagonal band, septum, amygdala, and hypothalamus—well placed to influence pleasure, aggression, and fear and to stimulate sexual acts.

Consider our own pair bond. Marriages, whether made in fits of passion, by parental design, or because of pregnancy—shotgun or perhaps spear-point weddings—have always been a mixture of successful and failed purposes. Given standard mortality rates, for most of human history "'til death do us part" meant about fifteen or twenty years, and divorce or abandonment was always a risk. As we saw in the last chapter, 80 percent of cultures allowed polygyny, yet most marriages in most societies have probably involved one woman and one man. In the industrial world we see marriages made for pragmatic or passionate reasons but sustained by social sanctions, economic necessity, and force of habit as well as friendship, sex, and love. Anthropologist Helen Fisher has shown that in many different countries the peak time for divorce is four years after marriage, which she believes is an adaptation for getting one infant born and nursed through weaning. But whatever the explanation, this four-year itch suggests a universal dynamic, a possibly profound insight into how marriages work—or don't.

Even in the West, where marriages often begin in romantic passion, few continue in that vein for more than a few years. Anthropological and historical evidence—not to mention everyday life—teach us that romance and pair bonding are different processes. Yet psychological research on adult attachment in recent years has all but ignored the distinction, naïvely blending romantic love, sexual relations, and pair bonding as if they were all one thing, and even as if they were a mirror image of attachment in infancy. The cross-cultural blinders and biological blankness of this work make it difficult to take seriously. The human pair bond, at best, includes sex, companionship, collaboration, and at least some hint of romance. But there are lasting marriages in which not only passion but sex have pretty much vanished. Sadly, there are marriages in which even affection is gone; people may hate each other and give each other unremitting misery yet be bound together for half a century by some force as mysterious and powerful as love. This, perhaps, is the irreducible essence of the pair bond, and whatever its physiology it is more than mere habit and it is not the same as love.

Fisher, following psychiatrist Michael Liebowitz, proposed a two-stage process for the physiology of pair bonding in cultures where falling in love is the entry path. First, the head-over-heels phase: monoamine neurotransmitters including norepinephrine (providing Sappho's racing heart and other sympathetic effects), dopamine (key to the brain's sense of reward), and perhaps phenylethylamine (also, incidentally, found in chocolate) are at high levels. As their torch dims, it is passed to the pair-bonding system, involving oxytocin and vasopressin, which ignite the cooler flame of lasting love. In the best marriages, an ongoing ability to recruit the chemistry of romance in the context of the pair bond somehow sustains the erotic friendship. Most couples can at least access the sexuality, fueled by estrogen and testosterone, within the long-term attachment made by oxytocin and vasopressin.

Half of American marriages end in divorce, and a good many others probably should as well. Yet worldwide, marriage is a great source of happiness. A seventeen-nation study published in 1998 showed that being married added to happiness in all but one country, Northern Ireland. In the others, fifteen Western industrial nations and Japan, "marital status was significantly related to happiness. Further, the strength of the association between being married and being happy is remarkably consistent across nations." In addition, "This effect was independent of financial and health-oriented protections offered by marriage and was also independent of other control variables, including ones for sociodemographic conditions

and national character." This study, like others in the 1990s, found that contrary
to common belief women as well as men benefit from being married. Also, being
married adds much more to happiness than just cohabiting does. This suggests
that commitment and social acceptance are important, and that the struggle for
recognition of same-sex unions, now achieving success in churches and courts,
will lead to something deeper than formal labeling.

Yet at the same time there has been a massive increase in the amount of
cohabitation in unmarried couples, a trend that, unlike the rising divorce rate,
continued into the 1990s. One third to one half of such households include children, and it is likely that half of all American children will spend part of their
childhood with couples whose pairing is not officially sanctioned. Of course,
none of these trends is independent of the most important social transformation
of our time, the reentry of women into the labor force. Still, something in human
nature causes people to pair off, independent of or even contrary to religious urgings. Clearly the human pair bond remains a biological mystery.

As for mothering, the first and most basic kind of mammalian love, it, too, is
physiologically complex. There are poorly understood forms of parental care in
invertebrates, fish, amphibians, and reptiles, but it is with the birds and mammals
that parenting becomes vital to reproduction for all species. In mammals, the
intensification of infant dependency and maternal care drove a great advance in
brain evolution, the elaboration of the limbic system. Hormones operating on that
system have historically been the first physiological factors studied. Not just oxytocin but estradiol, progesterone, and prolactin figure in the nest-building and
caregiving behaviors of rats, mice, rabbits, hamsters, ring doves, and canaries, but
they do so in different mixes and on distinct reproductive schedules. Prolactin,
which causes breast enlargement and milk-making in mammals, and the equivalents—crop and brood patch formation—in birds was at one time thought to be
the main mothering hormone. But it now seems that at least the onset of mothering is due to oxytocin, with estradiol and progesterone also playing a role.

This field has a long history. In pioneering studies, Howard Moltz of the University of Chicago showed that the best way to get a virgin female rat to respond
positively to pups is to put her through hormone treatments identical to those she
would have gone through in pregnancy and birth. That means a gradual, weeks-
long rise in estradiol and progesterone, an abrupt fall in both (such as normally
occurs just before delivery), and finally a prolactin surge. Psychologists Joseph
Terkel and Jay Rosenblatt of Rutgers University found out how, as it were, to cut
to the chase. They showed that if you hooked up the circulatory systems of a virgin female rat and a rat that has just given birth, the virgin would begin to behave
maternally toward pups. This raised the possibility that going through all the hormonal changes of pregnancy was after all unnecessary, and that the changes
immediately following parturition, combined with exposure to pups, would be
sufficient. Or, alternatively, as a result of the three-hormone program there was
some "Factor X" in the blood that could instantly make a mother out of a virgin.
Later, a student of Moltz, Michael Numan, and others showed that the "medial

preoptic area" of the hypothalamus—the same area where sex differences have been found—is essential for maternal behavior in rats, and that estradiol works by influencing this region.

Oxytocin, the hormone of parturition and milk let-down, was also a logical candidate for a mothering hormone, and studies began soon after it was chemically described. When Cort Pedersen and his colleagues introduced it into the brains of nonpregnant female rats, the rats became fully maternal in half an hour. Various treatments that block oxytocin prevent maternal behavior, but only if it hasn't yet started. Thus, the hormone is necessary for the onset of mothering but not for its maintenance. Oxytocin can't be the bloodborne factor suggested by the transfusion study, because it doesn't cross the blood-brain barrier. But its importance in kick-starting the process is increasingly evident.

How do these hormones work in the brain to produce maternal behavior? Numan has outlined "a core or elemental neural circuit that may be involved in maternal behavior in the rat," with an emphasis on pup retrieval. The medial amygdala, probably influenced by special olfactory cues, projects to the preoptic area of the hypothalamus. The preoptic area sends signals to the midbrain, especially the central gray and ventral tegmental areas, and oxytocin plays a role in this connection. Feedback from the midbrain to higher brain centers is also needed, and this occurs through dopamine neurons projecting to the ventral striatum and nucleus accumbens. Judith Stern, broadening the focus to other aspects of maternal behavior—licking, grooming, crouching over the pups, and nursing—has emphasized a tactile stimulation pathway in the brain separate from that of smell. The circuits she outlines include the preoptic area and the central gray but also the lateral septal area and especially the trigeminal nerve, which becomes more sensitive to pups under hormonal influence, much as the pelvic area becomes, during estrus, more sensitive to tactile stimuli from the male.

Finally, genes make a very large difference in maternal behavior, and several are under study. Jennifer Brown and her colleagues at the Harvard Medical School did groundbreaking research on a gene called *fosB*. This is one of several "immediate early genes," so called because they are promptly activated in neurons in response to certain kinds of stimulation. Repeated electrical stimulation, light, and the neurotransmitter glutamate are among the triggers for it, if the target neuron is right. The surface of the neuron is stimulated, and a chain of events activates a *fos* gene, which produces a transcription factor—a protein that can turn on other genes. But, important as they should be, it has been hard to figure out just how they affect behavior. Brown and her colleagues changed all that by creating *fosB* knockout mice. Mutants don't make the transcription factor, but they do survive and thrive.

What they don't do is mother. Despite a perfectly normal pregnancy, complete with all hormones and full mammary and milk development, they show so little nurturance that their pups die within a day or two of birth. If cross-fostered to a genetically normal mother, the pups do fine. Nothing is wrong with them, and there is nothing abnormal about the knockout mother's physical or hormonal state. And because in mice even virgins and males learn how to nurture pups with repeated exposure, some knockout virgins and males were also tested. Compared with genetically normal counterparts, they failed miserably. Their defect is in behavior, pure and simple.

But what *is* it? It isn't mental ability, since the knockouts pass complex tests with flying colors. It isn't the sense of smell, in which they are just like normal mice. It isn't a general defect in motivation or limbic system functions, because the mutants move around, explore foreign objects, eat, adapt to cold, have sex, get pregnant, and fight off intruders just like regular mice do. Their defect is specific to nurturing.

Now, when normal mice encounter pups and mother them, they have *fosB* expression in the preoptic area of the hypothalamus—a region essential for maternal behavior. It is also expressed in the olfactory system, but the preoptic area is a convergence point for sensory input from pups, including smell, touch, and taste. No one knows which gene the *fosB* protein binds to, but the promoter part of the oxytocin receptor gene has the binding site. This raises an intriguing possibility: oxytocin could cause *fosB* activation, and *fosB* binding could then stir up oxytocin-receptor production, making the neurons more sensitive to the hormone. This positive feedback would set up a virtuous cycle that might help explain the exceptional intensity of maternal behavior.

However, there are species differences in the biology of mothering, even among rodents, just as there are in pair bonding. Mice, we just saw, are easy touches where pups are concerned. Psychologist Elaine Noirot showed that priming—mere exposure to pups—can induce maternal behavior in virgin female mice in a matter of hours, with no hormonal or neurological treatment. Similar priming in rats studied by Rosenblatt took six to seven days, by which time the pups had to be kept alive artificially. And in hamsters the process, as studied by ethologist Martin Richards, took even more prolonged and repeated exposure, usually requiring the replacement of several litters killed by the female. Hamster virgins eventually get primed, but only at the cost of quite a few dead pups.

Testosterone effects are negative but consistent. Whether given to females around the time of birth, during pregnancy, or after delivery, it reduces maternal behavior in many mammals. In various species, castration of males *improves* their parenting. But the few mammals in which males play a natural nurturing role have only begun to be studied. As we have seen, paternal males like the prairie vole have brain adaptations for nurturing young, including receptors for oxytocin and vasopressin in critical brain regions. Zuoxin Wang's marmosets are similar. Darwinian theory predicts that such brain adaptations will be found in humans, too, if we look for them, and they may help us comprehend human dads, devoted or deadbeat. So if bad fathering is an illness, it may turn out to be treatable.

Human mothers go through hormone changes that are much more gradual but otherwise similar to those of rats: a slow rise of estradiol and progesterone, an abrupt fall of both before delivery (with a rising ratio of estradiol to progesterone, which falls faster), and a rise of prolactin. Not surprisingly, studies are under way of how these changes may influence women's attitudes and behavior, but they have so far produced more questions than answers. One finding, a correlation between cortisol level in mothers and their responsiveness to infants, echoes rat studies and suggests a role for subjectively experienced stress in triggering nurturance. (This might help to explain the besotted devotion of the new father described at the head of the chapter.) Putting the infant to the breast causes a surge in prolactin and oxytocin. These in turn suppress estradiol and progesterone, so that the entire profile of reproductive hormones, as well as the likelihood of resuming monthly cycling, is altered. This is as true of !Kung hunter-gatherers as of women in the West who nurse intensively, and Judith Stern showed that these changes influence mood and behavior, suppressing a woman's sexuality and promoting her responsiveness to the infant.

Still, there is a paradox for biologists: no hormonal changes occur in adoptive mothers, yet they are often as good or better than biological ones. So why attribute significance to the hormonal changes? Simply put, the cultural and social preparation for motherhood, important for any woman, is even greater for adoptive mothers, who feel that they are under scrutiny and who have often waited eagerly for years. In contrast, a significant minority of births throughout human history must have been unwanted; for those, hormonal preparation may have been crucial—a trick devised by natural selection to make good mothers out of reluctant party girls. Men, of course, go through no such changes (except for stress hormones), yet they care for children; there are now more than a million families in the United States headed by single fathers. Would average men be as inherently nurturing as average women, given equivalent environments of rearing? This is still a viable hypothesis; I wouldn't bet on it, but that doesn't make it wrong.

Traditional psychology has not made much of the ties among relatives outside the immediate family, yet it is increasingly evident that most complex species live by them. The problem is they don't talk about them. So you either have to hang around for years to find out who is whose uncle or grandmother or second cousin, or else use molecular methods to determine relatedness—a course more and more field biologists are following. Most *people*, on the other hand, at least in traditional settings, will talk a blue streak about kin, to one another and to anyone peculiar enough to listen. Kinship preoccupied anthropologists because it was a central concern of the people they studied. Relatedness by blood or marriage generates most of the social rules in such cultures, including rules of marriage, group membership, exchange, inheritance, authority, and descent. But what is the basis of the kin tie, or the feeling that "blood is thicker than water"? Psychologically, it could be merely a diluted or mixed form of parent-child, child-parent, sibling-sibling, or heterosexual affection. Structurally, kinship systems are just a way of ordering the social world; after all, you can't give gifts to or marry or take orders from everyone. Both these explanations surely have a piece of the truth, but recent advances in sociobiology have brought *genetic* relatedness into exquisite focus.

The great English geneticist J. B. S. Haldane put the problem this way. Suppose my brother was drowning. How great a risk could I run in saving him before natural selection would work against me—or rather, work against the "altruistic gene" that made me go after him? Well, all else being equal, I could run a maximum 50 percent risk, because the likelihood that my brother carries the same gene by common descent is 50 percent. That is, over the long run of evolution, considering many such acts of heroism, the frequency of the gene for altruism would not decline at that risk. If the hapless drowner were only my nephew, however, I could only run a 25 percent risk of death without reducing the frequency of the altruism gene. And if he were, alas, only my first cousin, I could only take a 12.5 percent chance of dying to save him.

Modern sociobiologists, notably the late William D. Hamilton, took this notion seriously and made a scientific revolution. Hamilton applied the theory to ants and honeybees, which have special patterns of genetic relatedness and special patterns of altruism; witness the wasp or bee's rash self-evisceration in the act of stinging an intruder—say, us. Because of these extremes of cooperation, such insects are called *eusocial*, or truly social. Hamilton also generalized the Haldane model to cover all altruistic acts. They should be favored by selection if and only if the altruist loses less (in fitness, or reproductive success) than is gained by the beneficiary (same units) *divided by their percentage of shared genes*. So I should help my full brother only if the benefit to him is more than twice the cost to me. Of course, this kind of calculation would not apply to my gift of flowers and dinner to an unrelated lady; that isn't altruism but mere self-serving courtship.

One consequence is that nepotism should rule social systems, and it does in animals and in most traditional human societies. But there are at least two ways for altruistic acts between unrelated individuals to be selected for: reciprocity and cooperation. Reciprocal altruism, proposed by Robert Trivers in 1971, can evolve in animals that have the ability to recognize and remember whom they have helped, and that are long-lived enough to be paid back. Monkeys and apes are ideal in these respects. They evolved subtle abilities to distinguish and store memories of faces, as well as to recall being aided, and they live long enough to benefit from the reciprocal gesture. Whales, dolphins, and elephants also fit the bill. Altruism and repayment alike might come by way of aid in fights, and such coalitions could easily trump a bully who never could be beaten one-on-one.

Unfortunately, it is difficult to make such a system resistant to the evolution of cheating. The Prisoner's Dilemma model, whose evolutionary implications were most fully worked out by Robert Axelrod, accounts for this by making the reciprocity simultaneous—in effect, cooperation. The game consists of a situation in which two prisoners must either cooperate or not cooperate (defect); the reward is greatest if you defect while the other fellow cooperates. However, if the game is repeated again and again, and the other fellow has some learning capacity, he will not continue to cooperate. When you both defect, which you will soon be doing repeatedly, you both gain less than you would have if you had both cooperated. It is not obvious what you should do in this situation, assuming there will be many trials, but computer simulations show that the most successful strategy is tit for tat—doing what the other fellow did last time—rather than consistent defection. It may be to your advantage occasionally to test the negative pattern, to cooperate once and see if your partner is also playing tit for tat. A shift could make you much better off.

One can go a surprising distance with this type of theory. The kin selection model began to be applied to research on higher primates almost immediately. Jeffrey Kurland, studying Japanese macaques, found that grooming and other helpful and cooperative behavior favored kin among hundreds of possible partners. This was one of the first tests of Hamiltonian theory. Paul Sherman's research on ground-squirrel alarm calls became a classic study of risk taking for kin, although this proved to be an inadequate explanation for a seemingly similar species. Under Sherman's leadership, a talented interdisciplinary group literally brought to light the extraordinary social life of the naked mole rat, an underground rodent sometimes called the "saber-toothed sausage." The life of this creature is as close as mammals are going to get to being eusocial. An exceptional degree of inbreeding in colonies leads to extraordinary sacrifice and cooperation as in the societies of ants, wasps, and bees.

Sometimes, however, this line of thinking is stretched in an effort to explain away every act of human or animal courage or generosity as based on nothing more than raising or lowering the frequency of "the altruism gene." No such gene has been found in complex animals, but it doesn't need to be, really. All we need is for one or more genes to enhance altruism in some way or other, and this is highly likely. In fact, identical twins resemble each other in altruistic inclination more than do fraternal twins. Genes *affecting* altruism will eventually be identified and cloned; the flaw in the argument is not in the nonexistence of such genes.

But there *is* a problem here: the theory seems to tell us about a mechanism when it really doesn't. That is, it obscures the fact that there are many possible paths to the goal of distributing generous acts as adaptation demands. Even in ground squirrels the basis of kinship has been experimentally shown to be due in part to being reared together, independent of genes. And Sarah and Harold Gouzoules have pointed out that "the most important mechanism whereby kin are recognized in primates is probably association during development." In captive rhesus monkeys, for example, the quality of relationships among females could be predicted from the relationships their mothers had had with each other, probably because of learned associations.

In humans, of course, such developmental association is the basis of lifelong kin relationships. Describing the stable power of these ties has been the bread and butter of generations of anthropologists, and Robin Fox in particular has brought the power of Darwinian theory to bear on traditional kinship questions. For example, in *The Red Lamp of Incest*, he proposed that our remote ancestors had some complex calculations to do, and even more complex alliances to make, if they were to find mates and at the same time avoid incest. This may be a key selection pressure for brain evolution.

But even in industrial states, kinship is still vital to social life. Carol Stack and others have shown the immense importance of kinship among poor urban blacks, while Alice and Peter Rossi have demonstrated its role in the lives of middle-class whites: "The great gift of increased longevity in terms of intergenerational relationships is . . . that the parent-child relationship persists for 50 years or more, and that for more than half of those years the child is also a parent." Although their research confirmed the greater importance of closer kin, they "were surprised to find that even in so highly structured a domain as kinship normative obligations, we could demonstrate several significant effects of early family experiences: 'Those who grew up with affectionate parents, in cohesive, intact families, and who had particularly good relationships with relatives they loved or admired, tend to internalize high levels of commitment to civic duties and to higher levels of felt obligation to a variety of kin, and even to nonkin.'"

The point about civic duties and nonkin is consistent with all we know about how humans work and must stem from the causal vagueness of genetic programs in our ancestors. For example, suppose a human hunter-gatherer were programmed by nature to save drowning people according to how well he loved them. The resulting distribution of risk across drowning victims might be a good approximation of what kin selection predicts. Yet the love guiding his efforts might come from none other than the usual processes of maturation and learning. And the same hunter-gatherer, having been thrust over the course of a few thousand years into a new kind of social world full of nonrelatives, would still learn to love.

Even when love is not the issue, helping others can be reflexive, as Haldane understood: "But on the two occasions when I have pulled possibly drowning people out of the water (at an infinitesimal risk to myself) I had no time to make such calculations. Paleolithic men did not make them. It is clear that genes making for conduct of this kind would only have a chance of spreading in rather small populations where most of the children were fairly near relatives of the man who risked his life." Many species have a specific chemical signal enabling them to recognize their kin by smell, but humans recognize kin by association and emotion. In this sense, in today's world, nonrelatives are kin.

It is possible that a simple signal exists in humans, too—although it would probably be something quickly and strongly learned rather than something innate, and it might be a seen rather than smelled signal. Indeed, this may be the original adaptive basis of our keen interest in faces—the "Are you related to so-and-so?" reaction. In a social world thoroughly sorted into kin groups, that ability may have been highly valued. And that same intense interest and fine discrimination may have helped males detect cuckoldry in the faces of their imputed offspring. For better or worse, it may also figure in our strong sense of what constitutes facial attractiveness and, negatively, in our penchant for bigotry.

Last but not least, the face contains our basic natural equipment for communicating affection. There is of course the smile, and laughter, provided the joke is on someone or something else; the suitably timed blush; the dilation of the pupils, an automatic sympathetic nervous system reaction to sympathetic attention; and the properly timed look away. Anthropologist Irven DeVore used to say that if two people look into each other's eyes for more than about six seconds, they are either going to make love or kill each other. (I would add that if one of them is very small, they are probably parent and infant.) As for the earlier stages of courtship (it used to take more than six seconds), ethologist Irenäus Eibl-Eibesfeldt has filmed in widely separated cultures around the world what seems to be a universal flirtation display in humans. The gaze alternates from the face of the person being flirted with to the side and down, while the head is half turned away. This is a classic approach-avoidance display, and it may be an innate *human* pattern for communicating romantic interest.

But it does leave us with an unsatisfying sense of the simplistic. To escape it we turn again to a master of affectional communication, Henry James, in a novel called *The Golden Bowl*. Adam Verver, a rich, decent, generous, middle-aged art collector, long since widowed, has become enamored of and has tendered an offer of marriage to Charlotte Stant, an exquisitely beautiful, well-bred, and charming person about half his age, who happens to be his daughter's best friend. Charlotte has said yes, *but*—the but being linked to the daughter's and her husband's consent. Charlotte will not contravene her best friend's wishes. We see them as they are about to open a telegram from the couple, with their answer:

What he could have best borne, as he now believed, would have been Charlotte's simply saying to him that she didn't like him enough. This he wouldn't have enjoyed, but he would quite have understood it and been able ruefully to submit. She did like him enough—nothing to contradict that had come out for him; so that he was restless for her as for himself. She looked at him hard a moment when she handed her his telegram, and the look, for what he fancied a dim, shy fear in it, gave him perhaps his best moment of conviction that—as a man, so to speak—he properly pleased her. He said nothing— the words sufficiently did it for him, doing it again better still as Charlotte, who had left her chair at his approach, murmured them out. "We start tonight to bring you all our love and joy and sympathy." There they were, the words, and what did she want more? She didn't, however, as she gave him back the little unfolded leaf, say they were enough—though he saw, the next moment, that her silence was probably not disconnected from her having just visibly turned pale. Her extraordinarily fine eyes, as it was his present theory that he had always thought them, shone at him the more darkly out of this change of colour; and she had again, with it, her apparent way of subjecting herself, for explicit honesty and through her willingness to face him, to any view he might take, all at his ease, and even to wantonness, of the condition he produced in her. As soon as he perceived that emotion kept her soundless he knew himself deeply touched, since it proved that, little as she professed, she had been beautifully hoping. They stood there a minute while he took in from this sign that, yes then, certainly she liked him enough—liked him enough to make him, old as he was ready to brand himself, flush for the pleasure of it.

Here we have as complex a rendition of the affectional burdens borne by the human face as we are ever very likely to require. Despite the fact that the thirteen (they prove unlucky) words of the telegram are in a sense the center of the piece, words have virtually no place in it. The telegram is a token, a piece of information. All the communication in the scene itself is in the faces, and indeed the very absence of words heightens the emotion. The subtlety of communication is all but inexpressible in words—would be for most writers other than James. The scene could be played on the screen but not onstage; the face is for private communication. Yet some of the most important of all human expression is made in just this way, with language playing just this dry—however critical—a role.

Freud famously said that the goal of a healthy mental life is to love and to work, in that order. Yet real love is no Victorian ideal. As for Adam Verver, love comes to the poet in middle age, and yet it is physically palpable. Adrienne Rich wrote a series of romantic and erotic verses that may be the most beautiful love poems composed in English in the late twentieth century. "Did I ever walk the streets at twenty," she asks, "my limbs streaming with a purer joy?" What sets these apart from most love poems is that they record a romance between two women, a relationship baffling to almost every notion of biological orderliness in affection. To all such mysteries, to all such incomprehensible possibility, *brava*.

CHAPTER 14

Grief

To have in general but little feeling, seems to be the only security against feeling too much on any particular occasion.

—MARY ANNE EVANS (George Eliot), *Middlemarch*

In the 1980s, as the divorce rate reached 50 percent of marriages, an Episcopal priest and a clinical psychologist in Darien, Connecticut, began performing a new church rite for divorce, complete with a new, original liturgy. The service was primarily for the children, to emphasize that their parents still loved them and to mitigate their deep sense of loss. They were, after all, losing a family, an illusion, at least, of their parents' love for each other, and a certain sense of the basic reliability of the world—although to be sure this latter loss is one we incur somehow sooner or later. A Unitarian Universalist minister in Cherry Hill, New Jersey, had written a divorce rite as far back as 1966. It is in the church's "Great Occasions" book, although it has not been used often. The idea seems to be spreading, however, in secular as well as religious settings, as more and more people deal with the loss of their marriages.

Meanwhile, the Hemlock Society—an organization that explains how you can take your own life expeditiously, quietly, and painlessly—was gaining fame from London to Oregon. Named for Socrates's final beverage, this rather dismal club grew rapidly in England and soon gained many adherents in the United States. They approved suicide only for terminal patients, especially those in pain, and they claimed to be providing information that would prevent people from making a mess of it, but obviously they believed that there is such a thing as a right to die. They were the vanguard of a sea change in the way Western culture looks at death. During the 1990s Dr. Jack Kevorkian, a Michigan pathologist, used a drug-delivery "death machine" to hasten dozens of deaths. Although he did this openly and in some places where it was clearly against the law, no prosecutor could get a jury to send him to jail. So in 1998 he committed euthanasia on television—or at least on videotape, which was later televised—basically daring the state of Michigan to prosecute him for murder. They did so, and he eventually did go to prison. But by then Oregon and several other states had passed right-to-die referendums. In 1997 the Oregon legislature passed its Death with Dignity Act, casting into law the right to die.

We seem to be trying to find new responses to common losses—the end of marriage, the end of life. During the last half of the twentieth century, the percentage of all women currently divorced roughly tripled. This was part of a very long trend in which the likelihood of divorce increased tenfold, from 5 percent in the 1860s to 50 percent in the 1980s. Since marriage termination through death at younger ages was far more common in the past—"until death do us part" once meant about fifteen years—and since unknown numbers of abandonments also played a role, the change in marital commitment is not quite what it seems. Nevertheless, it is large. Between 1910 and 1970 a related statistic—divorced women as a percentage of those ever married—increased from less than 1 percent to almost 6 percent in forty-year-old women. The most dramatic historical trend: divorce increased much faster in older age groups than in younger ones. The divorce rate, as a percentage of all marriages, leveled off at around 50 percent in the mid-eighties, but a profound cumulative effect continued. In 1997 the median age of divorced people in the United States was fifty, and they made up about 10 percent of all adults. Anthropologists like Helen Fisher began to speak of "the evolution of human serial pair bonding."

For loss of life, the trend is the opposite. Fewer of us than ever are dying before "three score and ten," and the fastest-growing age group is centenarians. Yet the demand for a timely end to life has become a din. Suicide has risen in some age groups, perhaps alarmingly so among teenagers, more of whom are putting an end to life just as they first come to grips with it. On the whole, suicide rates for the United States are about the same now as they were in the early twentieth century. However, there were major decreases during each of the two wars—a poignant and instructive irony—which were made up for with post–war surges. Japan, struggling with recession as the century ended, experienced a major increase in self-inflicted death, a rise that may be consistent with Japanese traditions about suicide. And today in the United States there has been a cultural change that permits public advocacy of suicide and even euthanasia—not out of melancholy or derangement, not to save face or preserve honor after failure, but to affirm the dignity and decency of a life destroyed by illness. Just as the inscription on the Japanese suicide blade says, "It is better to die than to live without honor," so ordinary people in the West increasingly believe that it is better to die than to live without dignity. We believe we have a right, claimed by writer Charlotte Perkins Gilman in her final note, "to choose a quick and easy death over a slow and horrible one."

> Public opinion is changing on this subject. The time is approaching when we shall consider it abhorrent to our civilization to allow a human being to lie in prolonged agony which we should mercifully end in any other creature. . . . I have preferred chloroform to cancer.

Public opinion has progressed to where this gallant woman stood in 1935. But to say that we are accustomed to these losses and to others as well—the distancing of family members, the end of neighborhoods, the abandonment of traditions—is not to say that we have come to grips with the feelings they engender. Elisabeth Kübler-Ross was a psychiatrist famous for facing up to death—her characterization of the mental processes of the dying person essentially founded a new field. Yet in the end she became a figure of ridicule by committing herself (body and, as it were, soul) to the service of a homegrown, midwestern clairvoyant, supporting the adage that there are no atheists in foxholes. Still, few people have attempted to dwell on the process of dying in so concentrated a way; and no one, perhaps, ever focused so systematically on the emotions of the dying. The dying themselves do, of course, but they only do it once, and they do not have to go on living afterward. Kübler-Ross did it vicariously many times, and she ended in denial.

Which is where most of us begin. Even without traditional religious beliefs, we usually manage not to think about it. Ernest Becker's *The Denial of Death* places this process at the center of a theory of human behavior. Becker—an anthropologist who taught at Berkeley and who wrote his masterpiece while he himself was dying of cancer—attempted to account for much of human action as a response to the presence of death. This was not merely because it is in the nature of life to try to keep on living but in a specifically human sense, that gives a central place to our awareness of death, and to the resulting fear and anticipatory grieving. For Freud and his followers it was the "death instinct," for Søren Kierkegaard, "dread" or "the sickness unto death." Since awareness of death can lead to strategies to avert it, it is adaptive. But awareness can be heightened to the point of interference with normal functioning. How can we understand this peculiarly human balance?

Consider how other creatures respond to loss. Among ducks and geese, the death of one member of a pair may result in repetitive searching and maladaptive behavior for at least a few days. In one case, a mother goose lost one of her four goslings and searched so persistently for the lost one as to endanger the lives of the other three. We know that in groups of wild monkeys, a mother whose infant dies may clutch it to herself for days, exposing herself to infection and predation. Psychologist Leonard Rosenblum found that when he anesthetized either an infant or its mother in the lab, neither would abandon the other, despite the unresponsiveness. These situations recall his findings on punishment and infant attachment and suggest that mere proximity can be rewarding even when the loved one is lifeless or punitive. They contrast vividly with other, also valid, reports of animals that immediately jettison or even eat the newly deceased. Ambivalence is expectable.

But can animals grieve? This possibility has been known since Darwin's second greatest book, *Expression of the Emotions in Man and Animals*. He reviewed grieflike reactions in animals and in many human cultures; in vivid word pictures, drawings, and photographs—it was one of the first books to use them—he showed the remarkable natural continuity in emotional expression. Today we understand Darwin's prescience in detail because we have incontrovertible evidence of these continuities. Paul Ekman and his colleagues proved that facial expressions of sadness have a common form and a common interpretation in cultures throughout the world, a finding that has held up well for three decades. Ekman's illuminating reassessment of Darwin's classic on the subject—including a detailed annotation and updating—vindicates Darwin's original vision.

Irenäus Eibl-Eibesfeldt has for decades documented cross-cultural regularities in more complex human interactions, leaving no doubt as to the psychic unity of humankind—not only in the basic core emotions but in their social expression. Claims for the absence of grieving after major losses in certain cultural settings have proved to be unconvincing on further study. Unni Wikan, working in Bali, has demonstrated that masking emotion is not tantamount to an absence of it, and Karl Heider, comparing emotion and its description in three Indonesian cultures, has established that small differences in the way emotions are labeled do not negate cross-cultural continuities. Paul Griffiths, a philosopher in the Darwinian tradition, has extended the human continuities in emotion to a systematically assessed phylogenetic range that goes far beyond the human.

Of course, human expressions are also uniquely colored by culture, and some that seem quite basic may be uniquely human. As Darwin suggested, blushing may be one of these, and weeping may be another. Despite some anecdotes suggesting that elephants, seals, and other mammals do it, there has been no clear evidence that in those species it is connected to emotion. Tears as protective lubricants flow continuously and drain away unnoticed, but just how emotional activation in the brain produces the overflow of weeping is a mystery. In one of the first extended studies, Tom Lutz showed clearly that despite being a human universal, weeping is strongly subject to cultural guidance in its time and place of expression, its relation to age and gender, and even its interpretation and meaning. Sacred tears, tears of joy, and tears of martial victory take their place beside tears of grief and pity in the life and lore of many cultures. There is clearly a gender difference in the likelihood of weeping. Yet in the last half of the twentieth century in the West, rules forbidding men from public tears changed so that soldiers, athletes, and politicians could cry openly without being ridiculed.

But this is not the only unique feature of human grief. As John Archer has pointed out, even the most advanced animals seem to lack what kids develop by age ten or so: a clear understanding of the finality of death. Still, after the death of an offspring, parent, or companion, many animals show behavior that looks remarkably like grief. The process in elephants has been described poignantly by Cynthia Moss and has been seen in many other species. Death may not be, for them, a defined moment of irreversibility, but there may be a growing emotional awareness of the finality of the loss. The point is made most clearly in Jane Goodall's observations of grief among the chimps of Gombe.

An old female, Flo, had already had several youngsters when her last, Flame, was born. Flame's next older sibling, Flint, was almost five years old when his mother, six months pregnant with Flame, stopped nursing him because her milk dried up. Flint whined, moaned, followed on her heels, clambered on her, and threw tantrums when she didn't respond. After Flame's birth, Flint's behavior improved, and he was very solicitous of his younger sister, but he still occasionally showed the tantrum behavior of the typical chimpanzee (or human) weanling.

Baby Flame died at six months, of an apparent infection. Her death was not observed and her body was lost, but it was likely that both mother Flo and older brother Flint experienced some emotional impact; they seemed to take solace in each other. Flint returned to nursing, although his mother was now elderly and he was six years old—about equivalent to a human nine-year-old. Flo never became pregnant again, and she proceeded to treat Flint like a baby, which apparently suited him well. In Goodall's first book, *In the Shadow of Man*, she expressed concern for his future: "Whatever the reasons for Flo's failure [to wean Flint] . . . Flint, today, is a very abnormal juvenile. Will he gradually lose his peculiarities as he grows older, or will some traces of infantile behavior characterize him when he is mature?"

The question was answered two years later. When Flint was eight, no longer nursing but still tied to his mother's apron strings, she became ill and died. He stayed with her body, moping around with a dejected expression and incapable of doing anything else. When the body was removed for autopsy by the investigators, Flint returned repeatedly to the place where it had lain. His activity was severely limited, and although he had by this time learned to feed independently quite well, he was not taking care of himself. A few days later he, too, died. A veterinary autopsy revealed gastrointestinal infection and peritonitis as the physical cause of death, but it seemed likely that his severe emotional reaction had weakened his natural defenses. Goodall had at least to consider the possibility that Flint had died of grief.

The study of such reactions in human children has a long history. René Spitz, a psychiatrist who studied infants' fears, became interested in their responses to losing a parent and studied infants brought to foundling homes. Separation in the first few months of life, while it required adjustment by the infant, did not usually cause serious problems if subsequent care was good. In vivid contrast, infants eight to ten months old might respond to the loss with a long behavioral depression resistant to even the best substitute care. Some developed a condition Spitz called *marasmus*, a gradual, dangerous wasting away. It was obvious that there were physical complications, but the triggering factor was the loss. Since there are major changes in emotional capacity during the second half-year of life, especially in the realms of love and fear, it is no surprise that the response to permanent loss is altered. But Spitz was the first to see in this period the emergence of the capacity for grief. Since this grief could be life-threatening for some infants, Spitz recommended that if there were any choice about when to begin a major or permanent separation from a parent, it should be done in the first six months of life, not the second.

An unintentional and tragic experiment in nature has resulted from the severe deprivation sustained by infants and children in Romanian orphanages during the waning years of Communism. Many of these children, adopted by British and North American families, are thought to have experienced what Spitz would have called marasmus, although it is complicated by unknown amounts of illness and malnutrition in addition to psychological deprivation. Follow-up reveals a remarkable degree of recovery even in the most deprived groups *as groups*. But many *individuals* never catch up, and some remain unable to form normal relationships in later childhood. In strong support of what Spitz had found half a century earlier, children adopted before six months of age have a greater likelihood of full recovery than those adopted later. These unfortunate children may, as they grow, help us to understand the emotional needs of children raised under more conventional conditions. But what are those needs?

John Bowlby, who spent four decades studying the child's response to separation, inherited Spitz's mantle as the psychologist of love's inverse, loss. In a classic study, Bowlby and John Robertson looked at infants and children who had to undergo prolonged hospitalization. Since those days, owing largely to Bowlby's work, children's hospitals have instituted rooming-in for parents, but at the time hospital stays meant separation. There seemed to be four stages in prolonged separation or loss for most children. (A similar sequence occurs when a mother has to leave a child for a hospital stay herself—but in this case the reaction can be softened by substitute care.)

The first stage is protest, usually lasting a matter of days. Its onset is seen in the Ainsworth experiment, as a reaction against a brief separation. It is the phase of resistance, active searching for the lost mother or primary caregiver, and refusal to accept the loss. Indeed, at this stage the loss may not be long-term, and the protest may be adaptive, bringing the parent back. Then, too, it is often no mere wail of dissent but specific intense hostility: at the mother for leaving, at anyone trying to substitute for her while she is away, and at the mother again when she returns. So common are hostile reactions that Bowlby titled the second volume of his work on attachment and loss *Separation: Anxiety and Anger*. Rage is second only to fear as a child's initial response to loss.

Stage two is grief proper: with the energy of protest exhausted, the futility of searching is accepted. Activity is slowed, and dejection shows in the facial expression, along with quiet whimpering and an immunity to pleasure. Stage three is a sort of emotionless adaptation. The dejected mood has disappeared, and there is superficial evidence of recovery, but subtle experience with the child may reveal an inability to experience emotion, especially love. The child is in a self-protective mode. The final stage is real recovery; the loss has been accepted and the capacity for love and friendship emerges again. This scheme is, of course, oversimplified, even for a child who more or less follows the phases. They are rarely so clear-cut; overlap and variation are common. Some children skip one phase or another or become stuck in the grief phase or, perhaps more commonly, in the phase of emotionless apathy. Probably there is no child for whom the recovery, however good, does not include some measure of at least occasional protest, grief, and detachment.

Prolonged separation in infant monkeys has been a model for these and other forms of grief, and for treatment. But the studies of infant monkey separation and loss are very different from the isolation studies described earlier. Isolation rearing begins at birth; the monkey has had no chance to develop normal or even abnormal social relations. The *absence* of social contact is what is being studied, and the result is a syndrome of social, communicative, and emotional incompetence; the mildest form is ineptitude, while the most severe is something like autism. Fortunately, such deprivation is rare in human life, and most cases of human autism do not result from early deprivation. Early *separation* is quite different. Here the infant is given an opportunity to form a strong emotional bond, and that bond is broken. The infant's capacity for the affection and attachment has so far developed normally, and is intact. The response to the disruption in some species at least is very like what Bowlby and others have described in human children: protest followed by grief. And the latter response has been studied as a model of depression.

In evolutionary terms it is fundamental. Paul MacLean, the great theorist of the limbic system, placed maternal care and the response to separation at the root of mammalian brain evolution. The infant's cry or call upon losing contact with the mother is a hallmark of mammal behavior, and in studies of rats, Myron Hofer showed remarkable subtleties in the pup's reaction to separation, including patterns that look like both protest and despair. But species' differences in response to maternal loss are not trivial, even in monkeys, so we must be careful what we say about human infants. Leonard Rosenblum, who gave the first serious attention to such differences, studied affectional ties in two species closely related to the rhesus monkey and to each other: bonnet and pigtail macaques. Bonnets habitually keep close physical contact among adults and young, and growing infants can interact with everyone. Because of this pattern, the infant responds to the mother's removal with protest and sadness, yes, but also by initiating relationships with other adults, one of whom often adopts the infant. But in pigtails, adults ordinarily keep their distance from one another, and the mother-infant pair is relatively isolated. With the mother removed, the infant is not picked up by a solicitous adult; it curls up into a ball in the middle of the cage floor, ignoring and ignored by other individuals, cycling through protest, grief, and apathy. The infant may recover, but little thanks to other occupants of the cage.

Species differences in other kinds of monkeys showed that selecting any one model for human reactions would not work. We need instead a broad framework that will place the reactions of various species, including humans, in a comprehensible perspective of cause and effect, genetic and environmental. As we have seen, James Chisholm, Jay Belsky, Patricia Draper, Michael Lamb, and Sarah Blaffer Hrdy have developed models that link early childhood experiences of loss with later, adult social or sexual behavior. They suggest that several forms of early attachment can be adaptive. The infant's attachment system may be designed to detect the different conditions that children will grow up to live under—richer or poorer, predictable or not, more or less competitive or promiscuous, with higher or lower mortality—and to produce an attachment system best suited for those conditions. In this view of life history, early relationships are facultative adaptations, designed by evolution to let adult behavior be shaped by conditions detected during early life. This approach goes beyond the older deprivation models. If loss or multiple caregivers in infancy make you more likely to divorce or be promiscuous in adulthood, then perhaps that is just natural selection's way of telling you in infancy what kind of reproductive strategy you will need when you grow up. Empirical tests of these theories are under way, but for the present they remain speculative.

Meanwhile, we know that for human children the loss of the mother is sometimes comparable to those monkey species where the effect is dramatic. But keep in mind the resilient bonnets; they may reflect what happens when a child who loses its mother has other close relationships to fall back on. Recall the study by Emmy Werner, who followed a multiracial sample of 698 human infants born in Kauai, Hawaii, in 1955, until their forties. One third of them had had severe deprivations and losses in early childhood, yet one third of those—about one ninth of the total sample—developed into "competent, confident, and caring young adults." All of those had had at least one very nurturing caregiver in infancy, such as a grandmother, sibling, or babysitter, and as they went through childhood "found emotional support outside their own families, . . . an informal network of kin and neighbors, peers and elders, for counsel and support." Studies like Werner's make human children seem resilient, more like bonnets. Yet two thirds of her deprived subgroup—twice the number of her resilient one-ninth— did not do well. So is the glass one-third full or two-thirds empty?

It is both. Human as well as monkey infants grieve over early losses with attempts to cope that sometimes succeed and sometimes fail. But how do we know that these reactions represent grief—that the image of the lost mother or caregiver as a *specific* individual plays an important role? This is a legitimate question, since the infant is losing a wide variety of previously dependable things. How do we know that the grief is really *for* the loved one? The question can be raised with regard to the grieving reactions of very young children, but even for monkeys, Rosenblum proved that grief is specific. This proof lies in the response to the mother's return *after* recovery. The infant is playing happily—having gone through protest, grief, and perhaps an apathetic phase—and now has normal relations with other monkeys. Many of the general losses from the mother's departure—interaction, touching, play—have been restored. Yet when the mother is returned to the cage, the infant does not go to her but instead has an immediate relapse of the grief phase, curling up on the floor with a dejected face and severely depressed activity. This response, specifically tied to the mother at this stage, proves that it is *she* the infant has missed, *her* departure that has brought the pain; that, in short, it is she the infant has loved.

In addition to such vivid descriptions, we have a growing idea of the physiology of these reactions. As early as the 1970s a group working with psychiatrist Martin Reite at the University of Colorado Medical Center repeated the separation study using four infant pigtail monkeys, this time with extensive physiological monitoring. The phases of separation response observed by Rosenblum were seen again in all four infants. The protest phase, lasting several hours, included increased physical activity and plaintive distress calling, as well as elevated heart rate and body temperature; the agitation was physiological as well as behavioral.

During the first night of separation, all four infants had marked sleep disturbances. Since Reite and his colleagues were measuring the infants' EEGs, they could identify sleep stages. The infants showed a decrease of rapid eye movement (REM) sleep—dream sleep in humans—to less than one fourth of the 100 minutes seen on average nights before separation. One infant showed no REM at all. Meanwhile, body temperature was down from the usual nightly level and heart rate was down as well in several different sleep stages—just the opposite of what had occurred during the day, in the hours immediately following separation.

The depression phase set in the next morning. Heart rate and body temperature remained low, but now there were behavioral changes apparent only in the daytime. The two older infants engaged in less social play and more interaction with inanimate objects. The two younger ones showed all signs of the classic grief phase, including slouching or curling up, depressed behavioral activity, impaired coordination, and dejected faces. All four had the same physiological reactions even though the older infants did not show the most obvious behavioral changes. Perhaps they could dissociate physiological and behavioral responses to losses, dissembling even in the midst of strong physiological reactions.

More recent studies have been carried out in the more resilient bonnet monkey and also in the very different squirrel monkey, a small, South American tree-dweller. These studies, many led by Christopher Coe, show that not just general physiological measures and more specific hormonal systems but even the body's defenses against disease are depressed along with behavior in separated infants. Further, these changes are correlated with behavioral signs of protest, like the separation call, and with the slouched posture of grief; this suggests that the emotional response to loss causes the immunity problem. Both lymphocytes and antibodies are suppressed to some extent. This kind of change may help explain why Flint, the Gombe chimp youngster, failed to survive his mother's death. If even brief separations suffice to lower infant monkeys' immunity, what effects might permanent loss have? Squirrel-monkey studies have told us a great deal, too, about hormone changes in separation. Separation causes adrenal activation and a rise in cortisol but a decrease in thymosin, a hormone involved in immune resistance. It also reduces the ability of white blood cells to attack and destroy undesirable target cells. Similar changes may occur in human infants and children suffering negative effects of separation.

But first, consider the nature of sadness and depression in adults. The first big book on this subject, and still in some ways the most ambitious, was Robert Burton's *Anatomy of Melancholy*. Considered a classic of English literature, the revered twentieth-century clinician Sir William Osler called it "the greatest medical treatise written by a layman." First published in 1621, it went through many editions and revisions and was a best-seller for decades. More than a treatise on one disorder, it gives a comprehensive account of knowledge and belief about human nature. The sections "Anatomy of the Body," "Anatomy of the Soul," and "Diseases of the Mind" are in effect a seventeenth-century introduction to behavioral biology. But mainly the book is about melancholy, from which Burton occasionally suffered: "a kind of dotage without a fever, having for his ordinary companions fear and sadness, without any apparent occasion." Believing that melancholy can only be defined by enumeration, he offered a four-page chart worthy of the most obsessive modern textbook author.

Some symptoms? Fear and sorrow without just cause; suspicion; jealousy; discontent; solitariness; irksomeness; continual cogitations; restless thoughts; vain imaginations; bodily symptoms, including much waking heaviness and palpitation of the heart; and excessive humours—among many other, more specific, symptoms. The causes could be supernatural, coming from God or the devil, either directly or through messengers, or even from magicians or witches. Or they could be natural: old age, heredity, temperament, nurses, education, terrors and affrights, scoffs, calumnies and bitter jests, loss of liberty, servitude, imprisonment, poverty, want, "a heap" of other accidents, physical diseases, or bodily inflictions, death of friends, and loss. Symptoms could also stem from one of many more "particular" causes. Among these he included "Love of learning, study in excess, with a digression on the misery of scholars," a subject that must have been close to Burton's heart.

One wonders if Burton as a boy might have seen *Hamlet*, and begun musing on "the thousand natural shocks that flesh is heir to." In any case, what he and his age meant by melancholy was something much broader than what we mean by it or by depression today, including mental states ranging from schizophrenia to lovers' moping. Still, Burton's list is acceptable today as an account of possible causes of depression. And most of them, aside from heredity and temperament, have something to do with loss. Severe depressions can be precipitated by losses, and there are high rates of suicide and suicide attempts in depressed people. But this association can be an oversimplification. Schizophrenia, a major psychosis, does not characteristically involve depression, yet it also has heightened suicide risk. Also, the acute-onset form of schizophrenia can result from life events very similar to those that provoke depression in other individuals.

But how do we separate depression from normal grief or loss, a natural reaction with its own character and symptoms? William James called depression "a positive and active anguish, a sort of psychical neuralgia wholly unknown to normal life." It can be part of what used to be called neurosis—milder mental illness treated without hospitalization—or an aspect of more serious mood-related psychoses. These fall into two groups: unipolar, characterized by recurring or continuous severe depression; and bipolar or manic-depressive illness, in which the depression alternates not just with middling mood but with a usually short-lived phase of wild and crazy elation. The subtleties are challenging. In the words of psychiatrist Gerald Klerman, "Feelings of sadness, disappointment, and frustration are a normal part of the human condition. The distinction between normal mood and abnormal depression is not always clear." We can, however, list the features of a pathological state:

> impairments of body functioning, indicated by disturbances in sleep, appetite, sexual interest, and autonomic nervous system and gastrointestinal activity; reduced desire and ability to perform the usual, expected social roles in the family, at work, in marriage, or in school; suicidal thoughts or acts; disturbances in reality testing, manifested in delusions, hallucinations, or confusion.

But with the broader array of symptoms, the continuum with normal experience is much more apparent:

> reports of feeling sad, low, blue, despondent, hopeless, gloomy, and so on; inability to experience pleasure; change in appetite, usually weight loss; sleep disturbance, usually insomnia; loss of energy, fatigue, lethargy; agitation; retardation of speech, thought, and movement; decrease in sexual interest and activity; loss of interest in work and usual activities; feelings of worthlessness, self-reproach, guilt and shame; diminished ability to think or concentrate, with complaints of "slowed thinking" or "mixed-up thoughts"; lowered self-esteem; feelings of helplessness; pessimism and hopelessness; thoughts of death or suicide attempts; anxiety; bodily complaints.

Few are those who never experience any of these symptoms—or indeed most of them—at some time or other. Most people do not experience them continuously, but then neither do most depressives.

Continuum or not, there is now incontrovertible evidence for strong genetic influences on sadness, depression, and suicide. Even in the normal range—whatever that is—studies of happiness versus sadness are unequivocal. Identical twins, whether raised apart or together, are 4 to 5 times more alike on standard measures than are nonidentical twins. For clinically significant depressions, the studies are more numerous and equally clear. Even the depressed person's susceptibility to stressful life events has a strong genetic component. Broadening the question to include the most serious mood disorder, manic-depressive illness, echoes the other findings. In seven twin studies of serious mood disorders, 65 percent of identical twin pairs were concordant, compared with only 14 percent of nonidentical twins.

So, unquestionably, depression genes are there to be found, but the search for susceptibility genes in mood disorders has been inconclusive. In the late 1980s a false lead about a gene for manic-depressive illness among the Amish led to widespread disappointment. Since then various studies have found putative genes for manic-depressive illness alone on chromosomes 6, 13, 15, 18, 21, and X. Obviously clarity is not on the near horizon. Because sadness and depression are

undoubtedly multigenic in most sufferers, the search for quantitative trait loci—
chromosome locations that contribute only small amounts to each mood variation—may be more efficient.

There have also been extensive family and twin studies of linkages between
depression and other emotional processes. Neuroticism measured by personality
tests runs in families with major depression, and the familial link between depression and anxiety is strong, in children as well as in adults. Mouse and rat models
of depression have been extremely valuable, but they have so far been mainly
experience-induced. Experiments that explore the differences among strains as
well as those that use knockout mice are growing in number—although it is difficult to identify depression convincingly in mice and rats. With more research,
and with greater understanding of the biology of mood, believable genetic animal
models will emerge.

But whatever genes are found, the power of monkey, rat, and mouse models
of depression *caused by the psychological environment* proves the importance of
nongenetic factors. So do the genetic studies themselves, which invariably leave a
large proportion of mood disorders unexplained—one of the most decisive proofs
of the power of environment. While we would like to have more carefully controlled human studies, a role for stress has been found so often that it cannot be
ignored. There is also a new understanding of how stressful life events induce the
biology of depression.

But first consider one extraordinary study. Angela Favaro and her colleagues
at the University of Padova studied 51 Italians who had been deported to Nazi
concentration camps for political reasons in the mid-1940s, and compared them
with 47 veterans of the anti-Nazi Italian resistance movement. All the Resistance
members also reported traumatic experiences during the same war. The lifetime
rate of post–traumatic stress disorder among the deportees had been 35 percent,
and of major depression, 45 percent. The comparable rates for former partisans
were *4 and 6 percent.* There was almost an order-of-magnitude difference between
the concentration-camp victims and those who voluntarily joined the Resistance
and had very bad experiences but not the ultimate trauma. The differences were
almost as great for *current* diagnosis of the disorders as they were for lifetime
prevalence. As the authors calmly conclude, "Concentration camp internment,
even for political reasons, appears to have severe long-term psychiatric consequences."

How can we make adaptive sense of permanent psychiatric damage from
stresses that have long since ended? The conventional explanation is that they are
not adaptive at all, just unfortunate by-products of attachment or of general trust
in life and the world. Love and trust are the adaptations; the reactions to stress and
loss are not. In the newer view, taken by Hrdy, Chisholm, and Belsky, early losses
sensitize the system to adapt better to uncertain environments, so adaptation over
the life course is optimized. Unfortunately, as Chisholm puts it, "there are as yet
no studies of the relationship between early attachment and adult reproductive
behavior that use any direct measures of early attachment."

Evolutionary theory also contributes a set of predictions about grief following deaths. The models derive from the great evolutionary geneticist Ronald Fisher's concept of reproductive value: "We may ask . . . about persons of any chosen age, what is the present value of their future offspring. . . . To what extent will persons of this age, on the average, contribute to the ancestry of future generations?" This is basically a lifetime graph of how an individual looks from the viewpoint of two factors: how much has been invested in her so far, and how much reproductive potential she has left—discounted by the risk that she will die before reproducing. Reproductive value approaches zero in old age, but it is also low in infancy, building slowly to a peak at the onset of reproduction.

Fisher went on to point out that the typical human death rates are the inverse of the reproductive value curve—lowest in adolescence. At first glance these models seem insensitive—"the present value of the future offspring of persons aged x is easily seen to be given by the equation"—but if they throw light on human reactions, they can eventually advance clinical understanding and treatment. Fisher understood that older people also contribute to the welfare of the young, and that reproductive value is not a simple matter of reproductive capacity—a suggestion that echoes through ongoing debates about "the grandmother hypothesis." Seven decades after Fisher's formulation, the evidence suggests that human grief reactions can be predicted from his equation. John Archer summarizes studies of the intensity and duration of parental grief after the death of a child:

> [P]utting together the findings at specific ages indicates a steady overall increase in the intensity of grief from early pregnancy loss to loss of an adult. . . . An increase in grief with age is consistent with the increase in reproductive value of the offspring from conception to early adulthood, and the decline in parental reproductive value, identified by the evolutionary perspective.

Finally, mood disorders themselves are subject to adaptationist explanation and prediction. Depression has long been seen as offering an opportunity to recover from losses. As evolutionary psychiatrist Randolph Nesse wrote, "Some negative and passive aspects of depression may be useful because they inhibit dangerous or wasteful actions in situations characterized by committed pursuit of an unreachable goal, temptations to challenge authority, insufficient internal reserves to allow action without damage, or lack of a viable life strategy." In *Darwinian Psychiatry*, Michael McGuire and Alfonso Troisi offer several possible adaptive explanations, including the appeal for help, the searching that improves self-awareness, the ability to withdraw from society for a time, and a possible genetic link to traits that are otherwise advantageous.

This last explanation almost certainly applies to manic-depressive illness. A connection between mood disorders and creativity has been hypothesized since ancient times. In Aristotle's exaggeration, "All extraordinary men distinguished in philosophy, politics, poetry and the arts are evidently melancholic." Traditionally,

psychiatrists have been skeptical, but several excellent studies, all pointing to similar conclusions, have laid most doubts to rest. Nancy Andreasen, a biological psychiatrist with a Ph.D. in English (her thesis was on John Donne) published the
first of these studies in 1987. Thirty faculty members in the Iowa Writers' Work-
shop—the nation's most distinguished writing program—were compared with a
group of occupationally varied controls, including lawyers, hospital administra-
tors, and social workers, who matched the writers in age, sex, and education.

Andreasen had hypothesized that creativity would show some relation to
schizophrenia. Many had likened the thought processes of this severe isolative
mental disorder—often including hallucinations and delusions—to bursts of creative genius. But she found no such relationship. Instead, there was an unexpectedly strong link between creativity and another common severe mental illness:
mood disorder, both unipolar and bipolar. In the depression phase of either,
despair can be so total as to prevent all action, and hospitalization may be
required. In the manic phase of the bipolar form, elation may give way to
extreme, even delusional, risk taking. Drinking problems can go with either form,
and the three disorders—depression, manic-depressive illness, and alcoholism—
afflict close relatives at a frequency much greater than chance. Of the 30 writers,
24—80 *percent*—had experienced one or another type of affective disorder at
some time during their lives. Forty-three percent had had bipolar illness to some
degree, and another 30 percent, alcoholism. Of the 30 control subjects, only nine
(30 percent) had had any affective illness; three had been bipolar, two alcoholic.
All three differences were statistically significant, but as Andreasen wrote sadly,
"two of the 30 [writers] committed suicide during the 15 years of the study. Issues
of statistical significance pale before the clinical implications of this fact."

Other studies provided strong support. Kay Redfield Jamison, a psychologist
at the Johns Hopkins Hospital and a leading authority on manic-depressive illness, studied 47 eminent British writers and artists. Thirty-eight percent had been
treated for mood disorder, three fourths of these with medication or even hospitalization. The poets had been worst off; 55 percent had received some treatment for
affective illnesses, of which all but 5 percent received medication, electric shock,
or hospitalization, not just psychotherapy. Two of eight novelists had taken medication for depression, as had one of five biographers and one of eight artists. Apart
from formal psychiatric treatment, 30 percent of the total sample reported severe
mood swings. Kareen and Hagop Akiskal, of the University of Tennessee, conducted similar studies—in Paris, of twenty painters, sculptors, and writers, and in
Memphis, of twenty-five blues musicians born in the South. About two thirds of
each group had "hyperthymic and cyclothymic temperamental disturbances"—
intermittent periods of agitated activity and elation, or marked mood swings.

I must emphasize that most creative people are *not* mentally ill, and most of
the mentally ill—the seriously mentally ill—do not function well enough to do
sustained creative work. But there is now no doubt that the overlap between the
two categories is too great to be explained by chance; there must be an intrinsic
link. Some hypotheses have been advanced: the brooding solitude and hypersen-

sitivity of depression leads to special insight, then the transforming energy of mania leads out of that depressed state to a productive one. And there are often flights of imagination, together with just that degree of grandiosity needed to push forward an innovative and ambitious project. Reasonable though they may seem, such explanations await investigation.

In the meantime, systematic psychiatric research continues. Jamison's remarkable book, *Touched with Fire: Manic-Depressive Illness and the Artistic Temperament*, reviews other research supporting the connection. Jamison lists 84 poets and 42 other writers unquestionably in the Western literary canon who were "greatly impaired by their mood disorders," including Blake, Byron, Coleridge, Poe, Shelley, Hart Crane, Goethe, Balzac, Virginia Woolf, and Ernest Hemingway. Similar illnesses affected more than twenty major composers, including Handel, Schumann, Berlioz, Rachmaninoff, Mahler, Rossini, and Tchaikovsky. In addition, Jamison was able to reconstruct detailed family trees for some of the most eminent Western composers, poets, and artists. She found in these family histories many cases of madness that bear the signature mood swings of manic-depressive illness. They are also characteristically scattered with suicides.

The link between creativity and alcohol abuse is more confusing because of strong cultural influences, but such a link clearly exists. Research psychiatrist Donald Goodwin described an epidemic of alcoholism among prominent, early-twentieth-century American writers. The frankly alcoholic ones include F. Scott Fitzgerald, William Faulkner, Ernest Hemingway, Eugene O'Neill, John Steinbeck, Malcolm Lowry, Tennessee Williams, Jack London, Truman Capote, Thomas Wolfe, Wallace Stevens, and Robert Lowell, among many others. Goodwin estimates that *at least one third* of twentieth-century American writers of stature were or are alcoholics by current reasonable standards, and that over 70 percent of American Nobel laureates in literature were alcoholic as well. Alcoholism may be on a genetic continuum with depression, and indeed there is evidence that some alcoholics become that way because they are using this over-the-counter drug to medicate themselves for mood disorders, just as others use it for anxiety. But in addition to any biological disposition, twentieth-century American writers had a hard-drinking subculture, which undoubtedly worsened the problem.

Both unipolar and bipolar disorders occur about twice as often in women: 8 to 10 percent of men and 16 to 20 percent of women have serious mood disorders at some time in their lives. For *severe* disorders, the incidence is closer to 1 and 2 percent for men and women, respectively. In contrast to these figures for the general population, the incidence in first-degree relatives (parents, children, and full siblings) of people with diagnosed mood disorders is 15 percent, about 10 times the usual risk. This fact, combined with a concordance rate of about 75 percent for identical and 20 percent for nonidentical twins, makes strong genetic influence virtually certain. Yet the genes have been hard to find. Studies linking bipolar disorder to genes on chromosome 11 and the X chromosome have not held up well on restudy.

Other genes for mood are under investigation and will be discovered. Given the evidence of a genetic basis for mood disorders, it is fair to say that the link between them and creativity is likely to be in essence a biological one. This is not news to everyone, however. The Akiskals quote "Born with the Blues," by Memphis Slim:

My mama had them, her mama had them
Now I've got them too . . .
You just got to inherit the blues.

The song goes on to name ten other famous singers affected by the blues. As the Akiskals point out, Slim "demonstrates an insight deeper than that of many psychiatrists." However, the 75 percent concordance rate for identical twins also tells us that there are environmental effects. Here we have two individuals with identical genes; one has a mood disorder, yet the other has a one chance in four of escaping it. Why?

Not surprisingly, stressful life events can cause depression, especially if the stress is loss. In one study, 25 percent of depressives and only 5 percent of controls had experienced the loss of a loved one shortly before the illness began. There are also early-life influences. Psychiatrist Aaron Beck has been a leading authority on depression and its treatment for half a century. He found in an early study that there were more people who had lost a parent before age sixteen among depressed than among nondepressed adults. Such losses were most common in the severely depressed—despite the decades that had elapsed between the early loss and the illness. This was echoed in studies of rhesus monkeys: the response to separation from a peer at adolescence was worse in individuals that had experienced separation from the mother in infancy. Rather than inuring the subject to separation (as in the adaptationist theories), early separation may make later losses worse. That would probably be maladaptive in any environment.

Whether or not they are preceded by childhood deprivations, separation and divorce are among the most common severe losses in adult life. As Diane Vaughan has put it, linking two kinds of losses, "Mourning is essential to uncoupling." And yet, "in the course of life, we do move on. Ultimately, we all are bound to fall out of love, divorce, separate, lose loved ones through death, or die ourselves. We are constantly 'uncoupling.' . . . We leave behind jobs, parents, coworkers, churches, neighborhoods, hospitals, mentors, schools and universities, prisons, clubs, children, and friends. And they leave us behind."

Because of the sense of guilt and failure, divorce may be the most difficult of these losses. But as Vaughan meticulously shows, it has remarkable common features: "Uncoupling is . . . woefully chaotic and disorderly. Yet, despite the dubious gifts of confusion, anger, sorrow, and pain bestowed on both partners, there is an underlying order that appears across all experiences, regardless of sexual preference, regardless of the unique characteristics of the partners and their relationship. Amidst diversity and disorder, patterns and natural sequences of behavior prevail." The "confusion, anger, sorrow, and pain" experienced by adults uncoupling echoes the melded phases and feelings of the separating child, but with a great, sad resonance of insight and guilt.

Still, whether the cause is genes, childhood experience, adult loss, or a combination of these, some change in brain chemistry must underlie or accompany depression. For decades the reigning model was the catecholamine theory that the level or turnover of the neurotransmitter norepinephrine (a catecholamine by chemical structure) was high in manic, elated episodes and low in depression. There are several kinds of evidence for this chemistry. Some drugs that help in depression influence norepinephrine in the brain. One group of effective agents blocks monoamine oxidase, an enzyme that removes norepinephrine from the synapse. This blockade increases the amount of the transmitter available to stimulate the next nerve cell, which would tend to elevate mood. Amphetamines, among other effects, make neurons release norepinephrine; they had been used but were less effective.

Because of the side effects of these early antidepressants, physicians in the 1970s and 1980s turned to tricyclic antidepressants, named for their three-ringed structure. They work in part by blocking the reabsorption of norepinephrine from the synapse by the nerve cell that secreted it in the first place. Since this sponging up is the normal means of removal of about 90 percent of the norepinephrine released, the slowing down of reuptake by the tricyclics—Elavil (amitriptyline) and Tofranil (imipramine) are common examples—increases the amount of norepinephrine stimulating the next cell. Other evidence, too, pointed to norepinephrine's role. For example, electroconvulsive shock therapy (ECT), which despite some people's doubts is an effective treatment for severe depressions, as well as a surprisingly safe one, raises norepinephrine levels in laboratory animals. Salts of the metal lithium, a remarkably simple and effective treatment for manic-depressive illness, work on the electrochemical wave that travels down the nerve, but they ultimately also change norepinephrine functioning. Finally, compounds that deplete norepinephrine, interfere with its synthesis, or poison neurons that make it cause depressionlike symptoms in monkeys, rats, and other animals.

The catecholamine theory still has a piece of the truth, but beginning in the late 1980s the picture changed dramatically. Several drugs that affect norepinephrine had long been known to affect another neurotransmitter, serotonin, in similar ways. One interesting difference, however, was that the tricyclics increased the response to norepinephrine at a pace consistent with the rather slow response of many depressed patients to the drugs, over a period of three weeks or longer. The effect of serotonin reuptake blockade was faster. It was hoped that patients suffering from depression might one day be partitioned into a group with a norepinephrine problem and another with a serotonin problem, treatable with two different drugs.

Enter Prozac, an antidepressant brand that by the mid-nineties became a household word. Followed in rapid succession by other new drugs—Zoloft, Paxil, Effexor, Wellbutrin, and Remeron, to name a few—Prozac was to change decisively the way we think about the meaning and treatment of depression. Some of these drugs have a common action, parallel to but distinct from the older antide-pressants: they block the transporter for serotonin—the chunk of molecular machinery in the nerve ending that sucks the neurotransmitter back into the cell. Others, especially Effexor, Wellbutrin, and Remeron, had broader or more obscure effects. But Prozac, Zoloft, and Paxil proved that the specific blockade of serotonin reuptake, with very little norepinephrine effect, could counteract depression. So without knowing what embodies mood, we nevertheless do know this: block the transporter, slow the sponging process a fraction of a hundredth of a second, and the extra serotonin hanging around in the gap will, in many patients, lift the depression. It doesn't work for all depressed people, and many others get better on their own. But while placebo-treated depressions have a 30 percent improvement rate, serotonin reuptake treatment yields a doubled improvement rate of 60 percent. New drugs with broader actions may be even more effective.

But by the early 1990s something new was noticed. Far more people were being treated with these drugs than had ever been treated with antidepressants before. Why? Two hypotheses were fielded. First, it might be just that the side effects were milder. Depressed people taking the classic tricyclics had to suffer clas-sic symptoms: dry mouth, constipation, hesitant urination, erection problems. This burden had made tricyclics—largely norepinephrine reuptake blockers—unlikely to be used except when depression was very serious and resistant to prolonged psy-chotherapy. In addition, such symptoms as dry mouth were so steady that the patient always knew the drug was there. Side effects constantly tapped you on the shoulder, saying, "Friend, you may feel better, but *you are not yourself.*"

The new drugs were different. They did suppress libido, and that loomed larger in people's minds as treatment continued, but patients felt normal most of the time. There were no constant physical reminders. Also, the new agents worked not just against depression but against the pervasive anxiety of our tense, neurotic age. People—were they still really *patients?*—began to say that they felt more like themselves than ever, that they became themselves for the first time. Some psychotherapists, such as Peter Kramer, author of the remarkable *Listening to Prozac*, were saying that these medicines changed personality, and that people liked the change. Now, new possible effects of these drugs appear so frequently that the stuff begins to sound like snake oil. *Runner's World* reported testimony from runners who say they have experienced improvements in their running time after taking Prozac or Zoloft. It is likely, though, that most of these people were depressed; depression can hurt running or any other kind of performance. Alberto Salazar, a world-class marathoner who had been suffering from listlessness and various minor symptoms, began taking Prozac and made a world-class comeback. Yet he later said, "I didn't really care how fast I ran at Comrades. . . . The only thing that's important to me is that Prozac has helped me lead a normal life again."

Critics worry that drugs will substitute for psychotherapy, personal effort, friendship, learning, even ethical reflection. They are making untenable distinctions. A study by Columbia University psychoanalyst Steven Roose has shown that a majority of certified analysts have prescribed psychiatric drugs to patients *in psychoanalysis*. Psychoanalysts have traditionally been the high priests of "working through," once rejecting drugs for all but the sickest patients. Depression, anxiety, and the like were considered essential to motivate recovery through psychological work. No longer. Today many psychotherapists find that drugs can facilitate this work and that the prescription may be only the beginning of a long, complex, positive process of learning and personal change. But this is not the most prevalent form of antidepressant use. In fact, the great majority of prescriptions are written by physicians other than psychiatrists, or by psychiatrists who see the patient only for 15 minutes or half an hour for occasional checkups, just as another kind of specialist might for an asthma or blood pressure follow-up.

This is not necessarily bad, since not everyone can afford or benefit from psychotherapy. But as Tanya Luhrmann showed in her account of psychiatry in the late 1990s, the success of psychopharmacology combined with the egregious pressures of managed care has beaten talk therapies into retreat. Yet excellent studies—for example, by Ellen Frank and her colleagues at the University of Pittsburgh—prove that psychotherapy and antidepressants are more effective than the drugs alone in treating severely depressed patients. Trying to control mood disorders without some form of psychological intervention is like trying to treat a broken leg without a cast, or at least trying to treat diabetes or heart disease without education about diet and exercise. Drug therapy alone is still worth doing, but it is not good medicine. Aside from treatment effectiveness, the trend in psychiatry at the turn of the millennium is a setback for the examined life. That will matter little to some, but it will be a great loss for many others.

Still, whatever can be done must be done. In his memoir, *Darkness Visible*, novelist William Styron writes of the pain of depression as few have done before. But he also writes about recovery, comparing it with Dante's emergence from the infernal regions: "For those who have dwelt in depression's dark wood, and known its inexplicable agony, their return from the abyss is not unlike the ascent of the poet, trudging upward and upward out of hell's black depths and at last emerging into what he saw as 'the shining world.'" Styron concludes with Dante: "And so we came forth, and once again beheld the stars." Those who have emerged at last from years of despair, weakness, darkness, and pain look back at those past travels and shudder—before sighing with immense relief. Something was no doubt learned on the infernal journey, but once is enough, and they will never be persuaded to go back.

If drugs can intervene in these circuits, why not, in extreme cases, look to the surgeon's knife, or at least the neurosurgeon's pinpoint suction tool? Direct intervention with psychiatric neurosurgery has been tried in extreme and intractable depression, just as in intractable violent fits, obsessive-compulsive disorder, and extreme phobia. These interventions logically involve interruption of one limbic system circuit or another, but they are very uncertain and their use must proceed with caution. As neuroanatomist Walle Nauta once said when this subject was debated, "We are digging around with a hand spade among the bulldozers, trying to figure out what the bulldozers are doing." The treatment methods—even the research methods—of brain science are weak compared with the forces that act on the brain in mental illness.

One of the hand spades is a very modest operation called cingulotomy, done during the 1970s and 1980s by Massachusetts General Hospital brain surgeon H. Thomas Ballantine. He placed electrodes in the right and left cingulum bundles, a paired cable connecting the forward part of the cerebral cortex to the limbic system, and made a one-cubic-centimeter lesion on each side. A completely independent research group at MIT did extensive mental testing and psychiatric interviewing before, immediately following, and years after the cingulotomy in 85 patients—26 of whom suffered from chronic pain and 59 from mental illnesses, mostly depression. The duration of illness averaged almost eleven years for the pain patients and fifteen years for the psychiatric patients. Thirty standard psychological tests were done, testing overall intelligence, frontal lobe dysfunction, and memory, among other functions.

In terms of short-term therapeutic outcome, moderate or marked improvement following cingulotomy occurred in 75 percent of the persistent-pain patients and in 61 percent of the largest and most successfully treated group of psychiatric patients, those with depression. Continued research with longer follow-up convinced the MIT group that no detectable deficits result from cingulotomy. With regard to efficacy, however, they were more skeptical. Follow-up is difficult, and some of the worst-off patients may be least likely to come in for study. A study without random assignment of patients or sham operations, both unfeasible in this clinical setting, cannot rule out placebo effects or spontaneous recovery, although the latter is less likely because of the duration of these chronic conditions and the failure of many other treatments.

Not surprisingly, Ballantine and his colleagues, including psychiatrist Anthony Bouckums, were more optimistic. In a 1987 report they summarized twenty years of clinical experience. They had performed 696 cingulotomies on more than 400 patients, with no deaths, two cases of partial paralysis resulting from hemorrhage (always a risk with brain surgery), and a 1 percent incidence of seizures (also a standard neurosurgical risk), which were effectively controlled by medication. In an uncontrolled clinical study using a sample of 198 patients out of the 273 who had received cingulotomy for psychiatric illness, the team found that 62 percent of patients improved considerably after surgery. The results were best in depression and other mood disorders. While there may be biases here, that has been true in the early stages of clinical research throughout the history of modern medicine. And these were not run-of-the-mill mental patients; each had been certified by a three-physician institutional review board (IRB) appointed by the hospital—the neurosurgeon, a psychiatrist, and a neurologist—as extremely resistant cases. "Cingulotomy was recommended only if the IRB concluded unanimously that all reasonable nonoperative treatments had failed." With Ballantine's death in the mid-nineties, few cingulotomies are being done in the

United States, and for now research on psychosurgery will have to be done elsewhere.

Ongoing experience in Britain and Ireland suggests that this kind of treatment will continue to have a role. In addition to cingulotomy, neurosurgeons there do a procedure called the subcaudate tractotomy. A small, precise lesion is made in the white matter connecting the cingulate cortex, near the frontal lobes, to the limbic system. As we will see, brain imaging independently shows that this area is hyperactive in depressed patients. So it is not surprising that this may be a useful operation in severe, intractable depression. Unfortunately, the British and Irish studies are uncontrolled, which they justify by saying that the patients have already tried all other treatments without success. But in any case, their reports suggest that the operation is quite safe and that 40 to 60 percent of these severely depressed people are able to lead normal lives after surgery, with suicide rates around 1 percent, as opposed to 15 percent for such extremely depressed people without treatment.

Meanwhile, a number of laboratories are creating images of the brain during sadness and depression. In Robert Post's lab, eleven healthy women were asked to enter moods of sadness. PET scans showed that the low mood activated limbic and related structures on both sides of the brain, including the cingulate cortex, the middle prefrontal cortex, and the inward-facing part of the temporal lobe, as well as subcortical areas. In contrast, transient happiness in the same women showed no areas of significantly increased activity but was associated with widespread *lowered* activity in the cortex, especially in the right prefrontal and the temporal-parietal areas on both sides. However, these changes were not just opposite activity in identical brain regions; happiness and sadness are not simple opposites in terms of brain function. Another study, though, using functional magnetic resonance imaging, compared healthy men and women and found that the transient mood of sadness had more focused associations in the men's brains, centering on the amygdala. Given what we know about the amygdala's role in fear and anger, this could mean that for men sadness is a more mixed emotion.

But what about serious depression? Some studies have implicated the amygdala and the frontal lobes in it as well. A group led by Helen Mayberg, one of the world's leading investigators of depression and the brain, used PET scans to compare depression with sadness. They found considerable convergence: increased blood flow in limbic circuits but decreased flow in higher cortical association areas, especially the prefrontal cortex. With *recovery* from depression, the reverse pattern was seen—limbic *decreases* in activity and cortical increases. In particular, there seemed to be a reciprocity between the forward part of the cingulate region and the outer part of the frontal lobe. Mayberg believes "that these regional interactions are obligatory and probably mediate the well-recognized relationships between mood and attention seen in both normal and pathological conditions. The bidirectional nature of this limbic-cortical reciprocity provides additional evidence of potential mechanisms mediating cognitive ('top-down'), pharmacological ('mixed'), and surgical ('bottom-up') treatments of mood disorders such as depression."

Mayberg also emphasizes the dynamic nature of brain moods. As with Parkinson's disease, we are not looking merely at damaged *centers*, but more importantly at a changing balance of power in reverberating neural loops. It is not incidental that circuit means circle, and the physiology of depression resonates through long, looping paths that cross much of the brain. However, Mayberg found a focus in the anterior, or forward, part of the cingulate cortex, which is very near the region damaged in Phineas Gage, the famous railway foreman. Gage was uninhibited, not depressed, and he and others like him had trouble making judgments in life. But the Gage cortex has proved to be a crossroads between thought and emotion in more ways than one. Depressed patients had many abnormalities visible on PET scans, but among the cortical regions, only the lower anterior cingulate, known as area 25, was hyperactive in depressed patients *along with the lower limbic brain regions.* The outer frontal lobes, deeply involved in working memory— touch your forehead a couple of inches above your temples and you'll be almost touching the region—is underactive in depression and has a reciprocal, contrary relationship to area 25. Image the brains of people *recovering* from depression and you will see increases in the activity of that higher frontal cortex and normalization of the hyperactivity of both cingulate area 25 and the lower limbic areas connected to it. Working memory is vital to thought, since it allows us to hold elements of a problem actively in mind while we manipulate them and try to put them together. Hyperactivity in the limbic system, including the limbic cortex, interferes with thought. Successful treatment reverses the balance, giving thought its rightful place again.

Perhaps the most compelling part of this story is that the reversal occurs in depressed patients as they recover, regardless of *why* they recover. Prozac causes the reversal, but only in patients who respond to it, so it isn't the drug, it's the recovery. Cognitive therapy, without drugs, produces the same brain changes in people who respond to *that* treatment. Transcranial magnetic stimulation of the frontal lobes for depression produces similar effects and, lo and behold, depressed patients who get better on *placebo* go through the same reversal of balance between the activity of these brain regions. "Getting better," Mayberg has said, "is about adjusting the balance between limbic and cortical areas." Her latest work even provides insights into the relationship between depression and normal sadness: If you take normal people and ask them to remember a sad experience in their lives, they go temporarily into the same imbalance that depressed people are in all the time.

For severe, intractable depressions, the future may bring methods that combine the best of chemical manipulation with a greatly improved version of psychosurgery—minor surgery for the delivery of a drug to a small set of nerve cells in a strategically defined part of a circuit. This could be better than destroying part of a circuit, and possibly better than systemic drug treatment, because it would not deliver a potentially harmful chemical to irrelevant and innocent parts of the brain and body. It is not an easy thing to contemplate, but neither is intractable mental illness. As the old medical adage goes, desperate maladies may require desperate remedies.

But most depressions can be understood and treated without such desperate measures, and many seem, biologically, a part of normal life. A case in point is a syndrome often called "postpartum blues," which affects a large minority of women who have recently given birth. In a classic study by psychiatrist David Hamburg, two thirds of a sample of normal women had one or more crying episodes—at least five minutes of crying, often with no evident cause—during the first ten postpartum days; 28 percent cried for over an hour. This was a much higher incidence than during the pregnancies or months after the birth.

Yet postpartum blues are not the most serious threat to mental health during this period. Studies in mental hospitals show that the risk of psychosis, obsessive-compulsive disorder, and other mental breakdowns is 5 or more times higher the three months after birth than in the last trimester of pregnancy. More than 10 percent of all women suffer depression within the first year after becoming mothers, which seems a greatly elevated level of risk for a mental illness with a *lifetime* prevalence of 20 percent. The presence of roughly similar rates for women in the United States, Italy, France, Scotland, England, and Uganda strongly suggest that this cannot be a purely culture-bound syndrome.

However, there is also no doubt that social conditions and cultural beliefs play a large role in influencing the degree and form of expression of postpartum distress. Women in Kenya, studied by Sara Harkness, and women in Fiji, studied by Anne Becker, have specific cultural strategies for supporting women in this period and protecting them from distress. Research by Patrizia Romito and her colleagues in Italy and France shows strong effects of relationships and employment on depression in the first year postpartum. Working outside the home is protective in these women, but even more important is that the mother have a confidante. A striking finding is that having a man in your life is a plus only if your relationship with him is very good or good. If it is only fair or worse, you are better off alone.

It would seem almost obvious that the physiological transformations of pregnancy and birth combined with the social transformations of motherhood would make this a unique time of psychobiological risk. However, three decades of research by psychologist Michael O'Hara and his colleagues have cast doubt on whether this depression is different from any other. Leaving aside immediate postpartum blues, depression in the rest of the first year was not more likely than in the last trimester of pregnancy, nor was it higher than in control subjects who were acquaintances of the mothers, of similar age and life circumstances but who were not going through pregnancy and delivery. The strongest predictors of depression in the first postpartum year are a personal and family history of depression and the presence of life stressors in the postpartum period. O'Hara and some others in this field conclude that prolonged postpartum depression is a manifestation of the general vulnerability of young women to depression, not something specific to childbirth or its aftermath.

Still, even O'Hara agrees that postpartum blues are related to birth itself, with a more than threefold increase in risk compared to acquaintance controls and at least some relationship to hormonal changes. Rates of distress were 26.4 percent for women in the first postpartum week compared to 7.3 percent for acquaintance controls. Many social and cultural factors are no doubt relevant, but the hormonal changes occurring around delivery must be considered as well. Estrogen and progesterone in the mother's blood can fall to 1 or 2 percent of their late-pregnancy highs in less than a day. These are placental hormones: remove the placenta and they crash.

In the laboratory, progesterone decreases the excitability of nerve and muscle tissue. The fall at this time changes the ratio of estradiol to progesterone, increasing nerve and muscle excitability and, perhaps, the irritability of the brain. Losing 98 percent of a soothing hormone in a distraught day full of pain, along with a massive role change that itself might make a newly minted mother cry her eyes out, creates a psychobiological crisis. Beta-endorphins, the brain's equivalent of morphine, are secreted in late pregnancy and especially during labor—perhaps an adaptation to manage pain in a situation where pain, fear, and helplessness chase one another in circles. Endorphins may help explain the postpartum exuberance experienced by some women. But they decline steeply when birth ends, so exuberance is probably a combination of things: endorphin infusion, putting the pain behind you, and falling in love with your baby.

A parallel (although much smaller) drop in progesterone and rise in the estradiol-progesterone ratio occurs each month just before menstruation. As we saw in chapter 6, even if depression or irritability associated with this time of the monthly cycle affects a small minority of women, it is a real problem for them, and hormonal analysis will contribute to a solution. In animal and human studies, mental functions, physical activity, seizure susceptibility, pain sensitivity, and symptoms of some neurological diseases are affected by hormone changes in the menstrual cycle. The idea of a *universal* premenstrual syndrome is indeed a sexist myth, but for those women seriously affected, cultural explanations don't help.

Another hormone known to figure in depression is the stress hormone cortisol. Early research by psychiatrist Edward Sachar of the New York State Psychiatric Institute found that a subgroup of severely depressed patients have excessive cortisol secretion, especially during the night. Depressed patients experienced more daily variation in cortisol, with their early-morning values dropping nearly to the levels of the control group. As patients improved with treatment, cortisol dropped. This work opened a line of research on cortisol in depression, culminating in the contributions of psychiatrist Charles Nemeroff and his colleagues. Mounting evidence has led to an important second focus for research on the neurobiology of depression, outside the monoamine synapses where so much has been learned. This other focus is called the "corticotropin-releasing factor hypothesis of depression."

Recall that cortisol, a hormone that mobilizes fuel for fight or flight, is released from the cortex, or outer part, of the adrenal gland. Behind it is ACTH, a pituitary hormone also known as corticotropin, because it makes the adrenal cortex grow. ACTH in turn is controlled by CRF, corticotropin-releasing factor, made in the hypothalamus and slipped to the pituitary through its private blood supply. Now, it's no surprise that this cascade of commands—the adrenocortical axis—would be activated in depression. This is a main axis of stress response, and depression is certainly stressful. Confirming Sachar's discovery, recent studies show that depressed patients have chronically high cortisol in blood and urine, that they do not respond to dexamethasone as most people do, by suppressing cortisol secretion, and that normal life stresses cause them to make more cortisol.

What is new is a growing conviction that CRF itself is intrinsic to depression, based on several lines of evidence. For one thing, CRF influences the brain, not just the pituitary. Introducing CRF into the brain reduces interest in food or sex, causes withdrawal from strange environments, and promotes activity in familiar ones. Depressed patients in many studies show heightened levels of CRF in their cerebrospinal fluid—probably a reflection of brain levels—and successful treatment with various methods brings the levels back down. At autopsy the brains of depressed patients have substantially greater expression of the CRF gene as well as many more CRF-producing neurons. Not surprisingly, depressed people have a blunted cortisol response to CRF injection They are just too accustomed to stress to react normally to this challenge; the adrenal axis is exhausted.

We don't know for sure yet if these changes are intrinsic to the illness or just consequences of depression's chronic stress. But growing evidence shows that traumatic or even difficult early experiences can produce enduring changes in the stress response axis. Paul Plotsky and his colleagues have shown that rat pups removed from their mother repeatedly before weaning grow up with greater CRF gene expression and more CRF receptors in several brain regions. This suggests an unfortunate synergy in which both the lock and key to the brain's stress axis are permanently enhanced. Yet giving the rats a serotonin reuptake blocker—standard antidepressant therapy today—restores normal function to the axis. Plotsky's work strongly suggests that early experience can upset normal neurobiology. In chapter 13 we saw how variations in food availability to monkey mothers can permanently affect the behavior of their young as they grow. Those same young had high levels of CRF in their spinal fluid. Clinical investigators have long believed that traumatic experiences in early childhood predispose us to depression as adults. Today the best evidence for such effects is in animal models.

Plotsky's group has also shown that different degrees of separation have very different effects, both immediately and in the long term. Rat pups from two to fourteen days of age are removed from their mothers daily for either fifteen minutes or three hours and kept warm and comfortable in an incubator. Fascinatingly, neither group shows increased corticosterone (the rat's version of cortisol) *during* the separation, suggesting that this in itself is not terribly stressful. The divergence begins when they are returned to the mother, when only the three-hour group has a doubling of corticosterone.

The key seems to be a difference in maternal behavior. Fifteen minutes of separation doesn't seem to faze mother rats. When the pups return, Mom retrieves them, licks and grooms them, arches her back, and nurses them. She even does more licking and grooming than usual. But three hours of separation is something else again. Mothers can't take this in stride, and they become rejecting for a time, although the persistent pups eventually bring them around. This lag in the mother's nurturing appears to be the main stress in these experiments. From that point on, these longer-separated rats react to *other* kinds of psychological stress—restraining them, for example, or startling them with an air puff—with more ACTH and corticosterone secretion, and *this excess increases as they grow to adulthood.* As in depressed patients, the excess cannot be effectively suppressed by dexamethasone, which readily suppresses it in normal people or animals. Throughout life, the three-hour-separated rats continue to show this hypersensitivity in the adrenal axis, starting at the top: they have elevated production of CRF mRNA in the brain. This means that brain genes are working overtime to manufacture CRF on a steady, day-to-day basis.

These rats also resemble depressed patients in other ways. They show less preference for sweetened water than normal rats do, and they require a larger current before they begin self-stimulation in the brain's reward areas. This suggests a kind of anhedonia, or dampened ability to experience pleasure. They show a greater preference for alcohol-laced water, and their preference grows stronger under stress, suggesting that they are medicating themselves for anxiety, a symptom frequently associated with depression. And they are more anxious and fearful in general, as are many who suffer from depression. Finally, treatment with standard antidepressants such as Paxil reverses much of this syndrome, although it returns substantially after the drug is withdrawn.

These findings suggest that some depressed people have a similar abnormality of hormone and neurotransmitter balance, genetically or environmentally caused. Genetic research shows that there is often an underlying vulnerability. It is also clear, however, that these chemical imbalances may be environmentally caused. Edward Senay remarked many years ago that in some cases "it would be correct to speak of an affective theory of catecholamine disorder rather than the reverse." Echoing this outlook recently, Nemeroff and his colleagues "propose that early abuse or neglect not only activates the stress response but induces persistently increased activity in CRF-containing neurons. . . . This effect in people already predisposed to depression could then produce both the neuroendocrine and behavioral responses characteristic of the disorder."

This notion is supported by the monkey studies: in our close relatives, separation and other forms of deprivation can induce syndromes that mimic human depression not only behaviorally but physiologically and biochemically. In fact, *the same symptoms* that have been produced in monkeys by norepinephrine depletion have also been produced in them by separation from the mother, and the same psychoactive drugs work in both cases. Sleep disorders and cortisol elevation are among the other symptoms that internally and externally caused monkey depressions have in common. One challenge for the early twenty-first century will be to develop brain models that integrate the monoamine theory of depression with the stress-axis theory centered on CRF.

Another source of knowledge about the physiological consequences of loss is research in psychosomatic medicine, or psychoneuroimmunology. Positive emotional states protect physical health in ways that range from enhancing healthy behavior to strengthening immune functions. Psychological stresses, including the stress of bereavement, can worsen, if not cause, serious medical diseases. Among the diseases that may be worsened by psychological factors are peptic ulcer, essential hypertension, bronchial asthma, and ulcerative colitis. Although we know a lot about their physiology, all are largely or partly mysterious in cause; there is some evidence implicating loss and other emotional stresses. Loss is also implicated in flare-ups of skin disorders, some heart conditions, and susceptibility to certain infections such as the common cold.

Neither bereavement nor or any psychological stress alone could account for these disorders. But a role for them is consistent with all we know about the physiology of loss, and the skepticism of some physicians is more to be wondered at than credited. We will learn much more about them over the next few years, partly from animal models. For example, psychological stresses can easily cause stomach ulcers in rats, and although it is difficult to say for certain, one such stress is a condition related to human depression. It is called "learned helplessness." The initial experiments were begun by Martin Seligman, a clinical and learning psychologist, using dogs as subjects. They were carried forward on rats by physiological psychologist Jay Weiss and his colleagues, who have spent more than three decades developing a model of how the brain generates depression.

The basic facts are simple. If you give dogs a series of electric shocks with no way to escape them no matter what they do, then later shocks *with* an opportunity to escape will find them very slow to learn. They have given up. They are convinced that they can do nothing to better their situation, and they proceed to do nothing even when they can. Weiss showed that there are clear physiological consequences of shocking rats without allowing them to escape (as compared with rats getting the same shocks but allowed to turn them off). The helpless ones have more gastric ulcers and depleted brain levels of norepinephrine, just as appears to be true of depressed people. The general idea is that severe stress, whether from learned helplessness, prolonged swimming, restraint, or other causes, triggers repeated release of norepinephrine until the neurons that make it are exhausted, and they make less forever after. Now, this may seem counterintuitive given the power of stress to sensitize us to further stressors, as we saw in some of the monkey models. But the key seems to be how prolonged and repetitive the stress is. Make it severe enough and repeat it relentlessly, and you can exhaust the system. Instead of producing a jittery, fearful animal, you get one too depressed to react at all.

It turned out that the norepinephrine depletion was most dramatic in the brain stem, specifically in a tiny region called the locus ceruleus, or "blue spot," which contains a large proportion of the brain's norepinephrine cell bodies. These two or three thousand cells in the blue spot project their axons to activate the sympathetic nervous system, as well as to many higher brain centers. With their norepinephrine depletion after uncontrollable shock, activity in these centers may be toned down. New evidence, however, suggests another, indirect route to higher brain regions. Neurons in the locus ceruleus project to the ventral tegmental area of the brain stem and use both norepinephrine and another transmitter, galanin, to regulate *dopamine* cells. These, we know, promote both activity and pleasure in higher brain centers. Weiss's revised model works through this indirect, dopamine path as well as through a direct, norepinephrine one. This may explain why Effexor, which blocks the reuptake of dopamine and norepinephrine as well as serotonin, has special effectiveness against depression. Interestingly, Paul Plotsky and Michael Meaney's rats, raised with daily three-hour separations followed by maternal rejection, also exhibit changes in the locus ceruleus enzymes that make norepinephrine, as well as changes in norepinephrine's brain receptors. Excess CRF may be causing these changes, neatly tying together learned helplessness and maternal rejection—not to mention combining the stress-axis and catecholamine theories.

It is tempting to say that what infant monkeys or humans go through during the first, protest phase of separation is a learning of helplessness comparable to that observed in Seligman's dogs. Indeed "learned helplessness" often follows any major loss, whether of a loved one, a job, or our own faculties. All habits, all motives or moves in the old direction fall flat. In a moving and tragic study, David Hamburg reported on the parents of children who were dying of leukemia. He was involved in one of the first systematic efforts to provide psychiatric care for such parents, who after all must go on, taking care of their other children and each other, and who are at obvious risk for emotional breakdown. Their experience is now an active field of research.

In Hamburg's study, parents' reactions included universal initial shock, followed by intellectual but not emotional acceptance. Many sought to disbelieve the diagnosis, to doubt the doctor's description of the future course of the illness, or to hope for a dramatic, curative medical breakthrough they had been cautioned not to expect. Many experienced an awakening of religious feeling, while others insisted on doubtful scientific explanations—some of which led them to partial self-blame—rather than accept the illness as random or meaningless.

> As the disease progressed, hope diminished. . . . As the child became increasingly ill, they hoped only for one further remission. They no longer made long-range plans but lived on a day-to-day basis. They would, for instance, focus on whether their child would be well enough that evening to attend a movie.

Grieving in anticipation of the loss varied greatly and changed as the child's condition worsened. The death of a child on the ward affected the mood of other children's parents. Growing resignation was frequently accompanied by the wish that it would be over. The narrowing of hope and the gradual letting go were described by one mother: "I still love my boy, want to take care of him and be with him as much as possible . . . but still feel sort of detached from him." Yet she continued to be effective in caring for and comforting him.

Such anticipatory grieving seems adaptive: the few parents who did not grieve in advance were in greater distress after the child's death. All the parents had high rates of excretion of 17-hydroxycorticosteroid, a breakdown product of cortisol; they were under prolonged, measurable, physiological stress. These physiological measures mirrored the severity of expressed grief, and those parents who grieved in anticipation had their highest levels during that early period. With the exception of anticipatory grief, all these psychological and physiological indicators also occurred in studies of victims of polio and of disfiguring burns. Adjustment to the loss of part of one's self, to whatever extent it is possible, has something in common with the loss of a child.

These processes also resemble people's response and (partial) adjustment to their own terminal illnesses. Kübler-Ross, whose later religious conversion was mentioned, identified five phases or stages of response in people who knew they were dying. These stages are shock and denial ("No, not me."); anger ("It's not fair, why me?"); bargaining ("Yes, it's me, but at least if I can have five, or three, or one more year of life . . ."); depression (characterized by weeping, brooding, withdrawal, despair, and suicidal thoughts); and acceptance—a restful, weary, but not apathetic period of having come to terms with approaching death.

These phases have often been satirized, as they should be if taken too literally. For a time there were crude attempts by hospital personnel to lead patients through these stages, one after another, as if they were written on the sky and every dying person had to go through all of them in order. Still, their elements are meaningful, and they resemble what happens in parents of dying children and victims of polio or disfiguring burns. They also resemble the response to loss of the mother in children and infant monkeys. Think of them, then, as *aspects* of adjustment to loss, whether imminent or recent, but expect them to appear in a jumble of cycles, orders, and combinations, with some left out in many cases. People may die in a state of denial, anger, or depression, despite long preparation and intervening periods of apparent acceptance. It is not even clear that acceptance is an ideal goal for every person; in some illnesses, nonacceptance lengthens life. Studies show that most terminal patients want to know that they are dying, but some do not. Some, including a few who are highly intelligent, virtually insist upon dying in a state of denial. Sherwin Nuland, in *How We Die*— based on a lifetime of watching people do it, despite his best efforts to save them—wrote, "Every life is different from any that has gone before it, and so is every death." Not only did most of the deaths he saw defy pattern and schedule, very few of them were "good" deaths.

Life is dappled with periods of pain, and for some of us is suffused with it. In the course of ordinary living, the pain is mitigated by periods of peace and times of joy. In dying, however, there is only the affliction. . . . The peace, and sometimes the joy, that may come occurs with the release. In this sense, there is often a serenity—sometimes even a dignity—in the act of death, but rarely in the process of dying.

Giving up on the prospect of a pleasant death, Nuland wisely says, "The honesty and grace of the years of life that are ending is the real measure of how we die. It is not in the last weeks or days that we compose the message that will be remembered, but in all the decades that preceded them. Who has lived in dignity, dies in dignity."

Nevertheless, there is a culture that is devoted to death with dignity, and it appears to have succeeded better than we have in the West. It is the subculture of religious Hindus who, together with their families, make a last pilgrimage to die a good death in the holy city of Banaras. As described by anthropologist Christopher Justice, hundreds of people each year make the journey, having decided that they are close to death and wanting the death to be surpassingly pure. They enter sacred hospices, stop eating, and usually die within a few days. We might call it self-starvation, but they call it fasting for purification, and their deaths are on the whole quiet and easy. Some of them are no doubt premature, but few are unnecessarily prolonged, painful, or cruel.

As for anger, the poet Dylan Thomas praised his father's final fury in the transcendent villanelle that ends:

> Do not go gentle into that good night,
> Rage, rage against the dying of the light.

Given these varied feelings, it can scarcely be left to psychiatric authorities to prescribe a method of dying. But they can identify for us some common features of the process that are essential to our grasp of human behavior.

Because we are, after all, all dying. We are all long-term terminal patients. We exhibit, at various moments of life, all the known symptoms—shock, denial, anger, bargaining, apathy, acceptance. Like the terminal cancer patient or the parent of that patient but much more slowly, we are engaged in a seventy- or eighty-year jumbled or repetitive cycle of expression of these symptoms. And, just as for those patients, the death or illness of another heightens the conflicts.

Every known society has grief and mourning of some kind in response to the death of a loved one. In the words of Paul Rosenblatt and his colleagues, after a classic study of the ethnographic record, this is one of the few situations where "an ethnocentric perspective has often been productive. . . . People everywhere experience grief . . . people everywhere experience the death of close kin as a loss and mourn for that loss." Among their findings, drawn from careful rating and statistical analysis of seventy-eight representative traditional societies, some degree of anger and aggression, as well as of grief, are universal or near-universal components of bereavement. Men and women respond similarly, but where there is a gender difference, women tend to cry and mutilate themselves, while men tend to direct aggression outward. Ghost beliefs and cognitions are "probably universal" and "arise from the normal psychological residues" of the relationship. Societies tend to have either ritual professionals to help the bereaved control their anger or approved ritual methods of expressing it. Others have rituals that approve or prescribe but limit the expression of grief, and that knit up the tear in the social fabric. Some societies have "final funeral ceremonies" weeks or months after the death occurs, and those that do not tend to have more prolonged grieving than those that do.

Recall from chapter 7 the young woman in Bali who tried to suppress her grief over her fiancé's death. Ridiculed and called "crazy" because she wept openly at first, she was soon shamed into a public show of cheerfulness. Yet she borrowed money to travel to her young man's grave and had herself photographed prostrate on it. She later said that she was afraid to think of his death or she might go mad. Funerals in Bali are not held until years after the death, perhaps on the theory that the bereaved can then handle them without succumbing to uncontrollable grief.

For many cultures, beliefs about illness and healing involve not a rational or scientific analysis but a concourse and bargaining with death. Among the !Kung San, the healing ritual requires the healer to enter a state called "like death," and may require him to leave his body, enter the world of the spirits, and win the soul of the ill person back through insistent advocacy in the face of the gods' resistance. The trance dance, their main healing ceremony, encourages active hostility against the spirits, who are thought to be observing on the sidelines. Even the highest gods are not exempt from vituperation. Indeed, hostility in trance may sometimes be directed against human beings, and the control of that aggression is a major goal in the training of healers. Healers sometimes in effect assault themselves, by running through the bush in the dark, heedless of rocks and trees, or by heaping glowing coals on their bare heads.

In the Western tradition we have little provision for such overt hostility toward heavenly power. When Job curses God it is forgivable but far short of the cultural ideal—and indeed, in the story, it is a comfort to Satan. In moments of impending or actual loss and grief we are supposed to remind ourselves (if we are religious) of the ancient Hebrew saying, "The Lord gives, the Lord takes away," or (if we are not) of the inevitability and orderliness of all events in the universe. If it is true that grief almost always includes anger, one has to wonder what it is we do with that anger. The relationship between anger and grief is still an unsolved problem in psychology. Late in his career Freud conceived the notion that they are expressions of one and the same "death instinct," an inevitable component of human and animal biology. Other depth psychologies appear to accept some version of the belief that the two emotions are intimately conjoined, that depression may really be "anger turned inward," and that it may be treatable by encouraging expression of suppressed anger. What basis is there for these assertions?

In the chapter on rage we saw many instances of anger provoked by loss. Some men murdered or assaulted the woman they said they loved; some form of grief was entwined with the rage in their homicidal emotions. The first phase of response to loss, even in small children, is protest, which is an agitated condition. The impulse to *do* something is frequently very strong, and the helpless feeling that enhances the despair can only be intensified by doing nothing. "Fight back" is certainly something to do, and it often comes to mind even where it is inappropriate, which must have to do with the circumstances under which we evolved. There have always been situations in animal life—predation, for instance, or aggressive competition—that confer advantage on those who fight during or just after a loss. And in many human societies vendetta has been a dark option after a loss through violence; thus, one violent loss brings another, for generations. Among the !Kung, vendetta is not pervasive, but it is a mainstay of social control, and in many !Kung homicides the main motive is revenge for another homicide.

But the grief that turns into rage is not always a vendetta. Renato Rosaldo did a classic ethnography on Ilongot headhunting, and subsequently rethought the problem in the light of his own grief and rage at his wife's tragic death during fieldwork. He showed how older Ilongot men, subject to major personal losses throughout the middle years of life, emerge from grief with a rage that, mixed with younger men's always potent anger, can stimulate a headhunting expedition. The link is convincing and powerful.

Why such violence might be a tonic for depression or grief is another question, and the answer involves presently unknown mechanisms. Violence is thrilling and involves a total systemic mobilization of an organism, a transcendence of social rules, and perhaps a triumph over danger. It involves the secretion of adrenaline, a cascade of impulses in the sympathetic nervous system, a flood of blood and nutrients to the muscles, a flushing of the skin—there is every physiological reason for it to be invigorating. It is also one of the easier ways to activate human beings. National economies are often at their strongest when the nations prepare for war, and economic depressions have sometimes ended in such preparations. So it is easy to believe that a depressed or grieving person might feel better after being provoked to anger and expressing that anger openly. That does not prove, however, that depression is anger "turned inward" or that it is causally related to anger in most cases. It may be that many potentially depressed individuals who do not show up in psychiatrists' offices are treating themselves by expressing anger at the wrong targets, at the wrong times, and in the wrong places—which may be invigorating for them but counterproductive and dangerous.

Still, we all have to deal with grief somehow, because we do not only deal with all the losses, large and small, of the normal course of life. We are also, as conscious creatures, dealing with the loss of ourselves, of our own lives, gradually but inevitably. For anthropologist Ernest Becker, this was the central fact of all human experience. However sedate our lives may be, we have a constant if not quite conscious flirtation with death and loss, which can come at any time and in some sense comes all the time. *We are the creature that anticipates death.* Most cultures show a steady awareness of death and its relationship to life. We have, for example, the !Kung healing ritual in which the gods are castigated, and which may take the healer out of his body into death and back again. Of the dead they may say, "Sky ate her," yet still describe her as alive and well in the village of the spirits. The latter, for their part, are jealous of the living, and they often take someone who is not being well cared for on earth. During an illness, such spirits are berated or cajoled, especially in a trance dance, when they may visit and linger on the outskirts of the watching human circle. At such moments a wide gate opens between the worlds.

Among the Australian aborigines, a person's life is a continuing relationship with the "sky world" or "sacred dream time," which he or she left temporarily at the moment of conception. At initiation, at the start of parenthood, and during various rituals, this relationship is confirmed, as is the transience of mundane life. Finally, at death, the burial ritual allows a return to the sky world, a passage more natural than life itself. Among the Bororo of central Brazil studied by Claude Lévi-Strauss, several weeks of funeral ceremonies follow a death. During this period, "every day is the pretext for negotiations between Society and the physical universe," by which they mean what we would call the spiritual universe, but which to the Bororo is unutterably real:

> The hostile forces in the physical universe have done harm to the human world, and that harm must somehow be put right: that is the role of the funerary hunt. Once the dead man has been avenged and redeemed by the hunters, as a group, he must be admitted to the society of spirits. That is the function of the *roiakuriluo*, the great funeral dirge.

And in some societies, the most fundamental events of life demand an intimate concourse with death. Among the Ndembu, a Zambian agricultural people, an infertile woman or one who has several stillborn children must be cured by means of a ritual called *Isoma*. Studied in exquisite detail by ethnologist Victor Turner, it requires the patient to pass through a tunnel, from a blocked place in the earth, near a fire that resembles the grave, to a cool place near a river that symbolizes life. Her physical body emerges into symbolic life and so can go on to create life. Rituals like these in many cultures suggest that for most human beings the boundary between life and death is vague, and that progress through life often requires grappling with death.

Children, at least in our society, come to grips with it only slowly. From age three to five they consider it reversible, resembling a journey or sleep. After six they view it as a fact of life but a very remote one, and they may deal with it by personifying it as a person or ghost. By age ten or so they have accepted its irreversibility and universality with a somewhat gloomy finality. And by fifteen they are already at significant risk for suicide, which is certainly one way to cope with death. The average annual suicide risk grows throughout life, but the steepest increase occurs before age twenty-five. Thus, it rises fastest during the growth of adult consciousness, the time of our first encounters with life's realities. Flirtation with suicide, like flirtation with violence, is a way to remember from day to day the slender boundary between life and death.

Suicide is related to grief and depression, but not strictly or simply. Mourning in many cultures entails suicide risk, and some even expect the bereaved person to follow the beloved—as in the tragic Hindu practice of *suttee*—burning a widow on her husband's funeral pyre. Many, although not most, clinical depressions that come to psychiatric attention involve suicide risk. In any case, inflicting this ultimate loss upon oneself is frequently an attempt to respond to or offset other losses. And it always involves exploring the transition from life to death.

Critics of assisted suicide for terminal cancer patients say that the distinction between them and the rest of us can never be clear enough to draw a believable line. To some extent this must be true. Certainly the suffering of the severely depressed patient can be very great, and it is not surprising that suicide rates are exceptionally high among them. In Burton's seventeenth-century phrasing, "In such sort doth the torture and extremity of the melancholic's misery torment him, that he can take no pleasure in his life, but is in a manner enforced to offer violence unto himself, to be freed from his present insufferable pains." Or in the plainer, twentieth-century words of Isaiah Berlin, "The logical culmination of the process of destroying everything through which I can possibly be wounded is suicide."

It is not easy to understand, if we take seriously all the known facts about it. They just do not form a simple pattern. Still, a brief focus on them may be useful. Some do relate comprehensibly to the argument of this chapter, namely, that grief is fundamental, inescapable, and risky. As for the other, less obvious parts of the pattern, well, at least they let us know how little we know.

In any given year, in a randomly selected group of 10,000 adults in the United States, about one will die by suicide, using the most conservative definition—probably a substantial underestimate. This exposes each of us to a lifetime risk of about half a percent—lower for some groups, much higher for others. The most common methods in the United States are firearms and explosives, poisons and gases, and hanging and strangulation, in that order, and it is well established that the availability of firearms in homes increases the risk. It is 3 times more common in men than women, despite the fact that both depression and suicide *attempts* are much more common in women. Not just depression, but manic-depressive illness, panic attacks, and other psychiatric disorders increase the risk.

The French sociologist Emile Durkheim knew in 1912 that married people are at much lower risk for suicide than those who are single, divorced, or widowed; that is still true in a variety of cultures. Elderly Americans of Chinese and Japanese descent have had steep increases, while older Americans of European descent have had more gradual increases. During the twentieth century in the United States suicide rates fluctuated, especially for males; there were major declines during each of the two World Wars and a peak during the Great Depression—a level not approached today. Natural disasters like severe floods, earthquakes, and hurricanes are like economic depressions rather than wars, causing increases in suicide for several years. Post–traumatic stress disorder carries an increased risk, regardless of whether the victims are depressed or not. Tragically, victims of torture tend to use suicide methods that resemble the torture. There is also evidence that both suicides and motor vehicle accidents, some of which are thought to be disguised suicides, rise significantly during a week after a widely publicized suicide—a kind of copycat effect. This leaves no doubt about the power of culture to encourage suicide, as occurs in Japan and Scandinavia. Scandinavian suicide is sometimes glibly pointed to as a price they pay for their low homicide rates, although in fact their excess of suicides over ours does not even begin to balance our excess of homicides over theirs.

Still, there may be some validity to the old psychodynamic idea that suicide is aggression turned inward, as is clear from the research of Marie Åsberg, who—perhaps not coincidentally—did her research in Sweden, at the Karolinska Institute. Åsberg, a biological psychiatrist, examined 5-HIAA in the cerebrospinal fluid (CSF) and made a chemical connection between suicide and aggression. CSF fills the spaces in and around the brain and spinal cord, cushioning them against shock. But it is also a confluence of the brain's secreted and excreted substances. It may even be a sort of underground river for distant transport of chemical cargo among brain ports of call. At a minimum it is a first-line filter, catching and concentrating brain chemicals before they are dispersed throughout the body.

In the CSF of people who will later take their lives is a low—we might say inadequate—concentration of 5-HIAA. We have already encountered it as the major breakdown product of serotonin. Serotonin is made in certain brain neurons and released across the synapse. Critical as it is to get that pulse of serotonin across, it is equally critical to remove it. Precision of function depends on the brevity of the pulse, and it doesn't do to have even as useful a molecule as serotonin indolently hanging around, causing further stimulation that wasn't explicitly called for. So it is changed by an enzyme to 5-HIAA, and the latter reflects serotonin turnover—the activity of serotonin neurons. The Åsberg finding, repeated at least four times, links suicide to a low level of 5-HIAA. Those depressed patients who had high levels of 5-HIAA in their cerebrospinal fluid—whatever they looked like clinically—were not really at risk for suicide.

But if 5-HIAA was low, a patient had either tried suicide or was likely to try it during the year after the measurement. To look at it differently, of thirty-six patients *below* the median level of 5-HIAA in one of Åsberg's study groups, six, or 20 percent, died by their own hands within one year of the measurement. For patients *above* the median, the risk of death within one year was *zero*. Later studies were not all so unequivocal but have strongly supported the association. Some of these "protected" patients did in fact attempt suicide—but by drug overdose, a method known from other studies to have a much lower success rate than the violent methods used by the "unprotected" group. In fact, none of those overdoses succeeded. Never has a chemical measurement in psychiatry had anything like this kind of discriminatory power. With it, the main source of mortality in psychiatry could be dealt with more effectively, while those who are not really at risk can be freed from meddlesome precautions. Thus, a molecule so small that it can be dwarfed by even a single DNA base pair may make the difference between life and death.

The question is *how*. More studies by Åsberg, Frederick Goodwin, and others produced major refinements. For example, individuals who do not have classic mood disorders but who are at risk for suicide because of personality disorders—chronic maladjustments rather than time-limited illnesses—can be divided into the same two groups: those with low 5-HIAA levels who have high suicide risk, and those with high levels who do not. Thus, the relationship is fundamentally to suicide rather than to a subtype of depression. An independent group of investigators in France found that violent suicide attempters indeed had lower 5-HIAA levels than controls had, but they added a further subtlety by testing the patients with an impulsiveness scale. The *impulsive* suicide attempters had such low levels of 5-HIAA that they alone accounted for all of the difference from controls, despite the fact that all the suicide attempts in this study were violent.

The cause of altered 5-HIAA levels in CSF is not clear, but it probably reflects compromised functioning of serotonin neurons in the brain. Anatomically, some serotonin neurons end not at other neurons but in the wall of the ventricles that hold the CSF. These cells dump serotonin into the river, and levels there can reflect what serotonin cells in general are doing. More to the point, direct study of the brains of suicide victims has found evidence of altered serotonin function, compared with brains of those who died of other causes. A repeated finding is increased binding to one subtype of serotonin receptor in the frontal lobes, an abnormality that may be specifically linked to suicide. There are changes in other neurotransmitter systems, especially the dopamine system, but the most common findings involve serotonin.

Further facts about serotonin now become interesting. A high serotonin level in the brain makes an animal more tolerant of pain, while an artificially lowered level makes it more sensitive; the possible implications for suicide are obvious. Equally intriguing, high serotonin levels go with normal sleep. Sleep disorders are common in psychiatric illnesses, and this may go beyond ordinary insomnia—it may be basic to such illnesses. Perhaps the ones who, like Hamlet, seek "that sleep of death" are those who have had a troubled sleep in life. Finally, people low in 5-HIAA (again, whether depressed or not) are more likely to commit not only suicide but also nonsuicidal violence. As psychodynamic theorists claim, inward and outward violence may be two sides of the same coin, two expressions of a unified aggressive urge.

Which brings us back to Hamlet—a play that, somewhat contrary to its hero's formulation, is less about "to be or not to be" than about "to act or not to act." It is mainly outward violence he is musing on, although in the course of it he may lose his own life. Some simple brain process may determine whether we "take arms against a sea of troubles/And by opposing, end them," yet without quite determining whether we try to end them with inward or outward violence. Procrastination has its time and place, and even for Hamlet it is not clear that he should have

acted sooner. Impulsiveness—at least the sort that leads to homicide or suicide—
is not a luxury we can casually indulge in.

Yet perhaps the whole business seems less of a puzzle if we change the focus
of the problem: given the tribulations of human life, given our consciousness of
life's inevitable end, given the available opportunities for self-release, why aren't
there more suicides than there are? "What youthful mother," wrote William Yeats
in "Among School Children,"

> Would think her son, did she but see that shape
> With sixty or more winters on its head,
> A compensation for the pang of his birth,
> Or the uncertainty of his setting forth?

Or, to paraphrase T. S. Eliot, another genius of twentieth-century poetry, it
might not have been worth it after all.

Most of us have similar doubts about our own futures on at least some occasions. And we may be thinking them, even when we are saying something quite
different. All those lame remarks on those inexorable birthdays: quarter-century,
whew! Well, now it's half of those threescore and ten. Forty, that's the big one.
Never thought I'd make it. Here we are half a century. How'd that happen? Two
thirds. Still get it up? Sure. You ain't changed a day. Never felt better in my life.
Enjoy this one and many more. Still, not getting any younger. Not exactly. Each
birthday a little death, each *rite de passage* another passage toward it. There is a
sense in which life is a continual condition of bereavement, during which we
mourn the loss of ourselves. Think of the anger, the affectless acceptance, the
denial, the shock, the colossal, ineffable sadness. Is it any wonder that we become
more selfish with age, abandon certain youthful ideals, and come to believe that
living well is the best revenge? Is it strange that the world's religions, great and
small, with their venerable, contradictory narratives, their competing gods and
spirits, and their occasional taste for holy war, have a grip on so many minds?

Of course, the birthdays are also celebrations. One must think, too, of the
growth, the gains, the strengths, the learning, the tests passed, the challenges mastered. The circle of loved ones can enlarge throughout life, and even death is
marked by a funeral-rite of passage that, whatever else it may do, has the function
of sealing the tear in the social fabric, bringing together those who remain to comfort one another. It is a strange and wondrous fact that, despite the losses of aging,
despite even the recent increase in elder suicides, older people in Western nations
express more satisfaction with life than do their younger counterparts. As poet
Louise Glück, not yet old but no longer quite young, wrote recently:

> Surely spring has been returned to me, this time
> not as a lover but as a messenger of death, yet
> it is still spring, it is still meant tenderly.

And religion is more than myths and holy war. Even the great, organized, frequently dangerous religions have much in common with small-scale primitive ones. Their functions for individual emotions are manifold. At least, they offer balm in the face of great pain; at best, they can lead to humility, and to a concomitant concern for something outside of and larger than oneself.

In "Sunday Morning," perhaps the greatest poem in English, Wallace Stevens attempts another sort of stand on the relationship between death and life and on how we can go through life in death's shadow. The claim is that life has its own validity in the absence of God or heaven, a validity paradoxically enhanced by our intimacy with loss.

> Death is the mother of beauty; hence from her,
> Alone, shall come fulfillment to our dreams
> And our desires. Although she strews the leaves
> Of sure obliteration on our paths . . .
> She causes boys to pile new plums and pears
> On disregarded plate. The maidens taste
> And stray impassioned in the littering leaves.

The poem leaves no doubt that it heralds a new religion. It is Sunday morning, and the heroine, unashamedly not at church, is relaxing in her nightgown with "late coffee and oranges in a sunny chair." "Why," the poem asks, "should she give her bounty to the dead?"

> Shall she not find in comforts of the sun,
> In pungent fruit and bright, green wings, or else
> In any balm or beauty of the earth,
> Things to be cherished like the thought of heaven?
> Divinity must live within herself:
> Passions of rain, or moods in falling snow;
> Grievings in loneliness, or unsubdued
> Elations when the forest blooms; gusty
> Emotions on wet roads on autumn nights;
> All pleasures and all pains, remembering
> The boughs of summer and the winter branch.
> These are the measures destined for her soul.

There follows an exquisite, mildly sardonic meditation on how very boring life in heaven must be, without death:

> Is there no change of death in paradise?
> Does ripe fruit never fall? Or do the boughs
> Hang always heavy in that perfect sky,
> Unchanging, yet so like our perishing earth,

With rivers like our own that seek for seas
They never find, the same receding shores
That never touch with inarticulate pang?

And then the proposed religion takes shape:

Supple and turbulent, a ring of men
Shall chant in orgy on a summer morn
Their boisterous devotion to the sun,
Not as a god, but as a god might be,
Naked among them, like a savage source . . .
They shall know well the heavenly fellowship
Of men that perish and of summer morn.
And whence they came and whither they shall go
The dew upon their feet shall manifest.

It is a primitive, even a savage religion, but one without fictions. A religion that takes as sufficient for an orderly, decent, human life the particulars of life itself: the experience of nature in its changes, human moods, the affections, even the knowledge of loss and death become part of what makes life so eminently worth living. For this we can take the risks of loving, acting, losing, going forward, even without convictions about heaven.

But acknowledging ultimate loss is dangerous. Freud came to believe that death and violence were the same, that both were the death instinct, violence turned inward. But the true relationship between them is at once less intrinsic and more dreadful: violence is what we do from our fathomless anger against death. If we can kill or hurt someone else, especially at our own risk, it takes us out of ourselves for a while, and for that little while we don't have to grieve. What we can hope for, with Stevens, is some sort of recognition that the grieving is part of what makes life precious, that we would not love life nearly so well without it. We could perhaps be less angry at it. We could try, at least, to stop taking it out on each other. We could get together one morning, perhaps, and shout turbulent praises at the sun.

CHAPTER 15

Gluttony

You citizens named me Ciacco, The Hog; and for the gross crime of
gluttony I languish, as you see me, in the rain . . .

— **DANTE**, *Inferno*, Canto VI

There is only one way to lose weight, and that is to grow accustomed to feeling hungry. This simple fact, known to most people in affluent countries, seems
somehow lost on the authors of the diet, weight-loss, and exercise books that find
their lucrative way through the drugstore book racks. Two questions, then: Why
do they fail to mention it? And why is it so?

The first is easily answered. Few fortunes or reputations are made by saying
something simple that people already know. Since little is new in weight loss—
except for new drugs, so far not very effective—the art of writing about it becomes
that of making what is old seem new, or what is trivial seem important. After
decades of exhortations and serious research, Americans are fatter than ever. A
third are obese. More than 9 times out of 10, successful assaults on body fat are
followed by weight rebound, leaving the problem as bad or worse than it was
before the diet—and cyclically reinvigorating the appetite for diet books. Nation-
wide, we are fatter in every age group than we were twenty years ago; in some age
groups the gain is more than ten pounds. This, despite all the research, fads,
crazes, tennis, jogging, diets, yoga classes, obesity encounter groups, sugar substi-
tutes, appetite pills, and all the determination of various government agencies
keeping an eye on our health. In the year 2001, the Centers for Disease Control
sounded an alarm about a growing epidemic of obesity *in children*—a setup for
later diabetes and heart disease.

Worldwide trends lag the United States but tend in the same dismal direction. In Britain, obesity rates jumped from around 7 to around 16 percent
between 1980 and 1994. Canada and Australia are a few percentage points lower
but resemble Britain in the overall trend. In the developing world there is more
variability, but the trend is upward wherever it has been followed. Strangely
enough, many countries now have serious problems with *both* starvation and

obesity. In Western Samoa, most people are now obese. In Southeast Asia obesity rates are very low but increasing. In Africa and Latin America, women's weight has increased. As one author put it, "The World Health Organization (WHO) and the International Obesity Task Force [IOTF] have declared an obesity epidemic on a global scale." The IOTF stated that "obesity . . . constitutes one of the greatest threats to human health and well being as the 21st century approaches."

We can almost hear the fat cells laughing at us, as if each effort to dissolve them succeeds only in making them multiply. This is not the case, of course, nor are all these solutions illusory; most have a piece of the truth. But to understand where they work and why they fail, we must understand appetite—what turns it on and off, and why we don't stop eating soon enough.

Something signals us to start eating; something else tells us to stop. We use the words *hunger* and *satiety*, but these have become ambiguous, since we sometimes say, "I ate it even though I wasn't hungry." Much research on body weight is directed at giving scientific meaning to that sentence. Why do we eat when we don't need food? A sort of creature that stopped eating too soon would not be around long enough to cause even a blip in the path of evolution. Yet an animal must stop eating eventually, and any animal does, abundance notwithstanding. Eating, like all behavior, is a product of the brain, and the brain must get a clear signal that enough has been eaten before it brings the behavior to a halt. What is the signal, and why doesn't it work more reliably?

As we eat, a stream of messages, first neural and later bloodborne, go from the mouth and gut toward the brain. Theoretically, any of them could help tell the brain how much has been eaten; actually, most of them do. One clue is that we feel satisfied and stop eating long before a meal is digested. This is as it must be, since digestion takes hours, and if we habitually ate for hours at a clip we would be grossly obese in no time. It tells us to search for signals that can work during and right after food intake, rather than signals keyed to the body's long-term needs. Another clue is that after a big meal we feel stuffed, and the thought of another slice of roast may leave us numb, but we may still eat a dessert as calorie-rich as the roast. So whatever the satiety signals are, they can be defied by the right stimuli; to put it another way, satiety may mean different things to different parts of the brain.

In addition, satiety and weight regulation are very different things. There are many signals to tell us when a meal is over, on a time scale of minutes. But there is also an overarching system of controls that operate on a time scale of days or longer and that change the brain's *sensitivity* to the meal-ending signals. These controls depend on the size of the body's fat stores *in comparison to a set point,* or what size fat stores the body "wants." No one really knows why this set point stays where it is, but we know that it differs from person to person, and that weight rebound after weight loss is extremely common. It is so common, in fact, that the people who succeed in keeping lost weight *off* are being entered into a registry and actively studied.

Food intake is complicated enough, but it is only half the story. Energy out-put is reduced after even a small (10 percent) weight loss, deliberate or not, so "a formerly obese person requires approximately fifteen percent fewer calories to maintain a 'normal' body weight than a person of the same body composition who has never been obese." This is due to an 18 percent decrease in energy use at rest and a 25 percent decrease in energy use while active. Weight regulation, as one group of researchers put it, is "highly integrated and redundant," and we have "a robust system for defending fat stores [that] may have conferred a survival advantage during human evolution." However, due to genetic differences, many people in our food-rich environment defend fat stores that are incompatible with health, and defend them as efficiently as if it were optimal. This is what we most often struggle with, and it operates on a time scale of years.

A large proportion of young people are overweight or obese in childhood, and this problem worsened dramatically during the last decades of the twentieth cen-tury. According to the Centers for Disease Control, "some 6 million American children are now fat enough to endanger their health. An additional 5 million are on the threshold, and the problem is growing more extreme even as it becomes more widespread." And this is by a standard more lenient than that applied to adults. For example, a five-foot-four-inch teenage girl would have to top 145 pounds to be considered at risk, 170 pounds to be considered overweight. Even by these lenient standards, the proportion of overweight children increased from 5 percent in the 1960s to 13 percent in the 1990s.

After childhood, weight gain continues relentlessly. Between age twenty-five and fifty-five the average American puts on an additional twenty pounds; all this takes is an energy intake that exceeds energy output by *one third of one percent*. So on a daily or weekly basis the system only has to be off by that much—a per-centage that the conscious mind is incapable of noticing—to gradually produce obesity. Which means that the unconscious forces easily win. Small surprise, then, that as the century turned, more than half of the adult population in the United States was overweight, nearly a fourth clinically obese. This insidious weight gain depends on small adjustments in the meal-ending signals and espe-cially in the set points—the targets for weight and body composition that the sys-tem seems to be aiming at. We see them as maladaptive, but that is because they are adaptations to a different environment from the one we live in now. Here is how they work.

Some basic facts have been known for half a century. Meal-ending signals are located throughout the gut from the mouth to the small bowel and beyond. They report over nerves or through chemicals sensed by brain cells, beginning in the mouth. Rats will press a lever putting food into their mouths longer than they will a lever that pours food directly into the stomach. This is because, in the early stages of a meal especially, taste and smell stimulate eating. But suppose the food goes through the mouth and instead of filling the stomach is drawn out again from the esophagus. A dog "sham eating" in this way will eat more than usual, but not indefinitely. It will still organize its food intake into meals, but with shorter intervals between them. This means that signals to the brain from the mouth can stimulate or quash hunger, but that the satiation from such signals alone is short-lived.

In the reverse situation, where an animal presses a lever to deliver food *directly* to the stomach, eating stops *sooner* than normal; this means there must be strong satiety signals coming from the stomach itself, or from the first part of the small bowel. The "hunger pangs" we feel in the stomach when it is empty are real—thirty-second contractions much stronger than the usual muscle-tone waves in the stomach wall. Since injections of *non*nutritive fluid, or even pumping up a balloon inside the stomach, will stop an animal from eating, it is clear that mechanical filling alone can work. But it takes more *non*nutritive fluid than nutritive fluid to end a meal, and the classic balloon experiment produced a degree of stomach distension that does not occur in eating unless we really gorge ourselves. Still, later studies showed that lesser degrees of distension do make a difference in both rats and men. Rats lick food less enthusiastically as their stomachs fill mechanically, beginning at only a fourth of the stomach volume. And twenty normal men who were given a milk-based drink before lunch ate less when the drink volume was larger, even though the energy and nutrient content was identical in all portions. Not only that, but their appetite was still suppressed hours later, at supper.

Yet even people who have their stomachs removed to prevent the spread of cancer get hungry and do not eat without stopping, so mechanical filling is not the whole story. Chemical signals, direct and indirect, account for most of the rest. Relatively little digestion and absorption take place in the stomach, which is mainly a reservoir; that is why some cancer patients can live without the stomach. In the small intestine, where most digestion takes place, fat reduces appetite even if it is introduced directly, bypassing mouth and stomach. Yet the same fat injected into the bloodstream has little effect. This "fat paradox" means that something in the intestine sends a satiety signal. The major nutrients are broken down into small, absorbable molecules—carbohydrates into glucose and other simple sugars, proteins into amino acids, fats into fatty acids. These seep into the bloodstream, first passing that great gatekeeper, the liver, which sends signals to the brain; but the brain also detects some breakdown products directly when they cross the blood-brain barrier.

The first theory to take advantage of this fact was the glucostatic theory, named by analogy to the thermostat. In this model, still plausible as *part* of what is happening, receptors in the brain sense the blood level of glucose and signal "full" when the glucose tide is high enough. Physiologist Jean Mayer suggested that the lower or ventral hypothalamus at the base of the brain could sense high glucose levels and send nerve signals adding up to satiety. It was later proposed that the lateral portion of the hypothalamus on either side could respond to *low* glucose levels with hunger signals. Indeed, destruction of the ventral hypothalamus produced a rat that grew grossly obese through overeating, a condition mimicked by some people with tumors in the same place. And destroying the *lateral* hypothalamus produced an animal that neglected food and lost weight—as if it had lost the impulse of hunger.

It turns out that many subtle and indirect effects of these lesions contribute to the changes. For example, lateral hypothalamic damage interferes with the rewarding effect of food by interrupting the mesolimbic dopamine path, part of the brain's great river of reward, so poisoning dopamine neurons alone has a similar impact. Also passing through the lateral regions are sensory pathways for the face, so damage here dampens sensations in eating. Ventral lesions, on the other hand, *heighten* food sensations. Furthermore, it is not just the eating itself or the rewards of eating that are changed but the body's set point for weight as well. Thus, if rats have been force-fed and fattened before the ventral lesions, they don't gorge themselves as much after surgery, whereas skinny rats that have been food-deprived will bulk up even after lateral lesions—but only up to a weight that is still less than normal.

It turned out that the ventral hypothalamus did have special receptors for glucose, but unfortunately for simplicity, poisoning those receptors did not make animals insatiable in the way that destroying the tissue did. Worse for the glucose theory, fluctuations in eating and stopping showed little relation to the level of glucose in the blood or its availability to the brain. This seemed to rule out regulation based mainly on brain glucose sensors, at least for a short-run cap on meal size. Still, pouring glucose water into the gut turns off prey catching in toads within about fifteen minutes, suggesting a basic vertebrate mechanism that somehow conforms to Mayer's theory. We now know that, in mammals at least, the brain senses satiation mainly by an indirect method, involving peptide hormones released in response to meals in the stomach and gut. Of the shorter-term chemical signals of satiety—the meal-ending signals—two overall mechanisms are prominent.

The first is glucose detection by the liver and the brain. Because nutrient-rich blood pours into it from the small intestine and stomach through the great portal vein, the liver is first to know when a meal is being eaten and first to know when it ends. Experiments show that the liver itself can detect glucose coming from the stomach and small intestine, and that it responds with signals over nerves, particularly the vagus, connecting the liver to the brain. These nerves fire more when glucose is low, and frequent firing causes hunger. Severing the nerves causes a long-lasting reduction of food intake to about one third the normal level, while injecting a drug that suppresses glucose activity into the liver's blood supply will cause an immediate tripling of food intake. As long as the vagus is intact and glucose is pouring into the liver, the sense of satiation will remain.

The second meal-ending mechanism involves gut hormones. These hormones, mainly peptides, were long thought to be involved only in stimulating digestion. Now we know that they play a role in feeding regulation by indirectly signaling the brain. The most important of these hormones, cholecystokinin, or CCK, is a gut peptide hormone that directly or indirectly affects the brain. It does have local effects—for example, it slows stomach emptying and stimulates the gallbladder; but it also helps to terminate feeding by way of nerves from the gut that eventually inhibit the ventral hypothalamus—what Gerard Smith, a pioneer in CCK research, calls the "information highway" from gut to brain. The main highway is the vagus nerve, a primary independent monitor of body states. A paraplegic with a spinal cord severed at the neck still receives this internal information, but cutting the vagus nerve blocks it completely. Nor does CCK involve higher brain centers; an animal with an intact brain stem can receive the CCK signal and stop eating. CCK injections make normal people voluntarily reduce their meal size by 20 percent; needless to say, drug companies are aggressively pursuing a version that could be taken as a pill. As for laboratory animals, Smith points out that "all kinds of interventions make rats eat less," including fenfluramine and amphetamine. But the normal rat behaves predictably after eating its fill, ambling and exploring for two to five minutes, then sinking into slow-wave sleep. CCK injections produce, after *less* food intake, a type of satiation that triggers this normal pattern. Amphetamine does not. Thus, a future CCK-like drug might be more "normal."

A seemingly parallel but actually quite different type of theory, based on fat detection by the brain (called "lipostatic" for lipid control), was also fielded early on. Mice can detect their own total body fat, summing various organ stores, and adjust their food intake to stabilize that level. If a piece of fat is transferred from one mouse to another, it will wither away; but if some of the recipient's own fat is removed first, it will be accepted. A mouse with damage to the lower hypothalamus, however—having a badly damaged satiation mechanism—will accept the fat graft yet continue to gain weight. Related, although less direct, evidence on humans comes from the life work of Rose Frisch at the Harvard School of Public Health. Frisch showed that a certain level of fatness in a young girl is needed before puberty and first menstruation can be triggered. She reasoned that this portion of the hypothalamus detects total body fat in the "lipostatic" effect. Such fat assessment could play a role in long-term regulation of body weight.

But the theory that the brain can assess the amount of fat in the body was not thoroughly vindicated until the 1990s, with the discovery of energy-regulating molecules released by fat cells themselves. Two are prominent among these: the old standby insulin and the relative newcomer leptin (from the Greek word for "thin"). Obese people have high levels of leptin, probably because they have less sensitive receptors and need more of it to feel satiated. Leptin is made by fat cells and acts in certain areas of the hypothalamus, where it affects circuits that put a stop to eating—with the help of a cascade of other brain peptides. It works in rats and chickens, suggesting an ancient molecular mechanism, and it is in effect an anti-obesity hormone. Both leptin and insulin are released by fat cells *in direct proportion to the mass of fat in the body*. Lay down more fat, and insulin *and* leptin levels increase; sensors in the hypothalamus know it, and cut appetite. They do this by holding back the release of neuropeptide Y, which works in the lateral hypothalamus to stimulate hunger. In fact, if injected, neuropeptide Y "produces ravenous, almost frantic eating," but insulin and leptin normally suppress it.

Neuropeptide Y is only one of several hormones and neurotransmitters modulated by leptin, and any of these may lead to new treatments. Insulin and leptin also increase energy output and affect the way meal-ending signals are perceived by *their* sensors in the lower brain stem. This is extremely important, because it means that *the meal-ending signal system is framed by the long-term fat-regulation system.* Unfortunately, it works both ways. Go on a diet, reduce your fat mass, and pretty soon your meal-ending sensors don't hear so well anymore. Now glucose, CCK, and the other satiety molecules have to raise their voices to make an impression. Translation: you stay hungry until you eat more, and you eat more until you gain back the fat.

Many attempts have been made to intervene chemically at one point or another between the temptation and the spare tire, but most have had mixed results at best. The older agents were usually stimulants like amphetamine, which increase energy output and may also suppress appetite. These are no longer used in weight control because of their effects on the heart and blood vessels. Others have been approved by the Food and Drug Administration and later withdrawn. Fenfluramine and dexfenfluramine block reuptake of serotonin from certain synapses, reducing hunger. The combination of fenfluramine and phentermine (fen/phen) was frequently prescribed in 1996 and 1997, and was effective, but it caused many cases of heart-valve leakage. Fenfluramine was withdrawn from the market, but another drug, sibutramine (Meridia), was soon approved. Like fen/phen, Meridia blocks the reuptake of both serotonin and norepinephrine from synapses but appears to be safer. It just doesn't work very well.

The synapses where these reuptake blockers act are probably not the same ones where antidepressants act, although this is still under study. Molecules that work in completely different ways are also being studied. Many brain peptides affect hunger and satiety, but the catch is that they cannot be given orally, since they are digested like any other peptide; nor can they be injected, since they are too large to cross the blood-brain barrier. Nevertheless, new drugs will emerge, as pharmacologists search for smaller molecules that mimic or influence some function of these peptides. But perhaps the most interesting agents on the horizon are analogs of leptin. The search is on for new drugs that will enhance leptin in some way, stimulate its receptors in the brain, or make them more sensitive.

Other agents, probably less risky, interact with the digestive system instead of the brain. Orlistat (Xenical) reduces the efficiency of fat digestion by blocking a pancreatic enzyme that breaks down fat. About a third of the fat eaten is excreted again, and since fat supplies more than twice the calories, gram for gram, that proteins and carbohydrates do, reducing fat digestion is a plus. Xenical's relative availability through the World Wide Web has increased its use. Finally, some weight-loss aids are officially food additives, not drugs. Olestra, a fat substitute, gives potato chips and other foods the taste and texture of fat without the health risks or excess calories. It passes through the gut without being absorbed or digested, so it has the occasional side effect of loose stools, but most people find the tradeoff acceptable.

Despite purists, who insist that obesity must be solved by diet and exercise alone—*Willpower!* is their battle cry—the scientific search for drugs and additives is excellent and necessary. Critics are often people who drew a lucky hand in gene poker. They have no idea what it feels like to play the same game with a lousy hand, and those who do should welcome any new pharmaceutical fix that is proven effective and safe. But no drug on the horizon, nor even any likely combination of drugs, will completely solve a problem that in the end comes down to human weakness. Of all motives, the impulse to eat is perhaps the most forceful, and the response to inadequate food intake—the response, in short, to starvation—is highly developed and overdetermined. That is, there are many mechanisms, and when something deflects one of them, whether that something is a new drug or a fad diet, others take over. Obesity expert William Bennett in 1987 wrote, laconically, that "dietary treatments for obesity are overdue for ethical as well as scientific reevaluation." This remains equally true in a new century. But how do psychological forces counter our best efforts at dieting? At any given time an estimated 15 to 20 percent of Americans are attempting to lose weight. Why are most of us destined to bounce back or just plain fail?

First, various factors can override the satiety produced by all the best efforts of the system. One is taste for variety: given an ordinary single-flavor rat diet, a rat will regulate its body weight; but given the same diet artificially flavored in four different ways, it will eat a lot more, in some cases twice as much as usual. Second, social factors, such as seeing another eat, enhance eating after normal satiety in various species. Social eating somehow induces false hunger signals. Both these effects work in humans as well as in rats, so a round of dinner parties is ideal for wrecking a diet in at least two different ways.

An equally interesting factor is stress, which in several species increases food intake to abnormal levels. Physical discomfort such as tail pinching will cause rats to eat more. But Michael Cantor and his colleagues showed that satiated rats or people can be made to eat more as a kind of nervous adjunct to a challenging task. They found that tail pinching and loud noise are somewhat interchangeable in triggering eating in rats, but simply putting the rat through a training process has similar effects. Furthermore, young men trying to track a spot on a rapidly rotating disc will dart their hands out to a plate of nibbles as the task gets more challenging.

In people eating freely without hunger or restraint, psychological distress can reduce food intake, but in people on diets, it *increases* intake. Even subliminal threats involving loneliness and anger can cause nervous eating. These fascinating findings support the hypothesis, already seen in the chapter on joy, that evolution left mammals with a motivational system that is just not specific enough, so that we often do not know what we want. Various different motives are handled in part by a *generalized* arousal system, using the lateral hypothalamus. Stress induces arousal, so it is not surprising that some kinds of stress can enhance eating, even though severe stress such as grief can diminish it. Depending on context, other strongly motivated behaviors such as hoarding may also be enhanced by stress.

And since anxiety is a form of arousal produced by vaguely perceived or imagined stresses, it stands to reason that we might feel like eating when we are anxious.

The very obese may or may not tell us much about ordinary overweight, but there are parallels between people who suffer from obesity and rats that become fat after damage to the ventral hypothalamus. For one thing, both are finicky; they eat much more of good-tasting food than the nonobese will, but much less of food adulterated with a touch of quinine. For another, obese humans and rats eat less if they have to work to get their food but eat much more if they do not have to work to get it. Moreover, there is a syndrome of hypothalamic damage in humans, called Froelich's syndrome, which has among its effects increased appetite and obesity. Yet while this syndrome parallels, in a real, anatomical sense, the syndrome of rats with hypothalamic damage, it is very rare. There is no *obvious* abnormality of the hypothalamus in more than a tiny fraction of the overweight population.

Still, stress experiments in animals may well be related to human overweight. It is a strange fact that in industrial countries rich people tend to be thin and poor people fat, whereas in poor countries the reverse is the case. This paradoxical inversion could be explained by an intersection between the stress-induced eating on the one hand and a seemingly obvious fact of life on the other. Consider the odd first sentence of the section on "Obesity" in the *Merck Manual of Diagnosis and Therapy*, a standard handbook for physicians: "The incidence of obesity coincides with the availability of food, obesity being conspicuously absent during famine." Obvious? Not quite. If obesity arose from an overwhelmingly strong drive, some obese people might somehow manage to stay that way by procuring extra food even in famine. Or, if the condition were due to a simple and overwhelming physiological defect—making much fat out of little food—then, too, we would see some obesity even under food-deprived conditions. But it turns out that when food is very hard to get, obesity disappears. And even under controlled conditions, fasting is one diet that works.

So of course poor people in the poorest countries are not going to be overweight, at least not for very long. Two of the three remaining groups—rich people in poor countries and poor people in rich countries—are motivated to overeat. Food supply is ample, but, perhaps unconsciously, these people are not quite sure how long this will be so. Other stresses, coming from general economic instability in the one case and subordination in the other, probably favor anxiety-induced overeating. But this is too simple; the quality of the diet must also play a role. Both those groups eat very starchy diets—starchy grains or tubers supplemented in many cases by alcohol and "junk" foods, with their high-caloric-density sugars and fats. Also, unlike the poor of poor countries, their diets are usually low in bulk or fiber. They might conceivably be striving to eat enough protein, vitamins, or other specific nutrients, and in the course of this effort put away enough starch to grow obese.

There have been a few demonstrations of such specific hungers. Rats being fed a sodium-deficient diet develop an appetite for sodium and seek salty foods.

This is an innate response, but there is also learning of specific food preferences: if a vitamin or mineral deficiency makes an animal sick, then whatever food contains that substance will reward it by making it feel better, and will therefore become preferred. As for the major nutrients, deprivation of glucose or lipids alone has to be pretty severe to cause hunger, but deprivation of both at the same time triggers substantial eating much more readily. Still, there is plenty of obesity among people who have no specific dietary deficiencies, so other factors must be operating.

Cultural forces can either enhance or contravene the general rules about social class and obesity. For example, although the poor are fatter than the rich in most industrial countries in the West, in Germany in the 1960s this was true only of women; rich German *men* were fatter than poor ones. Anthropologist Alexandra Brewis has shown that Samoans, who traditionally favor plump bodies, are even today very tolerant of weight gain. In West Africa, in certain traditional kingdoms, the ideal of "fat is beautiful" for noblewomen was carried to the extreme of incapacitating obesity. The Annang of Nigeria used a fattening room to plump up pubertal girls by restricting their activity and overfeeding them as they awaited marriage. Today in Niger, in the village of Maradi, young women encourage each other to gain more weight, even resorting to pills to increase appetite, in the name of beauty. At the other extreme, the cultural wave of anorexia nervosa in the West has had great power over young women's bodies. Remarkably, both groups of women—those in Niger who want to gain, and the emaciated ones in the United States who want to lose—use the same language: *Just another pound or two and I'll stop.*

So culture is powerful. But until we have better explanations, the anxiety theory seems helpful in understanding the obesity of both the industrial poor and the nonindustrial rich. As for rich people in rich countries, they do not escape overweight, but they keep it under control more successfully. They certainly suffer anxieties, but these are perhaps less strong or are assuaged in other ways. And, unlike the West African nobility, these rich are faced with two strong cultural counterweights. First, and probably less important, they are nowadays made continually aware of the increased risk of disease and death associated with being overweight, especially seriously overweight. Second, once the poor can afford to be fat, then a "healthy plumpness" is no mark of social distinction. A wealthy person had best stand out by slimming—showing that she is so sure of the quality of her environment that she can risk keeping her body close to the bone. Thus the "social X-rays" of the American and European nobility.

Can everyone do that? Evidently not. A quarter century of accumulating evidence has laid to rest the notion that willpower is all. In a classic study in England, twenty-nine women who claimed they could not lose weight were isolated in a country house and fed on a restricted diet. They ate only fifteen hundred calories a day for three weeks, little enough for any of them to lose weight given their body size and activity. Yet nine of them did not lose weight. These people had in common a long history of dieting, a low basal metabolism (fuel burning at rest), and a low energy-burning rate during normal activity. In other words, for some people, the complaint that the body seems to adjust its metabolism to countermand the dieter's best efforts is clearly true. The authors conclude "that among a group of would-be slimmers who claim to be unable to lose weight there will be some who have become metabolically adapted to a low-energy diet and others whose inability to lose weight is illusory." Of the two possibilities in the paper's title—"Resistance to Slimming: Adaptation or Illusion?"—both are real.

In another pioneering study, obese people showed reduced energy use at the cellular level. As measured by the ion-transport-pump activity of the red blood cells—theoretically representative of all cells—obese people used energy at less than 80 percent efficiency, burning less and storing more than the nonobese controls. Such studies have multiplied enormously, making genes for obesity an accepted reality and supporting the belief of at least some overweight people that they are physiologically different.

We could have guessed as much from animal studies. Genetically obese strains of rats and mice with quite simple mutations have been known and studied for decades. The Zucker "fatty rat" has a homozygous recessive mutation—a double dose of a bad gene—that makes a defective leptin receptor. Of two strains of severely obese mice, one has a defective receptor like the Zucker rat and remains obese when leptin is given; the other has a defect in the leptin itself, and injecting the molecule cures it of obesity. These are natural mutations, and there are several others in rodents alone. But in addition, transgenic mice—"knock-out" and "knock-in" animals, with specific genes removed or added—have been created and studied.

As for *us*, 40 to 70 percent of human variation in plumpness is genetic, as measured in various ways in many studies. In a classic adoption study led by Albert Stunkard, 540 adult Danish adoptees were compared with their biological and adoptive parents. The subjects ranged from thin to extremely obese, and this dimension was highly predictable from the fatness of either biological parent but not from either or both adoptive parents. "We conclude," the authors reasonably wrote, "that genetic influences have an important role in determining human fatness in adults, whereas the family environment alone has no apparent effect." Twin, adoption, and family studies have confirmed this view so many times that today debate about whether this is so has given way to the search for specific genes.

Undoubtedly these will turn out to be numerous, affecting every aspect of the control of hunger and satiety as well as energy expenditure. Experiments by Claude Bouchard and his colleagues at Laval University in Quebec have shown that identical twins resemble each other to a high degree in weight and fat-mass changes even when kept on an overfeeding or underfeeding regimen. The HERITAGE Family Study, a remarkable collaboration among five centers in the United States and Canada, is gradually separating the genetic contributions to different aspects of body composition. The research assigns abdominal visceral fat to a major chromosomal locus but links fat-free mass to the mitochondrial genome, which passes only from mothers to daughters. Also, there is substantial overlap between the inherited components of visceral fat and blood insulin level, which suggests a physiological mechanism for the family resemblance.

From this and other research we know that numerous genes are involved in weight and fat regulation. Every chromosome except 21 and Y has been implicated, most at more than one locus. Still, some genes are more important than others and "anonymous" major genes account for about 40 percent of the variation in fat mass, with many minor genes also contributing. Groups of people under study for obesity genes include some, like the Pima Indians of the American Southwest, who have severe epidemic obesity, and others, like the Old Order Amish and Mennonites, who tend to obesity and have large extended families ideal for genetic research.

At least nine individual people in various studies around the globe have been found to have mutations corresponding to rodent ones, including defects in both leptin and the leptin receptor. But this is only the beginning. By scanning the whole human genome in large groups of people, gene hunters have traced important obesity genes to specific regions of chromosomes 2 and 8. The chromosome 2 neighborhood includes the gene for a large protein, pro-opiomelanocortin, or POMC (pronounced "palm-see"), that is the parent molecule for a vital family of hormones, including endorphins, ACTH (the pituitary pilot of the adrenal gland's stress response), and beta-lipotropin, which helps govern fat stores. Any of these functions could theoretically be involved in promoting fatness, but there is direct evidence implicating melanocortin in feeding.

If rats are overfed through a tube in the stomach until they are 5 to 15 percent over control body weight, they reduce their food intake drastically when the overfeeding stops. At the same time, the genes expressing messenger RNA for POMC are turned on in the arcuate nucleus, a strategic part of the hypothalamus. In fact, the mRNA shows an 80 *percent* increase. Remarkably, if you block melanocortin with specific antagonists delivered into the brain, the overfed rats go right on eating at two or three times the rate of control rats. Thus, melanocortin helps ratchet feeding down when body weight is too high. This experiment confirmed other studies showing that in the hypothalamus and brain stem, melanocortinlike molecules slow or stop eating, while blocking them promotes overeating and fatness.

The relevant stretch of chromosome 8, on the other hand, contains the gene for an adrenaline-type receptor. This would influence obesity through energy output rather than through intake. For example, under conditions of arousal, the sympathetic nervous system is activated to prepare for fight or flight, and adrenaline is secreted from the core of the adrenal gland. This should mobilize energy stores and promote the burning of fat. But if a gene change makes the receptors less responsive to sympathetic neurons or adrenaline, the burn will be slower, leaving more fat stores unconsumed and making ideal body weight a more distant goal. Thus a few glimpses into energy regulation by the human genome. More genes will be found soon, since at least eight genome-scanning efforts directly or indirectly relevant to obesity were under way as the twentieth century ended.

So for some of us who can't stay slim the problem may just be willpower, but for others it is also metabolism. This is not an attempt to separate behavior from biology, but there is a real distinction. For many would-be dieters, physiology wrecks their best efforts by urging them to eat when food is superfluous; for some there is yet another genetically guided system that slows their fuel use to prevent weight loss. Those who have neither problem are a small, lucky minority of our species—they can live in the midst of plenty without conscious dieting and without tending to roundness. Those who have *both* problems are in a sort of physiological double jeopardy. Overall, the system seems badly designed. Here we have a species—so far a highly successful one—in which one of the most basic regulatory systems routinely malfunctions. The system does not regulate body weight—or to be precise, it regulates it at a level too high for ideal health and activity. Unwilling to accept malfunction on such a broad scale except as a last resort, an evolutionary biologist begins to cast about for a logical adaptive explanation: something, in a long-ago time, must have made this stubborn system advantageous.

Actually, we don't have to look far. There has always been talk about "natural" diets, but now there is real research that can raise that notion above the trivial: the ethnographic record of what people ate (and eat) in hundreds of nonindustrial societies—hunters, gatherers, fishers, gardeners, herders, and farmers. In a representative sample of 186 societies around the world (out of thousands known), 13 subsisted mainly by gathering wild plants; 14 by hunting; 17 by fishing (including shellfish collecting and aquatic-mammal hunting); 15 by herding domestic animals for meat, milk, or both; 70 by gardening or orchard keeping; 56 by large-scale agriculture; and 1 by exchange. Except for that last, the subsistence method also describes the food base. Ninety-two percent of these economies were studied between the years 1800 and 1965, so they represent worldwide variation during that period. They do not, however, reflect subsistence during most of human evolution: at say, 10,000 years ago, a comparable worldwide sample would have been at least 99 percent gathering, hunting, and fishing; and 20,000 years ago, 100 percent. Yet the traditional subsistence types have much in common.

Marjorie Whiting, a nutritional anthropologist, surveyed 118 nonindustrial societies with good research on diet. The alarmist view of some nutrition experts about primitive diet doesn't fit what she found, but neither does the idealized view held by some anthropologists. Of the 116 societies where quantity of food could be rated, just four had merely "minimal" food, the only rating below subsistence. As might be expected, health was poor in all four. Fourteen more had "subsistence" levels—barely enough to go around. Still, this left 98 societies (85 percent) with food supplies that were adequate or plentiful. Furthermore, diet *quality* in the sample was high. Averages for fat and carbohydrate fell within recommended ranges, while the percentage of protein was twice the recommended minimum. For the 85 percent of societies for which food supply was adequate or plentiful, the diet was probably superior to that in the United States today. Some other aspects of American diet (high intake of refined carbohydrates, excessive food quantity, and low intake of fiber) underscore the superiority of the "primitive" diets. Vitamins and trace elements present a more serious data-quality problem, but in any culture where a lot of wild plants were eaten (probably most of them), natural variety buffered against specific deprivations.

Archaeologists George Armelagos and Mark Cohen provided the first deep history of diet in their pioneering research beginning in the 1970s. They and others examined the bones of people who lived before and after the transition to agriculture, whether it occurred 10,000 years ago as in the ancient Near East or only 1,000 years ago as in northeastern North America. The remains showed something counterintuitive: health *declined* as hunting and gathering came to an end. The bones show increasing iron-deficiency anemia, growth-disrupting nutritional stress, and most surprisingly a decrease in average height, all linked to the rise of agriculture. These trends began even earlier, however, suggesting that agriculture was not a great voluntary advance but a forced transition for people running out of wild foods.

A group of us in Atlanta, led by physician-anthropologist Boyd Eaton, extended these findings by looking at the diets of recently studied nonindustrial cultures. Our first joint paper was "Paleolithic Nutrition," an analysis of hunter-gatherer diets. Using data from ecological anthropology, archaeology, and paleontology, we proposed a model of what people ate during most of human evolution. This had been tried before, but thanks to Eaton's scholarship, we did it with believable data amassed during decades of modern research. We compared the model Stone Age diet with that of modern Americans, on the one hand, and that recommended by medical scientists on the other.

The results were revealing. We estimated that daily calories in the Paleolithic were 34 percent protein, 45 percent carbohydrate, and 21 percent fat. The corresponding percentages for the current American diet were 12, 46, and 42 percent, and in the recommendations of the U.S. Senate Select Committee on Nutrition and Human Needs, 12, 58, and 30 percent. Thus, Americans ate *twice* as much fat as the hunter-gatherers, and even the recommended level was 50 percent more than our ancestors got. For the ratio of polyunsaturated to saturated fat, the estimates were 1.41 for the Paleolithic, 0.44 for the current American, and 1.00 for the recommended diet. For sodium, the estimates were 690 *milligrams*, 2 to 7 *grams*, and 1 to 3 grams, respectively. And for fiber, using various sources for the estimates of modern diet and recommendations, they were over 100 grams, 20 grams, and 30 to 60 grams. The exceptionally high fiber intake among hunter-gatherers also implies low intake of simple sugars. In each comparison, the Paleolithic diet met or exceeded the most stringent *recommended* standards. Although the ancestral plan had shortages, most were mild or moderate, preventing obesity without producing starvation.

In a burst of optimism, we sent the paper to the *New England Journal of Medicine,* arguably the most prestigious medical journal in the world. They had, we reasoned, a quick turnaround time, so the paper would not age much while we waited for them to reject it. Amazingly, they did not, and after a few minor revisions it was published. I was not prepared for the reaction. Much that appears in the *New England Journal* is news in a very practical sense, and science journalists follow it closely. But such intense interest in our offbeat paper? We began getting calls from newspapers and broadcasters days before we saw the *Journal* itself, not only from science and health editors but from food editors, too. Several reporters predictably coined the phrase "Caveman Diet" and went on to use it despite our insistence that it was misleading and insulting. "Cavemen Cooked Up a Healthy Diet," "Cave Man Takes a Healthy Bite Out of Today's 'Civilized' Diet," and "Check Ads for Specials on Saber-Toothed Tigers" were some headlines. Our prize went to the *Fort Lauderdale Sun-Sentinel*, which ran a group of "paleolithic" recipes and a full-color photograph of a stooge grotesquely done up as a caveman—skin, club, tooth necklace, and all.

America's more distinguished journalistic institutions were not above this sort of humor. A *Washington Post* editorial predicted that we would write a best-seller "inevitably" entitled *The Cave Man Diet*. It went on to say: "Some day in the near future you'll look out at daybreak and see people all up and down your street come loping out of their homes wearing designer skins and wielding L. L. Bean stoneaxes while every dog, cat and squirrel in the neighborhood runs for cover." Ellen Goodman, the *Boston Globe* columnist, ridiculed us in a piece accompanied by an etching of savages dancing, captioned "Make mine mastodon." The column bristled with the resentment of a noncompliant patient sermonized at once again by high-minded, pesky doctors. "But I am convinced," she concluded, "that the average Paleolithic person was the very role model of good health when he died at the ripe old age of 32." Some weeks later, *The New York Times* editorialized, "Did the people of the early Stone Age eat more healthily than their urban successors? The issue is being vigorously chewed in the *New England Journal of Medicine*, and it tastes like the myth of the Noble Savage."

But we knew that for any serious person who wanted to evaluate what we had to say, there would always be that tightly reasoned argument in technical prose, with five densely packed tables and eighty references to other scientific publications, sitting on some shelf in every medical library. This knowledge was comforting, and it still is. Eaton and his colleagues have gone on to extend, refine, and support the model ever since.

But recall the major flaw in traditional diets: all the cultures had shortages. In 29 percent, these were rare (every ten to fifteen years); in 25 percent, occasional (every two to three years); in 23 percent, annual ("a few weeks preceding harvest, anticipated and expected, recognized as temporary"); and in 23 percent, more than once a year. Generally, frequent shortages were mild and occasional ones more severe. Twenty-nine percent of the cultures had severe shortages ("comparable to a famine, deaths occur . . . many persons are desperate for food, emergency foods are exhausted"); 34 percent had moderate ones ("real suffering and deprivation, a few persons are hungry and incapacitated, weight loss may be considerable, food stores exhausted, emergency foods sought"); and 37 percent experienced only mild shortages ("fewer meals per day than usual . . . less activity, no great hardship experience, people may lose weight, food stores are used"). To take one example, a study of annual body-weight changes with the agricultural cycle in a peasant village in northeast Ghana showed fluctuations of about 5 percent a year—comparable to the frequent weight changes of many American dieters.

Still, the data show what is wrong with primitive economies. These shortages do not negate the advantages, but they do dull the edge of romanticism and they give us our first glimpse into the evolution of obesity. Since mild to severe deficits were common under natural conditions, *selection favored those who stored calories in good times.* For most cultures, such stores would be called on at least every two to three years. We did not adapt to continuous surplus because it never existed; there was no disadvantage for those tending to obesity. On the contrary, in the modern world our bodies "think" that the surplus is going to end and are storing up against that time, but it never arrives—hence endemic obesity. In addition, in traditional cultures food is harder to get, requiring a bigger energy output, and for the most part it is also harder to eat and to digest. Exercise for our ancestors was not a matter of choice, because all work was physical, and there was no free lunch.

The cultural ideal of beauty in most societies—including our own until the twentieth century—was plump: a Titian model, not a fashion model. Beginning with the Venuses of Willendorf and Laussel, the ideal of female beauty was one that seems to us, and to health authorities, overweight. Annelise Pontius showed that of 64 female figurines found in Paleolithic Europe from Spain to Russia, 41 showed below-the-waist obesity predominantly and 7 showed trunkal or upper-body excess. Such women were evidently idealized if not worshipped. The countless nudes painted during the Renaissance in Europe, many depicting classical goddesses, may not have been as generously endowed as the Paleolithic figurines, but they were plump by current standards.

Not one of these sumptuously gorgeous women, adored by Botticelli, Raphael, Titian, Rubens, or Correggio, would dare, today, to show her face in an upscale gym. None would be nearly thin enough for a *Playboy* shoot, let alone a strut down a high-fashion runway. Yet great artists immortalized them twice: by naming them goddesses and by drawing them into images that would last for centuries, enchanting the eyes of millions. Of Raphael's *Three Graces*, art historian Kenneth Clark wrote: "These sweet, round bodies are as sensous as strawberries." The nude in Titian's *Sacred and Profane Love* is "that sweet fruit, the human body . . . generous, natural and calm . . . a glowing panoply of flesh." In his *Venus*, "the female body, with all its sensuous weight, is offered in isolation, as an end in itself."

This ideal of beauty lingered for centuries. In "the miracle of Rembrandt's *Bathsheba*," painted in 1654, "the nude is of a young woman, clearly intended to be physically attractive." Simon Schama refers to her "dewy, roselike beauty" and calls her the most beautiful nude of Rembrandt's career although also "a vessel of tragedy." For Clark, "this ample stomach, these heavy, practical hands and feet, achieve a nobility far greater than the ideal form" of some Renaissance nudes. Ingres at age 82 painted the *Bain Turc*, a "whirlpool of carnality" full of exquisitely soft women "in attitudes of relaxed sensuality unparalleled in Western art," exuding "languor and satiety." Renoir's style grew beyond the "fair, round girls" of his early career, at last "creating a new race of women, massive, ruddy, unseductive, but with the weight and unity of great sculpture."

Clark means that they are unseductive in their poses, not that they are not beautiful. In fact, it doesn't take much gazing at these images to feel a subtle shift in one's own ideal of female beauty, and a growing embarrassment at the pubescent linearity of the bodies—childlike, really—that we now hold up to the light. Twenty-first-century women—real women, obsessed with and oppressed by unattainable, meager girlishness—should surround themselves with reproductions of these great paintings and drawings. If you can't change the superficial way the world sees you, at least look at yourself through Titian or Rembrandt's eyes. Or through the words of Nobel laureate Wislawa Szymborska in her poem "Rubens' Women":

> O pumpkin plump! O pumped-up corpulence
> inflated double by disrobing
> and tripled by your tumultuous poses
> O fatty dishes of love!

The male ideal in Western art was always more muscular, but in Asia and Africa the middle-aged man's paunch has long been a sign of health and power. In ancient Chinese medical theory a rounded abdomen signified health. Among the Massa of Cameroon, the ideal for men was such that some were fattened deliberately on milk and red sorghum, to the tune of a staggering 13,000 calories a day, gaining as much as twenty kilos in a relatively lean season when others lost or maintained weight.

As for women, if you calculate the extra energy stored in fat by a typical Renaissance siren, one of those "fatty dishes of love," it is just about the amount required to carry through a pregnancy and a year or two of breast-feeding. Approximately 80,000 calories are needed to take a woman through a successful pregnancy, while breast-feeding demands between 500 and 1,000 calories per day above a woman's normal intake. At a conversion rate of 9 calories per gram of fat, the pregnancy alone would burn almost 9 kilos, or over 20 pounds of fat. Thirty pounds would take her through the first few months of breast-feeding.

Of course, in almost all cultures women continue to work for a living through much of pregnancy and lactation, so they don't need to cover energy needs completely by burning stored fat. Instead, we are looking—or rather, our male ancestors were looking—for a cushion against the harshness of preindustrial life. The twenty-first-century actresses and models we idolize would be a poor bet to bring a healthy infant to term, much less to lactate. But the women whose images grace our great museums had just the ripeness needed. Among our ancestors, the woman who matched this ideal would have enough energy in the bank to make childbearing safe and healthy without slowing down a very active life. The men attracted to her were no fools. What appealed to them—unconsciously, since they didn't do the calculation—was reproductive readiness. Why moderns like women who look like girls remains a mystery and a tribute to the role of culture and learning in setting sexual responses. Women now signal high status by looking as close to the edge of starvation as they can—pubescent girls for the first half of their lives, social X-rays for the second—they are that sure of their future wealth and comfort.

Most of our ancestors, unfortunately, could not keep enough weight on to attain the Titian ideal. Even the !Kung are no exception to the rule about shortages. In the 1960s and 1970s economic anthropologist Marshall Sahlins made a habit of referring to them as "the original affluent society"—a strange way to describe a group of people with a 50 percent childhood mortality rate, resulting in a life expectancy at birth of thirty years. To be fair, Sahlins was referring to their apparent dietary sufficiency, their seemingly adequate leisure time, and above all their sense of satisfaction with their lives. But all three claims are controversial.

Excellent studies by Richard Lee showed that the !Kung spent just a few hours a day, a few days a week, in the food quest; that they had many leisure activities; that their diet was well balanced; that they did not exhaust their environment's food supply; that their caloric intake was just above the minimum needed for their size and weight; and that they did not aspire to the more well-to-do herding and agricultural life of their Bantu neighbors. Subsequent work, however, called some of these findings into question. Spending just a few hours a day and a few days a week in the food quest is impressive, but many more hours are spent making tools and weapons, curing skins, preparing and cooking food, making clothing, and planning future hunting-and-gathering expeditions—none of which was included in the initial research on !Kung work. If what lawyers and judges do is work, then when the !Kung sit up all night at a meeting debating a hotly contested divorce, they are also working. If what psychotherapists and ministers do is work, then a !Kung man or woman who spends hours in an enervating trance trying to cure people is working as well. Furthermore, the !Kung are often ill, with physical complaints apparent to anyone who visits them. They suffer endemic diseases including malaria, gut infections, parasites, and tuberculosis, among others. Most women spend the years from nineteen to forty-five either pregnant or nursing, a further major drain. Considering these facts about physical condition, we must also ask whether some of what looked like leisure to earlier investigators was perhaps just not feeling well. When people are feeling poorly they may not work, but that doesn't qualify as leisure.

As for available food left unused, that claim also requires scrutiny. Palatability and ease of access enhance eating, especially in the obese but also in normal people. Mongongo nuts are tasty and nutritious—they are the !Kung staff of life—but even a !Kung can eat only so many of them. If a woman who has eaten little else for a week straight declines an opportunity to take yet another ten-mile trek to the farther mongongo groves in the heat, carrying a child, and even chooses to skip a meal that day instead, this is not necessarily evidence that she is affluent. Perhaps she has merely made a cost-benefit analysis that allows the nuts to rot on the ground. Shortages of food were probably seasonal, and Edwin Wilmsen, who studied !Kung diet in the 1970s and 1980s, concluded that annual shortages result in significant weight loss (five to ten pounds) just as in Gambian farmers. In the end, both Lee and Wilmsen have a piece of the truth: Lee helped correct the widespread impression that hunting-and-gathering life was an unremitting, desperate search for food, but Wilmsen showed that it is not ideal.

Nancy Howell, a demographer at the University of Toronto, analyzed the !Kung population and found that food shortages help to explain its very slow growth. Her model of infertility draws on that of Rose Frisch. According to this widely accepted theory, fertile ovarian cycles are unlikely below a certain minimum level of body fat. Although the !Kung picture is not this simple, caloric insufficiency probably plays some role in lowering their fertility, by helping to lengthen birth spacing to four years. And then there are the mortality figures. How Sahlins could call such people affluent seems puzzling, but the argument goes something like this: the !Kung have lived in these same circumstances for thousands of years. Their continued existence in their present ecological situation would be impossible without high mortality, and they are used to it.

I do not buy this argument, and neither do the !Kung. Marjorie Shostak's book *Nisa* documented the life of a !Kung woman from her own narrative at age fifty-five, supplemented by Shostak's annotations. It was perhaps the most intimate life narrative ever collected from a "primitive" person. Together with the follow-up study, *Return to Nisa*, which adds another fifteen years to the story, the account achieves unprecedented insight into the !Kung view of their own lives. Clearly they were not satisfied with their lot. They are neither at peace with nor inured to the many losses those bleak mortality curves deliver, and they are quite envious of people who are better off. Still, they are tough, good-humored, resilient, self-possessed, and generous. They are not self-pitying and they do not allow their poverty or the conditions of stress they endure to destroy their joy in life. The !Kung, with far greater challenges, generally whine much less than the average upper-middle-class American does in a mild recession or even during a gasoline price bump. To provide some idea of the absolute differences in these circumstances, there is little doubt that perhaps not the poorest 5 percent of Americans but the next poorest 5 percent would seem to the !Kung to possess fabulous wealth, comfort, and safety. Imagine sleeping in a bed! Imagine eating fruit that has more flesh than pit! Imagine a 95 percent chance that your child will live!

For the first few months after returning from my two years with the !Kung, I used to hear a phrase in my mind, in the !Kung language, one that would often have been on the lips of a !Kung, if one had been with me: "Rich people, everywhere rich people." I remember being in Harvard Square—one of the busiest corners in the world—on an ordinary autumn day, watching someone get out of an ordinary car in ordinary clothing. I stared and said it aloud: "Rich people, everywhere rich people." For years every time I scraped a plate into the sink—from the most modest of meals, and meals that by American standards were quite thoroughly eaten—I would hear one of my !Kung friends asking me, "Are you a person who destroys food?" It was hard to throw out orange peels; !Kung women saved them to make perfume.

And yet, if they were to change places with the American poor, or even for that matter with the American rich, there is truth in the thought that they would not know what they were giving up. They would be trading a life of resilience and mutual support under the hammer of environmental exigency for a life of relative safety and mutual isolation, in which the quintessential social act is comparing oneself with someone else. As for the notion that because of their cultural background and upbringing they would not be utterly changed by the economic transformation, it is quite implausible. A !Kung family, transplanted to Cambridge or Atlanta, would be capable of leaving their old life behind and starting to focus their hopes and dreams on the difference between themselves and the movie stars; capable, even, of keeping up with the Joneses.

The !Kung gave every evidence of this in their traditional circumstances. Selfishness, arrogance, avarice, cupidity, fury, covetousness, all these forms of gluttony are held in check in their traditional situation in the same way simple alimentary gluttony is: it doesn't happen because the situation doesn't allow it— not, as some suppose, because people are somehow superior. Once a !Kung man— about forty years old, a father, well respected in the community, a good and substantial man in every way—asked me to stash the leg of an antelope he had killed. He had given away most of the carcass, as stringent custom required. But he saw a chance to hide some of it for later, for himself and his own family. Ordinarily, there would be no place to hide it in the entire Kalahari; it would either be prey to vultures, hyenas, and other scavengers or unsafe from demanding distant relatives. But the presence of foreigners provided an interface with another world, and he wanted to slip the meat, temporarily, through a chink in that invisible wall, into the only conceivable hiding place. I let him see my discomfort, but even knowing all I knew, I could not refuse him. We stuck it in the fork of an acacia tree near our hut, in plain view, and no one asked about it. It was in what anthropologists call a liminal space, a twilight zone between two worlds. He had schemed his way around a cultural norm that gave no quarter to greed.

Ordinarily, though, !Kung culture prevents individual ambition from getting out of hand. Economic anthropologist Polly Wiessner has shown, not only for them but for foraging peoples generally, that "meat hunger" confers a certain degree of status on the best hunters. This confirms Megan Biesele's demonstration, in *Women Like Meat*, that !Kung hunters are praised in song and story, as well as dramatized and occasionally mocked. She refers to "the peculiarly intimate identification between hunter and prey in Bushman belief," a relationship also evident in rock paintings in Southern Africa. This echoes the themes of the great cave paintings of Lascaux, Altamira, and Chauvet. The phrase "women like meat" came from the !Kung themselves, explaining why a certain old bachelor was unlucky in love. Summarizing this and other evidence, ethnologist Lorna Marshall wrote: "A boy who never killed any large meat animal would not be given a wife."

Among the Efe Pygmies of Zaire, studied by Robert Bailey, "Men spend more time hunting in the forest than in any other subsistence activity, and the meat they acquire on hunts is a very significant food item. . . . Meat contributes approximately 9 percent of the calories and 48 percent of the protein." The central role of hunting in their culture is vivid and clear. Efe men gain a modest advantage in economic exchange through hunting prowess, and this may facilitate finding a bride.

For the Aché of Paraguay, studied by Kim Hill and Magdalena Hurtado, hunting provided more than three-fourths of the calories, a reversal of the most common pattern for tropical hunter-gatherers. Yet "resource redistribution within Aché bands is more widespread than that documented for any other group of foragers. Hunters do not usually eat from their own kill, and nuclear family or kin are no more likely than unrelated band members to receive a share of meat." For the Sharanahua of eastern Peru, the densest part of the Amazon jungle, Janet Siskind found a certain ambivalence: "Sharanahua like to bring meat home, but the men are not enthusiastic about hunting. 'We hunt,' they say, 'because our wives and children cry for meat.'" Although there are often differences of opinion between the sexes about hunting and meat, the importance of meat is not tied to maleness. Even among the Agta of the Philippines, the one culture where women do about half of the hunting, meat has a central ecological and psychological importance.

The Hadza of Tanzania, like most hunter-gatherers, "attach particular value to meat and honey," and objective studies by Kristen Hawkes and her colleagues show that, despite the importance of plant foods, big game has a big nutritional function. Although restrained from boasting, Hadza men don't hesitate to sing of their adventures:

> You, big zebra, come here!
> You, in back,
> Just run
> To the water.
> So I can kill you.

This song seems to express frustration in the course of hunting, as does another:

> I am stalking the big giraffe.
> But he recovered.
> I hit him right in the lungs.
> I hit him right in the lungs.

And as with the !Kung, Hadza stories often dramatize the adventure, skill, danger, and humor of the hunt.

Still, Wiessner showed that, in twenty-five foraging societies, "when status is accorded to good hunters, its privileges are strictly curtailed to prevent it from being turned to dominance." There are clear rules of distribution, but there are also devices like forcing hunters to borrow other men's arrows, so that official ownership of the kill falls to the arrow lender. Meat distribution is characterized by a sense of duty, and the hunter's efforts are often ignored or belittled. Encouraging a sense of indebtedness in those who receive meat is out of the question. Meat is given to cohunters and kin, and nothing is expected or accepted in return. The hunter and his family do not get a larger or more preferred portion. Of course, they can rely on reciprocal treatment under the same rules. As Wiessner says, "The hunter spends his life hunting for others, and others spend their lives hunting for him." And Megan Biesele says of the !Kung:

> Reasoned, ritually disciplined, highly social hunting involving a lifetime of learning and of sharing information, with shared arrows denoting shared protein needs, underscores the necessary social relationships within small groups of hunters whose members have absolute reciprocal dependence upon one another.

Although hunting ability is not distributed equally, and although this may have consequences for regard, respect, and reproductive success, it has no consequences for the accumulation of wealth, which in foraging cultures is virtually impossible. In this way our ancestors, simple hunter-gatherers for most of our species' history, kept a lid on greed and on the exercise of power. But with the rise of *complex* hunter-gatherer societies—such as those in Europe during the late Paleolithic or among the Native Americans of the Northwest Coast in recent centuries—the resource base became rich enough to allow accumulation. Whether based on an abundance of caribou or mastodon on the Ukrainian or Iberian plains or just the vast harvest of mussels and other shellfish from the Pacific, combined with the thick rush of salmon every fall, these cultures could support dense populations. The strong could disadvantage the weak and keep their emotional distance. Social strata emerged, as did subordination and in some cases even slavery.

In these cultures food was used as a tool of domination. Grand feasts were held by those wealthy enough to throw them, with elements of ritual, self-aggrandizement, competitive display of wealth, and a spreading incursion of indebtedness. The classic example is the potlatch of the Kwakiutl Indians, in which competitive displays of wealth eventually reached the extreme of destruction of large amounts of property. With the advent of horticulture and pastoralism, feasts of various types become a way of life, a device for transforming the reciprocity of simple foragers into something much more hierarchical. In New Guinea, the "big man complex" rests on feasts and preparation for feasts, in which a man who aspires to any political influence must recruit many supporters, who pool their pigs and assign him the right to redistribute and at times slaughter them to feed allies and relatives. Wiessner and New Guinea colleague Akii Tumu, in a book that is a model of anthropological history, documented the transformation of Enga society over two or three centuries before European contact in 1939. The introduction of sweet potatoes as fodder greatly intensified pig production, leading to shifts in population and alliance, warfare and ritual on a vast new scale, and exchange networks that encouraged political hierarchy and challenged a previously egalitarian world. And in the classic ethnographic case of the Trobriand Islands, the rules of yam distribution after the harvest, as well as those for the flow of precious heirlooms, jewels, and other valuables around the South Sea islands of Melanesia, provide an opportunity for able men to establish social influence, as well as for those from important lineages to consolidate their positions. Wulf Schiefenhövel and his colleagues at the Max Planck Institute propose this as a basic mechanism in the rise of political society.

Recall that at the receiving end of each of these cultural systems was a poised array of hungry bellies. The cases mentioned so far represent the power of food in producing social complexity *before* the rise of the state. As agriculture became more centralized, depending on irrigation and the hierarchical organization of work groups, the power of small elite groups of men over those hungry bellies relentlessly grew. As is evident from the Bible, Homeric epics, and other ancient texts, ritual feasts were a mark of great men and a way to display and consolidate wealth and power. But that was just the beginning. Taxation, corvée labor, and other systems of redistribution became the groundwork of states and the funding devices of armies. Conquest subordinated more surrounding peoples as the states became empires, and alliances among landowners, merchants, generals, and priests controlled the lives and deaths of millions. These processes came later in the New World than in the Old, but with such pervasive and predictable similarity that anthropologist Marvin Harris called it "the second earth," a testing ground for hypotheses in the science of social complexity.

The use of food as a weapon pervaded the twentieth century, and in the developing world it persists in the twenty-first. Famines are relative, not absolute, and their functions are often political. But the direct control of food is only the most obvious continuity between the ancients and ourselves. The arcane legacy of ten millennia of combat and domination over agricultural land still shapes the world today, pervading every aspect of international politics. Geopolitics is hopelessly burdened with the habits of thousands of years. Biologically and psychologically, we may be simple hunter-gatherers. But culturally we are Babylon, Rome, Shang China, and the Kwakiutl potlatch. We love abundance and the growth abundance gives us, but it finally drives us places we may not want to go.

So in retrospect, the !Kung situation tells us much about gluttony, both the literal and the figurative kind. Natural selection could not provide us with an effective mechanism for keeping our weight down in times of abundance for the simple reason that it was giving us quite the opposite, a system that piles on excess fat in times of abundance, stores to draw on during shortages. Since there was rarely *continuous* abundance during the whole of human evolution—and certainly not combined with physical indolence—natural selection cannot have prepared us for such conditions. Similarly, with the metaphoric forms of gluttony, we were built to pursue targets of opportunity, to recognize our wildest material fantasies, our instinctual wishes of the soul, for what they were—dangerous and impossible 99 times out of 100, but perhaps doable once in a blue moon. And what happens when they become doable 10 or 20 or 30 times in 100? More or less what we see now—the brightest florescence of narcissism since imperial Rome. The obesity epidemic is merely the most obvious result of self-stuffing, in which our wish fulfillment piles flesh on our bones. As instinctual wish fulfillment goes, it is hardly the most dangerous.

The *dangerous* gluttons stuff themselves with land, money, power, with other people's happiness and safety. "Greed is good," says Gordon Gecko in Oliver Stone's film *Wall Street*. The model for Gecko, Ivan Boesky, said something almost identical in a commencement address at Stanford. Boys and girls just completing the best education in the West cheered loudly. But Boesky later landed in prison because his greed had gone over the line, as did Michael Milken's, the billionaire inventor of junk bonds, and Charles Keating's, the worst offender in the savings and loan scandal. All had found ways to separate countless small people from their money, and all paid a price. But scores of other not-quite-so-big fish swam away comfortably with their gains. And even as capitalism's biggest crooks served their prison terms, the worldwide fall of communism became a rationale for an even more heartless brand of capitalism, and for a safer, more widespread, equally "good" sort of greed.

For a while at the turn of the millenium, the triumph of capitalism over communism looked as if it might become a Pyrrhic victory. The constant, cumulative waves of crisis in the Russian economy pounded at distant American shores and made stock exchanges seasick the world over. The Asian economic miracle became the Asian contagion, and the promise of Latin America's dynamic, growing markets likewise weakened. As the century turned, left-of-center governments ruled in almost all of Western Europe, something few would have predicted in 1990. And in the year 2000 the Western technology bubble burst, dissolving trillions of dollars in wealth and tipping the world toward recession. In the United States even conservatives had to claim to be compassionate. Was capitalism stumbling? Were the great Western democracies rebuking it? Of course not. But capitalism was learning a lesson or two about the invisible hand and its limits in shaping a new world order. The invisible hand builds and heals, but it also swipes blindly about and causes great pain.

Economist Joseph Schumpeter called it creative destruction, but he neglected to mention that what is destroyed has a face and a name. As for the great democracies, they—*we*—were asserting something simple: we would not let the devil take the hindmost. We would sometimes guide or hold back the invisible hand. We would not indulge, at least not too much, in the vanity of guarding our privileges in isolation from the suffering of others. This, in addition to being morally discomfiting, was in a connected world increasingly impossible. We would acknowledge that greed was necessary if not exactly good, but we would harness it, not let it run wild. We would insist on capitalism with a human face. By the new millennium the whole world had decided: a furnace of greed must burn in the economic house. But that doesn't mean it can be allowed to burn the house down. And the great question remains: How high do we build the floor under the poor?

So far, not very high. Grain rots in the storage silos while millions starve in . . . somewhere. Can we suppose that as the green revolution contrives to provide a glut, we will automatically get it efficiently distributed to the hungry? When war comes, men will be men, we will meet the occasion and cover ourselves with glory. Haven't we always? History offers us an endless confirmation of such courage in the face of the flood, in the very eye of the hurricane. But can we rise to the call of human decency, when that call seems very faint and we hear it every day, when even to stop and cock an ear seems to risk a slip on whatever ladder we're mounting? Can we cease to think of ourselves even long enough to hear that the call, however faint, is anguished?

Consider one of the pinnacles of high-consumption culture. On the beach at Malibu are miles of tinsel mansions that look much as we imagine them. Until they were built, the French had no perfect use, perhaps, for their old phrase "newly rich." The verandas are high on stilts above the beach, to evade the water. Most are empty, even in summer, since their owners have other mansions—at Nice, in Manhattan, in Riyadh, in Beverly Hills. One can peek in some of the windows—from a proper distance of course—and see the glitter. The decor is plasticky palatial. A framed gold record clings to a living room wall. In one of the homes, they say, an original Rembrandt drawing hangs in a bathroom. On the beach the gulls are mottled, drab, as if to offset the opulence. A pair of young lovers runs through the dusk, angry, shouting, at terrible odds with each other. An empty bottle on the beach is Mum's champagne.

Every once in a while the sea rises up and slaps those houses down. No rational person would suppose such an event to be the sea's answer to human arrogance, and no decent person would take comfort in it. Nevertheless, one can scarcely escape the metaphor. Here are structures that embody in a reified, real world the whole effortful chaos of human motive, of human hunger and arousal and desire run wild, cast in an earthly form and held up so that all may gaze, and hope, and wish, and even hurt for wanting, so that even the very well-to-do may compare themselves and feel poor, and the average person burn with envy. And as for the poor and hungry, whether in Brooklyn or Gaborone or Laredo or Calcutta, they may go to the movies, and dream of a favorite star in one of the houses in Malibu, or look at pictures of it in a magazine and imagine themselves there.

As for the people who occasionally live there, they might feel at least embarrassed; perhaps they do. But they know that even all their wealth carved up and given out among the dreaming millions would not make a dent in the world's problems. And if their embarrassment should distress them, they have special doctors, just as kings had priests, to give them the world back as it really must be. And most of us in their shoes would be just like them; they are merely our own puzzled selves writ large.

Psychiatrist Robert Coles's study *Privileged Ones*—the fifth volume of his epic *Children of Crisis*—glimpses the inner lives of the scions of some of America's richest families. Perhaps the glimpses are not surprising. We see the richest children in the United States wishing for things they do not have. We see them in difficult moments consoling themselves with a fabulous array of toys and the deep conviction that they will one day have better ones. We see them comparing themselves to others who they presume to be wealthier than they are, and feeling anguish at that thought. We see them—expectably, yet somehow unbelievably—unhappy. We see them easing into an ingrained sense of entitlement. We see them wanting more, and encouraged to do so. Coles quotes a black maid who pities her employers; he pities them, too, and wants the reader to. But as the saying goes, it is better to be rich and unhappy than poor and unhappy. Coles had a physician's responsibility to express compassion toward these children; but for the social critic who wrote four volumes about children bent and sometimes broken by poverty, the transition is uneasy. They are taught that they are better, deserving, right to want more. And of course they are taught how to get it.

They feel compassion for people as for animals. One boy felt sorry for the men who labored in his father's mine but felt that his father, the owner, worked as hard or harder than they did. A statement by a twelve-year-old girl about one of her family's servants is typical:

> We had one maid, and she said we spent more time with the animals than she does with her children. I felt sad when she told me that. She has no understanding of what an animal needs. She was the one who was always telling me I was beautiful, and so I didn't need any lotion on my skin. I wanted to give her the lotion. She needs it. Her skin is in terrible shape. It's so dried and cracked. My mother says you can be poor and still know how to take care of yourself. If our maid stopped buying a lot of candy and potato chips, she could afford to get herself some skin lotion. And she wouldn't be so fat!

This may sound like a spoof of *Gone With the Wind*, but it was the urging of an all-too-real twelve-year-old in the New South in the late twentieth century.

The vast American middle class now practices a version of these child-rearing methods of the very rich. We withhold physical comfort, responsiveness to distress, even love from infants and small children for fear of spoiling them; but when the same children are eight and ten and twelve we spoil the very sense out of them—the sense of proportion without which life is a chaos of desperate wantings. Nature has enabled them to judge between want and need, between fear and need, between anger and need; but these judgments slip because of the uneasy legacy of human evolution, which makes us think that things are never good enough. By encouraging children to want more, by giving them whatever they want, by letting them believe they are superior and entitled, we blur the distinctions for them. We leave them languishing, like Ciacco, in the rain; a Dantesque fate in which the source of life's plenty supplies only torment without end.

PART THREE

Transforming Behavior

Effects vary with the conditions which bring them to pass, but laws do not vary. Physiological and pathological states are ruled by the same forces; they differ only because of the special conditions under which the vital laws manifest themselves.

—CLAUDE BERNARD, *Introduction to the Study of Experimental Medicine*

Freedom does not consist in the dream of independence of natural laws, but in the knowledge of these laws and in the possibility this gives us of systematically making them work toward definite ends.

—FRIEDRICH ENGELS, *Anti-Dühring*

CHAPTER 16

Change

We cannot adapt to everything, and in designing a way to the future we would do well to examine again what we are and what our limits are.

—JEROME BRUNER, "The Nature and Uses of Immaturity"

In the late 1990s a self-trained developmental psychologist named Judith Rich Harris created a scholarly stir that reached the national media. In a prizewinning article published in one of psychology's most intensely competitive journals and in a book called *The Nurture Assumption*, Harris challenged one of the most enduring and cherished beliefs of psychologists and parents. The article began: "Do parents have any important long-term effects on the development of their child's personality? This article examines the evidence and concludes that the answer is no."

Harris was a maverick, but her conclusion was respected. She won a prize for best paper of the year—in a crowning irony, named for the professor who had kicked her out of Harvard's graduate school decades earlier. Although not exactly embraced when the book came out, Harris did not stand alone. Many psychologists supported her. As the devoted mother of two adult daughters who were apparently doing well, she was not inclined to discount her own contribution. But she felt forced by the facts to a simple conclusion: "Parents matter less than you think and peers matter more." The argument is surprisingly hard to refute.

The last public debate of this kind was begun in the 1970s by Jerome Kagan, then as now one of America's leading child psychologists. Kagan's view was dramatically affected by a cross-cultural study he did with Robert Klein in Guatemala, in which they showed that on many basic mental tests, relatively deprived village children in that country eventually caught up with American children, in spite of a time lag. This was about universal cognitive functions, not (as in Harris's book) individualized personality. But the conclusion and the public impact were surprisingly similar, although separated by a quarter century: child-rearing practices matter much less than you think. I eventually accepted much of Kagan's reasoning, but only after painfully rethinking my own basic beliefs.

By chance, on a late summer night in 1998, while driving through a Pennsylvania thunderstorm, I heard Harris and Kagan debate on National Public Radio. Like Harris's prize, this was a delicious irony. Kagan was saying, almost word for word, the same things I (and many others) had said to him during the 1970s, and he was not doing well; Harris was all over him. About the best he could do was adopt a wise professorial tone while Harris ticked off the evidence, or rather the lack of evidence, for strong parental effects. In that intervening quarter century, genes had become common currency (Kagan himself was studying genes for timidity) and Kagan's perspective from the seventies had seeped into psychology and the wider culture. Kagan, in a sense, was now forced to debate with the echo of his own youthful skepticism.

Harris is not right, exactly, but she has a piece of the truth. Psychologists have irresponsibly touted parental influence as the major determinant of almost everything important. In their consistent bias ("the nurture assumption") they have belittled the importance of universals of psychological development, of genetic contributions to individual variation, of cultural factors other than parents, of environmental factors other than culture, of formal chaos (sensitivity to initial conditions) in development, and of luck. Kagan, in the early seventies, was calling attention to universals, driven by underlying brain growth; Harris, in the late nineties, was calling attention to genes and, for the environmental component, to peers.

Harris called her approach group-socialization theory, and her claim was that most of the nongenetic contribution to individual personality comes from the way children take after peers and the way peers respond to them. Such effects certainly have their impact, but this is not news. My youngest daughter's elementary school has a program called "Peer-Proof," based on fear of that influence. Peer behavior is among the strong predictors, along with hormones, of the onset of sexual intercourse in both boys and girls. As for good peer influences, psychologists such as Willard Hartup of the University of Minnesota have been documenting them for decades, and Ann Kruger has proven experimentally that peer interactions *enhance* moral reasoning in young children. Recall, too, the classic monkey studies: deprived infants with only peer-play opportunities grow up in better psychological shape, in some ways, than those given only mothers and no peer contact.

Group socialization is central to what anthropologists call culture, and they have for generations documented its independent, persistent power. Yet despite the coherence and durability of culture, anthropologists enter their exotic worlds on tiptoe, frightened that they might inadvertently change something. Indeed, television antennas have sprouted from thatched roofs throughout the world. So we have a paradox: How can something supposedly so durable also be so changeable? One possibility is that parents and teachers provide most stability and peers provide most change. This could give researchers the erroneous impression that peers do much while parents do little, since difference is what research usually tries to explain.

But what really allows the underestimation of parental influence—and indeed of all environmental forces—is that the families studied by psychologists are so similar to one another. As any basic-statistics student can tell you, if two causes influence an outcome, then holding one of them constant highlights the other's effect. So it is not surprising that when family environments in a sample are very similar, the effects of the genes look larger. Family environments cannot appear to have a strong effect if the variety between families is trivial. That is why cultural anthropology was invented. By enormously widening the spectrum of learning environments, it highlighted the impact of their effects.

Ignoring culture can make biologists look silly. Claude Fischer, a sociologist at the University of California, wrote a wise letter to *The New York Times Book Review* in 1998, in response to Pulitzer Prize–winning science journalist Natalie Angier's review of the book *Twins* by Lawrence Wright. The review ended by quoting Wright: "There has simply been nothing on the environmental side to counter the power of twin and adoption studies" in understanding "who we are." Fischer's answer:

> Consider that Americans living today have essentially the same genetic makeup as their grandparents and yet lead immensely different lives, hold starkly different values and even display different cognitive skills. Or consider the social and psychological gaps that develop between immigrants and the kin they leave behind—or even between immigrants and their own children.
>
> The effects of environment are so powerful and omnipresent that we fail to "see" them. . . .

We had better see them, since despite all our gene talk, they remain our best hope for the future. In fact, one can extrapolate the point backward for another five millennia. Significant genetic change in the species in that time frame is unlikely, but the behavioral transformations have been no less than astounding. And during the twentieth century alone, average IQ scores—which some naïvely believe reflect pure genetic potential—have actually increased, quite steadily, almost three points per decade. This phenomenon, named the "Flynn effect" for the psychologist who discovered it, is the most mysterious in the field of mental measurement. It stands as a deep challenge to our understanding of how human minds adapt and change.

The problems are stronger still. Sociobiological theory, for which I often give two cheers, has so far failed to resolve the sharpest challenge to its models: the demographic transition. In every country that has undergone even partial modernization, the rise in standard of living has brought about a fall in death rates, followed after a generation or two by a fall in birthrates. So, contrary to Darwinian predictions, people in better and better circumstances are steadily lowering their birthrates. Unfortunately, an overall decline may not come soon enough to save the human species. But aside from that concern, how do we reconcile the infamous selfish gene with the inconvenient fact of self-limiting reproduction? One possibility is that the transition is not over and that the genetic lines in the best circumstances are just biding their time, destined to once again be fruitful and multiply in some future generation.

There is evidence that the wealthiest Americans have more children than the basic middle class does. But recent research failed to support the biding-time hypothesis. People in the West just seem to be reducing their fertility. Of course, the difference in the future, in evolutionary terms, may be between having one child and having none. That alone will give selection immense power. But for now we have to admit we have a paradox. Rich countries like Italy and France, where people have real choices, have birthrates far below replacement. Theories of why abound, but they all show the power of culture over genes. *Ideas* about a better life, about focusing on one or two children, and about the desire to stay youthful have swamped any tendency to produce large broods. It may be a sociobiological strategy, but it is being realized through massive cultural change—behold the power of culture over 4,000,000,000 years of biological evolution. This is what makes human behavior hard to study; there is nothing more complex in the known universe.

Consider a system that is constantly changing, according to laws both known and unknown, from causes both internal and external, in a manner both cyclical and progressive, by processes both reversible and irreversible. Allow it to pass through an inconceivable number of states, and to come to rest for varying times in any of them. Give it many potential reactions to a given input, including changing, ignoring, and terminating that input. Enable it to reproduce itself through functions that have entered the design exclusively because they serve that purpose, though often indirectly and at the cost of other purposes. Endow it further with a trajectory of fixed maximum length (say, ninety years), which carries the system, predictably, through a series of potential or actual states from nonexistence to final cessation of function, with a possible termination of function at any earlier time. Finally, build in a sensor that can detect where the system is in the trajectory, assess the chance of continued functioning, and react, as far as possible, to change that chance for the better, except—and this is a crucial *but*—where that conflicts with the goal of reproduction. We now have something approaching, at least in its outlines, the complexity of the human behavioral system. We must add, of course, the potential for malfunction that is common to all systems, whether because of design flaws or unpredicted stresses.

And we must add, too, the prospect of change. If I did not believe in change, why would I write or teach? I would find a more lucrative, less taxing line of work. Indeed, it is impossible to summarize even a book like this, steeped as it is in biology, without a strong conviction that behavior can change. The question is how. "Behavior can be changed by changing the conditions of which it is a function," Skinner said. True enough. In fact, our future depends on it. But contrary to his narrow credo, those conditions range far beyond his own tiny county of reinforcement learning over the whole map of behavioral science. For it is not just the dolloping out of scheduled rewards that shapes behavior but also a host of other forces that can in a trice, in the right "conditions," swamp the effects of years of careful training.

The processes outlined by learning theorists beginning with Pavlov are real, widespread, and powerful. They emerge early and become more complex as we grow. A newborn baby startles at a noise, but if it is repeated without danger, she *habituates*—the startles soften, until she can sleep soundly through the thunder. When she wakes, brain stem reflexes direct her gaze to the general shape of a face, but as she stares at one or two very important faces, her brain *associates* their elements into wholes engraved in the circuitry. Reflexively at first, she turns toward the breast when it brushes her cheek, but in time the turning becomes *conditioned* to the mother's voice and smell. Crying gets results that *reinforce* it for a time, until brain growth replaces it with smiling and cooing, later reinforced in related ways. Early in the first year *mimicry* is possible, and in the middle of the second year, true *imitation*.

As the brain grows, the child experiences changing reinforcement patterns, and instead of being baffled and held back, she *learns to learn*, adapting quickly to new reward conditions by transferring a general skill, a trick, of *how* to learn. In early childhood, *language*, including internal, self-directed speech, becomes a vehicle of reward and imitation, shaping behavior toward expectations. *Identification*, an emotional focus on a person or kind of person the child wants to resemble, provides a deep basis for ongoing imitation. And after age seven or so, the uniquely human process of *collaborative learning* takes a mental or moral challenge out of the realm of a single brain, placing it as it were in the air between two children, as they comment on each other's ideas and invent a solution neither could have come upon alone.

On these processes, our collective fate depends. There will be no magic genetic fix, no technological sleight of hand, no escape from the thin skin of this planet. We can use new technology—in itself a process of learning—but ultimately we must rely on ourselves and on one another. We can take control of our evolutionary future not mainly by altering genes but by changing the conditions under which we will evolve. Indeed, if we don't learn massively and collaboratively soon, we will complete the mass extinction we have started and thoroughly stunt our own evolution. Collaborative learning, a uniquely human ability, is our best hope for the future, as we jointly attend to our environmental crises and try to create a livable world.

This is the sweeping overview, but as usual the excitement is in the details. Most fundamental, perhaps, is associative or perceptual learning. In the simplest laboratory setting, an animal encounters a set of stimulus elements, presented together again and again, and in time *constructs* them together in its brain. In the real world, this is how a melody becomes recognizable, how a perfume evokes a certain romantic bond. Through associative learning, an infant configures her mother's face as a consistent, coherent percept and ties it in her mind to a certain gentle voice.

A second kind of learning is classical conditioning; we pay homage to it whenever we mention Pavlov's dogs. Start with an inborn, or at least a very well-established reflex—say, the production of saliva in response to the smell of food. Ring a bell when the food comes, and in time the dog will salivate to the bell. This is the same conditioning process that brings a woman to the point at which she may, after a few days of nursing, wet her blouse with milk upon merely hearing her baby cry, or that brings some men to orgasm through fantasy alone. Or, to take a more common example, if a person has consistently approached us with tenderness, that person's mere approach may in time become sufficient to make us feel what, at first, only the tenderness itself could make us feel. The process may begin with habituation of fear, but it goes on to classical conditioning of warmth, love, joy.

Yet another kind of learning—the one Skinner focused on and which he grandly thought could encompass all behavior—is operant conditioning. Here we begin not with a stimulus but with a naturally occurring act. Skinner called it an operant, but it is simply whatever the animal may do. A rat in a box will walk around, lift its head, sniff, occasionally rear up—all operants. If there is a lever around, sooner or later the rat will press it. Any of these behaviors can be made more or less frequent by following them with certain stimuli, called reinforcers. Those that increase the frequency of operants are called rewards, those that decrease it, punishments. We know a lot about how various schedules of reinforcement affect the rate and stability of learning. Some of these findings are obvious: more consistent reinforcement produces faster learning. Some are not so obvious: less consistent reinforcement produces a behavior much more difficult to extinguish, that lasts longer after reinforcement is stopped. This may help explain addictive gambling.

Some of the laws, at least—just as Skinner predicted—apply to humans and pigeons alike. The first successful treatments for the autistic and the retarded have been based on such knowledge. Certified behavior analysts, trained in the Skinner tradition, can provide therapy through which patients become not mentally normal but capable of toileting and feeding themselves, often for the first time. Behavioral psychotherapy for milder psychological conditions, also derived from Skinnerian principles, is a widely accepted form of treatment. The evidence that it works in some situations is quite good—about as good as for other forms of psychotherapy. These cases include fear of flying, fear of spiders, fear of blushing, aggression in retarded children, and auditory hallucinations in schizophrenia.

But these specialized applications barely scratch the surface. There is not a single process having to do with the expression of the emotions—and not a single behavior—that is not subject to some degree of operant conditioning. Under most circumstances, when people we approach reward us in some way, we will approach them again and again; and when people punish us, even with a word or a glance, we avoid them. Paychecks make most of us work steadily, and bonuses strongly reinforce our underlying commitment. Harsh and kind words from coaches and trainers shape champions.

But current learning theory goes much farther. Duane Rumbaugh of Georgia State University, who has been a leader in animal psychology for five decades, has begun to reconcile learning theory with cognitive science. Rumbaugh has shown that with brain evolution there is an increase in mental power that overshadows Skinner's universals. Extending Harry Harlow's concept of the learning set—the ability to apply what we have learned to new and different tasks, making new learning easier—Rumbaugh developed a transfer index of animal mental abilities. He showed a clear correlation between relative brain size—extra neurons—and the ability to generalize over different learning tasks. Some mammals with simpler brains actually do worse in new learning challenges if trained in other, related tasks before. The prior experience *interferes* with later adaptation, because they can't get off the track laid down by past training. But for monkeys, and even more for apes, the prior experience is a clue. It's as if they think, "Oh yes, I've seen this kind of thing before, but I can't just keep doing the same thing. It's different. I've got to figure out something new." They step back from the specific task and begin learning again. They have learned to *learn*, the first step in what psychologists call metacognition, thinking about thinking, brought to fruition in the human mind. So Rumbaugh's transfer index measures the evolution of mind and the fact that in brain evolution, *more* is *different*. Incremental change eventually produced a qualitative increase, laying the groundwork for human thought.

But simple processes don't disappear when complex ones are added. Habituation, long a research focus of ethology and neurophysiology, is the gradual waning of an unconditioned reflex or at least a narrowing of the stimuli that trigger the response. A young frog at first darts its tongue at any speck that drifts past, but gradually it comes to ignore wind-borne debris and to target insects. A person will startle at thunder when a storm begins, but on subsequent thunderclaps the response habituates and dampens. A woman may well be wary of a man who approaches her with obvious sexual interest; but mere time spent in his presence without an untoward event can do a good deal to extinguish the fear, which finds itself, like the frog's tongue, misdirected—at least in the short run. Much of the emotional change of the early part of courtship is due to this dampening of fear. Habituation, in effect the converse of Pavlovian conditioning, weakens even instinctive responses that prove superfluous in facing reality; it is a ubiquitous kind of behavioral change.

Exercise, play, and practice strengthen associations or stimulus-response chains already in place, whether due to genes or learning. The process here is association *within* the organism—the linking of responses to each other rather than to external cues. Every complex behavior can be smoothed through exercise, which strengthens links from response to response, from response to stimulus to response, and within the neural circuits that manage the sequence. For example, when a baby's newly matured toddling is exercised, she is forming a series of links between the sight and feel of the floor at certain moments and the action of the legs and feet at the next—but also between this move of the ankle and the next move of the hip; most impressive, she is also smoothing and strengthening the neural circuit that steers walking from the spinal cord and brain. Something related is going on when we learn to play or sing a new melody, or for that matter when we become better at infant care, infantry combat, or lovemaking.

But there are several well-documented ways in which an internal, mental process intervenes to alter the stimulus-response connection. For example, observational learning occurs in many mammals and birds. The chance to observe a member of one's own species (preferably a familiar one) perform a response speeds subsequent learning by trial and error. It is one basis of formal training in many cultures, and of informal learning in all of them. And observation, like other learning processes, can produce undesirable results, as in the effects of advertising and peer pressure on teen smoking. Of course, imitation does not always involve learning. Also, it is difficult in some cases to distinguish it from social facilitation—the basis of a wave of yawns or coughs passing through a theater audience. Here a strong tendency already exists, and observation facilitates its release in a kind of social contagion—not really imitation, certainly not learning. When a female guppy takes an interest in a male because she has seen another female courting with him, is that imitation? Probably not, but it is an extragenetic influence on behavior. Some well-studied cases, such as social enhancement of overeating or the tendency of children to pummel a large doll after seeing an adult on film do the same, are probably intermediate points on a continuum between social facilitation and imitation.

Farther along that continuum is true imitation—the way a professional dancer learns a new dance from a choreographer or a medical student learns surgery. Ann Kruger and her colleagues at Georgia State University have persuasively argued that this kind of learning is distinctively human, because it involves intersubjectivity on both sides. That is, the learner pictures herself as the teacher and vice versa. In true imitation, the learner does not just watch the activity—say, weaving a shawl—but gets inside the head of the teacher, imagining the teacher's hands and eyes as her own, seeing and feeling the warp and woof as the teacher sees and feels them. And in true teaching, the teacher not only sees as if with the child's eyes but then adjusts her own performance of the task to make the next, observable step perfectly child-size. This is scaffolding, and it requires a kind of intersubjectivity that evolved with the human brain.

So observational learning at its most advanced is possible only in humans. Yet as simple a creature as the oystercatcher—a striking, black-and-white seashore bird with pink legs and a red-orange beak—depends for its very food supply on a related process. For two years, the young watch as adept adults open oysters with an intricate, studied routine. Gradually, the young shape their own behavior to meet the stringent demands of survival. But it is not likely that they do it through intersubjectivity. It may be that hanging around the parent for two years during feeding affords the young the opportunity to learn by trial and error, without true imitation but with an assist from observation—a socially facilitated learning.

An equally impressive process occurs in a number of songbirds. Males sing seasonally, as we have seen, under the influence of testosterone. But every species must have a distinctive song, to signal the right females. In some species the song is completely wired-in, but in others the brain has only a vague template of the properly formed song. Deprive the hatchling of experience, and you get only a crude, useless sketch or chunk of the song. In the white-crowned sparrow, for example, two stages of experience are indispensable: first, the youngster must hear his father sing; later in development, he must hear himself practice it. Then and only then will it be the fully formed species-specific song. Not only that, but sparrows along the northern California coast have dialects, based not on genetic differences but on local traditions of slightly different songs passed down from father to son.

Closer to our own learning is the classic case of Japanese monkeys developing the habit of washing sand off of potatoes before eating them. A monkey "genius" invented the procedure, which other females and juveniles, but not older males, gradually copied. Despite this omission, the behavior became established in the troop, an elementary kind of "protoculture." Something similar happened with the habit of floating a handful of grain on a water surface to let the sand sink away from it. Yet it is not clear whether this is real observational learning or merely social enhancement of trial-and-error learning. At a much higher level would be chimpanzee termite fishing, in which the young not only get the chance to practice but stare at older chimps, studying the process. The chimps seem to make a mental effort that oystercatchers and other species do not. Baboons sometimes watch the chimps, but do not learn. And if we move from termite fishing to fly fishing, we are talking about a subtle and intricate *human* cultural process, in which true teaching and imitation provide a quantum advance before trial and error begin.

Most of these forms of learning are straightforward and mundane, but the laboratory has quantified them and formulated laws. More important, it has revealed phenomena that are not obvious at all, and many of the resulting subtleties figure in the emotions. Recall, for example, the helplessness studies. Two rats in adjacent chambers are shocked simultaneously, for the same length of time. But one of them sees a warning light and can turn the shock off. The same act turns off the partner's shock, so both rats get shocked exactly equally. The rat in control is learning *active avoidance*: it gets a signal and *acts* to avoid a punishment. Meanwhile, the other rat learns helplessness. Its natural tendency to do something decays as it realizes that doing accomplishes nothing. This second rat develops more stomach ulcers, loses more weight, has higher levels of the stress hormone corticosterone in its blood, and is more fearful afterward. Within the brain stem, a small but powerful group of cells, called the locus ceruleus, changes its firing pattern: less norepinephrine enters these cells, and they fire more rapidly, irritating other parts of the brain. This rapid-fire function is linked to depression. Elsewhere in the brain, there is a deficit of another neurotransmitter, serotonin. And in the brain's reward system, which carries positive impulses from the midbrain to the limbic system, dopamine is depleted, blocking the normal experience of reward.

Since the rat in charge avoids these depletions, it can't be shocks alone that take this complex toll. In fact, the relatively unscathed rat undergoes classical conditioning, learning a suite of physiological responses to the warning light, just as Pavlov's dogs learned to salivate at the bell. These conditioned responses some-how protect the stomach wall. Along with this Pavlovian learning goes a simpler process of association, linking the warning light with the shock. And if the rat that could not control its shock could somehow watch its more fortunate partner, we might see an eager form of observational learning. Thus, one simple experiment in laboratory rats exemplifies every major form of learning. But they can all be illustrated with examples from human experience.

Simple habituation is at least part of what goes on when an infant makes the transition from smiling cheerily at everyone at age four months to smiling at very few people at six months; when a child finally ceases to cry for a parent who never responds; and when a young man practices desensitization to postpone premature ejaculation. Simple associative conditioning is part of what goes on when we learn to anticipate a haughty social mien from a person with a certain carriage of the head and shoulders; when a boxer learns that a certain facial expression of his opponent predicts a good right cross; and when someone we know very well inad-vertently teaches us that a characteristic respiratory rhythm anticipates by a few seconds the start of an orgasm. Classical or Pavlovian conditioning is at least part of what enables a toddler to use the toilet reliably or makes a child's stomach churn just hearing about food. Operant conditioning is at least part of the expla-nation for why we perform in school or go to work in the morning. And observa-tional learning explains why a boy and girl on prom night look the perfect carica-ture of North American masculine and feminine.

But these are only a few paths of change. On the fringes of the standard learn-ing laws are forms of special learning, such as the one-trial learning of Martin Seligman's "Sauce Béarnaise syndrome"—the strong tendency to avoid a certain food after one or two bouts of nausea. More broadly, there is mode-specific learn-ing, or the Garcia effect—differences in the rate of learning an association between a sound and nausea, as opposed to one between a taste and nausea. Other examples are the processes called imprinting—events ranging from the baby chick's quickly starting to follow its mother during the first days of life to a mother goat's learning the identity of her newborn kid, within minutes, by smell. These may differ from what happens to a rat or pigeon in a Skinner box, but they are learning. They modify behavior through of their own laws.

Early-experience effects are a type of special learning. Although they remain scientifically controversial, they have been taken for granted by most thoughtful people throughout the history of Western civilization. *Give me a child until he's seven, and he'll be mine forever,* said Lenin and Kierkegaard, echoing Jesuit edu-cators and many others as well. This belief is what Jerome Kagan caricatured as a sort of tape-recorder metaphor for the brain—the notion that young gray matter faithfully records sense impressions for later advantage or disadvantage, that noth-ing escapes it. This idea has shaped psychotherapy and education.

It is not without danger. As the century turned, we saw a surge of enthusiasm for early-experience effects that seemed a repeat performance of the 1970s. Back

then, those who insisted that early experience was overwhelmingly important—
Burton White of the Harvard School of Education was prominent among them—
caused a corresponding deemphasis on *later* experience. This led the rich to
lobby the "right" preschools on behalf of talented toddlers, lest they lose the path
to Harvard, and to teach babies to read. For the poor, the consequences were far
more serious: many people, even policy makers, concluded that if a child was
deprived for the first three to five years, there was little hope later. Societal neglect
of education for the poor had found its "scientific" justification. Today the same
beliefs, once again "new," seem to be buttressed by brain research. But Carla
Shatz, head of the neurobiology department at Harvard Medical School and one
of the researchers whose work led to these new exaggerations, said that "a break in
logic occurred" when research was publicized after a White House conference in
April 1997. "No neuroscientist ever got up there and said . . . that 0 to 3 was the
most important time."

The damage of this sort of hype is at least as great as it was in the seventies. In
the 1990s the governor of Georgia seized on the early-experience claims to do an
end run around the state's dismal history of investment in children and schools.
Inspired by a weak and unsupported study that supposedly showed that Mozart's
music improves the newborn brain, he sent tapes of Mozart to the mother of every
newborn baby in the state. No serious child psychologist believed this study, but it
was a grand gesture, and a cheap one—a temporary parry to the credible charge of
statewide child neglect. For the poor families of Georgia it was a double affront: it
belittled their concerns about the state's disgraceful record in child health and
education and, adding insult to injury, it asked them to play music they neither
liked nor understood, accepting that the state knew what was best for their babies.
If the people have no bread, let them eat cake, said the queen of France. Two centuries later the governor of Georgia declared, *If the children have no decent
schools, let them hear Mozart.*

John Bruer, in *The Myth of the First Three Years,* refers to "the basic infant-
determinist premise," which the Myth presents as fact. "The Myth rejects strong
genetic determinism in favor of strong *early neural-environmental determinism.*"
The consequences of both are similar: abdication of responsibility. We know this
because we have been through it before. In the 1970s the notion that poor nutrition in the first three years of life irreversibly damages the brain confirmed the
belief that people in the developing world must be intellectually inferior. This was
not exactly racism, since it was not based on genes, but it was close, and the belief
supported the easy way out in terms of policy: benign neglect. The truth is that
what happens during the first three years is very important, but so is what happens
during the next three, and the next. Small interventions at age eight can sometimes produce large changes in behavior. The environment during adolescence
shapes adulthood as much or more than anything that happens earlier.

While we cannot do experiments on children, animal models can tell us
much about how experiences change the growing brain, both at early and later
ages. These studies do provide support for the idea of sensitive periods, but only in

limited domains, and they also prove the power of later intervention. We have seen many examples, but it is useful to gather some of them here. Together they make a powerful story.

If you close one eye of a developing rhesus monkey for a few days before it is six months old, you will get an adult monkey with little or no depth vision. To see depth depends on certain cells in the visual part of the cortex—cells that in a normal animal respond to a flash of light on either eye. These binocularly responsive cells become linked up to both eyes during the first six months of life—you can close one eye for years after that age without much effect. The strange fact is that at that moment the two eyes are actively competing for those linkages. Take one out of the competition for even a few days, and the other eye takes over the binocular cell completely. This ends forever the cell's chance to respond to both eyes, integrate their slightly different information, and allow the monkey to see depth.

These studies on vision are the most elegant, but they are far from the only ones. We've seen that rhesus monkeys raised in social isolation for the first six months of life almost inevitably grow up to behave abnormally—they are socially and sexually inept and withdrawn, with inappropriate fear and anger. If females are forcibly impregnated, they are often neglectful, even brutal mothers. Most attempts to treat this syndrome, by restoring social contact after the deprivation, fail. Even less extreme deprivations—partial isolation, rearing with peers only, rearing with the mother only, and isolation for shorter periods—have lasting effects. An intervention as minor as two six-day separations from the mother during early life has measurable effects on the monkey's emotional behavior at age two. What happens in the brain during separations like these? A team led by James Rilling at the Yerkes Regional Primate Center investigated the question. Six young rhesus monkeys were separated from their mothers for only two hours and showed increased activity in their right frontal lobes, with a mirror-image decrease on the left. They also activated other right-brain regions, and the increased brain activity was related to higher levels of blood cortisol. Rilling's findings were consistent with other research on the brain areas active during anxiety and other mental stresses.

Although the biology of isolation rearing is by no means understood, we know that stress hormones are different in these animals. They also grow up with neurotransmitter imbalances, in the norepinephrine and serotonin systems. These imbalances make them vulnerable to despair, aggression, and self-injury. Other studies find changes in serotonin byproduct 5-HIAA, in brain glucose metabolism, and even in the function of the serotonin transporter gene—all tied to behavioral changes like increased aggression and alcohol consumption. Longer and more severe early social deprivation also alters dendrite branching in the cerebellar cortex. The cerebellum has been traditionally considered an organ of movement, and the restricted subjects also have much less physical activity, but we now know that the cerebellum plays a general role in learning. Finally, as Mar Sánchez and her colleagues at Yerkes have shown, rearing monkeys in social deprivation changes the corpus collosum in ways that relate to mental ability.

In rats, too, rearing changes the brain. Environmental enrichment and impoverishment, whether social or inanimate, can alter the anatomy of the rat cerebral cortex. The greatest changes are in the visual cortex, in the rear of the cerebral hemispheres. But even *blind* rats, when enriched, change in this "visual" area. We may need to rethink the functions of this region; parts of it may serve broader mental functions. Changes include variation in the weight and thickness of the cortex, the number and size of synapses linking nerve cells, the complex branching of dendrites that get incoming signals, the density of the spines on those dendrites (where most connections are made), and the activity of enzymes that process acetylcholine, a neurotransmitter involved in learning.

All this occurs in the cerebral cortex, the best part of the rat's brain or ours. But if you think you need an advanced brain to make such changes, consider the lowly jewel fish. Pyramidal neurons in these competitive, courtly, sexual fish (the same type of cells that are most changed in rats) also change with rearing—in this case, social isolation. The "listening" sites on these neurons—the spines or thorns on the smallest dendrites—change in number, distribution, and even shape under social isolation. These changes are *individual* adaptations; for example, repeated stimulation of the neuron by social experience makes the microscopic spine shorter and thicker, after which it affords less resistance to incoming electrical signals. The jewel fish neurons in question are in a part of the brain called the tectum—an advanced area in fish but a primitive one in us. One implication is that obscure lower regions of our own brains—not just the most advanced—also change with experience.

As we saw in the chick, imprinting on the mother hen in the first few days of life rapidly and permanently changes not only behavior but also the pattern of protein made in the roof of the forebrain. Imprinting also changes the structure of cells in the hyperstriatum, another forebrain region; in other areas it changes neuronal firing rate and dendritic spine density. And while imprinting is limited to a highly sensitive period, the time can be extended by altering brain chemistry. Thus, the circuitry of attachment in birds is transformed by certain experiences, and the permanent impact of imprinting—not only following the mother throughout infancy but restricted mate choice in adulthood—may be due to just these brain changes.

Yet the brain is not the only changing organ. Beginning in the 1950s, experimental work in rats showed that a few minutes of stimulation or stress each day for the first three weeks of life—about half of a rat's "childhood"—permanently changes behavior. The intervention could be mild electric shock, brief cooling, being stuck in a tin can with some nesting material and shaken, or even being stroked and petted in the hand of an animal caretaker—an intervention that may not have seemed so tender to the rat pup. Call it stimulation or stress, the resulting adults were toughened up. They were less fearful in an open-field testing situation, exploring more while defecating less. They also grew faster, became longer and heavier, showed greater learning ability, and were more difficult to kill by starvation, drowning, injected tumors, and various other means.

Physiological studies suggested mechanisms. These early interventions affect the stress response system—the hippocampus, hypothalamus, pituitary, and adrenal cortex. For instance, in adulthood the adrenal weighed less after stress if the rat had been stressed in infancy—the gland had been exhausted by early experience. Hence, the relative calm of rats stressed in infancy when later faced with threatening situations. Adrenal activity matured faster in these rats, supporting the theory, and the rats stimulated in infancy released less corticosterone in the open-field test. Yet when both groups got a more severe stress such as an electric shock, the rats stressed in infancy were *more* responsive. So, early-stressed rats were not *always* less responsive to stress; they just reserved their resources until they really needed them. Soon it was shown that two hormones higher up in the stress axis— CRF in the hypothalamus and ACTH in the pituitary—could be changed by early stimulation. Also, the degree of difference between the two halves of the rat brain, known as lateralization, can be permanently changed by early handling. Furthermore, mice respond more or less as rats do to early treatment; but more intriguing, the same effect in mice can be achieved by having them reared with rat "aunts."

The late twentieth century saw a resurgence of interest in early experience, proving its power and complexity. Studies during the 1990s by Dong Liu, Paul Plotsky, Michael Meaney, and others confirmed an old hypothesis of Seymour Levine's: early handling and separation—as long as it is brief—changes the *mother's* behavior toward the pup, and this explains most of the long-term impact. Two findings proved this: first, rat mothers double their licking and grooming after pups are handled; second, mothers that naturally lick and groom more have pups that grow up like handled pups, even though they never receive the early stress— just the extra licking and grooming.

This doesn't rule out the possibility of a complex interaction. Victor Denenberg, one of the founders of this line of research, has done work with mice and rabbits that confirms the rat maternal findings in some ways but not others. The long-term physiological effects of infant stimulation differ from those of maternal separation, and daily separation for fifteen minutes has very different long-term effects than does separation for three hours, both effects being dependent on altered maternal behavior. Finally, lower stress responsiveness may be due, as Robert Sapolsky believes, to reduced hippocampal receptors for stress hormones. Fewer receptors mean greater sensitivity to low hormone levels, which, through the hippocampus, would keep the hypothalamus from triggering the stress response. And of course, as we have seen in the chapters on fear and grief, the more extreme interventions in infancy reverse these effects, producing *more* stress-responsive animals—to the point of maladaptation throughout life.

Finally, some of the most intriguing research on how experience changes the brain is in psychopharmacology. In many strains of mice, housing each mouse alone for three weeks after weaning changes the brain's neurotransmitters and produces behavioral effects, such as fighting, in isolated males. Recent studies help explain these effects. Overall, post-weaning isolation of mice and rats affects the level, turnover, production, use, removal, and enzyme activities of major neurotransmitters. Graham Jones and his colleagues showed that rats reared in isolation have an increased sensitivity to amphetamine in the nucleus accumbens, making them more sensitive to reward. Others have found changes in monoamine neurotransmitters, in blood corticosterone, and in dendritic spines in the medial preoptic area of the hypothalamus. Ongoing work examines subtler behavioral changes due to post-weaning isolation or earlier, briefer maternal separation.

But what are the human implications? In some cases the parallel is clear. For example, there is a human analog of monocular deprivation in monkeys: childhood strabismus, which prevents both eyes from focusing on the same point. The condition can be corrected at any age, and the eye muscles will then do their job. But if it is corrected too late, the child will never see in depth, despite being able to focus both eyes on one point after the surgery. The likely reason is the same as it is in monkeys: a critical period for the normal development of those cells in the visual cortex that are designed to respond, *jointly*, to both eyes. Strabismus derails that development, almost as if one eye were closed. If the critical period—about five years—passes while the two eyes are deviated, depth vision cannot recover.

As for early *social* deprivation in monkeys, the human parallel is fortunately rare, at least in its extreme form: total and prolonged deprivation in the hands of negligent parents. Tragically, a few children have grown up while locked in closets, tied to bedposts, or confined to basements; the effects are psychologically dire yet often partly reversible. There may well be brain or hormonal changes in children in such extreme social deprivation, but we don't yet know what they are. An ongoing and horrific natural experiment, the severe deprivation of large numbers of infants in Romania during the transition out of Communism, is helping to answer this question, however. Psychiatrist Michael Rutter and his colleagues have been following 111 infants adopted into families in the United Kingdom after up to two years of both social deprivation and infectious disease. The catch-up in both physical and cognitive development appeared nearly complete at four years for those children who came to the United Kingdom before six months of age. For those placed after six months of age, the catch-up was "also impressive, but not complete." Thus, the deprivation syndrome as well as the recovery are reminiscent of what we saw in the monkey studies. In the Romanian orphans there appears to be a sensitive period for certain kinds of damage, with a threshold at six months of age. But years of research will be needed to trace the boundaries of the sensitive period and the pattern of residual deficits.

Are there human analogs of lesser forms of monkey social deprivation, with similar lasting effects? Possibly, but we need to learn more about them, and easy answers are unhelpful. For example, based on folk wisdom and inferences from animal studies, many have predicted dire consequences for children due to the rise in the divorce rate. But the situation is complex, and anyone who works with chaotic families knows that divorce can be a healing balm. The negative long-term impact of divorce on children has been exaggerated by some psychologists, based on poorly controlled studies. Better research shows that most ill effects come from continuing post-marital conflict, loss of contact with one parent, and drastic economic losses, rather than from divorce itself.

Parallels to the early-stimulation studies in rats have also been hard to estab-lish. Recall that the first experiments involved petting and stroking by the human caretaker, in an effort to give rat pups extra "tender loving care." When the pups grew faster and larger, that was interpreted in terms of the early need for love and touch. But when being placed on a block of ice or shaken up in a tin can had sim-ilar effects, it began to seem that to the pup, human handling was stress, not love. Parents in some cultures stress children deliberately. For example, Native Ameri-cans in the Great Plains region believed that letting a baby cry would make him tough and strong, a view consistent with animal studies. But there is better evi-dence than folk belief. Psychologist Thomas Landauer and anthropologist John Whiting showed that in cultures with stressful infant-care practices—head bind-ing, circumcision, and ice-water baths, for example—adults were fully two inches taller on average than adults in cultures with no such early stresses.

Whiting and Sarah Gunders later showed that cultures with stressful prac-tices in infancy also have an earlier age of first menstruation (two *years* earlier if mother-infant separation is also common) than those without such practices. This could mean that early stress makes girls grow faster. All these cross-cultural studies are correlational, not controlled experiments. But they suggest a direction for fur-ther research, which could open a new window not only on psychological devel-opment but on body growth. They also fit life-history theory, in which early stress is seen not just as a negative but as a signal to the organism that life will not be easy. Thus, science sometimes works indirectly: cultures as different as the Plains Indians and the British upper crust decide that early stress might have positive effects; laboratory research confirms the notion in rats; cross-cultural studies con-firm it in humans; and finally, an evolutionary theory emerges, producing a framework for these effects.

Presently, at a moment when brain-altering drugs are for better or worse a growth industry, we could use a human analog to neurotransmitter changes in iso-lated mice. Most major forms of mental illness—schizophrenia, manic-depressive illness, depression, obsessive-compulsive disorder, attention deficit disorders, and substance abuse, among others—involve abnormalities in neurotransmitters. Such abnormalities can be due to genes, but the studies of isolated mice prove that similar brain chemistry, along with behavioral changes, can result from expe-rience alone. So some aspects of both mental illness and normal emotion could represent effects of early experience on neurotransmitters. And despite doubts about the role of nurturance, research in monkeys shows that creatures like our-selves have lasting changes in brain chemistry due to nurturance alone.

Yet ultimately—and contrary to our intuition—widespread, lasting, impor-tant effects of early experience in humans have not been conclusively proved.

Some of the reasons are obvious. Children cannot be manipulated in experiments, and correlational studies are subject to alternative explanations. Other reasons are subtler. For example, the same experiences affect different children in very different ways. Maternal deprivation may have devastating effects on one child, mild effects on a second, and bracing or strengthening effects on a third. Individual vulnerabilities might be associated with certain genes yet manifest themselves in something as subjective as mind-set. For human beings more than other species, the way a child thinks about an experience alters the nature of that experience.

Some effects may be too subtle to be detected by the usual measures. In a famous study by Myrtle McGraw in the 1930s, one of a pair of twins was elaborately exercised in skilled movements for the first few years of life, while the other twin just watched. McGraw mistakenly thought the twins were genetically identical. Still, they were brothers the same age, sharing half their genes, with similar family environments. Against this background, McGraw introduced her extraordinary exercises. Throughout early childhood there were major differences between the boys in motor development and ability, favoring the exercised twin. Yet when McGraw visited them at age twenty-two, the twins did not differ, in occupation or leisure, in any obvious way that showed a lasting effect of the early training.

But a look at the film she made of the two young men—she asked them to perform simple activities, such as sitting, standing, or walking along a narrow plank—tells a different story. It is clear to any observer that the twins differ markedly in grace of movement, the graceful twin being the one exercised in infancy. But how do we measure gracefulness? Certainly not by any available formal test. McGraw's experiment is just a systematic anecdote, but it suggests a lesson. Perhaps in emotion and mind there are also differences of grace and tone that would be equally important, even obvious, yet be missed by the usual mental measures.

In fact, it is always easy to find no difference—easy enough so that when early experiences are compared we can always be skeptical of the conclusion *no effect*. Among the many ways to find no difference when there really is one: measure irrelevant outcomes, fail to look long enough, use a blunt measuring tool, or omit specific situations—especially stressful ones—most likely to reveal a deep-seated effect. This is especially true of human beings, who can be so successful at concealment that important difficulties—anxiety, suicidal thoughts, sleep difficulties, phobias, sexual problems—may be never be revealed. We need to look beneath the surface, at what may be changed in the brain. The frontier of brain imaging is advancing with amazing speed, providing an unprecedented window into function. Within the first decades of this century, it will be possible to view human brain activity down to the level of small networks of cells. In the meantime, even more precise laboratory studies leave no doubt that the brain is changeable.

In simple systems, a slice of brain, a network of nerve cells, or a single neuron is isolated, kept alive, and studied under the microscope. Such work removes all doubt about plasticity and points to mechanisms that probably act on a larger scale in conventional animal and human studies. Neurons, it turns out, respond to repeated stimulation with structural, functional, and chemical changes. They instantly switch on "immediate early genes," which in turn produce key enzymes that make more neurotransmitters. They change the structure of membranes and the electrical flow over them, the essence of brain activity. They grow new connections and modify old ones in countless ways, even changing the size and shape of dendritic thorns and synapses. And they alter their chemical trade with non–nerve cells around them, including the glia, keys to their life support. Glia multiply faster than neurons in both evolution and development, and they may play a role in learning.

These are general processes in neurons throughout the brain. Some we share with creatures like the sea slug *Aplysia*, even though it gets by with no real brain at all, just a few thousand neurons arranged in a few ganglia. But of course, with the mammalian brain came structures adapted specifically for learning. In the temporal lobe, such dedicated circuits have long been known. The famous patient H. M., who tragically lost the hippocampus on both sides of his brain, was never again able to store new memories, although he had good recall of his life before the loss. Versions of this particular amnesia—called anterograde because it blocks memories after the injury—also occur in people with alcoholic brain damage, which can target the hippocampus.

Such patients, along with animal studies and ever-more-precise brain imaging, have led to a coherent account of how we record new experience. Impulses from all over the brain converge on the temporal lobe, where the entorhinal cortex integrates them and delivers them to the adjacent hippocampus. There, in a daylong chemical stewing called consolidation, they become linked together and, as they circuit the limbic system, imbued with emotions, like fear and surprise, sadness and pleasure, that make them more vivid and retrievable. Amazingly, it has become possible to follow this process in the human brain not only through functional imaging but with electrodes implanted for treatment of people with severe epilepsy.

These studies deal mainly with *declarative* memory—the kind that you can retrieve on command and even talk about. But there are many other kinds of memory. *Performative* memory—what we achieve through practice—is widely distributed across brain structures. So are the more passive kinds of mental association, clusters of sights, sounds, and other features that stand out from the stream of stimuli as recognizable objects. Such *implicit* memories seem to be independent of temporal lobe structures. So is *working* memory—items that briefly circle the landing field of consciousness, ready for deployment in ongoing mental operations. The outer prefrontal cortex plays an important role in generating such mental holding patterns, and it is dampened in emotional states like sadness.

Even structures that serve movement are involved in complex learning. Some neurons in the motor cortex fire sequentially with a series of repeated visual signals, preparing for actions learned in response to the signals. And the cerebellum is active during learning—not just motor but verbal learning—especially during the first few tries at a task, and decreasing as it is mastered. Not long ago, most experts would have doubted that the cerebellum has a role in language learning. But one patient with a cerebellar stroke was bad at verbal discrimination learning despite being good at many other cognitive tests. The cerebellum is larger than expected in humans as compared to apes, and its role in learning words may be part of the reason why. The bottom line is that the laying down of memories involves many brain regions. As for memory *retrieval*, it depends on the *frontal* cortex—even for memories *put* in storage by the temporal lobe.

Overall, studies in the nineties proved the naïveté of two old views of learning. One, often based on research with literally brainless animals, saw it as a single microscopic process, such as long-term potentiation of neuronal firing or the formation of new synapses. The other set it in one particular circuit, usually centered on the hippocampus. In the end, both had a piece of the truth, but only in a much larger framework. Long-term potentiation certainly occurs in the hippocampus. But learning also depends on many small changes in every neuron repeatedly activated. Neurons change independently or jointly depending on the task, and although they are widespread in the brain, the learning circuits are not generalized, undifferentiated neural networks. On the contrary, they are often dedicated modules with exquisitely specific adaptive functions—imprinting and the Sauce Béarnaise syndrome being prime examples.

Finally, nerve cells die, and in mammals they are only rarely born again. Unlike most cells in the body, which die and are replaced every few days, weeks, or months, nerve cells are not simply and straightforwardly replaced in humans, and few new ones are born by means of cell division. As we saw in chapter 4, *some* new neurons are made. Research in the 1990s by Elizabeth Gould and others showed that in several mammals, including monkeys and humans, new neurons are born even in adulthood. But it has been hard to prove that these have functional relevance, and they are few in number. Even a lifetime of nerve cell birth at these rates would compensate for only a fraction of the estimated neurons lost in everyday wear and tear. "No new neurons" is dead as dogma, but "few new neurons" may for practical purposes mean much the same thing.

So our lifetime collection of nerve cells in their interconnected webs is pretty stable, except for steady losses that might, up to a point, be viewed as streamlining. Perhaps this is why we can record experience and use it, adapting to our environments—like few other cells in the body, nerve cells are with us throughout life and are exactly as old as we are, ticking forward like the grandfather clock in the song. Most of them, that is: an estimated 15,000 die every day, never to be replaced. Considering that there are ten to a hundred billion, this is a small number—even at the overripe age of a hundred it would only amount to a small percentage of brain cells. Still, as neuroanatomist Walle Nauta used to say, after a certain age one starts to wonder if they aren't adding up. Each day a tiny death of the brain, an infinitesimal "stroke," an unknown loss of structure and function. At least Nauta saw humor in it.

Brain anatomist Paul Yakovlev also saw hope. He presumed that the death of nerve cells is not random—that it is the least used ones that die. In this view, the nervous system is actually sculpting itself by means of slow death, adapting anatomically to what Yakovlev called "the pathos of life experience." Richard Dawkins based a model of learning on this process. Because of it, perhaps, our actions become smoother, until we lose too much. Even people with Alzheimer's disease, losing many mental functions, often preserve skills—tailoring, carving, piano playing—that they have practiced all their lives. This overlearning may be due in part to selective neuron death. Because of this, too, we may develop a certain inflexibility. Some who study the neural basis of learning have focused almost exclusively on individual nerve cells, frequently losing sight of the forest of the brain. And the forest can only be mapped through the difficult, classical study of brain anatomy.

We covered this ground in chapter 4, so we need only mention it now. The honing process begins in the embryo and only ends with death. At least twice the number of embryonic neurons are born than are needed, and the half that die do not die at random. Recall Changeux's paradox. A hundred thousand genes create a hundred billion neurons with a hundred trillion connections. It can't be done from within, by a push-pull, click-click sort of genetic control. Some of the early cells make connections, and these connections supply nutrients that help keep the cells alive while others die. Then synapses, too, stabilize selectively. Most connections are pruned, and the ones that persist are the most active—*functional* synapses have an edge in the cellular struggle for survival. Embryos are active very early, reacting to noise, flexing their limbs, and sucking their thumbs; this activity shapes brain circuits.

Third, beyond cell death and pruning, Gerald Edelman's model of neuronal group selection holds that functioning groups of nerve cells—local circuits—form initially under genetic control with little need of experience. These are fairly fixed, adaptive modules for perception and action. But during growth these preadapted neuronal groups compete, so that experience shapes the brain by selecting among innate microcircuits. In the growing brain, nerves that bring sensory signals from the thalamus determine the final structure of cortical regions. Ferrets, born so immature that they are basically embryos, can have their brains rewired after birth; direct the signals from the eye toward the auditory cortex, and the auditory cortex will end up looking like the visual cortex. Not only that, but the ferrets using this weird apparatus will see light.

Finally, chaos and complexity models help explain brain growth in at least two ways. First, sensitivity to initial conditions can multiply small differences exponentially, as in the famous butterfly-wing effect on weather. Such chaotic indeterminacy gives *identical* twins different brains, even in the womb. But computer models of embryonic cells, as they divide, change, and affect one another, show that chaotic unpredictability does not fan out indefinitely. Patterns emerge, as chaos in the developing brain resolves of its own accord. This is order, to be sure, but it is not genetically determined order. Still, many patterns *are* genetically coded. Chaotic and emergent effects, like activity-dependent ones, occur throughout development but are constantly pushed, pulled, shaped, and molded by the genes.

The power of change may be greatest before birth, but later experience still has major effects. We have seen that closing one eye of a young monkey during a sensitive early period allows the nerves from the other eye to compete unfairly, and no matter how much later experience they have, these monkeys never develop depth perception. Similarly, removing a few whiskers from the snouts of mice before they are seven days old eliminates cortical barrels that should have represented those whiskers, while nerves from neighboring whiskers bulge their own barrels accordingly. But *breeding* mice for *extra* whiskers will produce extra barrels in the cortex. Genes make whiskers, and whiskers in turn make cortical barrels.

Even evolution relies on this device. In mice, rats, rabbits, cats, and seals, whisker length and density follow similar patterns, enhancing the whiskers' task of detecting objects. In the star-nosed mole, a large stretch of cortex evolved just to represent the snout, so the creature can burrow sensitively in soft, moist soil. Its cousin, the common eastern mole, which burrows in dry, rocky soil, lacks these adaptations. Some of these differences are coded by genes, but they also rely on brain plasticity. Selection can't make the genes do things the environment can reliably do, because as long as the environment builds an adaptation, genes that might do the same job are unlikely to be favored.

So much for the *young* brain. It is clear that the adult brain also responds to experience with striking changes. In classic experiments by Michael Merzenich and his colleagues, an *adult* owl monkey with a finger removed lost the brain representation of that finger, while nerves from nearby fingers annexed the territory. Even finger stimulation, with no surgery, changes the cortex. Such changes are guided in part by the nucleus basalis, a region attacked by Alzheimer's disease. Perhaps the symptoms of Alzheimer's result in part from the brain's inability to rearrange its own structure. In amputees, phantom limb pain decreases as the cortex is reorganized in the months and years following amputation. But in humans as in monkeys, stimulation alone works. A remarkable study showed that string players—six violinists, two cellists, and a guitarist—had larger fields of representation in the cortex for the fingers of the left hand than the rest of us do, and the excess depends on how young they were when they started playing. Similarly, people who have been deaf from birth but are fluent in American Sign Language have expanded their visual cortex, especially in the left hemisphere—the benefits of a lifetime of seeing meaning in signs.

Even growing new cells is not out of the question. Testosterone-besotted songbirds, we know, make millions of new neurons each spring, the better to sing with. Rats make new granule cells in the hippocampus long after birth. Since the hippocampus is involved in learning and memory, this finding was always full of intriguing possibility, but similar new cells were hard to find in primates until Elizabeth Gould's work. In rats, tree shrews, and marmoset monkeys, new neurons are born in the adult brain, and they may have functions. Putting a male rat or tree shrew in a strange cage with the resident male reduced the production of new neurons by 30 percent, whereas environmental enrichment increased it. By the year 2000, adult neurogenesis had been shown in rats, mice, guinea pigs, tree shrews, marmosets, owl monkeys, and rhesus monkeys, in several brain areas including the prefrontal cortex.

Finally, Peter Eriksson and his colleagues found new neurons forming in the adult *human* brain. Cancer patients had for clinical reasons received a radioactive substance that labeled DNA in dividing cells. Their brains showed incorporation of their label into dividing cells that also had the distinct chemical signature of neurons. This was clear proof of the birth of neurons in the human adult, again in the hippocampus. They were few among the brain's legions of cells, and it is not certain whether the new neurons actually function in *any* species.

But with or without new neurons, significant positive change in the brain is routine in adulthood. Even aged rats respond to environmental enrichment with changes in the weight and thickness of the cerebral cortex, the ratio of supporting cells to nerve cells, the number and size of synapses, the amount and activity of neurotransmitters and their enzymes, the complexity of higher-order branching of dendrites, and the number of spines crowded onto a given length of dendrite. The last change, we saw, goes back in evolution as far as the jewel fish. Vertebrate adaptation requires it: experience, by changing the structure, changes the action of cell, circuit, and brain.

And neurons are not the only cells that change. Glands, including those that pump out behavioral molecules, grow or shrink in response to use. Also, a cell's functional, chemical character can be changed, and such adaptability goes beyond brain and glands. The skin forms calluses in response to friction, and similar things may go on in the gut wall. Muscles build in response to use, and their tendons grow longer and stronger; these effects of exercise work into old age. Even tedious, uninspiring bone responds to the stress of use with an intricate structural flowering. Under the microscope a sliver of bone from a jogger's leg is noticeably more complex in its living form and blood supply than the bone of a nonjogger of the same age and sex.

What, then, of those famous genes, and the countless studies that prove their power? Even the good studies need more thought. One of the major sources of error in some behavior-genetic research (and of genetic studies in general) is the strange statistical specter called an interaction effect. This is not just the common-sense meaning of *interaction*—the notion that the gene and the environment influence each other. That is certainly true, but the statistical interaction effect is a subtler, somewhat more difficult concept that may have great importance.

Suppose we have the simplest Mendelian situation: single-gene control of the height of pea plants. Say the gene "Tall" produces twelve-inch plants, while "Short" yields eight-inch ones. Now suppose we try a new fertilizer, Growitall, on Short and get twelve-inch plants, abolishing the difference between Short and Tall. We then predict that Tall will respond to the fertilizer by growing to sixteen inches. Or if we are very clever, we presume that it may work not by adding a fixed amount to the height but by adding a percentage, in this case 50 percent. So we cagily predict a total height of sixteen to eighteen inches, and we advise farmers accordingly.

We should have done the second experiment instead of theorizing. We failed to take into account the possibility that those same genes may, by unknown chemical means, also produce a completely different reaction to the new fertilizer. The farmers who buy the fertilizer and use it on their Tall plants get ten-inch-high plants—the shortest of all—and we lose our job at the agricultural institute. Also, we had better not conclude from our sad experience with Growitall that Tall cannot be stretched by fertilizers, period. We may lose our next job—to a competitor who, realizing that the next new fertilizer has to be tried on its own merits, grows herself a three-foot plant.

This instance is fanciful, but there are many real examples. Richard Lewontin, a critic of behavior genetics, often cites an experiment done by the great geneticist Theodosius Dobzhansky. Fruit flies were incubated at 16.5, 21, and 25 degrees Centigrade and survival time was measured. The genetic strain the flies came from was as important as temperature in predicting survival. Yet *no* strains outsurvived others at *all* temperatures. On the contrary: three strains survived best at the high temperature, less well at the lukewarm temperature, and worst at the low temperature; seven strains did the opposite, dying faster as temperature increased; another seven strains survived equally well at all temperatures; four fared worst when middling warm; and two survived best when lukewarm and worse at both extremes. The seemingly simple and sensible question "How do fruit flies respond to temperature changes, and which is the best temperature for them?" *has no answer.* Prediction can't be done off the rack for fruit flies generally but must be hand-tailored for each specific genetic background.

This seems like an irrelevant subtlety until we realize that much of human behavioral genetics, especially the genetics of intelligence, has discounted such interaction effects or, in some cases, invented them. A classic bad example came at the end of Arthur Jensen's 1968 monograph attempting to show race differences in intelligence. Claiming to have shown that known race differences in intelligence were genetically based—he had not shown this, nor have others shown it since—he irrelevantly and dangerously went on to conclude that intervention programs in schools and preschools are pretty useless. Indeed the very title of the paper, "How Much Can We Boost IQ and Scholastic Achievement?," implied this unwarranted conclusion.

If you drew a graph based on his predictions, showing genotypes as separate lines, each would increase in intelligence in response to enrichment. But at the origin of the graph, the lines would be joined, suggesting that under conditions of total environmental impoverishment the groups would be equally low in intelligence, and with enrichment the lines would diverge. All genotypes in this fantasy world would rise in intelligence as the environment improved, but one of the lines would rise much faster. So the better the environment got, the bigger the difference would be. Thus, Jensen strongly implied a specific gene-environment interaction effect, according to which minority people in enriched environments would lag even more behind whites than they do in the poor and moderate environments. In fact, it is at least as likely that appropriate enrichment would make the lines converge at the top, as all genotypes press against the ceiling of maximum human potential.

Such deliberate assaults on efforts to improve the school performance of African-Americans have continued, notably with the 1994 publication of *The Bell Curve: Intelligence and Class Structure in American Life,* by Richard Herrnstein and Charles Murray. This book presented strong evidence that genes play a role in intelligence but linked it to the unsupported claim that genes explain the small but consistent black-white difference in IQ. The juxtaposition of a good argument with a bad one seemed politically motivated, and persuasive refutations soon appeared. Actually, African-Americans have excelled in virtually every enriched environment they have been placed in, most of which they were previously barred from, and this in only the first decade or two of improved but still not equal opportunity. It is likely that the real curves for the two races will one day be superimposable on each other, but this may require decades of change and different environments for different people. Claims about genetic potential are meaningless except in the light of this requirement.

What about evolution? Can we at least rely on *evolutionary* change being based on genes? Some features of animals, especially but not exclusively behavioral ones, can be stably passed on outside the genes. We have seen several examples: dialects in song sparrows, shell opening in oystercatchers, potato washing in Japanese monkeys, and status in rhesus monkeys. Evolutionary biologist John Tyler Bonner summarized many instances in wild animals in his graceful book, *The Evolution of Culture in Animals.* Later studies have considered "cultural" transmission in animals from fish to chimpanzees. But only a few of these cases meet strict criteria even for observational learning, much less for culture.

Chimps probably come closest to meeting the standard. They vary from group to group in tool making and other aspects of protoculture. Using stones to crack nuts, building temporary nests for sleep, stripping the branches off a twig to fish for termites, sponging water out of a deep crevice in a tree trunk—all these behaviors and more are passed on from generation to generation. Yet another feature resembles human culture: the exact technique used and passed on has a distinct local flavor in any given part of the chimps' African range. Although this kind of local variety occurs in sparrow and cowbird song dialects, in chimps the brain mechanisms are probably simpler versions of our own.

Behavioral transfer outside the genes has also been shown in the lab. Victor Denenberg, famous for work on infant stress in rats, showed in a classic study called "Nongenetic Transmission of Behavior" that the behavior of those animals is influenced by the experience their grandmothers had as infants, even if these grandmothers are only foster grandmothers. This grandmother effect was confirmed in the nineties, using more advanced methods, by Darlene Francis, working with Plotsky and Meaney. Transgenerational, nongenetic transmission of maternal behavior is real, even in rats. The evidence for some form of protoculture in various monkey species is extensive. And in a move that makes archaeologists squirm, Kanzi, the world's most famous bonobo, has become a remarkably effective stone-tool maker—although in his case it is likely that his success depends on being taught by Nicholas Toth and Sue Savage-Rumbaugh— humans, and exceptionally gifted ones at that.

Some theorists propose a definition of culture that is too stringent to apply to animals. Ann Kruger has developed a theory of cultural learning that has teaching and collaborative learning at its core. The essence in Kruger's model is intersubjectivity, as seen in scaffolding, discussed earlier in this chapter; this rules out culture in nonhumans, while making it a human universal. The idea has a conceptual precedent in the name for our species proposed by the Australian ethologist S. A. Barnett: *Homo docens,* or "teaching person," and in his plausible claim that we have an instinct to teach. But for most anthropologists and naturalists, culture is much broader—not only teaching and scaffolding but observational learning, fads, accustomed sounds and odors, even psychological dispositions. None of these require teaching or collaborative learning, although all can be helped by these strategies.

Many evolutionists have adopted Richard Dawkins's simpler concept of the meme as a unit of cultural transmission, roughly corresponding to the gene. By the turn of the millennium, a new field, *memetics,* had emerged, complete with many rival models. In some formal theories, like that of Robert Boyd and Peter Richerson, cultural evolution is independent of genes and buffers the genes from natural selection; this is basically the standard social science view of what should happen in a cultural animal. In other theories, like that of Charles Lumsden and E. O. Wilson, culture may amplify the impact of natural selection on the genes. In reality both are possible, and the interplay between genetic and cultural evolution is complex and subtle. As William Durham showed in his classic *Coevolution,* there are multiple relationships between cultural stability or change and natural selection. Buffering, amplification, and mutual independence are all simple ideas compared with the richness of these interactions.

But cultural evolution also has a life of its own. I happen to know a descendant of Darwin's. She admires him but knows not much more about what he had to say than the average intelligent person on the street, while I teach his ideas daily, passing them on to the next generation of students. I also happen to know a descendant of Ralph Waldo Emerson's. She found him unreadable, whereas I have taken comfort from his essays at some of the darkest times of my life. Who are the real descendants of Darwin and Emerson? Well, these women, of course, if you ask about hair color or red blood cell surface antigens, or even native intelligence. But if you consider another part of those two men's phenotypes—the things they thought and wrote and considered immensely important—I am the one who, more than a century later, is repeating them and gaining adaptive advantage from doing so; different descendants, by different modes of inheritance, for different characteristics.

Of course, there is far more to human culture than the units we call memes. Memes are not like genes, which are more or less discrete and of roughly comparable size and action. Memes, in contrast, include a fad like navel-piercing, a game like tic-tac-toe or *Who Wants to Be a Millionaire?*, a religious form like the Buddha or the crucifix, an idea like communism or relativity, a melody, a notion of who is pretty, and the phrase "a pretty girl is like a melody." A character like Hamlet, a sexual position copied from the Kama Sutra, the architecture of a silicon microprocessor chip, and a stabbing spear can all be memes. So can the sentence "Love thy neighbor as thyself," a portrait of a Campbell's soup can, a dollar bill, a smart military salute as a flag passes, or an obscene gesture. Not only are they less simple than genes, but they come in many different forms and sizes, show little respect for space or time, and have no common mechanism of either transmission or expression.

Also (and here they resemble genes in a genome) they are not independent, and their coherence, however temporary, is a good deal of what we mean by culture. For theorists like Marvin Harris and Eric Wolf, the coherence comes from a common response to environmental challenge, and subsistence and power are driving forces. For Claude Lévi-Strauss, it comes from mental coherence, like the mind's universal taste for dualities. For John and Beatrice Whiting, shared experiences of childhood, channeling universals of the child's emotional life, give a culture its character. For Victor Turner, culture is a "forest of symbols" we must find our way through as we undergo life's drastic physical changes. For William Durham, coherence comes from a gating process that admits only some innovations. For Bradd Shore, a culture arrays mental schemas according to an overarching plan. And for Dan Sperber, it is a particular distribution of representations shared by a group of people, using the universal modules of mind that are our evolutionary legacy.

These accounts are not mutually exclusive; each has a piece of the truth. All provide, as they must, both a mechanism for stability and a mechanism for change. Change is no vain dream, but some things change more readily than others do. When we speak of future shock, we are generally talking about technological change and its consequences, like the breakup of families and neighborhoods, or the acceleration of the speed of life. Earlier in the chapter I quoted sociologist Claude Fischer, who rightly said that today's Americans have the same genes as their grandparents yet lead immensely different lives. But Fischer's own excellent study of the impact of the telephone showed that a quaint initial attempt by its distributors to restrict the network's use to business and emergencies failed miserably. People immediately used it for socializing, gossip, and keeping families together. "Reach out and touch someone" eventually became the marketing strategy of the same corporations that once tried to ban emotional life from the wires. And twenty-first-century ads for cellular telephones emphasize relationships as much as anything. We often hear that the Internet changes everything. But like every other means of human communication since the first words were spoken, the World Wide Web is used for love, hate, sex, family, gossip, humor, news, altruism, domination, lying, trading, greed, conspiracy, violence, theft, and fraud. The Web changes some things in life, but it doesn't change everything.

And what about genetic change? In sociobiology it is common to talk about "altruism genes"; this is not unreasonable, though no such genes are yet known. But consider this speculation, no more fanciful. Suppose a hormone, "altruin," causes altruistic behavior in a hypothetical bird species by crossing the blood-brain barrier into the limbic system. Suppose further that "proaltruin," found in certain berries, is eaten by the bird and then converted to altruin by the enzyme "proaltruin mutase." Only then can it enter the brain and promote self-sacrifice. It is converted to "altruic acid" by the enzyme "altruin oxidase," removed from the brain, and excreted in urine. A gene could affect altruistic behavior by altering the structure of one of the fantasy enzymes, which is just the sort of thing genes do best. Suppose a gene mutation occurs, and the gene for proaltruin mutase now codes a slightly different enzyme. The affected flyer has 10 percent more proaltruin mutase activity, 7 percent more altruin, and an increase in altruism toward kin.

Unfortunately for simplicity, a new berry bush soon colonizes the neighborhood. The fruit of this bush turns out to have aphrodisiac qualities, which increase reproductive success for a few berry-besotted, oversexed chirpers. Sadly, it contains no proaltruin yet is rapidly replacing the former berry. The result is lower blood altruin, despite the gene that works the other way. The more active form of proaltruin mutase is still favored, though, and over several hundred generations replaces the former gene. But the new berry also largely replaces the old one, much more rapidly in fact, and the net change in the level of altruin in the blood (and of altruistic behavior) is a sizeable decrease, despite a mutation for increased altruism duly favored by natural selection. Meanwhile, another mutation in the gene causes a whopping increase in the activity of the same altruin-making enzyme. Like the first, it distributes altruism by relatedness, and so is favored under kin selection. But, unfortunately for the elegance of theory, a glacier is advancing. Ambient temperature decreases 13 degrees, and this cooling ionizes altruin more, reducing its ability to cross the blood-brain barrier. The net change in altruistic behavior, since the first mutation, is zero.

The point of this game is not to deny genes—it assigns them strong influences. Neither does it suggest that natural selection—and its component, kin selection—would have trouble changing genes that control complex behavior. What it suggests is something subtler: that in the real world, *nongenetic* changes may swamp gene effects. This does not negate genetic change, it just finds it slow and weak compared with other forces. It may therefore be unhelpful in our effort to explain what we see and predict what will happen next. In the altruin reverie, the nongenetic effects were just large enough to counterbalance the gene changes. But what if the real nongenetic forces are much larger than that? Ten, a hundred, a thousand times larger? What use would there be in taking the genetic change into account? What chance of even detecting it? As Richard Lewontin put it, if the nongenetic forces are large enough compared with the genetic ones, then we need worry no more about genes than about the force exerted on a rocket fired from Cape Canaveral by the gravitational field of a distant star—a real, quantifiable force, consistent with elegant theory, but a negligible one, in practical terms not even worth reckoning.

Consider that in the year 1900 perhaps 20 percent of the human species (I am guessing) showed the behaviors of reading and writing, and that as the millennium closed perhaps 80 percent did. Thus, in one century (three or four generations) the species changed a crucial behavioral adaptation. Consider, too, that the average number of children per female lifetime for the same species was in 1900 somewhere in the vicinity of four, while a century later it is closer to two—another major change in a few generations. Now we can *nod* toward the genes. It's genes that make people seize the advantage afforded by reading and writing, and genes that enable a population to lower its reproductive rate when conditions require it. We could also refer to genes more specifically. For example, genes for dyslexia slow down our species' sweep toward literacy and put some at a disadvantage. Likewise, whatever enthusiasm underlies churning out twenty children while the species drops its birthrate gives its owners an obvious lead.

These are facts, interesting in the abstract, but for practical purposes, in the time frame we care about—centuries—they are trivial, like the term that is dropped out in a mathematical derivation because it complicates things more than it is worth and is small enough to be safely ignored. Strictly speaking, there is a relativistic contraction of an automobile on a highway, but no engineer in her right mind would think of paying attention to it. To the question "How can such changes in the behavior of the species as the spread of literacy and the fertility transition occur in only a century?" the genes have little to say. That can only be answered by conventional behavioral science: the laws of learning, cognition, social psychology, economics, and culture change. That genes underlie all these laws is a truism, but in these cases it doesn't help at all—which is to say that it doesn't help us solve two of the greatest challenges facing the human species in our time.

Of these, the issue of family size is far more serious today, and we will come back to it in the next chapter. In a general sense, Darwinian theory has the answer: we are built to reproduce, so population control is far more difficult than it looks. Yet family size has declined worldwide for generations. It has not happened fast enough, but it has happened, and for an evolutionist it is puzzling. Some nations, like France, have pro-birth policies to retard it; others, like China, have severe measures to limit family size. Family size has declined under both kinds of urgings. Elsewhere, as in Thailand, rapid declines occurred in the absence of government policy. And in Latin America, where the Catholic church has held back family planning, declines have been slower but still dramatic.

It is called the demographic transition. Public health and nutrition improve, so the death rate drops. People no longer expect half of their children to die. They begin to see the possibility of a better future. Women become healthier, better informed, and more equal. Investment in individual children increases. Family-planning information becomes available. At last the birthrate drops.

It is a well-known story but a serious problem for the Darwinians. This oft-repeated process runs counter to all predictions from theory, which suggest—no, require—that reproductive success will increase as environmental conditions improve. No satisfactory solution to this problem has been offered. Some claim the reduction is temporary, and that some lineages, biding their time, will out-breed others in future generations. But there is little evidence that this is happening. A group led by Jane Lancaster and Hillard Kaplan, at the University of New Mexico, has studied multigenerational reproductive success in men in Albuquerque, specifically looking for some upturn in reproductive rate in the third generation. No such uptick has been found, and the researchers have fallen back on more standard explanations.

We have seen, however, that for such problems as child abuse, nepotism, sociopathy, and sexually transmitted disease, Darwinian theory has proved relevant. And for other problems, biology may nevertheless tell us something about how *not* to solve them. It may serve to make us a bit more cautious in our meddling. And it may steer us away from potential disasters of "human engineering," simply by giving us some notion of the raw material we are working with.

Consider a parable. It is about a hundred years ago, and a practical gentleman is about to design the longest bridge in the world. He has the good luck to be on an ocean liner—not the *Titanic*—with a metallurgist, who makes a fine living exploring the properties of steel. The designer has visions of sunlit days on the upper deck playing shuffleboard, lounging, and smoking cigars, all the while drinking deeply of his companion's knowledge of steel, and in the evenings going back to his cabin and the drafting table. But he is getting nowhere. His questions evoke rhapsodies: *Steel is getting better all the time. People underestimate its potential. Poor present knowledge of the physics of solids is limiting steel's tensile strength. New metals may be created. Why, in my own laboratory experiments are going on that may one day increase the weight-support potential of current girders by 20 percent or more.* The designer's efforts to get some practical information end in frustration. Finally he shouts in exasperation, "But steel, man, present-day steel! I'm going to build a bridge, don't you see? I have to know the properties of steel!"

Many modern conversations between social planners and behavioral scientists take a similar bad turn. Social scientists typically will not say anything about the limits of human potential. But the planners—and that includes all of us—cannot work without some knowledge of those limitations, at least in the short run. While we wait for human beings to be transformed—by some combination of science, spirit, and the very best of will—into the pliable raw material we all want them to be, we may lose our last chances to take practical action, to ensure that people are around long enough for that ultimate and happy transformation. To recognize the limits of human nature, and the dangers built into it, is also a basis of change.

CHAPTER 17

The Invisible Galaxy

Evolutionary humanism is a demanding ethics, because it tells every
individual that somehow he or she shares a responsibility for the future
of our species. . . . Every generation is the current caretaker not only of
the human gene pool but indeed of all nature on our fragile globe.

—ERNST MAYR, *This Is Biology*

Although it has been argued otherwise, there is undoubtedly life, including
intelligent life, elsewhere in the universe. By "undoubtedly" I mean not with
absolute certainty, of course—only what I might mean by, "The sun will undoubtedly rise tomorrow morning." I mean with the strange and wise combination of
common sense, science, and metaphor that is the closest we get to certainty from
where we sit on this awkward, lonely planet. We may never find the intelligent
life—it is probably too far away—but it is there.

Unimaginably strong force makes protons and electrons out of quarks in an
unimaginably small period of time. Before long there is not only hydrogen but
helium and other, larger elements. Gases condense into fusion-explosion balls.
Chunks of these spin off into globes of liquid fire. The incandescent spheres condense into rock, which may or may not be laced with water, itself a simple union
of two elements.

At the right distance from a star of the right age and temperature, in the likely
presence of carbon, phosphorus, nitrogen, and oxygen, life is inevitable. Given
the numbers of galaxies of stars—*billions and billions*, as Carl Sagan used to say—
and the time, which the universe has plenty of, life must appear not in one place
but in many thousands of places, and intelligent life must emerge in some fraction of those. Until the late 1990s it was reasonable to counter that astrophysicists
had not found any planets at other stars, but as soon as the measuring tools
became fine enough, planets began to multiply. And if we do find extraterrestrial
life, it will not be based *im*probably on, say, silicon, but *probably* on the same stuff
that sparked it on our own wet, whirling world. It is not likely to look like us, nor
much like anything imagined in science fiction, but it will be there. Although

there is truth in Stephen Jay Gould's claim that evolution is not directed anywhere in particular—certainly, we were not a necessary end to it—that does not mean trends don't exist. Wherever there are simple living things, they will tend over time to produce at least some larger, more complex, more adaptable, and more intelligent forms.

Life is a balancing act staged in the face of relentless entropy, and only information can make such order more probable, at least in a few local eddies, than disorder. Information, in turn, can be stored in only a few ways. One is in long, structured molecules with changeable components—what physicist Erwin Schrödinger called "aperiodic crystals" (as opposed to periodic or repeating molecules, like nylon) predicting the basic nature of life. The second is in collections of electrically active cells, whose actions and connections are changed by events outside them. And the third is in cultures, which use imitation, language, symbol, and other devices to store, transfer, and stabilize learned information far beyond the individual organism. How these three kinds of information machines work has been the main preoccupation of late-twentieth-century science. Since mid-century it has been clear that the secret of life lies coiled—triply coiled, in fact—in an immensely long molecule, in the center of almost every cell. Just one snippet of these sinuous masterpieces, appropriately placed, can uncoil majestically and make either multiple mirror-image copies of itself (the essence of life) or even more copies of transcriptions—usefully transformed versions of itself.

Each of these tiny spiral stairways from the past in turn makes one of the many thousands of huge molecules that build body and brain. Architect, messenger, builder: each a giant, godlike as molecules go. Yet each, crunched and twisted, would seem a minuscule blob, a just-visible speck, under the strongest light microscope. Array them in rows or strings under a scanning electron microscope, and they start to look substantial yet poignant, like forlorn, overdone pasta stuck to a pot bottom. Yet these dull-looking, invisible strings swirl and climb to form whole galaxies of atoms, ordered, informative, generative, harbingers of all we need, codes for all that ever lived or ever will, beautiful, secret worlds within worlds.

It is quite possible that short clips of these order-bearing strings fell together from nucleic acids and sugars—the ring-based carbon compounds that are their fundamental components. This is the conventional view of the origin of life, and it is a likely one, repeated in laboratory simulations of the chemistry of the earth's primordial seas. It is also possible that life began differently, as in Stuart Kaufmann's emergence model. Here a molecule speeds a reaction between two others, producing a new compound that in turn speeds another reaction, and so on in a daisy chain of catalyzed chemical operations until, after six or eight links, they become recursive, generating the original two compounds and their catalyst. This has worked well in computer simulations, and may be as probable as the conventional view. But either way, we have to end up with long information-storage molecules. Aperiodic crystals. DNA.

This strange stuff perpetuates itself—or, rather, the information stored in its sequence—by making mirror-image copies, which in turn make mirror-image copies, duplicating what was already a duplex form, a double helix. This takes energy, and, of course, the world's disorder is always trying to tear the helixes down. They can only prevent this by anticipating that disorder more and more deftly, using the information they store. In practice, this usually means that they make other stuff—RNA, and then protein—that in one way or another protects them from ruin. At first, this will mean simple devices for cloaking the ordered molecules, or for patching the little rips and cracks made by stray cosmic rays.

Pretty soon you have something a bit like a virus but better—independent, a string of protected stuff that can reproduce on its own. Whether you consider such a thing to be alive or not, it is clearly a self-replicating system; and as we know from sad experience, it can do very well making a living at the expense of much more complex creatures, once they exist. Still, complexity did not prevent such creatures from eventually being assembled—or, rather, from assembling themselves, using the directions stored in their own DNA. *These* strings of DNA, to be sure, had to be longer, store more information, and make more stuff from the soupy sea around them. The more information, the more stuff, the more complex the organism, the more elaborately it protects the information in its DNA. But however complex it is, it can only be there to let the replicators work, to let DNA make DNA, to stabilize the sequence, the knowledge, the order, over seconds, days, years, eons. We return to the indelible, simplifying epigram: *a gene's way of making another gene.*

Thus life, and the engine of its history. Thicker and more complex protoplasm, made up of ever more varieties of molecules, doing ever more different and more interesting things, including behavior and, especially, reproduction. Replicators replicate. Replicating replicators last. They collect stuff around them, becoming organisms. Organisms foster replication. Those that didn't are not here to show and tell. "Urge and urge and urge," wrote Walt Whitman in the 1850s, "always the procreant urge of the world." But the procreant urge of the world is neither more nor less than the insensible, blind tenacity of genes.

Replicators, however, are not always genes. In humans and some other animals they can be elements of culture—memes. Memes can oppose genes, promote them, or do neither. They can spread like molasses or like wildfire; they can even "infect" our brains with genetically maladaptive purposes. Empirical studies and mathematical models alike show that memes, however intangible, have aegis in ours, the most cultural of species. But they never strip the genes of their power. The genes abide with us, their clumsy vehicles, and what we are, feel, and do, including how we love, grieve, hope, and die, belongs as much to their purposes as to ours. Directly or indirectly, they are the source of all the beauty and ugliness of life, the blind, creative power behind life's transformations. By 3,500,000,000 years ago, the first cells; by 350,000,000 years, the first land animals; by 35,000,000 years, the first primitive apes; by 3,500,000 years, the first protohumans; by 350,000 years, the first of our species; by 35,000 years, the first great art and the dawn of true culture; by 3,500 years, the first large urban empires; by 350 years, the first modern science and industry; by 35 years, the genetic code. From the first gene to the first mastery of the gene, a giant, jagged, crude, recursive circle, completed in less than 4,000,000,000 years.

These are estimates, but they are close to the mark. The last few changes, perhaps including the dawn of art, did not demand new genes. Yet in a sense they are still genetic, as is every living thing. Once there were creatures like humans, there had to be culture; once culture, art; once agriculture, civilization; once civilization, industry; once industry, science; and finally, the growing control of mind over genes. "Had to be," because our genes provide the ambition, the intelligence, the creative impulse, and the population growth that created the need, forced the invention, and advanced or at least complicated the culture. Nothing that led up to us could have been predicted, except for slow, fitful increases in size, complexity, and intelligence. No all-seeing eye could have predicted the catastrophes—mostly repeated asteroid impacts—that wiped out the bulk of life on earth; it could have predicted that there would be impacts, yes, but the timing in life's history, no. Georges Cuvier, the great pre-Darwinian evolutionist, said it best:

> Existence has thus been often troubled on earth by appalling events; calamities which, in the beginning, may have shaken to a considerable depth the entire envelope of the planet. . . . Living creatures without number have fallen victim to these catastrophes. . .

Earlier forms survived each crash—some by chance, some by adaptation. But afterward, in an almost empty world, natural selection had a vast new challenge: fill the world with novel forms metamorphosed from the remnants of the old. Unlike the meteor impacts themselves, this post-extinction work was not random but largely deterministic and thoroughly Darwinian.

We humans, given our growing technical knowledge—joined to our generalized, low-key but constant paranoia—may contrive to predict and deflect the next unluckily-aimed asteroid, averting the next cosmic extinction. Since such an impact is very unlikely for the next few thousand years, we should be technically ready to take it pretty much in stride when it does happen, without the desperate cliffhanger of Hollywood's *Armageddon*. In fact, after the cold war ended, scientists leading the Strategic Defense Initiative program wanted to turn this "Star Wars" antimissile technology to the task of asteroid defense. (*When your only tool is a hammer . . .*) Sooner or later, though, it will be tried, and it will be possible. But this assumes that we survive the next few thousand years. If we can't, the fault will lie not in our stars but in our genes.

To understand how thoroughly our culture has accepted genes, consider three headlines. One appeared across two pages of *Newsweek,* in a special 1997 issue about children: "Scientists Estimate That Genes Determine Only About 50 Percent of a Child's Personality." To those of us who remember when genes were completely discounted, the word *only* speaks volumes. On September 2 of that year, *The New York Times* ran an article called "Some Biologists Ask, 'Are Genes

Everything?'" And the lead-in to an article by Robert Sapolsky, in the *Newsweek* of April 10, 2000, was equally plaintive: "It's Not 'All in the Genes': The Environment You Grow Up in Is As Important As Your DNA in Determining the Person You Ultimately Become." In a mere two or three decades American scientists and journalists had gone from explaining almost everything in our behavior by the environment to trying to defend its influence as roughly equal to that of genetic effects.

Whatever we do, the genes at least vaguely guide us. They are not charmed either mystically from without or vitalistically from within; they are just big molecules that tend to stick around. But to do so, they must toss off messengers that build protein products that make us do the things that keep them around—or, to be more exact, things that *kept* them around through eons of evolution in a long succession of past environments. Things like withdrawing from pain, tagging and destroying some microbes while tolerating others by the billions, laying down fat in every moment of abundance, designing tools, abandoning children, dying to save other children, stealing copulations, deceiving ourselves and others, declaring lifelong love, delivering it, breathing a continuous mild buzz of grief and fear, reacting to frustration with threat and counterthreat, showing tenderness, piling up unspendable wealth extracted from those with much less, smiling kindly, spawning new, needier lives that shorten the fuse on the population bomb, crowding out thousands of species every year, slowly cooking the globe, creating new opportunities for exotic, deadly viruses, spreading nuclear weapons over the planet, and waging, in poet Robert Lowell's words:

> small war on the heels of small
> war . . . the earth, a ghost
> orbiting forever lost
> In our monotonous sublime.

Genes abide with us in inner ways, too. Magazines tell us of genes for happiness, sexual orientation, anxiety, sadness, thrill seeking, intelligence, grammar, memory, obsession, and phobia; and there is a piece of the truth in most of these claims. But that piece of truth must lie in a protein, because making one is all a gene can do; gene (DNA) makes messenger (RNA); messenger makes protein. Thus, the "central dogma" of genetics. Yet unlike every other dogma, it is sorely tested and lives to tell the tale every day in a thousand laboratories all over the world. This protein—protective, portable, primal, and protean—is all a gene can cough up—except, of course, another gene. The rest is development.

But development is *big*. The protein has potential. It can be structural, like the collagen in skin. It can be a receptor for a hormone or neurotransmitter. It can be an enzyme that helps manufacture or remove one of those smaller behavioral molecules. It can be chopped up to make smaller strings of amino acids that themselves serve as hormones or neurotransmitters. It can be a tag on an embryonic nerve cell that drags it to a certain spot up- or downstage, destined for a starring role or bit part in the brain. It can be a sentinel on the tip of a nerve cell's tongue, seeking its molecular mate from among the almost endless embryonic possibilities, wiring up the soggy circuit boards of mind. Or, if it really wants to puff itself up in importance, it can turn on another gene, or even a suite of genes.

That is about all proteins can do, and so it is all genes can do. They cannot think, love, learn, or strategize. They cannot even hurt, want, or grunt. They can only copy themselves and make proteins. Yet that is enough to constrain everything—and more than constrain; it is enough to encourage us along some paths instead of others. Therefore, it is not surprising that the genes' contribution to a wide range of traits—normal or pathological, physical, biochemical, mental, emotional, or behavioral—amounts to some 50 percent of the variation, on average. Some traits are almost *all* gene—or at least will be until we find a way to foil the gene. Others reflect little or no genetic effect. But 50 percent is a seat-of-the-pants average worth remembering.

Come home at the end of the day and stand before a full-length mirror. Look at your face and body. Think about what you did that day, what you plan to do in the evening, what relationships you have, how happy you are, what risks you took, when you last had sex, when you are next likely to pray, what shirt or blouse size you wear, how anxious you are about owing your boss a call, how annoyed you are that a certain someone hasn't called you, whether you took work home with you or not, how well you can tune out the heavy-metal band blaring upstairs, whether you'd rather have more pay or more vacation time, what color you'd wear to a neighbor's party, whether you're better at music or drawing, how quickly you get bored at the ballet. In all these and countless other ways, your rank among all the folks on your block would be about half predictable from genes.

Now, there's the rub: *on your block. On your block* cuts out an awful lot of variation. On your block no one wears a grass skirt, worships household idols, gathers wild vegetables for a living, builds an igloo every winter, paints cave walls to make magic, speaks Penobscot, sends twelve-year-olds to die in holy wars, observes absolute silence for a decade, dances herself into a rain-bringing trance, sacrifices chickens to cure cancer, excises clitorises, grows commercial cocaine, collects human heads and shrinks them, weaves age-old patterns into cloth with wood shuttles flying over handmade looms, sends a boy into the mountains for days to hallucinate a personal spirit, gives birth alone in the backyard, pounds heart-stopping syncopated rhythms on skin-tight belly-high drums, apprentices his ten-year-old son ritually to fellate his teenage friends and thus imbibe their masculinity, fattens a girl for months before her wedding, builds a tunnel for her to crawl through as a cure for infertility, or slashes a long, deep gash on the underside of his nephew's penis to make the boy a man.

These are real and present cultural variations that create diversity against the background of a largely constant genome. Since the calculations of heritability in even the best studies *never* include such dramatic cultural variety, there is a real, statistical sense in which they always and inevitably overstate the power of genes. Potentially, we can create a culture of the future that, while perhaps not as exotic as the ones mentioned, could depart as dramatically and fundamentally from what we presently have on our block. If we do that, there is no telling how weak the genes may, in retrospect, finally seem. But in the meantime, given the culture we do now have—you and I, gentle reader, members of the planet's literate post-industrial aristocracy—the genes have great power. And in fact, if anything, we tend to understate their power when we look at individual variation. Most of what we are and do is what we *all* are and do, because we are human, *Homo sapiens*, and 98.5 percent of that is no different from what makes a chimp a chimp, or a bonobo a bonobo.

Although the main subject of cultural anthropology has been cross-cultural variety, it has always had an inevitable, if tacit, complement: things about human life that vary little or not at all. But "universal" has at least five different meanings. First are things like upright walking or social smiling, shown by all normal members of the species. Second are acts universal to a given age or sex—the sucking reflex in newborns, for instance, or the ejaculatory pattern in adult males. Third, some universals affect all groups but not all individuals, such as the sex difference in physical fighting. A fourth group of universals applies to culture instead of behavior, like taboos against incest and in-group homicide, or the varied but always definable bond of marriage. Last, there are things that are unusual yet are found at some low level in every population, such as homicide, suicide, depression, schizophrenia, or incest.

The lists in these five classes are long, much longer than most anthropologists would have predicted. Ethologist Irenäus Eibl-Eibesfeldt described many constants of communication, Paul Ekman codified facial expression, and Donald Brown compiled the most complete list of universals, building on others. The search for societies without violence, without gender differences far beyond childbearing, without mental illness, or even without the ability to make fire has been a vain one. Although the frequency or context of most behaviors varies, the large, stable core proves that human nature is real. Universals describe our nature.

Yet traditional cultural anthropologists, even when reluctantly accepting the existence of universals, show little interest in them, deeming them trivial or outside anthropology's proper subject matter. This is like being interested in differences in diet but not in the way the gut works—fine for a chef but not very good for a scientist. To be fair, cultural anthropology does not always claim to be a science, and its historical mission is to document and validate the great variety of behavior and experience. Yet finding universal features of behavior and even culture is also a central task of anthropology. Even many cultural anthropologists—Claude Lévi-Strauss and Victor Turner, for example—have tried to find universal symbol systems and mental structures to link widely different cultural forms in art, language, and ritual. Dan Sperber and Bradd Shore have independently developed theories of culture that draw on universal features of the mind, and Sperber's goes on to establish continuity with current theory in evolutionary psychology. This welcome approach in cultural anthropology resembles that of linguists, who find common features of all languages, transcending their obvious variety.

But if the deep structure of mind is universal and genetically determined, what is the role of variety after all? We just spent a whole chapter on change; higher animals are plastic in countless ways, especially early in life but even in adulthood. Given the number and variety of these findings, and the fact that they range over the whole vertebrate family tree, adaptation itself must entail deep flexibility. So how can we reconcile genes with change, universals with variety? The short answer is that flexibility does not imply a completely open program. More often we find a *guided* flexibility, a range of reaction, a finite channel of good possibilities.

Even in the sex life of the redwing blackbird, flexibility is evident. Males that sing in richer territories—areas with more sumptuous food resources—mate with several females instead of one. The mechanism of this flexible pattern, known as a facultative adaptation, must be quite different in blackbirds from the parallel adaptation in human beings, but it is obviously not under tight genetic control. To take a more general and powerful example, consider how various creatures achieve flight. The wings of insects come from the thorax, of birds from forearms, of bats from fingers, and of humans from airplanes. Four solutions to the same adaptive challenge; similar functions, different devices. The same is true of adaptations in social behavior.

Take incest. Sociobiologists (like classical evolutionists and geneticists before them) predict that incest will be avoided in most sexually reproducing species to avoid getting individuals with paired maladaptive genes. But for this to happen, individuals on the verge of doing what comes naturally must recognize and rule out close kin. In insects and some vertebrates, recognition depends on pheromones: sniff the air around a friend or stranger, and you can guess how closely you are related. But even in ground squirrels, learning during development plays a role, and in humans the implausibility of aromatic triggers led to a search for other explanations.

Anthropologist Arthur Wolf, motivated by questions outside of sociobiology, discovered another mechanism. Working with standard ideas about kinship, he studied traditional China, where little girls sometimes came to live with the families of their intended spouses, who were still little boys at the time. The resulting marriages had a much higher rate of failure and infertility than did other arranged marriages. Wolf even identified a sensitive period for the effect—contact had to start between ages three and six. A related discovery was the exceedingly low marriage rate among Israeli kibbutz age mates: having grown up in close proximity, they have little sexual interest in one another. Both lines of research support the "familiarity breeds contempt" hypothesis of incest: brothers and sisters avoid incest because growing up together dulls sexual interest. Edward O. Wilson wisely made this a centerpiece of his understanding of *human* sociobiology. With our fancy blend of dependence on and freedom from genes, *we* avoid inbreeding through a developmental process that depends on cultural choice—even though the ultimate evolutionary impact in most cultures is the same as in species that base their avoidance on odors. In analyses like this, psychology and sociobiology are joined, and the study of human behavior in general is much better served than by sterile debates about nature and nurture.

However, if we insist on casting the discussion in debaters' terms, a win-or-lose argument where we have to choose sides, nature has won. Educated people accept a role for genes in behavior and even an evolutionary origin of the mind, to an extent unthinkable until very recently. This is because we are able to think about genes and about evolution in ways that have never been possible before. John Alcock's book, *The Triumph of Sociobiology*, states the case succinctly and persuasively, as of 2001. The quarter century since E. O. Wilson's classic *Sociobiology: The New Synthesis* has left little doubt as to the basic validity of the theory in the minds of most scientists who understand the data. But applying it to people is the subtlest move of all, requiring the broadest range of knowledge and methods, so that instead of accepting stupid and dangerous results, as the Social Darwinists of the nineteenth century and the eugenicists of the twentieth century did, we can begin to make cautious use of smart and helpful results. There is still danger, but the motives and possibilities are changed.

And of course, genes themselves are now profoundly changeable, at least in mice. The "mouse ranch" at the Jackson Laboratory in Bar Harbor, Maine, produces millions of them each year for research; some are literally worth their weight in gold. Of the 2,500 distinct Jackson strains, most have so far had their genes controlled only by conventional means—sex, sex, and more sex, starting at age six weeks and limited to close relatives, until each strain consisted entirely of almost-identical "twins." But today's "mouse wranglers" increasingly knock genes in and out, manufacturing unique genomes to include or exclude highly specific strings of DNA. These "fuzzy test tubes"—models of obesity, Alzheimer's disease, immune deficiencies, and scores of other defects—set out for labs worldwide, where, heroically if unwittingly, they help advance the genetic revolution. Some are futuristic species hybrids: they bear inserted human genes and pass them on through any number of mousy generations.

For us humans, an early gain of the revolution was diagnosis. In 1983 gene hunters succeeded for the first time in locating a genetic disease, Huntington's chorea. *Chorea* comes from the Greek word for "dance" and describes the tragic, uncontrollable, dancelike movements of the limbs that mark the afflicted. Woody Guthrie, the great Depression-era American folk singer, died of it; as is often the case, his first symptoms were of emotional instability, as the gene began to take its toll on neurons in the basal ganglia and limbic system. James Gusella of the Massachusetts General Hospital led the team that localized the gene to a smallish region of the short arm of chromosome 4. Soon it became possible to diagnose the disease before it started, usually between ages thirty and fifty. The test involved checking for a marker—a known string of DNA that was so close to the Huntington's gene that no chromosomal recombinations could come between them; everyone who had the marker had the gene.

In *Mapping Fate*, Alice Wexler told the moving story of her afflicted family and how their lives have intertwined with research on Huntington's disease. Her mother, Leonore Sabin Wexler, died of it after many years of suffering that transformed the lives of her daughters. Their father, a psychoanalyst divorced from Leonore before the disease was diagnosed, began raising money for research on Huntington's. Both daughters, Alice and Nancy, followed in his footsteps, and Nancy Wexler became one of the world's leading scientists researching the disease. Her team, studying patients and their relatives in a large extended family in Venezuela, carried out the successful hunt for the gene.

The geneticists steadily closed in on it, and in 1993 they found "IT95"—IT for "interesting transcript." Indeed it was. Unlike the point mutations—one substituted base pair—featured in many bad genes, IT95 is full of triplet repeats, a sequence of three bases (C, A, and G) repeated again and again. Up to a point, repeats have no abnormal effect. But much above thirty, and your risk for Huntington's disease crosses a tragic threshold. Each CAG triplet codes for the amino acid glycine, and if the protein made by the genes—Gusella named it huntingtin—has too long a string of glycines, it gloms on to a nearby, larger piece of molecular garbage and takes it into brain cells, killing them.

Despite our increasing understanding of this story, as of mid-2001—eighteen years after Gusella first located the gene—there was no treatment for the disease. For Alice and Nancy Wexler, as for thousands of other daughters and sons of victims, there was only the two-edged sword of diagnosis. The possibility alone, not to mention the dreadful news that follows half the tests, has strong and unpredictable emotional effects. Even good news is not uncomplicated, leaving residues of guilt over having escaped the fates of other family members. Not everyone wants to know. Fetal testing raises moral questions about abortion, and about the ethics of gene control. In more ways than one, the Wexler family and others involved with Huntington's have pioneered the biology and the biological ethics of the future.

Even now we can all get partial gene profiles, sufficient to tell us something about our destiny. Some industries test prospective workers for genetic vulnerability to substances they may have to work with; careers hinge on the answers. Health and life insurers want to know all they can about our genes before they take a chance on us. Orthodox Jews, some of whom still have arranged marriages, discretely test young hopefuls for the gene for Tay-Sachs disease, a dreaded and fatal form of mental retardation specific to some European Jews. Two carriers will not make a match. But the same community is wary of collecting more genetic information—relating to cancer risk, for example. Remembering the "racial hygiene" of the Nazis, they fear genetic stigma. Iceland, which also has a relatively homogeneous population, has built a universal gene data bank for its 270,000 citizens, provoking intense controversy about the risks of surveillance. In the United States, private companies have attempted to recruit tens of thousands of people to join a gene data bank voluntarily, for research. Ultimately, knowing what genes we have will not only predict the future but enable us to change it, with exquisitely personal programs of diet, exercise, and drugs made to order for our own private genes.

But getting beyond diagnosis has been difficult. If we can insert genes into mice with impunity, why not do the same with humans? For one thing, it's harder. By 2000, a decade of gene therapy could boast only one clear success, under Dr. Alain Fischer of the Necker Children's Hospital in Paris: four infants with genetic immune deficiency had been injected with a replacement gene. Gene therapists avoid the "C word" (cure); nonetheless, these children had normal immune systems ten months later. Other researchers have had more modest results with hemophilia. But a slow pace of discovery has greatly reduced private investment in the field. More important, the tragic death of eighteen-year-old Jesse Gelsinger, in September 1999 during a gene-therapy experiment, caused a crisis of public confidence.

Jesse died a hero, volunteering to test a therapy for babies much sicker than he was. It should have been routine, but he had a fatal immune reaction to the usually safe cold virus carrying the gene. The appropriate result is that now all research on human subjects comes under even more stringent scrutiny than it did before. Two thousand gene-therapy scientists meeting in Denver a few months later stood in silent tribute to "the young man who has given his life in pursuit of an ideal treatment." Others there referred to "a new humility in the field." But without doubt the field will move forward.

And of course, the heroic approach of gene therapy is not the only way to control genes. Humans have always used sexual choice for this purpose, and now we make "test-tube babies"—at least 300,000 worldwide so far. Louise Brown, the first, turned twenty-two on July 25, 2000, and is doing nicely, thank you. She was made by having her dad's sperm joined to her mom's egg in vitro and planted in her mom's womb—in vitro fertilization, or IVF. Simple stuff today, but that was just the beginning. Now parents can use an egg donor, a sperm donor, and a surrogate birth mother, all perhaps different, adding three biological parents to the two making the contracts and planning to raise the child. When biology later rears its head, courts have a tough time with competing claims.

ART—assisted reproductive technology—continued to pose new legal and ethical challenges in the new millennium. Each year in the United States alone this $2 billion industry gives us 60,000 births from donor insemination, 15,000 from IVF, and 1,000 or more through surrogate pregnancies. All over the world, similar techniques are used. A South African woman, age forty-eight, gave birth to her own daughter's triplets, becoming the surrogate mother for her grandchildren. But doctors have pushed the envelope of age much farther. A sixty-three-year-old woman in Los Angeles lied about her age to fertility doctors and, after intensive treatment, became the oldest person on record to give birth.

Because it involves the implanting of multiple embryos to increase its chance of success, IVF has produced some famous sets of septuplets. Multiple births make up 37 percent of ART deliveries, compared with 2 percent for the general population, and some countries limit the number of embryos transferred. Extra embryos are routinely destroyed or frozen for long-term storage and have been the objects of legal battles just as babies have. IVF clinics may screen embryos for

undesirable genes, such as cystic fibrosis, and selectively implant preferred candidates. Sex may also be selected in this way, and the future may hold such questionable prospects as choosing eye color or personality.

Controversial as these techniques may be, they don't hold a candle to what's
coming at the other end of the life course. Hundreds of laboratories are trying to
find the genes that age us and flat-out conquer them, effectively ending the nor-
mality of death. If they do, people will still die from accidents, violence, and many
specific illnesses, but the overall life span could double or more, and old age
could become a thing of the past. We could achieve our fondest, age-old dream,
relaxing into an infinite future with no fear of death. Only taxes would be
inevitable. But without a complete suppression of reproduction, this will ensure
the planet's destruction. Yet how would we stop it? As University of Manchester
scientist John Harris put it, "How could a society resolve deliberately to curtail
healthy life, while maintaining a commitment to the sanctity of life? The contemplation of making sure that people who wish to go on living cannot do so is
terrible indeed."

Meanwhile, what sanctity does life have when it's patented? In the years
bracketing the millennium, scores of thousands of patents quietly piled up, cov-
ering genes, cell and organ types, methods of gene insertion and transformation,
and genetically engineered life-forms. Does a cell line belong to the person it
was originally drawn from, or to the company that figured out how to breed it
and keep it alive? What rights, if any, do the five people whose genomes were
sequenced have over the information about them that now belongs to the world?
Someone has even tried to patent a hybrid creature combining human and
chimpanzee—actually a chimera, or mixture of early embryos, called a
"humanzee"—supposedly for the express purpose of preventing anyone from
ever creating such a life-form. There are countless such legal issues that have no
meaningful precedent.

The 1997 cloning of a sheep—Dolly—from an adult cell of another sheep
stunned a not-quite-ready world. Other species followed, each involving different
technical challenges. "Pharming," the use of nonhuman mammals to produce
pharmaceuticals in their milk, including the products of inserted human genes,
was underway. Mice were cloned in substantial numbers, followed by calves and
then adult cows. "Swine Squared" was a television headline on August 16, 2000,
when CNN announced that pigs had been cloned. By 2001 at least six mammalian species had been cloned from adult cells. Monkeys had been cloned from
fetal cells, and Dolly herself had been cloned, starting a wave of copied copies in
various labs throughout the world

Exotic species were cloned to avert extinction. A baby gaur—a rare species of
one-ton Asian ox—was born to an ordinary Iowa cow, after the cow's-egg DNA was
replaced with a gaur genome and the egg was implanted in her uterus. Plans are
under way for cloning cheetahs and giant pandas, two of the world's most trea-
sured endangered species. The last bucardo, a rare Spanish mountain goat, was
killed when a tree fell on her, but Spanish scientists saved her cells, and they may

soon *reverse* the extinction. Woolly mammoths thawing out of Arctic ice may well be next.

Almost certainly the same could be done with human cells, and at least one official attempt is under way, a collaboration between an Italian fertility clinic and a group at the University of Kentucky. Although human cloning seems ominous—raising science fiction specters of multiple Hitlers, as in the film *The Boys from Brazil*, or merely a few harmlessly clumsy copies, as in the benignly funny *Multiplicity*—it may be with us sooner rather than later. Dolly herself took most scientists by surprise, because her creation was an unexpectedly large technical step, large enough that some scientists were initially skeptical that it had been done. The field moves fast. A couple of years later a Texas rancher named Sandra Fisher, in mourning for her favorite steer, had him cloned. "We loved Chance," she said. "The thought of having him again is really exciting." She named the clone Second Chance. Others lined up to have their favorite pets immortalized. A couple gave scientists at Texas A & M a $3.7 million grant in exchange for having their dog cloned. Imagine the power of such motives applied to human beings. It is not surprising that there is already a vigorous movement dedicated to making it happen.

In any case, the moral and religious debate is under way. The benefits of animal cloning would be numerous, but the motives for human cloning are different. The rationale for the Italian experiment is to help infertile couples who do not want to use other people's eggs or sperm, and there are many other possible reasons. We could create individuals with a known genetic complement but altered to eliminate a deadly childhood disease. Some people might want to try raising "themselves" to see how they would have turned out with better parents. Or, like the pet owners mentioned, they might want to clone a dying child. Then too, cloning can mean not only the duplication of whole human beings but also the creation of body parts. Having transplantable organs genetically identical to their recipients would be of immense value in medicine.

Objections to human cloning include claims that it is a departure from the natural order, that it amounts to playing God, that it assaults the idea of individuality, that it would reduce human variation, and that it would be subject to abuses. But religious and ethical answers to these objections are possible. Some religious authorities oppose cloning, but others do not. Some theologians see genetic and reproductive manipulation as not only permissible but a part of human responsibility—not playing God but fulfilling human purposes, serving God by joining and completing the work of creation. Natural order is not fixed, and while we are part of this order, it is also in our nature to change it in cultural ways. We are simply trying to soften the edges of a natural, but also capricious and harsh, reproductive roulette.

As for individuality, the blurring of the line between individuals is no more serious a problem for a clone than for an identical twin. Although it can be a challenge for twins and their parents to insure psychological separateness, it is quite possible to remain an individual, and in a subjective sense individuality is absolute. Cloning is different, of course. It involves deliberate choice, and it creates a "twin" who is younger, but the problem of identity need not be greater; indeed, the age difference might make the line between individuals easier to draw. As in most of human experience, cultural construction of such a child's situation would strongly influence how she felt about herself and others. A world in which such children were common would learn to welcome them.

It is easy to be alarmed by cloning if we think genes are overwhelmingly powerful, but of course they are only one source of identity. They set development in motion and regulate it throughout life. Yet because of their sensitivity to initial conditions, genetically identical embryos are highly individual long before birth. More important still are the shaping effects of environment before and after birth. To grant cloning the power to blur individuality is to grant genes far more influence than they have. In any event, cloning is likely to remain inefficient, expensive, and error-prone for a long time. Abuse is certainly possible, but to set in advance an arbitrary limit on what we can know and do is also a form of pride. It is not easy to think of instances in which science was successfully held back by fiat. We decided not to have taxpayers fund human cloning research, and perhaps we could even effectively have such research banned in American laboratories. But what of other governments and private corporations around the globe? Did we really think that we could control it all?

When the human genome was almost completely sequenced, Dr. Francis Collins, the leader of the government's arm of the project, was interviewed on the Charlie Rose television show. Rose asked him whether he thought somebody, somewhere, might be experimenting with human cloning. He replied that he didn't think so. Pressed for a serious answer, he said wistfully, "I hope not." No wonder people mistrust scientists. Whatever we think about the complexity of the question, *I don't think so* and *I hope not* are not adequate answers. At that very moment, scientists had already planned the Italian human cloning study, and less than a year later they were carrying it out.

In a black-and-white world, human cloning would just be right or wrong, but the real world is shades of gray. We might decide that cloning would not be good in general and yet that certain conditions justify it, and carefully screen those who request it—as we do for surrogate motherhood or sex change. Consider a man dying of cancer who is the last of his family line; or a lesbian couple who want daughters cloned from each of them; or a woman whose husband has sickle-cell anemia but who finds a stranger's sperm intolerable. Human cloning may be bad for society or the species. But to look into the eyes of a person who could be helped by it and say no on philosophical grounds—*No, you may not have this, even though I could do it and it would help you*—is something else again.

New technologies for shaping the destiny of embryos have been called "unnatural selection, and in some sense this is apt." But we know that almost all hunting-and-gathering societies abandoned or killed defective infants. Such selection must have begun at or near the dawn of human consciousness. The majority of traditional cultures also attempted in crude ways to terminate unwanted pregnancies. So we don't really have a new ethical crisis, except that new techniques serve the same intentions. The effort could be immoral, but it is not unnatural. Since the invention of stone blades and the mastery of fire, creatures like us have repeatedly faced the fact that what can help can hurt. Perhaps we would have been better off without stone blades, but we could not have avoided inventing them. We are stuck with our inventiveness, so in gene control as in all else, we may as well make the most of it.

And of course, mapping the human genome is a case in point. Despite some ethical questions, this has largely been an open international effort that exemplifies the power of collaborative learning, a type of cultural learning unique to our species. By 1998, a vast chromosomal map had been pieced together. At the time it was thought that it comprised fewer than half of our genes, yet mapping the rest was a foregone conclusion. Craig Venter, a former government-funded scientist, founded Celera, a private company built to beat the international consortium to the prize. With millions in venture capital and billions more after the public offering, Celera amassed sequencing machines in vast factories, clicking through the rest of the genome at breakneck speed. Government-funded labs throughout the world had to choose between cooperating with Venter or being scooped.

It was an offer they couldn't refuse, and the collaboration worked. On June 26, 2000, at a White House conference, the Human Genome Project and Celera jointly announced the sequencing of the first complete "draft" of the code. The president of the United States and the prime minister of Britain (over the "tele" from 10 Downing Street) spoke first at some length, shamelessly stealing scientific thunder. After all, history was being made. The next day, banner headlines proclaimed a new era in medicine: "Genetic Code of Human Life Is Cracked by Scientists," blared The New York Times in inch-high letters. These proclamations were largely hype, but the moment did seem worth marking. Not many years ago a gene was just an idea, an inference from patterns of inheritance. Then it was a term in the algebra of evolutionary theory. Now genes are physical and chemical realities, each with a place, a name, and a set of quite predictable consequences for disease, biology, and behavior. To appreciate how concrete the reality now is, consider just one gene per chromosome, each playing a role in one or another tragic condition.

Chromosome 1 has a gene related to Alzheimer's dementia; 2, one that causes tremors; 3, a gene that suppresses lung tumors; 4, a syndrome that includes harelip; 5, cri du chat, a tragic syndrome that, among other defects, gives a newborn baby a catlike mew; 6, a gene for epilepsy; 7, the gene for leptin, which, when defective, causes obesity; 8, the one for Werner's syndrome, which causes early aging through a defect in an enzyme that unwinds DNA; 9, another tumor suppressor specific to skin cancer; 10, a metabolic error that causes progressive blindness; 11, long QT, an abnormal heart rhythm that causes sudden cardiac death; 12, phenylketonuria (PKU), a cause of mental retardation now preventable by newborn screening and dietary change; 13, BRCA2, one of the breast cancer genes; 14, another Alzheimer's gene; 15, Marfan Syndrome, which made Abraham Lincoln tall, thin, and depressed; 16, Crohn's disease of the bowel; 17, the p53 tumor-suppressor gene; 18, a gene for pancreatic cancer; 19, the gene for the receptor for LDL, the bad type of cholesterol; 20, "bubble boy syndrome," the genetic, noninfectious equivalent of AIDS; 21, ALS, better known as Lou Gehrig's disease of the nervous system; 22, one kind of leukemia. Chromosome X has a gene that causes color blindness, usually in males, and chromosome Y has the gene for testis-determining factor, which sets in motion an embryo's progress toward maleness. Through unknown means it leads to such symptoms as refusal to ask directions and seizing the remote control.

Although the effects listed for chromosome Y are only half serious, the traits of the other gene defects are well known and all too real. Each has hundreds of scientists following the trail it leaves, a chain of chemical reactions leading from chromosome to symptom. As each trail becomes known, interventions become possible. Many identified genes affect the brain, and so affect intellect, ability, emotion. Some critics say these are genetic disasters, which have nothing to do with the normal range of mind and behavior. These critics are wrong. Disaster genes are just the easiest ones to find. Traits like timidity, novelty seeking, sexual orientation, happiness, and intelligence have also led to hints about gene location. Where the critics are correct is that even if these prove true, the genes in question will only be a small part of the story. Even some disasters are remarkably complex. Alzheimer's dementia is influenced by at least five different genes on as many chromosomes. The five affect thought through at least two different brain proteins disrupting different chemical pathways. And these are only the first steps in the story.

When the 1998 preliminary map—a "shotgun assembly"—was published, the authors were sixty-five scientists, and their names suggested a wide sampling of the larger human family: Delaukas, Rodriguez-Tomé, Lijnzaad, Morissette, Nusbaum, Stewart, Thangarajah, Vega-Czarny, Polymeropoulos, Nomura, Cox, and Wu, to name a few. "The integrated map is available at www.ncbi.nlm.nih.gov/genemap," they wrote. "In addition, two Web servers . . . permit anyone to map a new marker relative to this map." Anyone who has a couple of million dollars' worth of gene-tracing equipment, that is; but you'd be surprised how many people do nowadays, and where they are. As with the old British Empire, the sun never sets on the domain of the gene hunters. And anyone (really) with Web access can get to that spot in cyberspace within seconds, and gaze in wonder at the newest map of our genes.

Due largely to this openness, the speed with which things happened after that interim publication in 1998 astounded even the experts, and a completed draft sequence was announced two years later. But at the White House press conference, after the pols ran out of clichés—"language in which God created life," "superhuman task," "change the face of medicine forever," "common good of all humankind," and so on—the scientists finally got to speak. Dr. Collins, perhaps feeling Dr. Venter's breath at his back, talked of the pace of things:

Most of the sequencing of the human genome . . . has been done in just the last fifteen months. During that time, this consortium has developed the capacity to sequence one thousand letters of the DNA code per second, seven days a week, twenty-four hours a day. We have developed a map of overlapping fragments that includes 97 percent of the human genome, and we have sequenced 85 percent of this.

Collins also poignantly cited a personal tragedy:

Less than twenty-four hours ago, I attended the funeral of my beloved sister-in-law, a wonderful marionette artist who brought magic and joy to thousands of children with her art. She died much too soon of breast cancer. The hope and promise of understanding all of the genes in the genome and applying this knowledge to the development of powerful new tools came just too late for her.

Yet it is also worth noting that we have had some knowledge of the genetics of breast cancer for years, including two well-known genes, and this knowledge has so far had no effect on treatment. Dr. Venter—who, had he not cooperated, could have beaten the government consortium to the sequence and in effect have been standing there alone—was more circumspect about gene power:

Thirty-three years ago, as a young man serving in the medical corps in Vietnam, I learned firsthand how tenuous our hold on life can be. That experience inspired my interest in learning how the trillions of cells in our bodies interact to create and sustain life. When I witnessed firsthand that some men lived through devastating trauma to their bodies, while others died after giving up from seemingly small wounds, I realized that the human spirit was at least as important as our physiology.

Venter also emphasized the universal human implications of the sequence:

The method used by Celera has determined the genetic code of five individuals. We have sequenced the genome of three females and two males, who have identified themselves as Hispanic, Asian, Caucasian, or African-American. . . . In the five Celera genomes, there is no way to tell one ethnicity from another.

In February 2001, scientific accounts of the draft sequence were published by both groups and open to study by other scientists and physicians throughout the world. One revelation: we have only around 30,000 genes. There had been a kind of gene-scientists' betting pool in which the authoritative guesses ranged from 28,000 to 140,000 genes. This must be the ultimate lesson in humility. Mice have almost as many genes, and the roundworm C. elegans, barely visible to the naked eye, has 20,000. Though we are made up of hundreds of billions of cells, we have only one-third more genes than a creature with 959 cells. It doesn't seem to compute.

Worse, most of the DNA in the human genome isn't even included in those genes. Ninety-five percent of it is junk—duplications of DNA strings that pointlessly jumped around the genome, and scraps left behind by viruses and bacteria that infected our ancestors over the course of millions of years. Talk about bloated bureaucracies. As Irven DeVore said about species extinction rates, if this was some kind of grand plan, it wasn't a very efficient one. But the trick is to understand how those mere 30,000 genes interact. Even taken just two at a time, the number of possibilities becomes astronomical, and for practical purposes may as well be endless. What the published sequences reveal is an exquisite hierarchical structure *within* the genome that eventually will explain the difference between us and our half-millimeter-long distant cousin, the roundworm.

There is good news and bad news here. The good news is we're not just fancy roundworms. Twenty thousand genes making fewer than a thousand cells can completely determine the roundworm, almost molecule by molecule. Thirty thousand making a hundred billion cells in the brain alone—cells that in turn link up in perhaps a quadrillion connections—just can't be all that bossy. We have vastly more degrees of freedom. The bad news is we barely have a clue as yet which of the countless possible interactions among genes in fact occur, or how they work. Some hierarchical suites of genes have not changed much since our common ancestor with the roundworm, or at least with the fruit fly. With the mouse we have far more in common, and with the chimp almost everything. The secret of our lives is almost certainly contained in the hierarchies of regulatory genes, but we have barely made the first steps toward understanding them.

Some parts of the genome, however, are better understood—the Y chromosome, for instance. As women might have guessed, there isn't a whole lot there. We're looking at long, repetitive stretches of junk DNA punctuated by a few genes whose main purpose seems to be to gum up an otherwise sensible plan. Y-DNA does seem to be especially variable, in keeping with the suspicion that males are unpredictable and weird. Storing variation is a dirty job, but somebody has to do it. But as journalist John Whitfield commented, "The news that male-defining Y chromosome is highly repetitive and mostly non-functional will come as no surprise to the half of the population without one."

Perhaps the Y holds the secret of Venter vs. Collins. Still, the draft sequence is a great achievement. And it wasn't just a race between two brilliant men. Venter had 282 coauthors, many of them women, and Collins had a similar number in 23 centers around the globe. Every effort has been made to keep this enterprise open, although Celera will try to recoup its vast investment by renting information and even patenting genes. But the genome map is just a start. A transcriptome map is next, delineating the different messenger RNA transcripts that the genes make, roughly three each, for a total of ninety or a hundred thousand. Since genes are full of introns—more junk, spliced out as transcripts are made—

and have more than one start point or reading frame, mapping three transcripts to each gene is a whole new challenge.

Like Collins, Venter believes that in "the research that will be catalyzed by this information, there's at least the potential to reduce the number of cancer deaths to zero during our lifetimes." Let us hope so. But of course, now that the hype surrounding the "working draft" has quieted, the harder work begins. In addition to the RNA transcripts, this includes such mundane tasks as proofreading the sequence up to six more times to correct errors, filling tens of thousands of gaps, and rounding up the 7 percent of the sequence deliberately excluded because it was wrongly thought to contain no genes. These ongoing efforts would produce a 99.9 percent "finished sequence" by 2003.

Yet that is still trivial. The important part goes under the very modest name *annotation*—which means that once you've sequenced the genes, you have to figure out what the darn things do. Which genes code proteins, and what gets spliced out and thrown away in the process? What *are* those proteins—perhaps 50,000—and what might they do? (This is the subject of biology's new, new thing, proteomics.) Which genes are content to just make RNA, a vital task in itself, and so fly under the radar of current protein-finding technology? What all-important regulating sequences turn genes on or off? What molecules—promoters, copromoters, repressors, transcription factors, and so on—enter into the layer upon layer of regulatory gene control? And most important, what happens in development?

The string of genes is minor. The sequence that matters is that of development—the step-by-step process by which those tens of thousands of genes enter and exit the pathways that build and rebuild body, brain, and mind. Those steps form an astronomical number, one that will dwarf the number of genes and probably exceed even the number of base pairs. Cures for diseases? Yes, in time. As of February 2001, the number of identified disease genes was 1,112 and counting, and these eventually will lead to treatments. But we have known for years now the location, sequence, and protein products of genes for Huntington's disease, sickle-cell anemia, cystic fibrosis, and some forms of breast cancer. There is not even a new, effective treatment for any of them, let alone a cure, based on the new genetic knowledge. White House press conferences with patched-in foreign heads of state may generate false hopes, and when those hopes are dashed, the disappointment could deal a blow to publicly funded science.

The risk is especially great when the fanfare is so exaggerated and politicized. What was accomplished by June 2000 was certainly needed and ahead of schedule, but it was no more than an early signpost on a very long journey. Later on that famous day, Craig Venter said that the rest of the job of annotation—the *basic* analysis of how the genes work—will take most of this century. That is a long time to wait, and for the ill it is forever. If the gene hunters don't produce some new treatments in the next few years, their avid public relations may come back to haunt them. NASA put a man on the moon in 1969, supposedly opening vast new worlds of space exploration and untold benefits to humankind. More than three decades later, the moon shot seems an isolated achievement. Space exploration has not lived up to its early promise, and in many ways we have come back down to earth.

This will not happen in the same way in biology, but the risk of irrational exuberance is there. To get an idea of where we really are, consider this analogy. You have just been handed a shiny new phone book for a small city. Here, you are told, are the names, addresses, and phone numbers of most of the 50,000 residents. Your task: explain the city. The directory is 90 percent accurate, and you see the gaps and errors, but that is not the problem. The problem is that the list tells you nothing about the city. You do see a few annotations. One man, it says, is a mail carrier. One young woman packages apples in a supermarket. An older woman is a pediatric surgeon in the main hospital. Someone noted recently that she works less efficiently if she has too many forms to fill out. A boy works part time delivering pizza, but only in better neighborhoods. A very old man in a very large house does little now, but the note says that for forty years he manufactured typewriters. A woman in midlife draws designs for cell-phone towers. Four different people spend all their work time changing tires on cars.

That's about it. That's all you know. Now try to reconstruct the city. With this relatively complete directory of citizens, you should soon be able to deal with the nurses' strike, reduce crime levels, resolve the traffic-congestion problem, get the lead out of the house paint in poorer neighborhoods, and stop the spreading wave of corruption among public officials that threatens a vibrant urban life. After all, your directory contains, to a level of 90 percent accuracy, the names and addresses of all the people. It's just a matter of figuring out which ones are which, what they *all* exactly do, and why they don't always cooperate.

Don't laugh, because it's roughly where we are in the genome. Just as a list of residents tells you almost nothing about how a city works and how to fix its problems, the genome sequence, even at 99.9 percent accuracy, can tell us almost nothing about how the body works. But you can start telephoning the 50,000 residents; you can even visit them. Laboriously, you can find out what each of them does. And finally, you can begin to imagine how their efforts join to build, maintain, and repair the millions of functions the city performs every day. The genes are equally unknown, the body and brain equally unexplained. Annotation has barely begun. Intellectually, we are just out of the starting gate. As one geneticist said of the genome sequence, "Now we have to learn to use it to understand the biology of the organism."

Fittingly, Venter noted that his press release for when the sequence was published was dated on Darwin's birthday. The brash genetics entrepreneur quoted the grand master, from *The Descent of Man*: "False facts are highly injurious to the progress of science, for they often endure long; but false views, if supported by some evidence, do little harm, for every one takes a salutary pleasure in proving their falseness: and when this is done, one path towards error is closed and the road to truth is often at the same time opened." Those who would like to keep the new genetics, with its immense potential for medical applications, separate from the science of evolution, will find little comfort in genomics.

As for the open road to truth, the new race is on. For researchers, the urgency is palpable. You don't have to click on the Web site, or even study the journal papers. Just read the ads:

"Publish faster," says one.

". . . your fast and easy way to great transfections," claims another.

"Now your lab can win at DNA analysis," another promises, on a brightly colored game board reminiscent of Parcheesi.

"How to win the human race. . . . Run faster, you win. Slow down, you lose: facts of life. And of life science . . . because in some races, you just can't finish second."

"Pump Up your Gene Expression," over bulging biceps.

"NEED HUMAN SPECIMENS FOR RESEARCH?"

"50,000 Knockouts In Our Freezers . . . rapidly generate new mutant mice."

"Win the race to market with novel full-length genes."

"Transcribe longer. Run hotter . . . Do what hasn't been done before."

"I need a novel gene—fast!"

"Time in a Bottle. Expression-Ready Human Genes."

And, beneath a picture of a pretty woman whose bald head is doodled with gene diagrams, "The future of cloning is here. You may proceed."

We may indeed. Each of these come-ons is from a different ad, most published in *Science*, America's leading scientific journal. They sound scary, or at least science-fictional, but they are really just signs of enthusiasm, the childlike glee of thousands of smart people playing with toys that will change the world. Knowledge in this field does not just explode, it pulses with the sheer white heat of nuclear fusion. Money pours in from governments, then from venture capitalists, scientists go private, companies go public, acres of gene sequencers churn, supercomputers hum, noble ambition thrives, and genius is mobilized throughout the world. And in case you don't happen to have a multimillion-dollar lab, try the recipe for "PCR Soup" in *Scientific American*. Amplify genes in your kitchen, for just over a hundred dollars. When gene hunting becomes this affordable, it may be the dawn of genetic democracy—or at least a free-enterprise zone of sustainable genomics.

We *will* increasingly control those genes—knock them in or out, switch them on or off, block or change their protein products. We will clone the genome in whole or part, make artificial chromosomes, permanently change the genes in some family lines, and attempt, in our bumbling way, to make better people.

Admittedly, we humans have not done well with this sort of project, especially as a matter of national policy. In fact, such policies produced some of the worst crimes of the dismal twentieth century. Evolutionary geneticists agree that China's national eugenics policies accomplish little against the tide of evolution, altering only fundamental human rights. But what about when our own national policy does the opposite—prohibiting people from choosing their own children's genes? Now who is violating rights?

It is a fact of nature that since males were invented, female choice has guided evolution. Should we now take away a woman's right to choose the genes of a champion athlete, a cello virtuoso, or a Nobel laureate because he offered them at a sperm bank instead of her bedroom door? Or her right to reject the genome she is carrying? Or to choose a particular embryo for implantation, based on its genes? Or to replace a gene or two before the embryo is implanted? Or to clone her dying, childless sister so she can raise that sister's "twin" as her daughter? Now we have moved from the question of national policy to improve the genome to the question of whether to frustrate personal choices. What, in a democracy, would justify such bans?

Instead, we can do three things. First, encourage research in genetics. Second, get government out of the way of personal choices. Third, practice no deliberate eugenic policies but create conditions in which decent people of varied talents and inclinations can thrive and grow. That means standard conditions as free as possible of war, dictatorship, racism, genocide, famine, disease, inequality, and poverty—the human and inhuman plagues that have burdened our species for millennia. Contrary to the predictions of the Social Darwinists, current conditions do not result in survival of the fittest—or rather, they do, but only if we mean fittest for war, dictatorship, racism, genocide, famine, disease, inequality, and poverty. If we want, instead, those people who are fittest for peace, democracy, justice, health, equality, productivity, creativity, and prosperity, then somehow we must begin to create those conditions first.

And we have one more major task at this time in our evolution, one that runs counter to all we have had to do so far. Like any successful species, we have beaten the competition. We are far more successful than any other primate. We have routed our closest competitors, the apes, destroying several species completely and bringing the last four to the verge of extinction. We may yet be stupid enough to extinguish even those, erasing our best hope of understanding our own uniqueness. And while we are at it, we may also rupture the balance of the four classic elements—air, earth, water, and fire. This is no metaphor. That balance made life possible, then workable, then stable for us and for other animals. So here is the reigning paradox of our generation, poised at a pivot point in human history: to succeed in the future we must stop doing what made us succeed in the past, what has allowed every good species to succeed—breeding. There are forces afoot in the world—infectious, maladaptive viruses of the mind—that would lead us wrongly to think that this problem is solved. Nothing could be farther from the truth.

In 1798 Thomas Malthus, a British economist and man of the cloth, published an essay on population based on two facts: that we need food, and that "the passion between the sexes is necessary and will remain." He predicted a *geometric* progression of human numbers against an *arithmetic* increase in food. Result: catastrophe. Today many otherwise intelligent people believe that Reverend Malthus has been proved wrong. The population has increased geometrically in these twenty decades, but so has the food supply, and the projected disaster has therefore not occurred. Worldwide breeding has slowed, and the population "bomb" will be defused during this century.

These people are naïve and dangerous. They have missed three crucial truths. First, archaeologists have, since Malthus's *Essay*, definitively shown the catastrophe of overpopulation, diminishing agricultural returns, ecological destruction, and population crash through war, disease, and famine; these have happened over and over again in human history. Sites of great civilizations remind us that the process Malthus described is old, reliable, and real. We often find these civilizations buried under . . . nothing. In "Ozymandias," Shelley mocked their pretensions:

. . . Two vast and trunkless legs of stone
Stand in the desert. . . . Near them, on the sand,
Half sunk, a shattered visage lies, whose frown,
And wrinkled lip, and sneer of cold command,
Tell that its sculptor well those passions read . . .
And on the pedestal these words appear:
"My name is Ozymandias, king of kings:
Look on my works, ye mighty, and despair!"
Nothing beside remains. Round the decay
Of that colossal wreck, boundless and bare
The lone and level sands stretch far away.

The disasters that doomed such civilizations and covered them with sand were worldwide, continual, and predictable according to Malthus's reasoning. *Ah, but the species survived and grew!* True. Now think of the enormous suffering of the people in those dead civilizations: the hopes dashed, the dreams and fears they saw in the faces of their children, the sight of those children succumbing to sword, plague, famine, and flame—or, at a minimum, dragged from their homes to become paupers, misfits, slaves, strangers in a strange land. The Malthusian crisis was not just a model of future events but a summary of what had already happened many times.

The second crucial truth is that, since Malthus wrote, the process has widened and intensified. A few years after his book came out, the people of France, as Tolstoy put it, decided to go to Russia, hoisting Napoleon like a flag. Meanwhile, the people of Europe and Asia overflowed into land they conveniently but mistakenly thought was empty and, since it was not, proceeded to empty it by murdering its inhabitants or infecting them with deadly diseases. The English, French, Germans, Spanish, Portuguese, and Dutch went to Africa,

North and South America, and Australia, always acting as if they owned the place. In the nineteenth century the newcomers to North America brutally savaged one another in disagreement over the fate of people they had dragged there in chains from Africa. In the twentieth century, the people of Germany went to Russia and France twice, shouting (among other things) *"Lebensraum!"* The second time they murdered 11,000,000 civilians, including 6,000,000 Jews, which did leave them some extra living room.

The Russians built a seventy-year empire on a process of *self*-destruction, killing scores of millions of their own people. The Japanese went to the South Pacific, Korea, and China, and the French and the Americans went to Southeast Asia. From time to time the conquerers withdrew from exotic places they had gone to, leaving millions of ruined lives in their wake. Somewhere in the world tens of millions, if not hundreds of millions, of people were starving at any given time. Refugee populations throughout the world swelled to staggering proportions. Genocide or something like it took place, with hundreds of thousands or even millions of civilian deaths, in Indonesia, Uganda, Cambodia, Rwanda, and, on a smaller scale, in the former Yugoslavia. In the second half of the twentieth century, an era without a major, armed conflict, at least 50,000,000 people died in small wars, and at least 10 times that from starvation and pestilence.

Malthus was *wrong?*

Third and most important, the anti-Malthusian calculations are tragically naïve and misleading. In the words of Bill McKibben, author of *The End of Nature,* "very few of the estimates of carrying capacity . . . take into account the possibility that we will degrade the planet's support systems below their theoretical maximums." This includes most of the more pessimistic estimates as well as all of the dewy-eyed fantasies. For example, grain grown per capita increased from 250 kilos in 1959 to 350 kilos in the mid-eighties; but it shrank 10 percent by the mid-nineties. Why? Because there was little land left to expand into, because fertilizer had pretty much done what it could (its use began dropping), and because warmer summers cut corn and soybean yields. The soil is being mined and will be depleted. But the shrinkage is worse than it looks, because more people want to eat meat; this requires much more grain per capita than does eating grain. Exploit the oceans? Well, the fish catch per capita fell 9 percent between 1989 and 1997, because the world's fisheries have pressed the bounds of sustainable harvest. Economic statistics that show the developing world to be doing well fail to assess the depletion of environmental capital and gloss over the immense disparities of wealth and welfare that make national growth statistics meaningless.

Worse still, the carrying-capacity models ignore the logistics of getting food and water to where it is needed. Divide the accessible fresh water in the world by 8 or 10 billion—the minimum projected peak of human population—and you get enough for each person. Now the reality: there is already no way to get the water to the mouths, a situation that worsens daily. Twentieth-century dam building—monumental projects mounted throughout the world by every major head of state—have changed ecologies permanently. Water composition, soil minerals, water tables, agricultural potential, and resident flora and fauna have been dramatically altered since dozens of these immense dams were completed. This transformation of the planet's major rivers has irreversibly changed the equations of fresh-water supply.

Other misleading calculations induce complacency about the amount of fish or grain. In an ideal world, food and water go where they are needed; in the real world, every stubborn, ugly feature of human nature intervenes to block the flow. Greed does its work wherever corrupt officials can magically skim gold from sacks of rice. Worse, a new kind of small, relentless war is being waged throughout the world, churning grief, fear, and hatred in its roiling wake. But the worst threat is one deep, steady feature of our nature that the statisticians ignore: *desire*.

For the anti-Malthusian calculations to be remotely plausible, the aspirations of every one of the 6,000,000,000 people alive today—add 200,000 tomorrow, 80,000,000 by this time next year, a country the size of India in a decade, and another, up to a minimum of 8 to 10 billion—would have to stay as they are or be scaled down. People throughout the world would no longer be allowed to hope that the lives of their children and grandchildren would be any richer, in material terms, than their own lives are now. For the world to support us all, billions of dreams would have to end. But of course it is not in our nature to be satisfied for long, and if expectations for the future are no better or even worse, it *is* in our nature to feel frustrated, angry, afraid. This will not make it any easier to get the food and water (not to mention the fuel, entertainment, and vaccines) to the people the accountants find it available for.

So what will allow the earth to support 10 billion people? Will the hopelessly poor mass of them be forbidden to see the lifeguard serial *Baywatch*—alas, at the turn of the century, the world's most popular show—and told that their goals for their children must stop at full bellies, since those who "proved" that Malthus was wrong thought no farther than the food supply? Or will they just gaze, as if on a dream of heaven, at a world so idyllic that it is permanently beyond their grasp, beyond even that of their children? Do we think they will stop wanting it? And if they do not, what will they do to get it? And how will the thousands of violent acts on that same flickering box shape what they do with their envy and anger?

These are the real questions. This is the real world of human nature. A young man in a crowded mud hut somewhere within a thousand miles of the equator gazes at his village's only TV, eyes glazed over. Perhaps he wants to touch the silicone-enhanced breasts of the make-believe young woman in the modest, messy, three-room apartment. But even more, he wants the things in the apartment—a refrigerator and a TV with a VCR, a little makeshift bar near the refrigerator with two or three mostly full bottles of spirits. And the used subcompact car parked outside that can take a young couple to a movie, over a well-paved road, at a speed of at least thirty miles per hour. That is all the young man wants—a fraction of what the readers of this book take for granted. Now ask whether the earth can provide that apartment, that car, that basic, decent, modest, modern lifestyle—not to mention the hospitals, sewage, fire departments, earthquake protection, and other support systems that go with it—for 10 billion people, 5 or 10 times as many as enjoy that lifestyle now.

Everyone knows it cannot. That is why the optimists focus on food. There was a major international environmental summit in Rio de Janiero in 1992. Looking back, the most important fact about it was its refusal to discuss fertility reduction. But considering the things it *did* discuss, how is it doing? A follow-up five years later concluded, "The only bright spot is a decline in population growth," which presumably could not be credited to a summit that had categorically refused to address this issue. As for the things it did address, United Nations assessments found a continuing loss of topsoil and productive farmland and a growing scarcity of fresh water. In the late nineties a third of the world's people had inadequate fresh water, and this is expected to double to two thirds by 2020. Many future wars will be fought over water.

A year 2000 symposium edited by Mikhail Gorbachev, the former Soviet leader, describes the problem.

> United Nations Secretary-General Kofi Annan asks us to face up to the threat of a catastrophic world water crisis and to counter such bleak forecasts by adopting a new spirit of stewardship. To do otherwise would be nothing less than a crime—and history would rightly judge us harshly for it.

Annan goes on to say that we must "help humanity confront the threat posed by the unsustainable exploitation of water resources and the broader danger of living on a planet irredeemably spoiled by careless human activity." And Madeleine Albright, the former U.S. secretary of state, writes: "More than two billion people live in countries experiencing some degree of water stress. . . . We cannot afford to . . . simply stand by while competition for water becomes a threat to international stability and peace."

But the water shortage is only one threat among many. As we've seen, pollution and overfishing threaten the oceans along with the coastal peoples whose lives depend on them. Deforestation proceeds apace globally, with an area the size of Florida cut or burned each *year*. Two treaties signed in Rio were broken before the ink was dry. The Convention on Biodiversity, signed by 161 countries not including the United States, has not slowed what the United Nations deems an "unprecedented" rate of habitat destruction, by far the major cause of species extinction. There is a wide consensus among scientists that we are now in one of the most massive extinctions—perhaps the fastest in world history. A climate treaty signed by 166 nations in 1992 was without effect five years later; the major contributors to climate change—urban and industrial contamination, atmospheric pollution, and deforestation—are worsening in Latin America and Asia and improving nowhere.

In 1997 another treaty, the Kyoto Protocol, was drafted by 170 nations. Its purpose was to slow global warming. A follow-up conference at The Hague in November 2000 was supposed to finalize the treaty. Instead, "Disputes festered among blocs of countries large and small, rich and poor, over what rules, tools and penalties should be used to stem the flow of heat-trapping greenhouse gases." The meeting broke down and there was no agreement. But even if these negotiators could reach a consensus, no one knows how to fix the problem. Despite scientific agreement on the basic facts of warming, there is no proven way to slow its relentless progress, and "it is far from certain that the [Kyoto] strategy would work." Almost two trillion kilograms—one third—of the carbon we pour into the atmosphere each year is unaccounted for by present methods.

But it is thoroughly irresponsible to continue to say that warming itself is in dispute. New evidence for it accumulates continually. Even by the time of the Kyoto Protocol,

> a task force of leading climate scientists from 98 countries . . . had studied the problem exhaustively. . . . The report had been authored not by one or two leading researchers . . . but by 78 lead authors and 400 contributing authors from 26 countries, whose work was then reviewed by 500 additional scientists from 40 countries, and then *re*reviewed by a conference of 177 delegates representing every national academy of science on earth. The [task force] had been unequivocal in its conclusions that 1) warming is happening, rapidly; 2) human activity is causing it; 3) the warming is likely to unleash devastating weather disturbances ranging from unnaturally heavy storms and floods to heat waves and droughts; and 4) it is therefore urgent that carbon emissions be cut sharply all over the world, but particularly in the industrial nations where these emissions are worst.

As of November 2000, no industrial nation had signed the Kyoto treaty. In 2001 a new American president with no electoral mandate withdrew the United States from *discussion* of the treaty. Yet the treaty's demands were absurdly modest. It called on industrial nations to reduce carbon dioxide emissions to about 5 percent below 1990 levels by 2012, while to stabilize world climate would require a reduction of *60 to 80 percent*.

The industrial world can be blamed for wrecking the planet so far, but the developing world's teeming billions will soon take over the burden. Glimpses of a dismal future are emerging. In November 2000 there were riots in New Delhi, a city of 14,000,000, in which "mobs torched buses, threw stones, and blocked major roads." Why? Because the supreme court of the world's largest democracy had ordered the closing of more than 90,000 small factories that were polluting the city, and the factory closings put a million people out of work:

> The factory owners, on strike against the court ruling, have pulled down their metal shutters. Ashok Mehta, who owns a shop that makes industrial molds, led the way into the two dark, windowless rooms where his 10 men labor over three oil-spattered machines. Faded posters of Hindu gods are taped to the walls. The smell of oil fumes hangs in the close air.

One worker in this shop said that his family's survival depended on his $35-a-month wage. Another said, "If these factories are relocated or closed down, we will die with empty stomachs." Multiply this dirty little factory, polluting the air and water, by 90,000, and you have the problem in New Delhi. Multiply it by millions, and you have the developing world.

Back in 1976 demographer Nathan Keyfitz warned that future problems would emerge more from rising aspirations than from rising numbers. Today we quantify his warning. Vaclav Smil, a leading authority on the biosphere, reflects widespread scientific opinion: "[T]he projected rise in energy consumption would bring CO_2 emissions in the year 2020 to levels 30 to 60 percent higher than in the mid-1990s. . . . Most of the increase will come from modernizing economies in Asia, above all in China and India," where some two billion people are trying to enter the middle class. If you think that the first billion or so people who have become fully industrialized have hurt the planet some, wait till you see what the next two billion will bring. At every international convention, developed and developing worlds exchange accusations, each holier than thou. But when the planet finally falters, there will be plenty of blame to go around.

In December 2000, 121 countries met in Johannesburg for the last of a series of five meetings on persistent organic pollutants (POPs). They focused on twelve of these POPs—DDT, dioxins, and PCBs are among them—all of which spread throughout the world's environment and stay there indefinitely. POPs originating in Africa have been found in fish and in people in and around America's Great Lakes. Pest control, crop protection, and industrial waste generate these pollutants, and happily there is hope for controlling some of them. But there are many more, and we know little or nothing about most of them. Current levels may not cause cancer, but without control, future levels will.

Finally, the destruction of the earth's shield against ultraviolet light—the ozone layer—proceeds apace, despite strong international agreements to protect it. The ozone's brief history alone speaks volumes about our predicament. It was not until 1985 that we even knew there was a problem, and by that time decades of emitted chlorofluorocarbons (CFCs)—products of aerosol sprays, refrigerators, and air conditioners—had greatly reduced the value of the shield. The Montreal Accord of 1987 caused CFC emissions to drop 87 percent by the end of the century—the first real success of global environmental governance. But reducing emissions is far different from repairing the ozone or even stopping further damage. There is still a black market in CFCs, and more important, a half century's worth of accumulated gases continue to remove the ozone now. A year 2000 report from NASA estimated that the hole in the ozone above Antarctica was larger in that year than in either of the two previous years, and 13 times as large as in 1981.

Ozone damage (O.D.?) should blaze across our consciousness as a prime example of negative synergy. Trapped greenhouse gases warm the globe, but the stratosphere above them cools accordingly. Stratospheric cooling accelerates ozone damage. Ultraviolet light singes forest foliage, and leaf reduction favors global warming. Missing such mutually reinforcing downward spirals is one of the many reasons for our inappropriate confidence.

Still, the deepest circle of hell certainly must be reserved for those whom *Worldwatch* editor Ed Ayres calls "techno-optimists," the ones who proclaim "that human ingenuity will solve all problems in the future because it always has in the past." They believe this "partly because they live in places that have not yet been overwhelmed by floods, squatter populations, mafias, food shortages, electric grid failures, or epidemics. But history, which leaves us now with the legacy of a human population too large for Earth to support without liquidating its resources, does not support the view that our ingenuity is supreme." So what is the answer? It's a no-brainer: reduce population, reduce consumption, reduce pollution. That's it. Difficult? Too bad. Be grateful it's still possible. Ayres calls it "God's last offer." Take it or leave it.

So far, we show little taste for the challenge. Instead, we soothe ourselves with an ongoing Green Revolution and slowing population growth. But what of the wars over water? The modest nuclear exchanges between developing countries bulging with hopeless citizens that will pour radiation into the air and soil? The potentially disastrous results of global warming? We dither about whether the world will warm half a degree or only a quarter in this century. We divide projections of the number of mouths to feed into the number of calories the world's agricultural land can produce and ask if the dividend can prevent a child from starving. But the vast, dreadful famines of the twentieth century have not been about this planetary exercise in long division. They have been about bridging the distance between a bag of barley and an empty stomach. Time and again human nature, with its resonant depths of selfishness and fear-induced rage, has caused tens of millions to die for want of a bowl of rice.

The Malthusian nightmare, then, is not about species averages of grams of rice per belly, or computer projections of the number of ears of corn the world can yield if the Green Revolution continues. It is about human nature, just as it was for Malthus himself. It is about too many people in too little space with the wrong desires and the wrong beliefs about one another throughout too many years of frustration and resentment. It is about Cambodia, Bosnia, and Rwanda; about famine caused by politics; about war and terror and genocide. It is not about something that hasn't happened yet. It is about what happened to the ancient civilizations and the early modern empires. It is about what is happening now, and will continue to happen and worsen if we don't change our lives. Read the daily headlines. Malthus was right.

Have I gone on too long? I guess I have—environmental preaching bores me to tears, too. I would rather kick back, pop open a microbrew from Brooklyn, light up an illegal Cuban cigar, and watch *Baywatch* myself until my eyes glaze over, or at least until I nap and dream of being rescued. But my attention drifts during commercials, and pretty soon I'm wondering if there's another beer in the fridge, or whether my grandchildren are going to have a home planet they can actually live on. Woops—that's no fun—think about something else! But sometimes I just can't. Sometimes, for the whole next segment of *Baywatch*, I'll have intrusive thoughts about my grandchildren and their planet. My best guess at the moment is no, they won't have a planet they can live on. Sure, the planet and the species will still be here, coughing and spitting, gorging and starving, procreating and perpetuating genes, the old biological round interrupted every few years by hiccups of mass destruction. My grandchildren may still *exist*—at least somebody's will. But we were talking about a planet they could *live* on.

In 1936 Albert Einstein, already famous for thirty years and increasingly respected as a sort of world statesman without portfolio, wrote the following "Letter to Posterity":

> Dear Posterity,
>> If you have not become more just, more peaceful, and in general more sensible than we are (or were) today, then may the Devil take you! Respectfully expressing his opinion with this devout hope is (or was)
>>> Your,
>>> Albert Einstein

So far we have not, and something devilish is taking us.

Yet we have within us, too, a strong inclination to love nature and not just ourselves. Naturalist Edward O. Wilson has written movingly of what he calls biophilia, the love of life itself—not just of being alive but the wonder and beauty and value of all living things. "Natural philosophy," he writes, "has brought into clear relief the following paradox of human existence. The drive toward perpetual expansion—or personal freedom—is basic to the human spirit. But to sustain it we need the most delicate, knowing stewardship of the living world that can be devised." Theologian James Gustafson has written similarly of our "sense of the divine" in all things living. He derives a reverence for life from what he sees as our innate sense of piety, a feeling of awe that precedes and grounds—but in some of us is independent of—conventional faith. Thus, a wise biologist and a wise theologian converge on the same hope: that a love of life and living things, or a piety and reverence toward them, will somehow serve to protect life against us.

On a warm, wet planet adrift in the visible galaxy, about 5,000,000,000 years ago, there appeared an equally beautiful, invisible galaxy, which set itself against all the hostile forces of a random, disordered, relentlessly entropic world. Improbably, it prevailed. Eventually, it produced a species able to be awestruck by its own dazzling visions painted on cave walls, in the depths of the earth, by torchlight. Capable, too, of being impressed by its own incessant cleverness, so much so that it was not for hundreds of centuries that it began, a little, to question that cleverness. As the epigraph from Ernst Mayr reminds us, we are in charge of the species and of the planet now, and we need an evolutionary humanism to guide us. Our cleverness has even put us in charge, in a small way, of the minute, tenacious, fateful invisible galaxy—both directly, through technique, and ethically, through judgment. We are at last where we wanted to be, where it was all along our destiny to be: *in our own hands.* If in the end we let our light flicker out, who will grieve? No one, except, for a while, in the last throes of extinction, us.

But just who is this *us*? Anthropologists have spent two centuries searching for *Homo sapiens* in space and time, and we have found some pretty exotic things. But sometimes we have missed the common ground among all people. It is not hard to summarize what we know about that. There is an old song written by two great songwriters, George David Weiss and the late Bob Thiele, Americans of very different descent; a song immortalized by Louis Armstrong, a descendent of African slaves. I can always imagine Armstrong's big, deep, gravelly, strangely lovely voice, slowly but surely delineating the obvious, so that I hear it each time as if I had never heard it before. He sings of travels "all through the world and back . . . but no matter where you go you're gonna find/that people have the same things on their minds." The short list is explicitly male, but symmetrical things can be said about women. To be sure, it does not include everyone, but it links most men in cultures throughout the world. And it is remarkably close to what we anthropologists have found in our own travels:

A man wants to work for his pay.
A man wants a place in the sun.
A man wants a gal proud to say
That she'll become his lovin' wife.
He wants a chance to give his kids a better life.

Well, hello . . . hello . . . hello, brother.

PART FOUR

Human Nature and the Human Future

Woe unto them that call evil good, and good evil; that put darkness for light, and
   light for darkness; that put bitter for sweet, and sweet for bitter!
Woe unto them that are wise in their own eyes, and prudent in their own sight!
Woe unto them that are mighty to drink wine, and men of strength to mingle
   strong drink:
Which justify the wicked for reward, and take away the righteousness of the
   righteous from him . . .
Therefore is the anger of the Lord kindled against his people, and he hath
   stretched forth his hand against them, and hath smitten them: and the hills
   did tremble, and their carcasses were torn in the midst of the streets. For all
   this his anger is not turned away, but his hand is stretched out still.

   —Isaiah 5:20-25

CHAPTER 18

The Prospect

This is a present from a small distant world, a token of our sounds, our science, our images, our music, our thoughts and our feelings. We are attempting to survive our time so we may live into yours. We hope someday, having solved the problems we face, to join a community of galactic Civilizations. This record represents our hope and our determination, and our good will in a vast and awesome universe.

    —**President Jimmy Carter,** *Voyager 2* Spacecraft Message, 1977

Rechristened the Voyager Interstellar Mission (VIM) by NASA in 1989 after its encounter with Neptune, *Voyager 2* continues operations, taking measurements of the interplanetary magnetic field, plasma, and charged particle environment while searching for the heliopause (the distance at which the solar wind becomes subsumed by the more general interstellar wind). . . . *Voyager 2* is speeding away from the Sun at a velocity of about 3.13 AU [Astronomical Units]/year toward a point in the sky of RA = 338 degrees, Dec = –62 degrees (–47.46 ecliptic latitude, 310.89 degrees ecliptic longitude).

    —NASA *Voyager 2* Web site, last updated 1998

Editorial

This morning the *Galaxy Times* learned from Interstellar Network News of yet another "bottle message" from a remote island planet. This one, carried in the hulk of an ingenious if crude vehicle, managed to drift halfway across the Galactic Federation before it was intercepted by the debris patrol. It was about two hundred thousand years old.

    The geographic circuits of the *Times* photon library output more than we cared to know about the quaint planet, but by the third request the report was down to thinkable proportions. The planet—named by the early geographers, with their courtly sense of language, "the Blue Drifter"—followed a well-known planetary genesis. Intelligent life on the Blue Drifter entered its light-speed signal phase about one lifetime before the message. Intelligent life there extinguished itself three or four lifetimes later, by a typical process, recounted from other examples on every child's history chip.

As usual, light-speed signals over the Blue Drifter's surface were recorded by our ancient geographers with compulsive diligence. These billions of signals, machine-sifted, cataloged, and edited for noise and repetition, became the grist for a great number of archaeological and geographic dissertations. One of these, the pride and hope of a now long-forgotten young scholar, was titled "The First Sublightspeed Interstellar Message from 'the Blue Drifter,' Third Planet of Star 868-2893-41162-33: A Study in Contrasts."

The "contrasts" are hardly new; they were not new even then. Yet they are so vivid that they seem to bear almost endless examination. For example, photon-wielding creatures were sending a message at a tiny fraction of light-speed to an interstellar community of true civilizations. A little thought might have revealed to them that if we were so advanced, we would be detecting and unscrambling their light-speed signals, and that in about ninety years we would have the message from their descriptions of it to each other, two hundred millennia before their spacecart could cross the galaxy. Once within the Galactic Federation—which the casters of this "bottle" imagined with notable accuracy—it could only pose an admittedly trivial threat to spaceliners in interstellar corridors. So strong, evidently, was their impulse to shout, "I am here!" as to obscure the uselessness of sublight-speed messages in bits of crude hardware.

But there are larger contrasts. Following standard procedure, the dissertation systematically studied the contents of the message, a sober selection from life on the Blue Drifter as it was. Nothing was new in it, even then. Sounds of atmospheric conditions and smaller-brained creatures, as well as greetings and musical compositions—all well known to our geographers— were included and were innocent enough. But the representations of intelligence, including the "message" part of the message, were nothing but half-truths and lies. The scholar notes in his conclusion:

> These creatures expect us to take at face value their expressions of goodwill and cheerful greeting. Members of a community approaching extinction, they do not report their almost continual mutual slaughter, the widespread starvation on a lush planet, the perpetual fouling of their little sphere with chemical excreta, nor the incessant hoarding of wealth by the few, who use it largely to prepare for even more mutual slaughter. The module sending the message exists in utter unreality. This module, housing 5 percent of the species, consumes most of the planet's resources. Its members' hopes for material comfort are completely out of proportion to what the planet has to offer at this time, thus the inevitability of further conflict. Their self-indulgence, irresponsibility, and decadence beggar description. But lest it be thought

that this tendency is restricted to one module, almost all others on the planet show similar tendencies, especially among the dominating sectors of their populations.

We may resist the young scholar's arrogance; had he been older, he would no doubt have been more compassionate; billions of intelligent creatures were on their way to a sad end, doing the best they knew how. But we must ask, with him, *Why were they sending toys into space?* This was a moment when all energies should have been turned to the sciences of communication, learning, mental illness, interpersonal conflict, population reduction, ocean cultivation, desalination, photon harvest, and waste control. Most of them lived by myths that flatly denied their evolutionary past and probable destiny, and so they could not grasp that it was time for them to take charge of that destiny. Voyaging blankly around a barren interplanetary wasteland, they failed to take seriously the nearer explorations that might have saved them.

From prior research on primitive planets and his analysis of all relevant light-speed signals from the Blue Drifter, the scholar predicted "protocivilizational terminus in approximately 2.6 lifetimes." He was too pessimistic by almost a lifetime but otherwise depressingly right. And most poignant, as always in these cases, was that they were so close; it is always those last three or four lifetimes that make the difference between exodus and terminus.

A familiar tale, as expected by now as death, taxes, and the light-speed boundary. But perhaps, even today, we may read in the Blue Drifter's crude gesture a deeper message, about the limits of what we are pleased to call intelligence. Instead of simply mocking or ignoring these creatures' dilemmas, we might take occasion to pause and review possible flaws in our own sentient condition. Smugness, now as then, is ever the enemy of survival.

To make the point in as offensive a form as possible, there are grounds for holding that the attractive young upper middle class mother, driving a station wagon (nowadays often decorated with a peace sticker) full of happy sunburned children, represents a major threat to the prospect of a humane civilization, even one defined according to the purely negative criterion of reducing misery.

— BARRINGTON MOORE, JR., *Reflections on the Causes of Human Misery, and Upon Certain Proposals to Eliminate Them*, 1972

At this writing, in the spring of the year 2001, the young mother drives a sports utility vehicle, completely out of place on her suburban street. It is her day with the kids, and tomorrow her husband will be behind the wheel of the gas guzzler, carting the kids around the touch points in their tightly planned lives, while Mom goes to the office. The kids are sunscreened, not burned, and the sticker shows not the old peace symbol but a fresh new color photograph of the earth seen from space. Some details of suburban life have changed, but its impact is the same. "Think globally, act locally," the words on the sticker may urge, but the young family, for all its good intentions, is acting locally to bring about global disaster.

At this writing, too, there is a worldwide economic slowdown. The financial markets sulk and brood, except for energy, where shortages have made prices soar. These trends are no doubt temporary, a pause in an unprecedented, worldwide revolution in technology. In our heart of hearts, we believe that this will solve everything. The Internet and the genome, silicon and DNA, somehow, magically, will bring an end to thirst and famine, poison and pollution. *Globalization* is the word on every pundit's lips, and it helps us to ignore many vicious small wars. In the West we see top-heavy population pyramids, and we worry about the absence of children, but of course the West is not the world. The world, for decades to come, will see unprecedented numbers of children turn into young adults each year and bear children themselves.

At the moment, despite economic dislocations, the richest man in the world is still one William H. Gates III, a college dropout, silicon geek, entrepreneurial genius, corporate baron, and now philanthropist. In the 1970s, as a boy really, he saw the potential for great wealth in a set of instructions for desktop computers, then mainly of interest to hobbyists—instructions whose authors had simply been giving them away. This ambitious, brilliant young man went on to build the world's largest corporation, the first one based on nothing but ideas. He led a quiet, worldwide revolution that eclipsed the industrial economy, created many thousands of newly rich people, turned millions of skeptics into followers, and brought him into legal confrontation with the world's most powerful government—a conflict he has the resources to fight and one he may yet win.

But as the young man mellowed into middle age, although no less tenacious in the marketplace, he began to give way before steady criticism that he was hoarding his wealth. He became, almost overnight, the world's leading philanthropist. Enlisting the aid of William Foege, former head of the Carter Center and the Centers for Disease Control and probably the greatest living authority on world public health—not to mention one of its most tireless advocates—Mr. Gates has learned to intervene wisely in some of the world's most distressing health problems. But the causes he has espoused have little to do with technology. In fact, he has become a thorn in the side of naïve technological optimists.

In conferences, speeches, and interviews, he ridicules the notion that computing and Internet services can be brought to all of the world's billions. "I mean, do people have a clear idea of what it means to live on one dollar a day?" he said at a conference in the fall of 2000. "There's no electricity in that house, none." In his office at the world's largest technology company, he envisioned an African village that gets a computer. "The mothers are going to walk right up to that computer and say, 'My children are dying, what can you do?' . . . They're not going to sit there and like, browse eBay or something. What they want is for their children to live. They don't want their children's growth to be stunted. Do you really have to put in computers to figure that out?" He went on to say that "as a father of two children, thinking about the medicines I take for granted that are not available elsewhere, that sort of rises to the top of the list."

Thus, the world's richest, most ambitious, and most vocal leader of the technology revolution now calls into question the power of that revolution to solve the most pressing problems of the new millennium. Technology, he now clearly says, is not enough. And his unsurpassed entrepreneurial mind, the same that made computers pervasive in modern life, now paints this picture: an African mother vainly and uncomprehendingly begs a computer to feed her children. His picture could hardly contrast more starkly with that of the cheerful suburban mother drawn by Barrington Moore. It is these two pictures that we must see, side by side, every day, if we are to understand our world.

In 1846, Gustave Flaubert wrote, in a letter to the poet Louise Colet, "As a rule the philosopher is a kind of mongrel being, a cross between scientist and poet, envious of both." It is sad to say that many behavioral and social theorists are like Flaubert's philosophers. They are almost immune to criticism. When their facts or logic are challenged they hide behind a cloak of humanism. Yet they expect to be taken much more seriously than poets because they are not, after all, offering mere deftly worded sentiment but presume to draw on a large body of science. This is a pretty treacherous middle ground. It is clear that some behavioral and social theorists have negotiated it very well, and they are justly admired for that, even by those of us taking biological approaches.

Still, the late twentieth century saw dire confrontations between the two camps. The social and behavioral sciences of the last hundred years have made great advances in data gathering and analysis, but their theoretical progress has been slim. There are small theories to explain restricted bodies of data—for example, the law of distributed versus massed practice in learning theory, the (usually) inverse relation between inflation and recession, or the theory of demographic transition—but no grand conceptions. Or, to be more precise, the grand conceptions are nineteenth-century leftovers, and they taste more than a little stale. Biological approaches are threatening for three reasons. First, they have been politically misused and caused great harm. Second, proper evaluation of them requires a large body of difficult knowledge, and few behavioral and social scientists are willing to try to master it. Third, they might provide the grand conceptions social scientists know they lack.

This helps explain why social scientists protest the efforts of ethologists, geneticists, sociobiologists, neuropsychologists, psychopharmacologists, and other behavioral biologists to contribute something new to our common understanding of behavior. But there is a better reason for resistance: recent contributions of biologists undermine the philosophical structure of social science as it has existed for at least a hundred years. This structure rests upon two pillars, neither of which stands on solid ground. Both are matters of faith rather than knowledge, poetry rather than science. Each is pretty, but each is wrong.

The first is the metaphor of society as an organism. The units are individual people that, like the cells of the body, are built up into tissues, organs, and finally into the whole itself, all serving one goal: survival. Signs of conflict are pathological and can be corrected by restoring the system to balance. In its everyday functioning the healthy society is no different from the healthy individual: a collection of cooperative units with a single goal.

This metaphor is simply and demonstrably wrong, because it requires that society be a plausible unit of natural selection. This, at least in animals like ourselves, it has never yet been shown to be. Group selection works in computer simulations under certain conditions, and that is welcome news to those who find gene selection, individual selection, and kin selection philosophically troubling. But in the grand scheme of real evolutionary processes in complex animals, it plays at most a small role.

Contrast the individual in a group with the cell in an organism. Evolution has stripped the cell of its independence; consequently, it is devoted almost entirely to the survival and reproduction of the organism to which it belongs. But the purposes of the individual human being are wedded to the survival and reproduction of society only transiently. Indeed, the same process of evolution has designed the individual with a full charge of independence and a canny ability to subvert the purposes of society to its own. Every time a human being gets fed up with his or her society or church or club or even family, and voluntarily changes affiliation—or merely slacks off, defecting internally—we have another factual disproof of the central metaphor of social science.

The second pillar is an article of faith, which I call the "tinker theory." It holds that human behavior and experience are basically good and decent and healthy and warm and cooperative and intelligent but that something has gone a bit wrong somewhere. A fuse has blown in the child-rearing process, or a tube has overheated in the psyche, or an evil madman has taken over the social controls, or some bungler has ordered the wrong grade of concrete for the foundation of the economy (or at the very least the wrong glass for the windows). So what we need is to do some tinkering: change the teaching apparatus, apply the right kind of psychotherapy, kick out the king and queen, elect a conservative, institute socialism, rewrite the songs and TV programs, slash taxes, or at least print less money, and all will be well. If you can do more than one of these things and, preferably, also get rid of your present husband or wife, everything will be just fine, and after a certain period of near-optimal functioning, you may just ease into paradise on earth.

Now, it is far from my mind to discredit any of these estimable varieties of tinkering, since any of them can and will sometimes make life better. But what we *can* discredit is the twofold act of faith that makes people consistently overestimate both how often and how well these strategies will work. Such miscalculations are perilous because they ruin the risk-benefit analysis that is the basis of all intelligent action. Psychologists and social scientists of various schools encourage such misestimates, just as is done in commercial and political advertising—using similar means, for similar reasons, leading to similar disappointments. It is not charlatanism; that is too easy to decry. It is the everyday, ubiquitous, sometimes benign but often pernicious exaggeration that is common to almost every person who thinks of a possible solution to a human problem.

Everything will not be "just fine." After we make the change, even if it improves us, we will still be full of the flaws of the human condition. We will still be weakly responsive to the needs of those around us. We will still be beset by unnecessary motives that we do not comprehend and that are sometimes dangerous. We will still be too wary of threats real and imagined. We will still have to do things we do not like to do. We will still tire easily of things that should delight us. And of course we will still be dying.

Tinkerers don't deny these truths, but neither do they make much mention of them. They don't have to. They know perfectly well that they can rely on denial and self-deception in their listeners to sweep such matters under the rug. They are selling hope, and the people are taking their money out, and the rule of the market is *Let the buyer beware*. To be fair, they are in the main not cynics; they are merely wide-eyed optimists, reluctant to think dark thoughts just as most of us are. That is why they are not good poets. Poets not only think the dark thoughts, they dwell on them sensuously and build them into beautiful arrangements of words, in order to show us how to live in spite of them. Since the biologist's view of human nature is seen as a new departure from the venerable traditions of social theory, we do well to recall that there is another tradition older than either—that of classical literature—which is much more consonant with the biological view.

In Greek tragedy we have such a clear, consistent view of the dark side of life that when the chorus says it is better to die than to live, and best of all never to have been born, it scarcely causes a ripple of doubt. In Shakespeare the same tradition is carried forward and developed. In Hamlet's soliloquies, in Macbeth's "sound and fury" speech, in the Ages of Man passage in *As You Like It*, in Lear's ramblings, and in the pellucid decrees of his fool, we have the same tale of chaos, blindness, and despair repeated in many different mouths—all of it at odds with the faith of the tinkerers. And the sonnets, in the poet's own voice:

> Tired with all these, for restful death I cry,
> As, to behold desert a beggar born,
> And needy nothing trimm'd in jollity,
> And purest faith unhappily forsworn,
> And gilded honor shamefully misplaced,
> And maiden virtue rudely strumpeted,
> And right perfection wrongfully disgraced,
> And strength by limping sway disabled,
> And art made tongue-tied by authority,
> And folly, doctor-like, controlling skill,
> And simple truth miscalled simplicity,
> And captive good attending captain ill. . . .

For Henry James, life is "a slow advance into enemy territory":

Life is, in fact, a battle. Evil is insolent and strong; beauty enchanting, but rare; goodness very apt to be weak; folly very apt to be defiant; wickedness to carry the day; imbeciles to be in great places, people of sense in small, and mankind generally unhappy. But the world as it stands is no illusion, no phantasm, no evil dream of a night; we wake up to it again for ever and ever; we can neither forget it nor deny it nor dispense with it.

For Goethe, through the pen of that sad figure Werther:

All the highly learned schoolmasters are agreed that a child does not know why he wants something; but that adults, too, like children, stagger about on this earth of ours and do not know where they come from or where they are going; that they act just as little in accordance with true purpose and are governed just as much by biscuits and cake and birch rods—no one wants to believe that, and yet it seems to me that this is palpably so.

On the opening page of his *Antimémoires*, André Malraux tells of running into an old friend with whom he had been through the war. The friend had spent fifteen years as a country priest. The man of letters asked the man of the cloth, in solemn humility, what he had learned from fifteen years of hearing confessions, and the latter replied, after some thought, that he had learned two things:

"First, people are much more unhappy than one imagines . . . and then . . ." He raised his lumberjack's arms into the night full of stars: "And then, the bottom of everything, is that grown-ups do not exist."

One could go on almost endlessly with such quotations, but if we dwell on them long enough to disperse the mist of familiarity, we can look again at what they say. Let us invite these artists of the soul to a friendly gathering. On one side of the room a group of tinkerers banter cheerfully about various ways to make everything just fine. On the other side, a group of biologists are discussing, rather glumly, the unchanging facts of human nature. Which group would the artists join?

The great literary figures, past and present, have had much more in common with the view of human nature taken by biologists than with that of human potential taken by social scientists. Thus, it is almost as amusing as it is unjust for the latter to decry biologists as technicians while they try to hide behind the cloak of humanism; for them it is a tattered cloak indeed, affording scant cover and, in the long term, no safe disguise. Despite their dim view of human nature, many—though certainly not all—of the artists who wrote these things were critics of the societies they lived in and even of the governments that ran them. But they were skeptical of proposals for change, especially grand schemes. Their insight into human character made them less sanguine about how much change is possible, and so more resistant to the tinker theory. *Life is a battle . . . Adults stagger about like children . . . Grown-ups do not exist.* Like children, indeed, in our motives, in our confused yet hot arousals, in the willful working out of our personal chaos upon the world. But unlike children, who stumble endearingly over themselves before they wreak much havoc, we do the real-world work of our besotted, arcane emotions after slower deliberation, in loftier grandeur, using subtler indirection, out of a deeper, more vengeful vein of selfishness, and with more power.

> The dream of reason produces monsters.
> —GOYA, *Caprichos*

Human behavioral genetics is the most controversial of all pursuits in behavioral biology, as it should be. It was only yesterday that an explicitly genetic theory of human behavior resulted in, or at least strongly supported, the ghettoization, deportation, concentration, enslavement, and mass extermination of millions of helpless victims guilty of absolutely nothing. The genetic taint that the Jewish people were accused of was the product of deeply entrenched racism, but it was explicitly embraced by many German physicians and scientists. Doctors were the first professional group to support Hitler. Under the rubric of public health, a chain of medical institutes promoted racist lies. In principle, all the Jews murdered in every concentration camp had death warrants signed by physicians, in the name of racial hygiene. Any thoughtful person must take pause at this specter of mass murder, intertwined with decades of genetic pseudoscience, and tremble at its possible repetition, in some unpredictable form, at some unknown time in the future. Should this possibility not be enough in itself to keep us from meddling with such theories?

Unfortunately, no. Rejection of genetic theories has been no guard against the terrors of authoritarian violence. Deportation, imprisonment, virtual enslavement, and direct or indirect slaughter of as many as 20,000,000 innocent Soviet citizens was based on an ideology rejecting the stable effects of genes. As the Jewish victims of Hitler were transported and worked to death or murdered because of presumed defects due to unchangeable genes, so the Soviet victims of Stalin were transported and worked to death or murdered in order to change them, through "reeducation," in the name of malleability, or at least to change other people around them. As the lies of the anti-Semitic "geneticists" (who knew nothing about genetics) justified the one, so the lies of Trofim Lysenko and other antigeneticists justified the other. The "Jew-Free Europe" and the "New Soviet Man" were approached through partly similar means, despite stemming from irreconcilable theories.

Such things did not, of course, end in the mid-twentieth century. In 1965, in Indonesia, a mass slaughter of alleged "Communists" claimed more than half a million lives, including a far disproportionate number of ethnic Chinese, whose historic role in the country paralleled that of the Jews in Europe. In the late 1970s, in the formerly peaceable kingdom of Cambodia, a program of deportation and mass extermination that, by some estimates, was equal to a fourth of that accomplished by the Nazis—quite a feat for a technologically backward country—

was based entirely on "reeducation," and on a theory of the extreme plasticity of human behavior. In the mid-1990s, a mass murder of Tutsi by Hutu claimed more than a million victims, based on no ideology at all really, only a long history of mutual hatred combined with population pressure and periodic shifts in the balance of power. And just to bring our fears up to date, in the late nineties a smaller but equally ominous ethnic slaughter in the former Yugoslavia proved that Europe is still capable of large-scale savagery.

So there is no morally safe haven in ideology, culture, or history, unless it is an ideology of decency, a culture of respect, and a slowly, painstakingly earned history of fair play. The political experience of the century just concluded proves that almost any theory about behavior can be bent to evil purpose, up to and including mass murder, just as religious ideologies have been bent to similar purposes for many centuries. The source of large-scale wickedness, then, cannot lie in the minds of social theorists—however misguided they may be, however they may sometimes lend themselves to evil purpose. To insist that sociobiology and behavioral genetics are inherently bigoted or reactionary is simply slanderous, a tarring of thousands of dedicated scientists with a brush that should be reserved for only a few. It is the worst sort of guilt by association.

And it is simply not true that recognizing genetic influence must be an obstacle to change. In this, the Gene Age, finding the cause of a problem is not just compatible with but often essential to its correction. With glasses I have excellent vision, despite my myopic genes. Diabetes is partly genetic, but insulin, appropriately used, can make those genes weak. PKU, once a leading cause of mental retardation, is itself caused by a simple genetic defect in one enzyme. But the disease has an environmental solution—the removal of phenylalanine from the diet—that could never have been arrived at without prior knowledge of the genetic determination of the enzyme defect. Families that have endured depression for generations have found medication that has freed them from their ancestors' suffering. Finally, new reproductive technologies have given couples an unprecedented range of choices, yet they are only the beginning. Gene control is coming faster than we think, and it will make biology a servant of freedom, releasing us from countless age-old miseries.

But biology is only the beginning of understanding. What is needed is not determinism but what I call biological influentialism—the belief that biology, while not decisive, has an influence too strong to ignore. Consider the following plausible policy arguments based on biology:

1. Women and men differ, for biological reasons attributable to genes, in their tendency to violence. One simple measure to reduce worldwide violence would therefore be to replace men with women in positions of political and diplomatic power, in a strategic effort to dampen irrational sources of violent conflict. This would have to do more than just put a woman at the top of an almost all-male hierarchy, which does little to change the inherent violent tendencies of the system. Women would have to enter the system at all levels, even perhaps predominate. Social

and political systems with such pervasive representation of women would be buffered against irrational mobilization for war.

2. Some individuals of low normal intelligence are limited by genes. Research into the genetics of intelligence can offer them the hope of freedom from those limitations. Yet many well-intentioned intellectuals think that research on the genetics of intelligence should stop. This is like calling for a halt to research into the genetics of obesity. After all, this condition is really just one end of the normal range of human nutritional adaptation, and restrictive diets, faithfully pursued throughout life, can often prevent it. Also, eliminating obesity would tend to make everybody the same. But fortunately, there is little opposition to research on the genetics of obesity. Depriving the less intelligent of research that might help solve their problem is a policy lacking in compassion. If we can have a national institute devoted to alcoholism and drug abuse, why not have one devoted to learning impairments?

3. In all hierarchical societies, some people have wealth and power because of traits—such as intellect, cooperativeness, and compassion—that may be good, as well as traits—such as a tendency to authoritarian violence— that are indisputably bad. Democracies monitor that rise to the top and limit the power of those who complete it. They also limit what can be transferred to genetic offspring. Some of the rise to the top must be due to genes, but the genetic shuffle is unpredictable, and people generally transfer wealth and power to their offspring, even those showing few of their own good traits. So democracies should have very high inheritance taxes. If rich children cannot succeed with the advantages they have while their parents are alive, they have probably not inherited their parents' abilities and should cede wealth and power in open competition with more able contemporaries.

4. The reproductive imperative has shaped human destiny from the beginning, but we have now filled the planet and are headed for disaster. Future historians will not deal kindly with those who, in the twenty-first century, discouraged any reasonable means of birth control. Fortunately, we know that there are other imperatives. Human nature exaggerates the importance of certain stimuli. Suppose we were to take advantage of these weaknesses to combat population growth? Youth, health, fitness, beauty, sex, play, comfort, status—these powerful incentives make people abandon the goal of having a large family. A rational program for the developing world would build family-planning programs on a massive scale while aggressively promoting these incentives.

5. Reproductive technology allows people to choose to improve on their own genetic endowment within their own families. This is not coercion, it is choice. If parents feel that their genetically guided abilities or intelligence would not give their children the best life has to offer, why shouldn't they be able to improve that potential? Even now, without gene therapy,

choices are possible in the realm of artificial insemination or purchase of
donor eggs, but they are extremely expensive and not available to poor
couples who might choose them. Making these and future technologies
available to those who really want them could give them the opportunity
to lavish love on children with greater promise than their own genes can
provide. *Opportunity*, not coercion; *choice*, not chance.

Whether or not I subscribe to these arguments is not the point here. Neither is
the question of how convincing they are. All are fraught with problems and would
require many years of care and thought. The point here is that although they are
arguments arising from behavioral biology, they can scarcely be called conservative. Most recent practitioners of ethology, sociobiology, and behavior genetics
have done little or nothing to justify the charge that they are politically conservative, much less reactionary. With few exceptions, they are trying to find out some
truth or other about human behavior. Those truths are not only usable by policy
makers across the political spectrum, they are absolutely vital to their success.

Still, one must not be too confident about new proposed solutions, wherever
they may come from. There is a strange Goya etching, in the series *Caprichos*,
showing a man—probably an intellectual—asleep over some writing at his desk.
In a whirlwind-shaped cloud coming out of his head emerges as horrendous-
looking a collection of scary figures as ever were drawn on paper. The caption
reads, *El sueño de la razon produce monstruos.* "The dream of reason produces
monsters." It seems to have three meanings. In one, Reason is allegorical, person-
ified in the man; it sleeps at times, and its dreams produce monsters. In another,
reasoning as a process is a kind of dreaming, and that kind of dreaming is always
nightmarish. The third meaning uses the "wish" or "prayer" sense of the word
dream; the possibility of reason is seen as a vain hope, which in its predictable failure produces monsters—ones perhaps more terrible than any ever engendered by
mere passion.

Joan Didion, reviewing a book by V. S. Naipaul, voiced the two novelists'
common suspicion that ideas can be overrated. Instead, she wrote that one can
have a "sense of the world as a physical fact without regret or hope, a place of
intense radiance in which ideas may be fevers that pass." Another novelist, Leo
Tolstoy, wrote in the private notebook in which he set down meticulously the
details of his life: "As soon as man applies his intelligence and only his intelligence to any object at all, he unfailingly destroys the object."

So far we have applied our intelligence, and only our intelligence, to the
ordering of human life on earth. Not that I don't believe in intelligence; I do. But
we all have a tendency to believe in it too much, to serve it as if it were a god. I
have to remind myself of the words Brecht gave Galileo: "The purpose of science
is not to open the door to infinite wisdom, but to set some limit on infinite error."
In the twenty-first century we are going to hear an almost continuous stream of
ideas about the nature of experience and the solutions to our problems, ideas produced by human intelligence. In his great book *An Introduction to the Study of
Experimental Medicine*, Claude Bernard wrote: "Man is by nature metaphysical

and proud. He has gone so far as to think that the idealistic creations of his mind, which correspond to his feelings, also represent reality." If this was true of ideas in physiology, how much truer must it be in the realm of behavior?

It would be easy to say that the ideas we hear during the first few decades of the millennium will all be wrong. They will not. A few will be right, and I hope we will know them when we hear them. But even then, every idea, every trend is a double-edged sword. Globalization breaks down barriers, but fundamentalism uses fear to erect others. Technology shrinks the world but makes that smaller world more vulnerable. The Internet wires the world but leaves the poor behind; and like every form of communication, it is already a vehicle for the worst in human nature, allowing hatred, bigotry, destruction, greed, deception, and exploitative pornography, even as it fosters new levels of cooperation, altruism, democracy, education, and information exchange.

We face immense challenges. In this century, fresh water will run out. Vast populations will suffer war and famine. AIDS will devastate Africa and disrupt the economies of South and Southeast Asia. New viruses will emerge as forests succumb to blades and fire. Small, autocratic countries will make nuclear weapons, and if they cannot, they will wage even more destructive germ and chemical warfare. Billions of young people full of hope will see their hopes dashed and feel the frustration that follows. Selfish men with petty minds will lead them.

Yet human ingenuity will be at our disposal, as it has been since we first walked upright. Technology is no easy fix, but it allows us to try to countermand these problems. Birth control will bring about zero population growth and a subsequent badly needed population *decline*. There will still be crises caused by more new billions, but an economic engine of unprecedented power could perhaps draw those billions in its wake. What will *not* happen is automatic resolution. Technology alone has never solved our problems because those problems are, in essence, human problems. They require human solutions—ideas about how to make a better world out of the same old stuff of human nature. Silicon chips, fiber optics, fuel and photovoltaic cells, designer drugs, and gene control will all play their roles, but the future is human, made of flesh and blood. Without a deep grasp of human nature, no amount of technology will put us where we want to be a century from now.

Only a serious study of our nature, in the context of the rest of the natural world, will allow us to protect ourselves at this precarious evolutionary juncture. In the meantime, if you ask me how to set your sail in the storm of claim and counterclaim, of fact and lie and theory, of warning, prophecy, judgment, and exhortation, I do have a bit of advice that I earnestly believe in. It can be summarized in the one-word injunction: Doubt.

Placed on this isthmus of a middle state,
A being darkly wise, and rudely great . . .
He hangs between . . .
In doubt to deem himself a god, or beast . . .

—**ALEXANDER POPE,** "An Essay on Man"

In the old ape house at New York City's Bronx Zoo, among the highest primates, was a sign that read, THE MOST DANGEROUS ANIMAL IN THE WORLD. Having learned already that the fearsome-looking gorilla is not very dangerous, much less the chimpanzee or orangutan, the puzzled zoo visitor would lean forward to discover which cousin of these relatively benign creatures merits such a label. Above the sign was a set of ordinary cage bars, and behind the bars was a mirror. It was too set-up, too much of a cliché, to engender a big shock of recognition. *Of course*, one would think. *Still, cute*. Yet there was something about one's own image above that particular sign, and—however illusorily—behind bars, that somehow did not fail to give pause.

A recurring theme of this book has been the problem of human destructiveness, including self-destructiveness, and its discouraging intractability. If I have seemed at times to let it slip into synonymy with the simpler problem of violence, let me now clearly distinguish them. The human propensity to violence is there, is inborn, is—up to a point—enhanced or reduced by experience, is serious, and is, in a word, bad. But it is only one part of the problem of destructiveness, which is made up of much else besides. Biological explanations may have little to say about certain major transformations now under way in our species—for example, the remarkable progress from widespread illiteracy to widespread literacy, one of the most rapid and profound behavioral changes any species has ever undergone in so few generations.

But there are other challenges facing us, as we have seen, and about some of them sociobiology and behavioral genetics have a lot to say. The world's population will stabilize, but not soon enough. In the interim we will be at constant risk for international and civil conflict, with sweeping movement of peoples and struggles over resources. More important than simple population growth, however, are the continually expanding wants or needs—the distinction, always vague, undergoes a sea change with modernization—of the billions already here.

Although it will be small consolation to the hundreds of millions who will starve during the twenty-first century, sheer food productivity may not be a lasting problem. But the vision of people thrown into armed conflict over scraps of food in a starving world is naïve, because people do not have to be deprived of food to be drawn into violent confrontation. People make war not because they are starving but because they are paying more for gasoline, or because their national honor has been offended, or because they think it has, or because they want to prevent someone from making war on them.

We live in an energetic getting and spending—and not just of buyable things. Our motives are laced with a froth of anger that can easily churn and roil. When someone distant suffers, we have a feeling of pity; only when it is close is the feeling grief. *That is only natural*, say the wise, biologically accurate words. But in most lives there is evidently enough grief to make pity a poor clarion call to action. This, although never exactly admirable, worked for most of our history, when we all lived in small, face-to-face groups. But over the last 10,000 years we have piled ourselves into vast, dense aggregates, and the same set of motives and limitations has brought us to a condition in which, with regularity, we commit the most despicable acts in the whole long record of life on earth. And without blushing, in our respites from mutual slaughter, we prepare ourselves to try to commit more.

According to T. S. Eliot's poem "The Hollow Men," the world ends not with a bang, but a whimper. But suppose it were not about to end, neither in fire nor in ice, neither suddenly nor slowly, at any near time. Suppose the world were to go on just as it has for the last few hundred years, an unbroken extension of the past into the future, for as long as we please. Would that not be reason enough to despair?

In Chaim Grade's novel *The Yeshiva*, about the Talmudic tradition of Poland and the Jews who tried to follow it, the hero's first sermon says: "Man is evil from birth. But his nature prevents him from finding the evil in his supposedly good deeds. That is why the Torah was given to us—to teach us to lead an ethical life." This idea, in one form or another, is at the core of many religious systems. It is an idea more compatible with the recent discoveries of human behavioral biology than with the social science of the last hundred years. We could rephrase it: *Human beings are irrevocably, biologically endowed with strong inclinations to act in ways that our own good judgment tells us to reprehend—that is, if we are in the least capable of sympathy with the suffering of others, or have any sense of the joy and order and beauty of life.* The judgment, the sympathy, the sense of joy and beauty all evolved for other purposes than to save the human species from slow destruction. Yet there they are. Can we not now use them for this purpose?

The hero of Grade's novel suffers because he fails to see that people are good as well as evil, from birth. His foil, a rabbi named for the patriarch Abraham, tries to teach him to fight evil by drawing out good. He cannot, because he ignores good while insisting upon a relentless fight against evil.

The good, too, is deeply engrained. The English geneticist Conrad Waddington, near the end of a long life, argued

> that the choice of an ethical system is like the choice of a set of axioms on which to found mathematics. . . . But though in mathematics we are free to choose . . . when we need to deal with the world of objects that are about the size of their own bodies, we find that it is the Euclidian axioms which are by far the most appropriate. They are so appropriate, indeed, that we almost certainly have some genetic predisposition to their adoption built into our genotypes—for example, the capacity of the human eye to recognize a straight line.
>
> . . . If we wish to develop an ethical system which we can apply to human life as we know it, there are probably some ethical axioms which we are almost forced to incorporate. They would be the common ground which we find between all the major ethical systems of different religions and groups of mankind—such values as truth, respect for self, respect for others, and respect for something larger and more embracing than one's

own immediate experience . . . a built-in predisposition towards certain ethical values which have the same degree of general relevance to human society as do the Euclidian axioms of geometry to the material world . . .

Who knows what good may yet lurk in the hearts of men? In the hope of discovering it and bringing it to light, in the hope that some earthly nurturance can cause it to thrive and grow, we may well set our hearts and minds to a most momentous task. And, as a sort of amulet, an ornament of tradition, to speed us on our difficult way, we could repeat with the Psalmist, "Break thou the arm of the wicked; and as for the evil man, search out his wickedness, till none be found. . . . Lord, Thou hast heard the desire of the humble . . . to right the fatherless and the oppressed, that man who is of the dust of the earth may be terrible no more."
Amen. Selah.

PART FIVE

The Tangled Wing

Then beauty is nothing
But the start of a terror we're still just able to bear
And the reason we love it so is that it blithely
Disdains to destroy us . . .

—RAINER MARIA RILKE, "The First Elegy"

CHAPTER 19

The Dawn of Wonder

The most beautiful experience we can have is the mysterious. It is the fundamental emotion which stands at the cradle of true art and true science.

—ALBERT EINSTEIN

One of the most fascinating and least discussed discoveries in the study of the wild chimpanzees was described in a short paper by Harold Bauer. He was following a well-known male through the forest of the Gombe Stream Reserve in Tanzania when the animal stopped beside a waterfall. It seemed possible that he had deliberately gone to the waterfall rather than passing it incidentally, but that was not absolutely clear. In any case, it was an impressive spot: a stream of water cascading down from a twenty-five-foot height, about a mile from the lake, thundering into the pool below and casting mist for 60 or 70 feet, a stunning sight to come upon in a tropical forest.

The animal seemed lost in contemplation of it. He slowly moved closer and began to rock, while starting a characteristic round of "pant-hoot" calls. He grew more excited and finally began to run back and forth while calling, to jump, to call louder, to drum with his fists on trees, to run back again. The behavior resembled that observed by Jane Goodall in groups of chimps at the start of a rainstorm—the "rain dance," it has been called. But this was one animal alone, and not surprised by sudden rain—even if he had not deliberately sought the waterfall out, he certainly knew where it was and when he would come upon it.

He kept this up long enough that it seemed to merit some explanation, and he did it again in the same place on other days. Other animals were observed to do it as well. They had no practical interest in the waterfall. They did not have to drink from the stream or cross it. To the extent that it might be dangerous, it could be easily avoided, and it certainly did not interest every animal. But for these it was something they had to look at, return to, study, watch, become excited about—a thing of beauty, an object of curiosity, a challenge, a fetish, an imagined creature, a god? We will never know.

But for a very similar animal, perhaps 5,000,000 years ago, in the earliest infancy of the human spirit, something in the natural world must have evoked a response like this one—a waterfall, a mountain vista, a sunset, the crater of a volcano, the edge of the sea—something that stopped it in its tracks and made it watch, and move, and watch, and turn, and watch again. Something that made it return to the spot, though nothing gainful could take place there, no feeding, drinking, reproducing, sleeping, fighting, fleeing, nothing animal. In just such a response, in just such a moment, in just such an animal, we may, I think, be permitted to guess, occurred the dawn of awe, of sacred attentiveness, of wonder.

The human infant, for its first few months of life, is all eyes, in a way that no other animal infant quite is. It isn't just that its eyes are good, that it does a lot of looking; it's that it does so little else, really. It can suck, of course, and swallow, but the rest of what it does is very primitive, except for attentiveness. Even in the adult brain, a third of all incoming signals are from the eyes. In the infant, looking and seeing are way ahead of most other functions in development, with the possible exception of hearing. The infant is not a passive figure, or an active one, but what might be called an actively receptive one—eagerly, hungrily receptive, famished for sights and sounds; it possesses not a vague, fuzzy intelligence in a blooming, buzzing confusion but a highly ordered, if simple, mind with a fine sense of novelty, of pattern, even of beauty. The light on a leaf outside the window, the splash of red on a woman's dress, the restless shadow on the ceiling, the drubbing sound of hard rain—any of these may evoke a rapt attention not, perhaps, unlike that of the chimpanzee at the waterfall.

For most of us, that sense of wonder diminishes as we grow, becoming at best peripheral to the business of everyday life. For a few it becomes the central fact of existence. These few will follow one of two paths: analysis or simple contemplation. Either way the sense of wonder is the first fact of life, but the paths are different in most other ways. The analyst, or scientist, moved to reveal by explaining, breaks apart the image and the sense of wonder, focusing sequentially on the pieces. The contemplator, or artist, moved to reveal by simply looking, keeps the image and the sense of wonder whole. The artist contrives to keep the attention riveted without fragmentation, by means of high trickery. This trickery involves transmuting the image into human speech—whether in literary, plastic, or musical form—forever fixing in place the sense of wonder.

There is a photograph that by now has been seen by most people living in what we call the developed world. It was taken from an ingenious if crude vehicle traveling thousands of miles per hour, across a vast expanse empty of air, by men who had devoted their lives, courageously and at great personal cost, to the mastery of nature through machinery. This photograph cost a billion dollars, and it is worth every penny.

It shows an almost spherical object poised against a field of black. The object is partly colored a deep, warm, pretty blue, with many broken, off-white swirls drawn across it. At first it resembles a mandala, a strange symbol woven on black cloth. It looks whole, somehow, and rather small. But as we study it (it draws us in mysteriously) some red-brown shapes obscured among the swirls of white take on before our eyes the unmistakable images we first saw and memorized as children, encountering the geography of the continents. If the space program accomplished nothing else, we must be grateful to it for producing that photograph.

"Got the Earth right out our front window," said Buzz Aldrin, a medium-size mammal from a middling planet of a middle-aged star in the arm of an average galaxy, gazing at home. There was no excess of poetry on that mission. There was, of course, the stark poetry of aeronautics gobbledygook, and the arch, well-prepared, historic mot of Neil Armstrong setting foot on the Sea of Tranquillity. But "Beautiful, beautiful," "Magnificent sight out here," and "Got the Earth right out our front window," represent the level at which these unique first views of the natural world were transmuted to human speech. This was no fault of Armstrong and Aldrin; they were chosen for other talents, which they had in full measure. But it is intriguing that such spontaneous poetry was inspired by the machinery. "The Eagle has wings," one of them said as the lunar landing vehicle separated, after some difficulty, from the orbiting command station. The eagle, bold symbol of hope on the North American continent and, beyond that, of the hope of humanity in the mission, has wings, has the means to transcend technical difficulty and emerge, having mastered natural law.

But this stepping off the earth is an illusion. The mastery of natural law has proceeded no farther than the grasp of some elementary laws of physics. Compared with the uncharted, infinitely more intricate laws of biology and behavior that govern the human spirit and the planet Earth's future, this mastery is trivial, a mere conjurer's trick. The mastery of physical law cannot save us while we are grounded in ignorance of the natural laws that govern our behavior. In this sense, the eagle does not have wings.

When I was a young man in college a professor took me to the American Museum of Natural History—not to the exhibits, which I had often seen, but into the bowels of the place, among the labyrinths of cabinets storing bones and skins and rocks and impossibly ancient fossils. I was very much impressed by this chance to see the museum the way insiders saw it.

There I met a man who had devoted most of his life to the study of the skeletal remains of *Archaeopteryx*, the earliest tetrapod with feathered wings—as we now know, a descendant of nonflying dinosaurs feathered for warmth. I saw this man as he stood over the bent, vaguely birdlike shape, embedded in a 150-million-year-old Mesozoic rock. I was introduced to him, awed by him, impressed with his intelligence and wisdom. It was obvious that he wanted to impart to me some piece of useful knowledge gained from countless hours of squinting over that crushed tangle of bone and stone.

What he finally said was that he thought *Archaeopteryx* was very much like people. This puzzled me, of course, as it was calculated to do, and when I pressed him, he said, "Well, you know, it's such a transitional creature. It's a piss-poor reptile, and it's not very much of a bird." Apart from the shock of hearing strong language in those relatively hallowed halls, there was an intellectual shock to my young mind that filed those phrases in it permanently.

The dinosaurs ruled this planet for over 100,000,000 years, at least 100 times longer than the brief, awkward tenure of human creatures, and they disappeared almost without a trace, leaving nothing but crushed bone as a memento. We can do the same more easily, and in an ecological sense we would be missed even less. *So what?* seems an inevitable question, and the best answer I can think of is that we know, we are capable of seeing what is happening. We are the only creatures that understand evolution and that, conceivably, can alter its course. We see the possibility of self-extinction and are probably capable of averting it. It would be too base of us to relinquish this possibility.

It seems at times that we are losing the sense of wonder, the hallmark of our species and the central feature of the human spirit. Perhaps this is due to the depredations of science and technology against the arts and the humanities, but I doubt it—although this is certainly something to be concerned about. I suspect it is simply that the human spirit is insufficiently developed at this moment in evolution, much like the wing of *Archaeopteryx*. Whether we can free it for further evolution will depend, in part, on the full reinstatement of the sense of wonder. It must be reinstated not only in relation to the natural world but to the human world as well. We must once again experience the human soul as soul, and not just as a buzz of bioelectricity; the human will as will, not just a surge of hormones; the human heart not as a fibrous, sticky pump but as the metaphoric organ of understanding. We need not believe in them as metaphysical entities— they are as real as the flesh and blood they are made of. But we must believe in them as entities; not as analyzed fragments but as wholes made real by our contemplation of them, by the words we use to talk of them, by the way we have transmuted them to speech. We must stand in awe of them as unassailable, even though they are dissected before our eyes.

As for the natural world, we must restore wonder there, too. We could start with that photograph of the Earth. It may be our last chance. Even now it is being used in geography lessons, taken for granted by small children. We were the first generation to have seen it, the last generation not to take it for granted. Will we remember what it meant to us? How fine the Earth looked, dangled in space? How pretty against the endless black? How round? How very breakable? How small? It up to us to try to experience a sense of wonder about it that will save it before it is too late. If we cannot, we may do the final damage in our lifetimes. If we can, we may change the course of history and, consequently, the course of evolution, setting the human lineage on a path toward a new evolutionary plateau.

We must choose, and choose soon, either for or against the further evolution of the human spirit. It is for us, in the generation that turned the corner of the millennium, to apply whatever knowledge we have, in all humility but with all due speed, and to try to learn more as quickly as possible. It is for us, much more than for any previous generation, to become serious about the human future and to make choices that will be weighed not in a decade or a century but in the balances of geological time. It is for us, with all our stumbling, and in the midst of our dreadful confusion, to try to disengage the tangled wing.

Notes and References

CAVEAT: The Dangers of Behavioral Biology

The contents of this book are known to be dangerous.

I do not mean that in the sense that all ideas are potentially dangerous. Specifically, ideas about the biological basis of behavior have encouraged political tendencies and movements later regretted by all decent people and condemned in school histories. Why, then, purvey such ideas?

Because some ideas in behavioral biology are true—among them, to the best of my knowledge, the ones in this book—and the truth is essential to wise action. But that does not mean that these ideas cannot be distorted, nor that evil acts cannot arise from them. I doubt, in fact, that what I say can prevent such distortion. Political and social movements arise from worldly causes, and then seize whatever congenial ideas are at hand. Nonetheless, I am not comfortable in the company of scientists who are content to search for the truth and let the consequences accumulate as they may. I therefore recount here a few passages in the dismal, indeed shameful history of the abuse of behavioral biology, in some of which scientists were willing participants.

The first episode is recounted in William Stanton's *The Leopard's Spots: Scientific Attitudes Toward Race in America, 1815–59* (Chicago: University of Chicago, 1960). Such names as Samuel George Morton, George Robins Gliddon, and Josiah Clark Nott mean little to present-day students of anthropology, but in the difficult decades between the death of Jefferson and the Civil War, they founded the American School of Anthropology. This movement dedicated itself to proving the inevitable separate status of the races and to placing white supremacy on a scientific foundation. They attempted to do this through the study of skulls and brain volume, combined with some "obvious" observable facts of behavior and custom—"niggerology," as one of them privately called it (Stanton, *The Leopard's Spots*, p. 161). In its more dignified public guise it was called "polygenism," a reference to the supposed separate evolutionary origins of various races. (This, incidentally, was a view that Jefferson and his intellectual circle had rejected.) Two of the three (Morton and Nott) were physicians, but their conjectures were based on so little and such silly "evidence" that it is puzzling how they succeeded.

Yet succeed they did. When they came on the scene in the early part of the nineteenth century, the views of Samuel Stanhope Smith, according to which humankind had a single origin and a single biological plan, held sway. It was the view taken by Thomas Jefferson and his circle (see Daniel Boorstin, *The Lost World of Thomas Jefferson*, Chicago: University of Chicago, 1981) and is universally accepted today. But thanks to the efforts of the American School, by the 1850s the unity of humankind was an idea

effectively dislodged from favor, linked to atavistic, religious, antiscientific sentimental-
ism. Miscegenation was viewed as a threat to civilization, and slavery as the logical lot of
the Negro. Now no one would suppose the Civil War to have been caused by a handful
of anthropologists; but they were highly respected and popular writers and lecturers, and
it cannot be doubted that they deceived many. Meanwhile, their counterparts in Britain,
France, and Germany laid a foundation for scientific racism that would stand firm for
about a hundred years (Marvin Harris, *The Rise of Anthropological Theory*, New York:
Thomas Y. Crowell, 1968, ch. 4).

The second episode involved Social Darwinism, some of which was in fact pre-
Darwinian. It is recounted by George Stocking, in chapter 6, "The Dark-Skinned Savage:
The Image of Primitive Man in Evolutionary Anthropology," of *Race, Culture and Evolution: Essays in the History of Anthropology* (New York: Free Press, 1968) and by Marvin
Harris in chapter 5, "Spencerism," of *The Rise of Anthropological Theory*. In the latter part
of the nineteenth century, most social theory was "evolutionary," but in nothing like the
modern sense. Leaders of social and cultural anthropology, like Lewis Henry Morgan and
Edward Tylor, although they greatly admired the "primitive" tribes and races they studied,
nevertheless viewed them hierarchically, with the "less developed" or "less complex"
groups as essentially frozen relics of past epochs. Marx and Engels took over this view from
Morgan and made little attempt to conceal their own patronizing attitude toward preindustrial, especially pre-state peoples.

Darwin (see Stocking, *Race, Culture and Evolution*, p. 113) and his evolutionist predecessor Charles Lyell (see Harris, *Rise of Anthropological Theory*, p. 113) both predicted
the extermination of the "savage" races by the civilized ones, and did not seem to shed any
tears over the process. This in an era when some of their readers were doubtless pursuing
that very extermination. Morgan and Tylor's hierarchical arrangements of social and cultural forms went along with explicit presumptions of a corresponding hierarchy of mental
capacity; the more complex the civilization, the greater the native intelligence of its members. Progress through improvement was the inexorable motive force, and the pinnacle of
progress was the civilization of Victorian England.

How comforting these ideas must have been to the representatives of that and similar
civilizations just then engaged in the difficult work of subduing, enslaving, or, where necessary, exterminating those "primitive" peoples. It is not surprising that they were easily
convinced, despite the lack of evidence. Herbert Spencer, the leading exponent of social
evolution, cuts a rather sad figure against this background. Always claiming to be a friend
of the poor, abhorring war and the greedy rape of the underdeveloped world, Spencer was
viewed by many contemporaries, as well as by later scholars, as an apologist for the worst
that was going on. He, not Darwin, coined the phrase "survival of the fittest" and justified
the exploitation of the weak by the strong, on the grounds that the inevitable march of
progress is only held back by humane intervention in the struggle for existence. Spencer
explicitly apologized for the most unrestrained capitalism, and opposed socialism and all
forms of social welfare. It is not difficult to imagine his words in the minds of the Robber
Barons or of the legislators who voted against child labor laws. The progress of human
decency in the nineteenth century was no doubt a complex matter, but it is logical to suggest that ideas about the biology of behavior retarded that progress. (On evolutionary theories of social behavior and their consequences, see Stephan Chorover, *From Genesis to
Genocide*, Cambridge: MIT, 1979, ch. 5).

The third episode took place on both sides of the Atlantic between the beginning of
World War I and the end of World War II. The American side of the episode is recounted
in Daniel Kevles's *In the Name of Eugenics: Genetics and the Uses of Human Heredity*
(New York: Alfred Knopf, 1985), and in the works by Kamin, Chorover, and Stocking cited

above. Although Alfred Binet, the French psychologist who originated IQ testing in 1905, had intended it as a device for identifying children who needed mental improvement through training, it began to be used a decade or so later in the United States for very different purposes. Under the auspices of Lewis Terman of Stanford and Robert Yerkes of Harvard—two leaders of American psychology—it was explicitly used to reduce immigration. Both these men believed that IQ was largely genetic, and they saw a chance to provide a much needed social service—giving the U.S. government a good excuse to stem the rising tide of immigration. Vast numbers of potential immigrants were labeled as retarded and sent away after taking intelligence tests in a language they did not understand.

Meanwhile, the behavior-genetic theories of the nineteenth century had crystallized in a clear eugenics movement in the United States. With the approval and encouragement of leading psychologists, compulsory sterilization laws were passed by the state legislatures of Pennsylvania, Indiana, New Jersey, Iowa, California, and Washington, providing for the "unsexing" of an impressive range of undesirables. In upholding the California law, the attorney general of California explicitly used the language of behavioral biology:

> Degeneracy means that certain areas of brain cells or nerve centers of the individual are more highly or imperfectly developed than the other brain cells, and this causes an unstable state of the nerve system, which may manifest itself in insanity, criminality, idiocy, sexual perversion, or inebriety.

He went on to include "many of the confirmed inebriates, prostitutes, tramps, and criminals, as well as habitual paupers" in this class, all of whose members were potentially eligible for legal castration. *The Harvard Law Review* of December 1912—by which time all these state laws had been passed—argued that they would be constitutional, but only in the case of "born criminals" (Kamin, *I.Q.*, pp. 11–12).

Retrospective criticism of these lawyers and officials has been justifiably great, but they were influenced by psychologists, biologists, and physicians who gave them a false account of the facts. These experts provided what seemed to be definitive statements in a context fraught with uncertainty. They held out false hopes for great improvements in human welfare through eugenics, and rang loud, false alarms of racial degeneracy and eugenic disaster in the event that their advice was not followed.

Given these remarkable intellectual and legal developments in the United States, the parallel movements in Germany and elsewhere in Europe seem a bit less astounding. The ideas of eugenics and racial hygiene (*Rassenhygiene*) became respectable and established in German academic and medical discourse while Hitler was still a child. In 1895 the physician Alfred Ploetz wrote *The Excellence of Our Race and the Protection of the Weak*; in 1903 Wilhelm Schallmeyer won a national prize (given by the Krupp armaments family) for his *Inheritance and Selection in the Life-History of Nationalities: A Sociopolitical Study Based upon the Newer Biology*. *Politisch-Anthropologische Revue* and *Archiv für Rassen und Gesellschaftsbiologie* (Archive for Racial and Social Biology), two important scholarly journals concerned with eugenics and racial purity, began publication in 1902 and 1904, respectively. In 1920 a distinguished jurist, Karl Binding, and a distinguished psychiatrist, Alfred Hoche, published *The Release and Destruction of Lives Devoid of Value*, advocating large-scale, eugenic euthanasia.

It is critical to realize how very respectable these ideas were. They had nothing to do with brown shirts, breaking glass, goose-step marches, or diabolically energized mass rallies. They had only to do with respectable scientists, physicians, and lawyers communicating soberly through the usual means of discourse. Long before the Nazi party was founded, it was widely agreed that discoveries in social biology constrained scholars to certain beliefs. Civilization was the result of genetic determinants, and its future depended on racial purity and the relentless elimination of the unfit from the gene pool.

This was not a national but an international phenomenon. In 1923, a year before the publication of Hitler's *Mein Kampf*, a director of health in Zwickau wrote to the German minister of the interior urging the enactment of a program of eugenic sterilization: "What we racial hygienists promote is not at all new or unheard of. In a cultured nation of the first order, the United States of America, that which we strive toward was introduced and tested long ago." Still skeptical, the interior minister pursued the matter through the German Foreign Office, and after receiving an extensive report became convinced. Through the legal and judicial example set by the United States, eugenics became respectable government business in Weimar, Germany (Chorover, *Genesis*, p. 98).

Daniel Goldhagen's comprehensive and chilling account of the perpetrators, *Hitler's Willing Executioners: Ordinary Germans and the Holocaust* (New York: Vintage/Random House, 1996), shows how deeply German culture was steeped in anti-Semitism, not just in the folkview but in the highest intellectual circles. Yet ideas about the role of the Jews in what might be called "racial history" were also current in international discourse. The English historian Houston Stuart Chamberlain had argued, in such works as *Foundations of the Nineteenth Century* (originally published in German) and *Race and Nation*, that the fall and rise of nations could best be understood by reference to the introduction and removal of Jews respectively. Chamberlain's work was widely discussed among German students from the time it was first published. (See Lucy S. Dawidowicz, *The War Against the Jews, 1933–1945* [New York: Holt, Rinehart & Winston, 1975] for discussion and references.)

Alfred Rosenberg, Hitler's advisor during the early years of the Nazi movement, called Chamberlain's work "the strongest positive impulse in my youth," and prepared excerpts of *Foundations of the Nineteenth Century* (*Grundlage des Neunzehn Jahrhunderts*) for Hitler's easy study (Dawidowicz, *War*, p. 20). Heinrich Himmler, later and throughout the war the head of the SS and a key figure in all concentration and killing operations, read *Race and Nation* (*Rasse und Nation*) at the end of 1921, and wrote of it in his diary: "It is true and one has the impression that it is objective, not just hate-filled anti-Semitism. Because of this it has more effect. These terrible Jews . . ." (Dawidowicz, *War*, p. 95). The last sentence is almost poignant; it makes clear that reading Chamberlain gave Himmler an added measure of conviction.

Are the scribblings of intellectuals about behavioral biology really important in causing great and destructive social movements? We don't know in every case, but the truth is poorly served by a smug conviction that they are not. Certainly the Nazis relied heavily on racial "science," and on physicians who studied and practiced it, for the justification of their program. As shown by Robert N. Proctor, in *Racial Hygiene: Medicine Under the Nazis* (Cambridge: Harvard University, 1988), racial theories and "research," emanating from official medical and scientific institutes and journals, was of the utmost importance in giving Nazism credibility. Doctors and public health officials were a central part of the program from the beginning, and were numerically as well as intellectually the professionals most supportive of Hitler. In addition to Proctor's account, see Michael Kater's *Doctors Under Hitler*, (Chapel Hill: University of North Carolina, 1989) and Robert Jay Lifton's *The Nazi Doctors: Medicalized Killing and the Psychology of Genocide* (New York: Basic Books, 1986/2000).

Many people wonder why the Jews did not try to get out. Of course, they did, in much larger numbers than were able to do so. The rising tide of immigration to the United States after World War I was in part due to the recognition by Jews and other Europeans of ominous signs on the horizon. As mentioned above, American psychologists helped to stem this tide. Terman, Yerkes, and others, referring to very poor research, involved themselves in the perpetration of falsehoods that laid the foundation for a much more restrictive immigration policy, formulated in the Immigration Act of 1924 and other laws. A much-quoted study was Henry Goddard's report about IQ testing of immigrants at Ellis Island, which claimed that 83 percent of the Jews, 80 percent of the Hungarians, 79 percent of the Italians, and 87 percent of the Russians were "feeble-minded" (Kamin, *I.Q.*, p. 16). These statistics were due primarily to sloppy testing and language barriers. Robert Yerkes published the results of similarly poor, "confirmatory" research, under the auspices of the United States National Academy of Sciences, in 1921.

The result in immigration policy was formidable for many ethnic groups, but for Jews it was deadly. Because of the views of American psychologists and other behavioral biologists about the genetics of mental competence, many Jews were trapped in Europe, later to become Nazi victims. Speeches and writings by respected Americans like Henry Ford and Charles Lindbergh echoed the vicious anti-Semitism pervasive in Germany, but they would have had less credibility without the assent of scientists. (The definitive work on the Holocaust remains Raul Hilberg, *The Destruction of the European Jews*, New York: Holmes & Meier, Inc., 1985. See also Hilberg's *Perpetrators, Victims, Bystanders: The Jewish Catastrophe 1933–1945*, New York: HarperCollins, 1992; Lucy Davidowicz's *The War Against the Jews* and Daniel Goldhagen's *Hitler's Willing Executioners*, cited above; and Martin Gilbert's *The Holocaust: A History of the Jews of Europe During the Second World War*, New York: Holt, Rinehart & Winston, 1985.)

Incidentally, after 1920 the role of American anthropology in these intellectual currents became a very different, rather heroic one. (See Stocking's *Race, Culture and Evolution*, ch. 11, for details.) Franz Boas had established a new and completely different "American School" of anthropology, the main thrust of which was to break decisively with the racist and evolutionist past. He and his students (among them Alfred Kroeber, Ruth Benedict, and Margaret Mead) rejected all notions of cultural hierarchy, and Boas's book *The Mind of Primitive Man* broke down the notion that mental function was correlated with civilizational complexity. Anthropologists of the Boas school placed the concepts of culture and cultural relativism at the center of the field, stressing the dignity and independent validity of all ways of life.

In the arguments over IQ, race, and eugenics that raged during the 1920s and 1930s, they opposed the psychological testers and eugenicists, stressing the mounting evidence for cultural conditioning in all dimensions of ethnicity and for the universality of human mental functions. They traveled everywhere on earth searching for evidence, sifting and organizing it into a new science of culture. As Stocking put it:

> In the long run, it was Boasian anthropology—rather than the racialist writers associated with the eugenics movement—which was able to speak to Americans as the voice of science on all matters of race, culture, and evolution—a fact whose significance for the recent history of the United States doubtless merits further exploration. (*Race, Culture and Evolution*, p. 307)

But the taint of scientific racism lingered. Konrad Lorenz, who shared the Nobel Prize in medicine and physiology in 1973 for his work in behavioral biology, and who remained an active and distinguished investigator well into the 1980s, provided an uncomfortable link with the past. As noted by Leon Eisenberg (in "The Human Nature of Human Nature," *Science* 176 [1972], pp. 123–128) and by Chorover (in *Genesis*, pp. 104–105), Lorenz wrote an article in a scholarly journal in 1940, decrying miscegenation and racial impurity as leading to degeneracy in the genetically determined aspects of behavior and character. And he explicitly praised the Nazi state for its accomplishments against this danger. Lorenz deeply regretted and retracted these statements. He also paid a high personal price for his support of the Reich, spending years in a Soviet prison camp after his capture on the eastern front. Yet the watchword should not be "forgive and forget" but perhaps something more like "forgive and remember."

Statements made by Arthur Jensen, William Shockley, and other investigators in the late 1960s and early 1970s about race and IQ or social class and IQ rapidly passed into currency in policy discussions. Many of these statements were proved wrong, but they had already influenced some policymakers, and that influence is very difficult to recant. The sociobiology of the late 1970s was soon cited in support of neofascist movements. It must be said that there is nothing specific about these ideas that should be useful to neofascists; merely the highly visible statement that genes affect behavior, combined with an emphasis on the strict Darwinian sense of the word *fitness*. The National Socialist youth movement in Britain adopted a sort of sociobiological cant, quoting or referring to E. O. Wilson, Richard Dawkins, and others. To be sure, they had little understanding of what they read, yet they found it useful.

An early exchange of letters published in *Nature*—correspondence between Steven Rose and Dawkins, to this day English arch-rivals in the sociobiology controversy—is still of interest (S. Rose, *Nature* 289 [1981], p. 335; R. Dawkins, ibid., p. 528). Rose pointed smugly to the neofascist use of Dawkins's views, called on Dawkins to dissociate himself publicly from them, and said, almost explicitly, *I told you so.* Dawkins dissociated himself, and expressed amazement that anyone could have so misconstrued his views as to make use of them in a neofascist cause. He said explicitly that it never crossed his mind that this could happen. Now, Rose was ill-mannered, and one wonders whether he expects other scientists to conceal their findings when they turn out to be susceptible to misuse. But Dawkins's naive amazement was more distressing.

Early in the controversy, an article in *Time* magazine on sociobiology included a brief, innocuous quotation from me. I simply pointed out that not only bad human traits but also good ones such as altruism were part of our evolutionary endowment. I did not mention race or individual differences, and the rest of the article said little or nothing about either, focusing instead on universals of human nature. Yet I received a long, poignant letter from a woman who identified herself as African-American and who, despite being quite articulate, expressed thoughts and feelings that suggested mental illness. Among other things she wrote at length on the genetic and moral inferiority of African-Americans, attributing many of her own and her people's problems to this "theory." I had said nothing remotely related to the main theme of her letter, yet she had interpreted my little remark about altruism as support for her theory. She was writing in a strange spirit of collegiality and congratulation. Never since then have I underestimated the power of even a few words about behavioral biology.

What of the latest currents of thought? Are they likely to lead to, or at least encourage, further distortions of social policy? The indications are not all encouraging. Richard Herrnstein and Charles Murray published a book in 1994 clearly directed at policy, just as Jensen and others had in the 1960s and 1970s. *The Bell Curve: Intelligence and Class Structure in American Life* (New York: Free Press, 1994) teamed a psychologist with a conservative policy advocate to try to prove that both the class structure and the racial divide in the United States result from genetically determined differences in intelligence and ability. Their general assertions about genes and IQ were not very controversial, but their speculations on race were something else again.

Also in the 1990s, Phillipe Rushton has tried to couch racial differences in IQ in a theory drawn from evolutionary biology. This theory takes the concepts of *r* and *K* selection, crudely useful when applied to a vast range of living creatures considered on a continuum, and apply it to subtle differences in skull form, mental test results, and sexual behavior within our one species. This theory has no academic legitimacy and little relationship to real evolutionary theory, but it taints the whole Darwinian enterprise, strongly recalling the "scientific anthropology" of the era of slavery.

The reality is quite different. As argued by George Armelagos in his Presidential Address to the American Association of Physical Anthropologists ("Race, Reason and Rationale," *Evolutionary Anthropology* 4, 1995, pp. 103–109) race itself is a dubious concept for the human species. Obviously it is sociologically meaningful, but even in the social realm it is a constantly moving target with little or no core biological legitimacy.

The overwhelming genetic unity of our species becomes clearer all the time. We are, every one of the six billion of us, descended from a very small group of people who lived in Africa around 100,000 years ago. During almost all of that time, challenges to intelligence have been remarkably similar on every continent. There has been little or no opportunity for racial separation, and the physical variety that seems so obvious to us is just an intersection of geographic trend lines known as clines, each a gradient of variation along a particular dimension, such as nose shape, height, or skin color. You can point to any spot on earth, draw a circle around it, and call it a race, but all you will have done is arbitrarily label the local intersection of several of these clines.

The human genome project has draft-sequenced five people's genes, three women and two men, self-described as Hispanic, Asian, Caucasian, and African-American. Craig Venter, head of Celera Genomics and one of two main leaders of the project, said, "In the five Celera genomes there is no way to tell one ethnicity from another." (See chapter 17 for discussion and references.) Statistically, it has been repeatedly shown that the vast majority of human genetic variation occurs within, not between, ethnic groups (See Ryan Brown and George J. Armelagos, "Apportionment of Racial Diversity: A Review," *Evolutionary Anthropology* 10, 2001, pp. 15–20).

If this is so, why is there a persistent difference in IQ and school success among African-Americans and European-Americans? Here are a few of the reasons. Slavery was a devastating blow to African-Americans, destroying language and culture and gravely damaging family and identity. It lasted for almost three centuries, followed by another century of systematic deprivation. African slaves were virtually the only group of Americans not self-selected to come here. Twentieth-century immigrants from Africa and the Caribbean had the same racial background as the slaves but had an experience very similar to that of all other ethnic groups in U.S. history. This alone gives the lie to a genetic explanation of the problems of the descendants of former slaves.

Genetic explanations of group differences ignore the immense power of culture to govern motivation and performance. There is no culture-free test of intelligence, and mental tests have been constructed on which African-Americans outperform European-Americans. Studies have shown that even a hint of racial stereotyping in the setting of a test markedly diminishes the performance of those being stereotyped. Peer pressure is also powerful. Identifying test performance and school success as "White" has kept generations of African-Americans, especially boys, from doing well. If genes are the explanation for group differences, why do African-American girls do so much better than boys, while among European-Americans the sexes perform more similarly?

If genes are the explanation, why is there no correlation between the test performance of African-Americans and their degree of admixture of European genes? Because *sociologically* our racial designations are categorical, so that any noticeable degree of African ancestry—indeed, the label alone—exposes you to all the cultural risks associated with being Black. Why do African-American children adopted by Whites grow up performing as well as Whites? Because they are given most of the advantages of the dominant culture. Why do the illegitimate children of Black and White American servicemen in Germany have comparable mental test scores? Because they are all brought up by German mothers, sharing most of the same cultural opportunities and disadvantages.

These are only a few of the counterarguments against ongoing, twenty-first-century racial determinism. Race is the least interesting and least significant of biological categories, yet it continues to compel the attention of many people. The most likely explanation for this is not the intrinsic merit of the subject. It is the desire to simplify the world, to justify unfair treatment of minorities, and to shore up a weak identity with a false sense of superiority. Human beings characteristically dichotomize the social world, and much of what is wrong with the world stems from this fact of human nature.

The need for vigilance continues. Anyone who investigates or writes about behavioral biology without recognizing the potential for grave misuse of it, proven many times in the last two centuries, is either a dangerous charlatan or a dangerous fool. Since the Enlightenment gave science a central place in our lives, scientific ideas have been abused. But to those who think these studies should stop, there is a clear answer: closing our eyes to biological influences cannot make them go away or prevent other people from distorting them. In fact, the distortions are made more likely by such suppression. Will there be further abuses? Of course. But can we ignore a subject so central to self-understanding? I don't think so.

Behavioral biology is a strong, dangerous physic, potentially healing if used appropriately, poisonous if not. For the great questions of race and social class, it has far more relevance to the behavior of the oppressors than it does to that of the victims. It does not show that the oppressed are inferior, but it does help explain why the oppressors are selfish, greedy, and violent. Yet other, false claims will be made for it—claims that echo the worst errors of the nineteenth and twentieth centuries. Hence, this caveat, a sort of package insert for the book, warning of the known dangers of improper use of this kind of knowledge. I would not purvey such medicine if I did not think that the human species is in a critically ill condition, needing every kind of knowledge it can get. But it would be far better for behavioral biology to disappear from view than to be applied as carelessly, as stupidly, and as destructively as it has been in the past.

A NOTE ON THE NOTES

When my editor, John Michel, and I came to the end of three years work on this revision, he pointed out to me that the text of the book was more than five hundred pages long. But in addition, there were two hundred pages of notes. This seemed to both of us to be too much; it would add substantially to the price of the book, and would also risk intimidating readers who might otherwise find the book inviting. John made what seemed a radical suggestion: Print the text, but publish the notes, free of charge, on the World Wide Web. I hated the idea, but agreed to think about it.

The more I did, the more sense it made. I was nervous because neither of us could think of another book published that way, but then I thought that pioneering the strategy could be exciting. I was afraid that the colleagues who had generously praised the book in advance comments would be disappointed or even embarrassed. But as many of them were consulted and almost to a person welcomed the idea, I realized that the problem was more mine than theirs.

Certainly, there would be no real problem of access; anyone likely to care about the references for the book would be able to find and use the Web site. I knew that many readers, like me, would miss having the notes right there in the back of the book, where doubts could be resolved or questions followed up without delay. But those of us who need this process can print out the contents of the Web site.

To those inconvenienced, please accept my apology, but please also know that I am trying my best to make this book accessible to an audience of readers that goes beyond the academic community. At the same time, I would hate for academic and professional readers to miss the two hundred pages of notes, the meticulous preparation of which added at least a year to the project. Almost every statement in this book has at least one specific supporting reference, most no more than a few years old. I stake my reputation on the quality of those references, and I urge the interested reader, as strongly as I can, to consult them at www.henryholt.com/tangledwing/.

CHAPTER ONE: *The Quest for the Natural*

This chapter uses the !Kung San as one example of the hunting-gathering adaptation that is known to have played a central role in human evolution. The best current reference on hunters and gatherers generally is Richard B. Lee and Richard Daly's *The Cambridge Encyclopedia of Hunters and Gatherers* (New York: Cambridge University, 1999). From it, myriad paths lead to a large and rich literature on this vital and once-central human adaptation. Accessible introductions to the !Kung may be found in Lorna Marshall's *Nyae Nyae !Kung: Beliefs and Rites* (Cambridge: Harvard University, 1999), Richard Lee's *The Dobe Ju/'hoansi* (New York: Harcourt Brace, 1993), and Marjorie Shostak's *Nisa: The Life and Words of a !Kung Woman* (Cambridge: Harvard University, 1982). Shostak's *Return to Nisa* (Cambridge: Harvard University, 2000) offers a personal account of their situation late in the twentieth century.

Excessive emphasis on the !Kung has been rightly criticized; many other hunting-gathering adaptations have existed or still exist, some quite different from the !Kung. Studies of the Hadza of Tanzania by James Woodburn, Kristen Hawkes, and Nicholas Blurton Jones, of the Efe Pygmies of Zaire by Robert Bailey and Nadine Peacock, of the Ache of Paraguay by Kim Hill and Magdelena Hurtado, and of the Netsilik Eskimo by Asen Balikci are among the outstanding modern investigations cited below. Critics would do well to follow these examples and do research on hunter-gatherers while they are still around to be studied. In addition to the *Cambridge Encyclopedia*, good sources on a range of hunter-gatherer societies include *Hunters and Gatherers: History, Evolution, and Social Change*, edited by Tim Ingold, David Riches, and James Woodburn (London: Berg Publishers Limited, 1991) and *The Foraging Spectrum: Diversity in Hunter-Gatherer Lifeways*, by Robert L. Kelly (Washington, D.C.: Smithsonian Institution, 1995).

The chapter, focused on the !Kung, draws primarily on the works of Marshall *(The !Kung of Nyae Nyae* and *Nyae Nyae !Kung: Beliefs and Rites)*, Lee and DeVore *(Kalahari Hunter-Gatherers)*, Lee *(The !Kung San)*, Howell *(Demography of the Dobe Area !Kung)*, and Shostak *(Nisa: The Life and Words of a !Kung Woman* and *Return to Nisa)*, as well as on my own experience and research. These and other works are cited fully in the notes.

No account of the !Kung can omit mention of their present situation. After centuries of oppression at the hands especially of whites but also of blacks in southern Africa, some now find themselves choosing either near serfdom on Bantu farms or dependency on reservations in Namibia (formerly South-West Africa), while others struggle to maintain independence. We who study them must not forget that they are in the throes of an ongoing historical crisis, now complicated by AIDS. For further information see Lee's *The !Kung San* and Edwin Wilmsen's *Land Filled with Flies*, cited in the notes. John Marshall's film *N!ai, The Story of a !Kung Woman*, shown several times on national public television, as well as his numerous other films (see *The Cinema of John Marshall*, Philadelphia: Harwood Academic, 1993, edited by Jay Ruby), vividly illustrate !Kung life. Marshall has devoted many years to helping the !Kung survive into the twenty-first century. Lee, who is professor of anthropology at the University of Toronto, and Polly Wiessner, at the University of Utah, keep up with current developments.

CHAPTER TWO: *Adaptation*

An easy-to-read popular account of adaptation theory is Richard Dawkins's *The Selfish Gene*, New Edition (New York: Oxford University, 1989), and Robert Wright's delightful *The Moral Animal: Evolutionary Psychology and Everyday Life* (New York: Pantheon, 1994) brings the theory to bear on humans—including Darwin himself. John Alcock's *The Triumph of Sociobiology* (New York: Oxford University, 2001) thoroughly justifies its title with a cogent summary of current research and controversy. James and Carol Gould's *Sexual Selection* (New York: W. H. Freeman/Scientific American Library, 1989) vividly introduces the natural history of this core theoretical problem. More advanced are two books by George Williams that have bracketed a long and distinguished career: *Adaptation and Natural Selection* and *Natural Selection: Domains, Levels, and Challenges.* Edward O. Wilson's *Sociobiology* (Cambridge: Harvard University, 1975) is the best-known comprehensive text. Though widely criticized and now a quarter-century old, it is still a vital foundation. Wilson's *Consilience: The Unity of Knowledge* (New York: Alfred A. Knopf, 1998) argues for using evolution to integrate the human sciences with the rest of scientific knowledge.

The best general introductory textbook of sociobiology remains Robert Trivers's *Social Evolution* (Menlo Park, Calif.: Benjamin Cummings, 1985). Current professional summaries of research in specific areas are collected in *Behavioral Ecology: An Evolutionary Approach*, 4th Ed., edited by John R. Krebs and Nicholas B. Davies (Oxford: Blackwell, 1997). No one should criticize a new scientific approach without reading the original papers that have convinced practitioners. They are collected in T. Clutton-Brock and P. Harvey (eds.), *Readings in Sociobiology* (San Francisco: W. H. Freeman, 1978). The papers of W. D. Hamilton and Robert L. Trivers are especially noteworthy. Applications of the theory to human behavior are collected in Laura Betzig's *Human Nature: A Critical Reader* (New York: Oxford University, 1997). Darwin's *On the Origin of Species* is perhaps the only major work of nineteenth-century science that continues to be essential reading. The edition introduced by Mayr (cited in the notes) is authoritative. A modern textbook worthy of Darwin's legacy is Mark Ridley's *Evolution* (New York: Oxford University, 1997).

For a widely cited critique of adaptationist approaches see Richard Lewontin, "Adaptation," *Scientific American* 239, September 1978, and Lewontin and Stephen Jay Gould's "The Spandrels of San Marco and the Panglossian Paradigm: A Critique of the Adaptationist Program," *Proceedings of the Royal Society of London*, 1979, 581–588. A quarter century of controversy is analyzed, and should be laid to rest, by Ullica Sagerstråle in *Defenders of the Truth: The Battle for Science in the Sociobiology Debate and Beyond* (Oxford: Oxford University, 2000).

The basic work on the interface between theories of adaptation and of learning is Martin Seligman and Joanne Hager's *Biological Boundaries of Learning* (New York: Meredith, 1972). Various writings of psychiatrist David A. Hamburg helped set the tone for a generation of interdisciplinary research in anthropology, psychiatry, and evolutionary biology, and helped inspire this book. The banner of evolutionary psychiatry has been taken up by Randolph Nesse, Michael McGuire, and others in works cited in the notes.

What Emotions Really Are, by Paul E. Griffiths, is the best account of the emotions in evolutionary perspective. The subdiscipline of adaptationist cognitive psychology was founded by Jerome Barkow, Leda Cosmides, and John Tooby in *The Adapted Mind: Evolutionary Psychology and the Generation of Culture* (New York: Oxford, 1992). That of evolutionary social psychology is summarized in the book of that name, edited by Jeffrey A.

Simpson and Douglas T. Kendrick (Mahwah, N.J.: Lawrence Erlbaum, 1997). For a classic account of the theory and methods of ethology, see Konrad Z. Lorenz, *Foundations of Ethology* (New York: Springer-Verlag, 1981).

CHAPTER THREE: *The Crucible*

My favorite book on human evolution is by Donald Johanson and Blake Edgar, *From Lucy to Language* (New York: Simon & Schuster, 1996), a spectacular large-format assembly of life-size photographs of all the most important hominid fossils discovered up to the mid-nineties, accompanied by a brief authoritative text. For the later part of the story, Ian Tattersall's *The Last Neanderthal: The Rise, Success, and Mysterious Extinction of Our Closest Human Relatives*, Revised Edition (Boulder, Colo.: Westview, 1999) is a good choice. In *Lucy's Legacy: Sex and Intelligence in Human Evolution* (Cambridge: Harvard University, 1999), Alison Jolly redresses the male bias that has affected such studies for generations. For two brief, easily readable accounts by top fossil hunters, see Richard Leakey's *The Origins of Humankind* (New York: Basic Books, 1994) and Alan Walker and Pat Shipman's *The Wisdom of the Bones* (New York: Alfred A. Knopf, 1996). In exploring the role of hunting in human experience, Matt Cartmill's *A View to Death in the Morning* (Cambridge: Harvard University, 1993) is subtle and insightful. Craig Stanford's *The Hunting Apes: Meat Eating and the Origins of Human Behavior* (Princeton, N.J.: Princeton University, 1999) has persuasively revived a central role for hunting in our evolution.

Robert Sapolsky's *A Primate's Memoir: A Neuroscientist's Unconventional Life Among the Baboons* (New York: Scribner's, 2001) is a funny paean to a quarter century of fieldwork on the African plains, combined with some of the most insightful observations ever made on wild monkeys. Karen Strier's *Primate Behavioral Ecology* (Boston: Allyn & Bacon, 2000) is a superb, brief, theoretically sophisticated introduction to our closest relatives and their significance for our evolution. The strange and varied roles of males among those animals is detailed in *Primate Males: Causes and Consequences of Variation in Group Composition* (Cambridge: Cambridge University, 2000). Frans DeWaal's *Good Natured: The Origins of Right and Wrong in Humans and Other Animals* (Cambridge: Harvard University, 1996) reviews a neglected and important positive aspect of primate life.

An excellent advanced text on human evolution is Richard Klein's *The Human Career*, 2d Ed. (Chicago: University of Chicago, 1999), and the great modern account of the background to human evolution is Robert D. Martin's *Primate Origins and Evolution* (Princeton, N.J.: Princeton University, 1990). For the latter part of human evolution, especially the Neanderthals, there are books by Erik Trinkhaus and Pat Shipman, by Christopher Stringer and Clive Gamble, and by Ian Tattersall, cited in the notes. The oldest known paintings, discovered in the 1990s, are beautifully displayed and explained in *Dawn of Art: The Chauvet Cave*, by Jean-Marie Chauvet and colleagues (New York: Harry N. Abrams, 1996). John E. Pfeiffer's *The Emergence of Culture* (New York: Harper & Row, 1982), on the great advances of the late Paleolithic, remains a valuable and original synthesis. The serious student should also consult *The Cambridge Encyclopedia of Human Evolution*, published by Cambridge University Press in 1992.

CHAPTER FOUR: *The Fabric of Meaning*

Books on the brain fill libraries, and I can only mention a few. Excellent starting points are Jean-Pierre Changeux's *Neuronal Man*, 2d Ed. (Princeton, N.J.: Princeton University, 1997), a developmental approach, and John Allman's *Evolving Brains* (New York: W. H. Freeman, 1999), an evolutionary one. *The Human Brain Coloring Book*, by Marian Diamond,

Arnold Scheibel, and Lawrence Elson (New York: HarperPerennial, 1985), is a wonderful exercise for the learner. Walle Nauta and Michael Feirtag's *Fundamental Neuroanatomy* (New York: W. H. Freeman, 1986) is a classic introduction to structure; Gordon Shepherd's *The Synaptic Organization of the Brain* (New York: Oxford University, 1990), a cogent account of nerve cell and circuit function; and Jack R. Cooper, Floyd E. Bloom, and Robert H. Roth's *The Biochemical Basis of Neuropharmacology* (New York: Oxford University, 1996) is the best introduction to brain chemistry. Other books are more technical. There is no shortcut to the nervous system, but it is perfectly accessible to anyone willing to invest a few hundred hours. The standard comprehensive text is *Principles of Neural Science*, by Eric Kandel, James Schwartz, and Thomas Jessell (New York: McGraw-Hill, 2000). For visual learners, the beautifully illustrated and brilliantly designed *Human Brain*, 4th Ed., by John Nolte (St. Louis: Mosby, 1999), is a feast for eye and mind. And for higher-brain functions, the indispensable collection is Michael Gazzaniga's *The New Cognitive Neurosciences* (Cambridge: MIT, 2000).

CHAPTER FIVE: *The Several Humours*

Behavior genetics has found wide acceptance only recently. The definitive scientific summary (for now) is *Genetic Influences on Neural and Behavioral Functions*, edited by Donald Pfaff and colleagues (New York: CRC, 2000). Complementary introductions in lively prose include *Living with Our Genes*, by Dean Hamer and Peter Copeland (New York: Doubleday, 1998), and *Galen's Prophecy: Temperament in Human Nature*, by Jerome Kagan (New York: Basic Books, 1994). Central to our understanding of both human individuality and universality is the five-factor model of Paul Costa and Robert McCrae, presented mainly in articles (cited in the notes) rather than books. Other valuable accounts are David C. Rowe's *The Limits of Family Influence: Genes, Experience, and Behavior* (New York: Guilford, 1994) and *Behavioral Genetics*, 3d Ed., by Robert Plomin, John C. DeFries, Gerald E. McLearn, and Michael Rutter (New York: W. H. Freeman, 1997).

Three papers by Sandra Scarr, all in the journal *Child Development* (the first with Kathleen McCartney), are central: "How People Make Their Own Environments" (54 [1983], 424–435); "Developmental Theories for the 1990s" (63 [1992], 1–19); and "Biological and Cultural Diversity: The Legacy of Darwin for Development" (64 [1993], 1333–1353). *Schizophrenia Genesis: The Origins of Madness*, by Irving Gottesman (New York: W. H. Freeman, 1991), set a standard of how to think about these things. The field's intellectual pitfalls are cogently summarized by Richard C. Lewontin, in "Genetic Aspects of Intelligence" (*Annual Review of Genetics* 9 [1975], 387–405). Its ethical aspects, related to but separable from its intellectual accomplishments and pitfalls, are discussed in chapter 17 and in the section of the notes entitled "Caveat: The Dangers of Behavioral Biology."

CHAPTER SIX: *The Beast with Two Backs*

The best treatment of the behavioral differences is Eleanor Maccoby's *The Two Sexes: Growing Up Apart, Coming Together* (Cambridge: Harvard University, 1998). Worthy counterparts in psychological anthropology are by Beatrice Blyth Whiting and Carolyn Pope Edwards, *Children of Different Worlds: The Formation of Social Behavior* (Cambridge: Harvard University, 1988), and by Alice Schlegel and Herbert Barry III, *Adolescence: An Anthropological Inquiry* (New York: W. H. Freeman, 1991). Deborah Tannen, in *Gender and Discourse* (New York: Oxford University, 1994), collects some of her writings on male-female differences in discourse.

The evolutionary background to gender roles was greatly advanced by Sarah Blaffer Hrdy in *The Woman That Never Evolved* (Cambridge: Harvard University, 1981). Her themes of the centrality and flexibility of female roles in nonhuman primates were taken up by many investigators, as summarized by Shirley Strum and Linda Fedigan in their chapter of *The New Physical Anthropology: Science, Humanism, and Critical Reflection* (Upper Saddle River, N.J.: Prentice-Hall, 1999), edited by Strum, Donald Lindburg, and David Hamburg. Hrdy's 1999 book, *Mother Nature: A History of Mothers, Infants, and Natural Selection* (New York: Pantheon, 1999), is of the greatest importance in our understanding of women's reproductive roles. Two other major works on the evolutionary background of sex roles are Alison Jolly's *Lucy's Legacy: Sex and Intelligence in Human Evolution* (Cambridge: Harvard University, 1999) and Bobbi S. Low's *Why Sex Matters: A Darwinian Look at Human Behavior* (Princeton, N.J.: Princeton University, 2000). *Sexual Selection*, by James Gould and Carol Grant Gould (New York: Scientific American Library, 1989), is a graceful and beautifully illustrated account of the relevant evolutionary principles.

Biological constraints notwithstanding, little in anthropology's history is more important than its ongoing challenge to narrow views of sex roles. Margaret Mead's *Male and Female: A Study of the Sexes in a Changing World* (New York: William Morrow, 1949, 1967) remains an important document, along with later work on women cross-culturally: eds. Michelle Rosaldo and Louise Lamphere, *Woman, Culture and Society* (Stanford, Calif.: Stanford University, 1974); ed. Rayna Reiter, *Toward an Anthropology of Women* (New York: Monthly Review, 1975); Naomi Quinn, "Anthropological Studies on Women's Status," *Annual Review of Anthropology* 6 (1977), 181–225; and Carol Ember, "A Cross-Cultural Perspective on Sex Differences" in *The Handbook of Cross-Cultural Development*, eds. R. L. Monroe, R. H. Monroe, and B. Whiting (New York: Garland, 1981). Ember integrates cross-cultural findings with the psychobiology of sex differences. An illuminating account of how women in many cultures come into their own after menopause is Judith K. Brown's "Cross-Cultural Perspectives on Middle-Aged Women," *Current Anthropology* 23 (1982), 143–156. A recent essay on sex roles in seven cultures is Serena Nanda's *Gender Diversity: Cross-Cultural Variations* (Prospect Heights, Ill.: Waveland, 2000).

Interesting anthropological collections include *Beyond the Second Sex: New Directions in the Biology of Gender*, eds. Peggy Reeves Sanday and Ruth Gallagher Goodenough (Philadelphia: University of Pennsylvania, 1990); *The Other Fifty Percent: Multicultural Perspectives on Gender Relations*, eds. Mari Womack and Judith Marti (Prospect Heights, Ill.: Waveland, 1993); *Gender at the Crossroads of Knowledge*, ed. Micaela di Leonardo (Berkeley: University of California, 1991); and *Naturalizing Power: Essays in Feminist Cultural Analysis*, eds. Sylvia Yanagisako and Carol Delany (New York: Routledge, 1995). Unfortunately, the real knowledge about sex roles in this material tends to get buried in postmodernist rhetoric.

Good antidotes include Christina Hoff Summers's *Who Stole Feminism?* (New York: Simon & Schuster, 1994) and Warren Farrell's *The Myth of Male Power* (New York: Simon & Schuster, 1993). Camille Paglia's *Sexual Personae: Art and Decadence from Nefertiti to Emily Dickinson* (New Haven, Conn.: Yale University, 1990) is a powerful account of gender as depicted in Western art and literature. Lionel Tiger's *Men in Groups*, 2d Ed. (New York: Marian Boyars, 1984) and *The Decline of Males* (New York: Golden, 1999), summarize decades of his research. Two books about sex roles in the kibbutz—one by Tiger with Joseph Shepher, entitled *Women in the Kibbutz* (New York: Harcourt Brace, 1975), and one by Melford Spiro, called *Gender and Culture: Kibbutz Women Revisited* (New Brunswick, N.J.: Transaction, 1979)—raise questions about the extent to which sex roles can be engineered. David Gilmore's *Manhood in the Making* (New Haven, Conn.: Yale University, 1990) reviews masculinity in different cultures, and Gilbert Herdt's collection, *Third Sex, Third Gender: Beyond Dimorphism in Culture and History* (New York: Zone/MIT, 1994), displays the stunning variety of transgender roles in cultures throughout the world.

An excellent brief introduction to the biology of gender is Simon LeVay's *The Sexual Brain* (Cambridge: MIT, 1993). Other sources include *Behavioral Neuroendocrinology*, eds. Jill Becker, Marc Breedlove, and David Crews (Cambridge: MIT, 1992), and Randy J. Nelson's *An Introduction to Behavioral Endocrinology* (Sunderland, Mass.: Sinauer, 1995). How the sexes diverge as childhood ends is summarized in *Adolescence and Puberty* (New York: Oxford University, 1990), edited by John Bancroft and June Machover Reinisch, both former directors of the Kinsey Institute. A skeptical account is Anne Fausto-Sterling's *Myths of Gender* (New York: Basic Books, 1985), but its claims become more difficult to sustain as time goes by.

CHAPTER SEVEN: The Well of Feeling

The way we think about emotions and the brain has been permanently changed by Antonio Damasio's *Descartes' Error: Emotion, Reason, and the Human Brain* (New York: Putnam, 1994), at once an eloquent plea for the centrality of emotion in mind and a persuasive account of how it plays its role, extended and completed by his 1999 book, *The Feeling of What Happens* (New York: Harcourt Brace). Still, for an evolutionary account the only place to start is Charles Darwin's 1872 *Expression of the Emotions in Man and Animals*, annotated by Paul Ekman (New York: Oxford University, 1998). Ekman's annotations, introduction, and afterword make this great work absolutely current. Paul Griffiths's *What Emotions Really Are* (Chicago: University of Chicago, 1997) establishes the philosophy of emotion on firm scientific ground, and Aaron Ben-Ze'ev's *The Subtlety of the Emotions* (Cambridge: MIT, 2000) gives a fine, broad overview. Michael Lewis and Jeannette Haviland's *Handbook of Emotions*, 2d Ed.(New York: Guilford, 2000), is a standard reference for psychological research.

The first comprehensive neurobiology of the emotions is Jaak Panksepp's magisterial *Affective Neuroscience: The Foundations of Human and Animal Emotions* (New York: Oxford University, 1998). *The Neuropsychiatry of Limbic and Subcortical Disorders*, by Stephen Salloway, Paul Molloy, and Jeffrey L. Cummings (Washington, D.C.: American Psychiatric Association, 1997), is a more specialized attempt to establish psychiatry on a new neurological foundation.

Joseph LeDoux's *The Emotional Brain* (New York: Simon & Schuster, 1996) reviews his important research on fear and the amygdala and gives a clear idea of how brain scientists think about emotion. Karl H. Pribram, ed., *Brain and Behavior*, Vol. 4: *Adaptation* (Baltimore: Penguin, 1969) collects many of the classic papers mentioned in this chapter. Further classic accounts by neuroanatomists are Walle Nauta and Michael Feirtag, *Fundamental Neuroanatomy* (New York: W. H. Freeman, 1986), and Paul D. MacLean, *The Triune Brain in Evolution* (New York: Plenum, 1990).

For insight into the anthropology of the emotions, the most impressive scientific study is Karl Heider's *Landscapes of Emotion: Mapping Three Cultures of Emotion in Indonesia* (New York: Cambridge University, 1991). Theoretical integrations by anthropologists include Melford Spiro's *Culture and Human Nature* (New Brunswick, N.J.: Transaction, 1994) and Gerald Erchak's *The Anthropology of Self and Behavior* (New Brunswick, N.J.: Rutgers University, 1992). Humanistic approaches, deeply individual yet compelling in their evidence of universality, include Lila Abu-Lughod's *Veiled Sentiments: Honor and Poetry in a Bedouin Society* (Berkeley: University of California, 1986), Unni Wikam's *Managing Turbulent Hearts: A Balinese Formula for Living* (Chicago: University of Chicago, 1990), and Marjorie Shostak's two books, *Nisa: The Life and Words of a !Kung Woman* (Cambridge: Harvard University, 1981) and *Return to Nisa* (Cambridge: Harvard University, 2000).

CHAPTER EIGHT: Logos

A graceful and lucid book on the nature of language is *Lingua ex Machina: Reconciling Darwin and Chomsky with the Human Brain,* by William Calvin and Derek Bickerton (Cambridge: MIT, 2000), an exchange of letters and collaboration between a brain scientist and a linguist that attempts with some success to locate language in the brain. Steven Pinker's *The Language Instinct* (New York: Morrow, 1994) is a classic account of the universals of language, how language develops, and how it may have evolved. Equally influential for my thinking have been Eric Lenneberg's *Biological Foundations of Language* (New York: John Wiley & Sons, 1967), the first comprehensive statement of the biological approach, Terence Deacon's *The Symbolic Species: The Co-Evolution of Language and the Brain* (New York: W. W. Norton, 1997), and Sue Savage-Rumbaugh and Roger Lewin's account of ape language, *Kanzi: The Ape at the Brink of the Human Mind* (New York: John Wiley & Sons, 1994). Norman Geschwind's "Language and the Brain," *Scientific American* 226 (1972, pp. 76–83), remains a concise, readable statement of the anatomical essentials. For comprehensive and current treatment of the brain functions involved, see Brian Kolb and Ian Q. Whishaw, *Fundamentals of Human Neuropsychology* (New York: W. H. Freeman, 1996).

Roger Brown's *A First Language: The Early Stages* (Cambridge: Harvard University, 1973) is a classic account of language acquisition, and John Locke's *The Child's Path to Spoken Language* (Cambridge: Harvard University, 1993) contextualizes the process in the intensity of early relationships. *Language Socialization Across Cultures,* edited by Bambi B. Schieffelin and Elinor Ochs (Cambridge: Cambridge University, 1986), reviews child language in varied anthropological settings. The four-volume *Universals of Human Language,* edited by Joseph Greenberg, Charles A. Ferguson, and Edith A. Moravcik (Stanford, Calif.: Stanford University, 1978), is a monument to the common regularities of language throughout the world. Bernard Comrie's work, exemplified by *Language Universals and Linguistic Typology* (Chicago: University of Chicago, 1989), reviews and extends this research.

Noam Chomsky helped create modern thinking about language, and his most accessible account is *Language and Problems of Knowledge: The Managua Lectures* (Cambridge: MIT, 1988). An exchange of immense interest is *Language and Learning: The Debate Between Jean Piaget and Noam Chomsky,* edited by Massimo Piattelli-Palmarini (Cambridge: Harvard University, 1980). It includes views by Jerry Fodor, Hillary Putnam, Marvin Minsky, Jean-Pierre Changeux, and other leading theorists of mind, all commenting (civilly, no less) on one another's views. Charles Hockett's accessible article "The Origin of Speech," *Scientific American* 203:3 (1960, pp. 88–111), has influenced the outlook of most anthropologists. For a still valuable traditional presentation of the viewpoint of anthropological linguistics see Edward Sapir, *Language* (New York: Harcourt, Brace & World, 1949). An excellent current text is Alessandro Duranti's *Linguistic Anthropology* (Cambridge: Cambridge University, 1997).

The old nineteenth-century ban notwithstanding, worthwhile theories of the evolution of language include Bickerton's *Language and Species* (Chicago: University of Chicago, 1990) and Merlin Donald's *Origins of the Modern Mind* (Cambridge: Harvard University, 1991). Attempts to model language evolution anatomically include Patricia Greenfield's "Language, Tools, and Brain: The Ontogeny and Phylogeny of Hierarchically Organized Sequential Behavior," *Behavioral and Brain Sciences* 14 (1991, pp. 536–595) and Calvin and Bickerton's *Lingua ex Machina,* cited above. Finally, Irene Pepperberg's *The Alex Studies: Cognitive and Communicative Abilities of Grey Parrots* (Cambridge: Harvard University, 1999) provides the intellectually bracing, indeed humbling experience of trying to figure out how a creature far removed from the primates, with a very different brain, can achieve a flexible grasp of words and concepts and use them in conversations.

CHAPTER NINE: *Rage*

An excellent account of the biological mechanisms of aggression is Debra Niehoff's *The Biology of Violence* (New York: Free Press, 1999), while the best overview of psychological studies is *Aggression: Its Causes, Consequences, and Control* (Philadelphia: Temple University, 1993), by Leonard Berkowitz. *Demonic Males*, by Richard Wrangham and Dale Peterson (Boston: Houghton Mifflin, 1996), considers the evolution of aggression in our closest relatives, and Michael Ghiglieri's *The Dark Side of Man: Tracing the Origins of Male Violence* (New York: Perseus, 1999) complements it with an account of the human outcome. Sarah Blaffer Hrdy's *The Woman That Never Evolved* (Cambridge: Harvard University, 1980) is the classic evolutionary account of female roles, which include a good deal of aggression. Important papers by Barbara Smuts, cited in the notes, on male violence against females appeared in the 1990s. Violence research has been transformed by the work of Martin Daly and Margo Wilson, especially in *Homicide* (New York: Aldine de Gruyter, 1988), *The Truth About Cinderella: A Darwinian View of Parental Love* (New Haven, Conn.: Yale University, 1998), and many papers on domestic violence. Family violence cannot be understood without reference to their work.

Good collections with varied emphases include *Human Aggression*, edited by Russell G. Geen and Edward Donnerstein (New York: Academic, 1998), *Rage, Power, and Aggression*, edited by Robert A. Glick and Stephen P. Roose (New Haven, Conn.: Yale University, 1993), and *Family Violence*, edited by Lloyd Ohlin and Michael Tonry (Chicago: University of Chicago, 1989)—although the latter precedes the important work of Daly and Wilson. *The Anthropology of War: A Bibliography*, by Brian Ferguson and Leslie Farragher (New York: Harry Frank Guggenheim Foundation, 1988), is a good place to begin research on that subject, but the essential facts are set forth in Lawrence H. Keeley's *War Before Civilization: The Myth of the Peaceful Savage* (New York: Oxford University, 1996), a convincing plea for honesty about the archaeological record. R. Brian Ferguson's *War in the Tribal Zone: Expanding States and Indigenous Warfare* (Santa Fe, N.M.: School of American Research, 1992) deals with traditional warfare in tribes becoming states. *Sick Societies*, by Robert Edgerton (New York: Free Press, 1992), is an extended, empirical critique of Rousseau's naïve view of primitive cultures, and it should lay that view permanently to rest. Irenäus Eibl-Eibesfeldt's *The Biology of Peace and War* (London: Thames & Hudson, 1979) remains a valuable reference and is particularly good on the German ethnographic literature often ignored in English-language treatments.

J. van der Dennen's *The Origin of War: The Evolution of a Male-Coalitional Reproductive Strategy* (Groningen, The Netherlands: Origins Press, 1995) is a comprehensive comparative overview. David Hamburg, a psychiatrist whose papers are cited in the notes, pioneered the serious study of the evolution of human aggression. Joseph Popp and Irven DeVore provided an early sociobiological perspective in "Aggressive Competition and Social Dominance Theory," in *The Great Apes*, edited by David A. Hamburg and Elizabeth R. McCown (Reading, Mass.: Benjamin Cummings, 1979). Monkey and ape reconciliation behavior is summarized in Frans de Waal's *Peacemaking Among Primates* (Cambridge: Harvard University, 1989) and in his *Good Natured: The Origins of Right and Wrong in Humans and Other Animals* (Cambridge: Harvard University, 1996). Although not the viewpoint taken here, the thesis that human aggression is purely learned is defended by Ashley Montagu in *The Nature of Human Aggression* (New York: Oxford University, 1976) and by a group using the invented name John Klama, in *Aggression: Conflict in Animals and Humans Reconsidered* (New York: John Wiley & Sons, 1988). But as an antidote, see my review in *Nature* 333, June 2, 1998 (p. 405).

CHAPTER TEN: *Fear*

Joseph LeDoux's *The Emotional Brain* (New York: Touchstone, 1996) summarizes elegant experiments on the role of the amygdala in learned fear, and extends them to form a current account of the physiology of this and other emotions. Jeffrey A. Gray's *The Neuropsychology of Anxiety* (New York: Oxford University, 2000) is a more advanced account. A classic treatment of this subject is Gray's earlier book, *The Psychology of Fear and Stress* (New York: McGraw-Hill, 1987). It is outdated in some concepts of evolutionary biology but has the virtue of attempting a synthesis of all aspects of science pertinent to fear. *Fears, Phobias, and Rituals*, by Isaac Marks (New York: Oxford University, 1987), is a good introduction to the clinical phenomena, informed by evolutionary ideas.

Freud's short book *The Problem of Anxiety* (New York: W. W. Norton, 1963; also published as *Inhibitions, Symptoms, and Anxiety*) remains vitally illuminating, as does Donald O. Hebb's classic essay "On the Nature of Fear," published in *Psychological Review* (53, pp. 259–276) in 1946. One classic account of the fears of childhood is included in John Bowlby's *Attachment and Loss*, 3 Vols. (New York: Basic Books, 1970–1980). Modern psychiatric views of anxiety and its disorders are given by Donald Goodwin in *Anxiety* (New York: Oxford University, 1986) and by Aaron Beck, Gary Emery, and Ruth L. Greenberg in *Anxiety Disorders and Phobias* (New York: Basic Books, 1985).

Jerome Kagan's two decades of research on children's timidity has transformed our understanding of the genetics and development of this trait and its opposite, boldness. His *Galen's Prophecy: Temperament and Human Nature* (New York: Basic Books, 1994) summarized much of this work. Ned Kalin's 1993 precis, "The Neurobiology of Fear" (*Scientific American* 268:5, 94–101), takes a different experimental approach using nonhuman primates. Randolph Nesse, an evolutionary psychiatrist, has done interesting theoretical work on the adaptive functions of anxiety and phobia, cited in the notes, and Michael McGuire and Alfonso Troisi treat the same subject very effectively in their *Darwinian Psychiatry* (New York: Oxford University, 1998).

CHAPTER ELEVEN: *Joy*

Joy remains the most poorly studied human emotion, being as elusive for investigators as for everyone else. The burgeoning new field of positive psychology was the subject of a special issue of the *American Psychologist*, 55(1), January 2000, edited by Martin Seligman and Mihaly Csikszentmihalyi. The most original modern work is Csikszentmihalyi's *Flow: The Psychology of Optimal Experience* (New York: HarperCollins, 1990). An excellent and comprehensive set of psychological papers, including summaries of the relevant brain science, is *Well-Being: The Foundations of Hedonic Psychology*, edited by Daniel Kahneman, Ed Diener, and Norbert Schwarz (New York: Russell Sage Foundation, 1999). A companion volume, by Ed Diener and Eunkook M. Suh entitled *Culture and Subjective Well-Being* (Cambridge: MIT, 2000), presents the cross-cultural evidence. These books will greatly advance the field. A good collection focused on psychodynamics is *Pleasure Beyond the Pleasure Principle*, edited by Robert A. Glick and Stanley Bone (New Haven, Conn.: Yale University, 1990).

The indispensable book about human play is Brian Sutton-Smith's *The Ambiguity of Play* (Cambridge: Harvard University, 1997), although, as the title implies, play is not always joyful. Robert Fagen's superb evolutionary overview, *Animal Play Behavior* (New York: Oxford University, 1981), is a classic and still vital source, and Mark Bekoff and John Byers's collection, *Animal Play* (Cambridge: Cambridge University, 1998), updates it by two decades. Modern psychology has gone far in explaining the neurology of joy through brain-reward systems; James Olds's classic summary is cited in the notes along with recent studies.

Barbara Fredrickson's "broaden and build" model is developed in "What Are Positive Emotions Good For?" in the *Review of General Psychology* 2 (1998). A more specifically Darwinian view is Jerome Barkow's "Happiness in Evolutionary Perspective," cited in the notes. Important reflections on the functional value of even misguided positive emotions are found in Lionel Tiger's *Optimism: The Biology of Hope* (New York: Simon & Schuster, 1979) and Shelley Taylor's *Positive Illusions: Creative Self-Deception and the Healthy Mind* (New York: Basic Books, 1989). George Vaillant's books, *Adaptation to Life* (Boston: Little, Brown, 1977) and *The Wisdom of the Ego* (Cambridge: Harvard University, 1993), teach us by example that our power to make our lives happier and better, in spite of pain and stress, can be far more than mere illusion.

CHAPTER TWELVE: Lust

Lust and its behaviors have proved surprisingly accessible to scientific study. Simon LeVay's *The Sexual Brain* (Cambridge: MIT, 1993) makes a graceful, excellent starting point for the neuroscience, but the definitive book on mechanism is Donald Pfaff's *Drive: Neurobiological and Molecular Mechanisms of Sexual Motivation* (Cambridge: MIT, 1999). *Behavioral Endocrinology*, edited by Jill Becker, Marc Breedlove, and David Crews, (Cambridge: MIT, 1992) and *Reproduction in Context: Social and Environmental Influences on Reproduction*, by Kim Wallen and Jill Schneider (Cambridge: MIT, 2000) broaden the perspective.

A fine collection of studies on the effects of age and culture is *Sexuality Across the Life Course*, edited by Alice Rossi (Chicago: University of Chicago, 1994). The classic introduction to the evolutionary dimensions of the subject is Donald Symons's *Evolution of Human Sexuality* (New York: Oxford University, 1979). David Buss's *Evolution of Desire: Strategies of Human Mating* (New York: Basic Books, 1994) shows just how prescient Symons was. Any doubt about the applicability of Darwinian sexual-selection theory to human beings should be dispelled by the papers in two collections: *Human Reproductive Behavior: A Darwinian Perspective*, edited by Laura Betzig, Monique Borgerhoff Mulder, and Paul Turke (Cambridge: Cambridge University, 1988); and *Human Nature: A Critical Reader*, edited by Betzig (New York: Oxford University, 1997). Engaging evolutionary overviews are Natalie Angier's *Woman: An Intimate Geography* (New York: Anchor/Random House, 1999) and Jared Diamond's *Why Is Sex Fun? The Evolution of Human Sexuality* (New York: Basic Books, 1997).

The reports by Alfred Kinsey and his colleagues, *Sexual Behavior in the Human Male* and *Sexual Behavior in the Human Female* (Philadelphia: W. B. Saunders, 1953), remain classic accounts, but we now at last have a worthy modern counterpart: *The Social Organization of Sexuality: Sexual Practices in the United States*, by Edward Laumann et al. (Chicago: University of Chicago, 1944). This work was summarized briefly and with clearer charts in *Sex in America: A Definitive Survey*, by Robert Michael et al. (New York: Little, Brown, 1994). Philip Blumstein and Pepper Schwartz's *American Couples: Money, Work, and Sex* (New York: William Morrow, 1983) remains a superb account of the real lives of hetero- and homosexual couples. The best general overviews of sexual orientation are Simon LeVay's *Queer Science: The Use and Abuse of Research into Homosexuality* (Cambridge: MIT, 1997) and *Being Homosexual: Gay Men and Their Development*, by Richard Isay (New York: Farrar, Straus, & Giroux, 1989).

The most interesting six pages I have read on the emergence of sexual interest is "Rethinking Puberty: The Development of Sexual Attraction," by Martha McClintock and Gilbert Herdt, in *Current Directions in Psychological Science* 5 (1996, pp. 178–183). Herdt's collection, *The Third Sex*, cited in Chapter 6, is the definitive account of transgender roles cross-culturally. Nancy Etcoff's *Survival of the Prettiest: The Science of Beauty* (New York: Doubleday, 1999) is the best overview of this fascinating and increasingly rigorous field. The honest but often unsuccessful effort of Catholic priests to suppress their sexual instincts is detailed in Richard Sipe's remarkable book *A Secret World: Sexuality and the Search for Celibacy* (New York: Brunner/Mazel, 1990). Last but not least, no one with a serious interest in sex should fail to read William H. Masters and Virginia E. Johnson's *Human Sexual Response* (Boston: Little, Brown, 1966), a source of crucial information and a turning point in the history of sex research.

CHAPTER THIRTEEN: *Love*

The near-definitive work on the evolution of motherhood is Sarah Blaffer Hrdy's *Mother Nature: A History of Mothers, Infants, and Natural Selection* (New York: Pantheon, 1999). It complements John Bowlby's enduring three-volume classic, *Attachment and Loss* (New York: Basic Books, 1970–1980), which successfully integrated animal and human attachment in an evolutionary context for the first time. *The Handbook of Attachment*, edited by Jude Cassidy and Philip Shaver (New York: Guilford, 1999), brings the subject up to date. Important contributions to our grasp of attachment in animals have come from Konrad Lorenz, Harry F. Harlow, Robert Hinde, Patrick Bateson, Peter Klopfer, Leonard Rosenblum, and Stephen Suomi, whose papers are cited in the notes and described in the text. Harry Harlow's short book *Learning to Love* (San Francisco: Albion, 1971) is a brief overview of his lifetime of discoveries about attachment in infant monkeys. Two accounts of human infants' relationships that focus on interaction rather than attachment are Daniel Stern's superb treatise, *The Interpersonal World of the Infant* (New York: Basic Books, 1985), and Philippe Rochat's fine collection, *Early Social Cognition* (Mahwah, N.J.: Lawrence Erlbaum, 1999). These form a counterpoint to the notion that love is all you need.

On the subject of romantic attachment, there is no more beautiful or illuminating work than Ethel Spector Person's *Dreams of Love and Fateful Encounters: The Power of Romantic Passion* (New York: W. W. Norton, 1988), which combines a common-sense psychoanalytic approach with a remarkable literary and humane sensibility. Diane Ackerman's *A Natural History of Love* (New York: Random House, 1994) is more history than natural history, but it gives a good account of the annals of love in the West. For real natural history, a good place to start is Helen Fisher's *The Anatomy of Love: The Natural History of Monogamy, Adultery, and Divorce* (New York: W. W. Norton, 1992). Great strides in our understanding of the physiology of love and the pair bond have been made by Daniel Lehrman, Sue Carter, Stephen Porges, and especially Thomas Insel, in works cited in the notes.

Evolutionary theory has helped to explain altruism and cooperation in a relentlessly competitive universe. The applicability of this kind of theory to human life is best appreciated through Laura Betzig's fine collection, *Human Nature: A Critical Reader* (New York: Oxford, 1997). The original theoretical contribution was by W. D. Hamilton, "The Genetical Evolution of Social Behavior," *Journal of Theoretical Biology* 7 (1964), 1–52. Robert Trivers's classic paper, "The Evolution of Reciprocal Altruism," *Quarterly Review of Biology* 46 (1971), 35–57, provided a missing piece of the puzzle highly relevant to anthropology. An elegant extension is given by Robert Axelrod and W. D. Hamilton in "The Evolution of Cooperation," *Science* 211 (1981), 1390–1396, discussed accessibly in Axelrod's book of the same title (New York: Basic Books, 1984). Group selection remains an embattled addition to the theory, but most evolutionists now give it a minor role. A brief in its favor, widely criticized but still interesting, is Elliot Sober and David Sloan Wilson's *Unto Others: The Evolution and Psychology of Unselfish Behavior* (Cambridge: Harvard University, 1998).

CHAPTER FOURTEEN: *Grief*

Courageous and moving first-person accounts of mood disorders include William Styron's *Darkness Visible: A Memoir of Madness* (New York: Vintage, 1992) and Kay Redfield Jamison's *An Unquiet Mind:* (New York: Alfred Knopf, 1995). Charles Nemeroff's "The Neurobiology of Depression" (*Scientific American* 278:6 [1998], 42–49) offers a brief and authoritative account of the physiology, and Hagop Akiskal and Giovanni Cassano's collection, *Dysthymia and the Spectrum of Chronic Depressions* (New York: Guilford, 1997), is the best introduction to depressive personalities. *The Harvard Guide to Modern Psychiatry*, edited by Armand M. Nicholi (Cambridge: Harvard University, 1999), includes several chapters on the causes and consequences of feelings of loss and depression. *Manic-Depressive Illness*, edited by Frederick K. Goodwin and Kay Redfield Jamison (New York: Oxford University, 1990), is a landmark textbook on bipolar disorder. Robert Burton's *The Anatomy of Melancholy* (cited in the notes), one of the best-selling books of the seventeenth century, is of great historical and literary interest.

The definitive account of the relationship between mood swings and creativity is *Touched with Fire: Manic-Depressive Illness and the Artistic Temperament* (New York: Free Press, 1993) by Kay Redfield Jamison. An excellent overview of grief in the literal sense is John Archer's *The Nature of Grief: The Evolution and Psychology of Reactions to Loss* (London: Routledge, 1999). It brings the loss part of John Bowlby's *Attachment and Loss* (New York: Basic Books, 1970–1980) up to date through the 1990s, in both evolutionary and psychological terms. Ernest Becker's classic, *The Denial of Death* (New York: Macmillan, 1973), is a psychoanalytic and existential account of the pervasive effect of our knowledge of death on the way we live our lives. To understand the sad reality of dying itself, read *How We Die* (New York: Knopf, 1994), an unflinching yet eloquent account by surgeon Sherwin Nuland; it isn't pretty. A. Alvarez, in *The Savage God* (New York: Random House, 1970), provides an exceptionally literate, eloquent, and sensitive view of suicide.

On the nature of divorce and separation, one cannot do better than Diane Vaughan's *Uncoupling: Turning Points in Intimate Relationships* (New York: Oxford University, 1986), a common-sense, systematic, largely positive view of this increasingly common loss. The demographic benchmark is Andrew Cherlin's *Marriage, Divorce, Remarriage* (Cambridge: Harvard University, 1992). *Solitude: A Return to the Self* (New York: Free Press, 1988) is Anthony Storr's graceful demonstration of the therapeutic value of solitude, a state entirely different from loneliness.

CHAPTER 15: *Gluttony*

Gerard P. Smith's collection, *Satiation: From Gut to Brain* (New York: Oxford University, 1998), is a good place to start for the science of what ends eating—or fails to. Several major papers on the physiology of hunger and satiety appeared in the journal *Science* 280:5368 (May 29, 1998). "Neuroendocrine Responses to Starvation and Weight Loss," by Michael W. Schwartz and Randy J. Seeley (*New England Journal of Medicine* 336, 1302–1311), is an elegant overview of how the body foils efforts to starve it. A brief account of what makes an animal (or person) stop eating is provided by Gerard Smith, "The Direct and Indirect Control of Meal Size," in *Neuroscience and Biobehavioral Reviews* 20 (1996), 41–46.

A classic account of the psychology of obesity is *Obese Humans and Rats*, edited by Stanley Schachter and Judith Rodin (Potomac, Md.: Lawrence Erlbaum, 1974). Medical aspects of the problem are reviewed in Derek Chadwick and Gain Cardew, eds., *The Origins and Consequences of Obesity* (New York: John Wiley & Sons, 1996). Insights into "nervous eating" are given by Josephine Wilson and Michael Cantor in "An Animal Model of Excessive Eating," *Journal of the Experimental Analysis of Behavior* 47 (1987),

335–346. At the other end of the spectrum, Joan Jacobs Brumberg's *Fasting Girls: The History of Anorexia Nervosa* (Cambridge: Harvard University, 1988) provides an absorbing history of what is mainly a twentieth-century disorder. Félix Larocca's collection, *Eating Disorders* (San Francisco: Jossey-Bass, 1986), remains a good, concise introduction to the clinical issues.

Food and the Status Quest, edited by Polly Wiessner and Wulf Schiefenhövel (Providence/Oxford: Berghahn Books, 1996), is a highly original collection showing how people and apes use food to create relationships and exert control. Hunter-gatherers are central to our understanding of the human relationship to accumulation, and the foundation for modern studies was laid by Richard Lee and Irven DeVore in *Man the Hunter* (Chicago: Aldine de Gruyter, 1968). The field was advanced by Bruce Winterhalder and Eric Alden Smith in *Hunter-Gatherer Foraging Strategies: Ethnographic and Archeological Analyses* (Chicago: University of Chicago, 1981) and Francis Dahlberg's *Woman the Gatherer* (New Haven, Conn.: Yale University, 1981). The culmination of twentieth-century research is Richard Lee and Richard Daly's *Cambridge Encyclopedia of Hunters and Gatherers* (Cambridge: Cambridge University, 1999). For a view of life in societies where maximizing wealth is not the principal goal, see Marshall Sahlins, *Stone Age Economics* (Chicago: Aldine de Gruyter, 1972); a more balanced view is Stuart Plattner's *Economic Anthropology* (Stanford, Calif.: Stanford University, 1989).

For a view of childhood in a culture in which maximizing wealth *is* the principal goal, see Robert Coles's *Privileged Ones: The Well-Off and the Rich in America* (Boston: Atlantic-Little, Brown, 1977). Robert Frank's *Luxury Fever: Why Money Fails to Satisfy in an Era of Excess* (New York: Free Press, 1999) analyzes the background and human consequences of out-of-control consumption and conspicuous display of wealth in contemporary America. John Kenneth Galbraith's writings have, for over half a century, gently and gracefully called our attention to the excesses of corporate capitalism and the people it leaves behind — most recently, *The Culture of Contentment* (New York: Houghton Mifflin, 1993) and *The Good Society: The Humane Agenda* (New York: Houghton Mifflin, 1997).

CHAPTER 16: *Change*

Laws of learning persist intact, although research on them is no longer popular. For a basic account of those laws and the facts that support them, I like Michael Domjan's *Essentials of Conditioning and Learning* (Pacific Grove, Calif.: Brooks Cole, 1996). An approach through major theorists is B. R. Hergenhahn and Matthew H. Olson's *Introduction to Theories of Learning* (Englewood Cliffs, N.J.: Prentice-Hall, 1993). On learning in children, the classic by Albert Bandura and R. H. Walters, *Social Learning and Personality Development* (New York: Holt, Rinehart & Winston, 1963), is still worth reading. Another landmark work is *Biological Boundaries of Learning*, edited by Martin Seligman and Joanne Hager (New York: Meredith, 1972). Recent emphasis on cognition, linguistics, evolution, and genetics has led to a decreased emphasis on traditional studies of learning, partly due to past excessive claims for the power of learning theory. Perhaps fruitless controversy will now give way to a more synthetic approach.

Ulric Neisser's *Memory Observed* (San Francisco: W. H. Freeman, 1982) is a classic on the natural history of memory, and Charles A. Nelson's collection, *Memory, Affect, and Development* (Hillsdale, N.J.: Lawrence Erlbaum, 1993) introduces important perspectives on childhood memory and its role in identity formation. To some extent the effects of early experience are separate from traditional learning theory. Victor H. Denenberg's *The Development of Behavior* (Stamford, Conn.: Sinauer, 1972) is a collection of key papers in this field, introduced by a leading practitioner. For the contributions of anthropology to the effects of experience on development, see the *Handbook of Cross-Cultural Development*, edited by Ruth H. Monroe, Robert L. Monroe, and Beatrice B. Whiting (New York: Garland, 1981).

Work in the late 1990s promises to lead cultural anthropology out of its dark night of wandering in the postmodern wilderness. The best recent theoretical books are Dan Sperber's *Explaining Culture: A Naturalistic Approach* (Oxford: Blackwell, 1996) and Bradd Shore's *Culture in Mind* (New York: Oxford University, 1996). In *Culture: The Anthropologist's Account* (Cambridge: Harvard University, 1999), Adam Kuper offers the most cogent explanation of the concept by tracing its modern history. The definitive work on cultural evolution is William Durham's *Coevolution: Genes, Culture, and Human Diversity* (Stanford, Calif.: Stanford University, 1991).

Studies of the physiology of learning and memory abound. The advanced reader should consult sections on plasticity and memory in Michael Gazzaniga's collection, *The New Cognitive Neurosciences*, 2d Ed. (Cambridge: MIT, 2000). An accessible account of the modern science of memory is Daniel Schacter's *Searching for Memory: The Brain, the Mind, and the Past* (New York: HarperCollins, 1997). For a wisely skeptical summary of the effects of early experience in children, see John T. Bruer's *The Myth of the First Three Years* (New York: Free Press, 1999).

CHAPTER 17: *The Invisible Galaxy*

It is difficult to recommend books about the genome, they go out of date so fast. My turn-of-the-century favorite is John Avise's *The Genetic Gods* (Cambridge: Harvard University, 1998), although Matt Ridley's *Genome* (New York: HarperCollins, 1999) is also excellent. A fine textbook of genetics is Anthony Griffiths et al., *An Introduction to Genetic Analysis*, 7th Ed. (New York: W. H. Freeman, 2000). Up-to-the-minute information about the genome is on the Web at www.nhgri.gov.

There are many graceful, intelligent books about the fate of the earth and its inhabitants, but my favorites are Edward O. Wilson's *Biophilia* (Cambridge: Harvard University, 1986), James Gustafson and Frederic Blumer's *A Sense of the Divine: The Natural Environment from a Theocentric Perspective* (Cleveland: Pilgrim, 1994), and Ursula Goodenough's *The Sacred Depths of Nature* (New York: Oxford University, 1998). *An Essay on Population*, by Thomas Malthus (New York: Penguin, 1985, orig. 1798), remains a compelling read more than two centuries after its publication, easily justifying Darwin's admiration. Paul Ehrlich's *Human Natures* (Washington, D.C./Covelo, Calif.: Island Press, 2000) and Ed Ayres's *God's Last Offer: Negotiating for a Sustainable Future* (New York: Four Walls Eight Windows, 1999) are good antidotes to twenty-first-century population complacency. The science of the biosphere is gracefully explained in Vaclav Smil's *Cycles of Life: Civilization and the Biosphere* (New York: Scientific American Library, 2001).

For a recent assessment of worldwide poverty, see the World Bank's *World Development Report 2000/2001*, or visit www.world-bank.org/poverty/wdrpoverty. To join a forum on worldwide poverty, go to www.world-bank.org/devforum/forum_qog.html.

Recommended Reading

The following books published since 1985 have helped me to understand human nature, something I thought I knew a lot about before that year dawned. As my readers know, I put great stock in old books, so please read the bibliographic notes to find out which of *those* I found most valuable. The list below is a starting point; it is already obsolescing. If you read what is on it, you will pretty much know what I know—at least the part that can be learned from books. If you go on to read the books and papers to come, then in just a couple of decades you will know far more than I or anyone today can dream of knowing. If you are young and are interested in this subject, you should not be satisfied with anything less.

Adair, Virginia Hamilton. 1996. *Ants on the Melon: A Collection of Poems*. New York: Random House.

Akiskal, Hagop S., and Cassano, Giovanni B., eds. 1997. *Dysthymia and the Spectrum of Chronic Depressions*. New York: Guilford.

Alcock, John. 2001. *The Triumph of Sociobiology*. New York: Oxford University.

Allende, Isabel. 1988. *Of Love and Shadows: A Novel*. New York: Bantam Books.

Allman, John. 1999. *Evolving Brains*. New York: W. H. Freeman.

Archer, John. 1999. *The Nature of Grief: The Evolution and Psychology of Reactions to Loss*. New York: Routledge.

American Psychiatric Association. 1994. *Diagnostic and Statistical Manual of Mental Disorders*, Fourth Edition. Washington, D.C.: American Psychiatric Association.

Bancroft, John, and Reinisch, June Machover, eds. 1990. *Adolescence and Puberty*. New York: Oxford University.

Barkow, Jerome H., Cosmides, Leda, and Tooby, John, eds. 1992. *The Adapted Mind: Evolutionary Psychology and the Generation of Culture*. Oxford and New York: Oxford University.

Betzig, Laura, ed. 1997. *Human Nature: A Critical Reader*. New York: Oxford University.

Bloom, Harold. 1998. *Shakespeare: The Invention of the Human*. New York: Riverhead Books.

Brown, Donald E. 1991. *Human Universals*. Philadelphia: Temple University.

Buss, David M. 1994. *The Evolution of Desire: Strategies of Human Mating*. New York: Basic Books.

Calvin, William H. 1996. *The Cerebral Code: Thinking a Thought in the Mosaics of the Mind*. Cambridge: MIT.

Calvin, William H., and Bickerton, Derek. 2000. *Lingua Ex Machina: Reconciling Darwin and Chomsky with the Human Brain*. Cambridge: MIT.

Changeux, Jean-Pierre. 1997. *Neuronal Man: The Biology of Mind*, Second Edition. Princeton, N.J.: Princeton University.

Conway, Simon Morris. 1998. *The Crucible of Creation: The Burgess Shale and the Rise of Animals*. Oxford: Oxford University.

Cooper, Jack R., Roth, Robert, and Bloom, Floyd E. 1996. *Biochemical Basis of Neuropharmacology*, Seventh Edition. New York: Oxford University.

Csikszentmihalyi, Mihaly. 1990. *Flow: The Psychology of Optimal Experience*. New York: Harper and Row.

Dabbs, James McBride, and Dabbs, Mary Godwin. 2000. *Heroes, Rogues, and Lovers: Testosterone and Behavior*. New York: McGraw-Hill.

Daly, Martin, and Wilson, Margo. 1988. *Homicide*. New York: Aldine de Gruyter.

Damasio, Antonio. 1994. *Descartes' Error: Emotion, Reason, and the Human Brain*. New York: G. P. Putnam's Sons.

———. 1999. *The Feeling of What Happens: Body and Emotion in the Making of Consciousness*. New York: Harcourt Brace.

Deacon, Terence. 1997. *The Symbolic Species: The Co-Evolution of Language and the Brain*. New York: W. W. Norton.

de Waal, Frans. 1996. *Good Natured: The Origins of Right and Wrong in Humans and Other Animals*. Cambridge: Harvard University.

de Waal, Frans B., ed. 2001. *Tree of Origin: What Primate Behavior Can Tell Us About Human Social Evolution*. Cambridge: Harvard University.

Diamond, Jared. 1997. *Guns, Germs, and Steel: The Fates of Human Societies*. New York: W. W. Norton.

Edelman, Gerald M. 1987. *Neural Darwinism: The Theory of Neuronal Group Selection*. New York: Basic Books.

Edgerton, Robert B. 1992. *Sick Societies: Challenging the Myth of Primitive Harmony*. New York: Free Press.

Eibl-Eibesfeldt, Irenäus. 1988. *Human Ethology*. New York: Aldine de Gruyter.

Ekman, Paul, and Rosenberg, Erika L., eds. 1997. *What the Face Reveals: Basic and Applied Studies of Spontaneous Expression Using the Facial Action Coding System (FACS)*. New York: Oxford University.

Etcoff, Nancy. 1999. *Survival of the Prettiest: The Science of Beauty*. New York: Doubleday.

Frankl, Victor E. 2000. *Man's Search for Ultimate Meaning*. Cambridge: Perseus Publishing.

Fukuyama, Francis. 1992. *The End of History and the Last Man*. New York: Free Press.

Gazzaniga, Michael S. 2000. *The New Cognitive Neurosciences*, Second Edition. Cambridge: MIT.

Gilmore, David D. 1990. *Manhood in the Making: Cultural Concepts of Masculinity*. New Haven, Conn.: Yale University.

Glück, Louise. 1999. *Vita Nova: Poems*. Hopewell, N.J.: Ecco Press.

———. 2001. *The Seven Ages: Poems*. New York: HarperCollins/Ecco Press.

Goldhagen, Daniel Jonah. 1996. *Hitler's Willing Executioners: Ordinary Germans and the Holocaust*. New York: Vintage/Random House.

Goodall, Jane. 1986. *The Chimpanzees of Gombe: Patterns of Behavior*. Cambridge: Harvard University.

Goodwin, Donald W., and Guze, Samuel B. 1996. *Psychiatric Diagnosis*. New York: Oxford University.

Grandin, Temple, and Sacks, Oliver. 1996. *Thinking in Pictures: And Other Reports from My Life with Autism*. New York: Vintage.

Griffiths, Paul E. 1997. *What Emotions Really Are: The Problem of Psychological Categories*. Chicago: University of Chicago.

Harris, Marvin. 1997. *Culture, People, Nature: An Introduction to General Anthropology*, Seventh Edition. New York: Longman.

Hart, Josephine. 1996. *Damage: A Novel*. New York: Ballantine.

Hatfield, Elaine, Cacioppo, John T., and Rapson, Richard L. 1994. *Emotional Contagion*. Cambridge: Cambridge University.

Heider, Karl. 1991. *Landscapes of Emotion: Mapping Three Cultures of Emotion in Indonesia*. New York: Cambridge University.

Herdt, Gilbert, ed. 1994. *Third Sex, Third Gender: Beyond Sexual Dimorphism in Culture and History*. New York: Zone Books.

Hrdy, Sarah Blaffer. 1999. *Mother Nature: A History of Mothers, Infants, and Natural Selection*. New York: Pantheon.

Jamison, Kay Redfield. 1993. *Touched with Fire: Manic-Depressive Illness and the Artistic Temperament*. New York: Free Press.

Jankowiak, William, ed. 1995. *Romantic Passion: A Universal Experience?* New York: Columbia University.

Johanson, Donald, and Edgar, Blake. 1996. *From Lucy to Language*. New York: Simon & Schuster Editions.

Jolly, Alison. 1999. *Lucy's Legacy: Sex and Intelligence in Human Evolution*. Cambridge: Harvard University.

Kahneman, Daniel, Diener, Ed, and Schwarz, Norbert, eds. 1999. *Well-Being: The Foundations of Hedonic Psychology*. New York: Russell Sage Foundation.

Kandel, Eric R., Schwartz, James H., and Jessell, Thomas M., eds. 2000. *Principles of Neural Science*, Fourth Edition. New York: McGraw-Hill.

Keeley, Lawrence H. 1996. *War Before Civilization: The Myth of the Peaceful Savage*. New York: Oxford University.

LeDoux, Joseph. 1996. *The Emotional Brain: The Mysterious Underpinnings of Emotional Life*. New York: Touchstone/Simon & Schuster.

Lessing, Doris. 1989. *The Fifth Child: A Novel*. New York: Vintage International.

LeVay, Simon. 1993. *The Sexual Brain*. Cambridge: MIT.

———. 1997. *Queer Science: The Use and Abuse of Research into Homosexuality*. Cambridge: MIT.

Lewis, Michael, and Haviland, Jeannette M., eds. 2000. *Handbook of Emotions*, Second Edition. New York: Guilford.

Low, Bobbi S. 2000. *Why Sex Matters: A Darwinian Look at Human Behavior*. Princeton, N.J.: Princeton University.

Lykken, David. 1999. *Happiness: What Studies on Twins Show Us About Nature, Nurture, and the Happiness Set-Point*. New York: Golden Books.

Maccoby, Eleanor E. 1998. *The Two Sexes: Growing Up Apart, Coming Together*. Cambridge: Harvard University.

MacLean, Paul D. 1990. *The Triune Brain in Evolution: Role in Paleocerebral Functions*. New York: Plenum.

Mayr, Ernst. 1997. *This Is Biology*. Cambridge: Belknap/Harvard University.

McGuire, Michael, and Troisi, Alfonso. 1998. *Darwinian Psychiatry*. New York: Oxford University.

Nuland, Sherwin B. 1994. *How We Die: Reflections on Life's Final Chapter*. New York: Alfred A. Knopf.

Olds, Sharon. 1996. *The Wellspring: Poems*. New York: Alfred A. Knopf.

Olweus, Dan. 1988. *Aggression in the Schools: Bullies and Whipping Boys*. New York: John Wiley & Sons.

Ondaatje, Michael. 2000. *Anil's Ghost: A Novel*. New York: Alfred A. Knopf.

Paglia, Camille. 1990. *Sexual Personae: Art and Decadence from Nefertiti to Emily Dickinson*. New Haven, Conn.: Yale University.

Panksepp, Jaak. 1998. *Affective Neuroscience: The Foundations of Human and Animal Emotions*. New York: Oxford University.

Person, Ethel Spector. 1988. *Dreams of Love and Fateful Encounters: The Power of Romantic Passion*. New York: W. W. Norton.

Pfaff, Donald W. 1999. *Drive: Neurobiological and Molecular Mechanisms of Sexual Motivation*. Cambridge: MIT.

Pfaff, Donald W., Berrettini, Wade H., Joh, Tong H., and Maxson, Stephen C., eds. 2000. *Genetic Influences on Neural and Behavioral Functions*. New York: CRC.

Pinker, Steven. 1994. *The Language Instinct: How the Mind Creates Language*. New York: William Morrow.

Prunier, Gérard. 1997. *The Rwanda Crisis: History of a Genocide*. New York: Columbia University.

Ridley, Mark. 1997. *Evolution*. Oxford: Oxford University.

Robinson, Michael H., and Tiger, Lionel, eds. 1991. *Man and Beast Revisited*. Washington, D.C.: Smithsonian Institution Press.

Rochat, Philippe. 2001. *The Infant's World*. Cambridge: Harvard University.

Rossi, Alice S., and Rossi, Peter H. 1990. *Of Human Bonding: Parent-Child Relations Across the Life Course*. New York: Aldine de Gruyter.

Rowe, David C. 1994. *The Limits of Family Influence: Genes, Experience, and Behavior*. New York: Guilford.

Sacks, Oliver. 1985. *The Man Who Mistook His Wife for a Hat, and Other Clinical Tales*. New York: Summit Books.

———. 1996. *An Anthropologist on Mars*. New York: Vintage.

Salloway, Stephen, Malloy, Paul, and Cummings, Jeffrey L., eds. 1997. *The Neuropsychiatry of Limbic and Subcortical Disorders*. Washington, D.C.: American Psychiatric Press.

Sapolsky, Robert M. 1994. *Why Zebras Don't Get Ulcers*. New York: W. H. Freeman.

———. 2001. *A Primate's Memoir: A Neuroscientist's Unconventional Life Among the Baboons*. New York: Scribner.

Savage-Rumbaugh, Sue, and Lewin, Roger. 1994. *Kanzi: The Ape at the Brink of the Human Mind*. New York: John Wiley & Sons.

Shostak, Marjorie. 2000. *Return to Nisa*. Cambridge: Harvard University.

Simpson, Jeffrey A., and Kenrick, Douglas T., eds. 1997. *Evolutionary Social Psychology*. Mahwah, N.J.: Lawrence Erlbaum.

Sipe, A. W. Richard. 1990. *A Secret World: Sexuality and the Search for Celibacy*. New York: Brunner/Mazel.

Smuts, Barbara B. 1985. *Sex and Friendship in Baboons*. New York: Aldine de Gruyter.

Smuts, Barbara B., Cheney, Dorothy L., Seyfarth, Robert M., Wrangham, Richard W., and Struhsaker, Thomas T., eds. 1987. *Primate Societies*. Chicago: University of Chicago.

Spiro, Melford E. 1987. *Culture and Human Nature*. Chicago: University of Chicago.

Stern, Daniel N. 1985. *The Interpersonal World of the Infant: A View from Psychoanalysis and Developmental Psychology*. New York: Basic Books.

Storr, Anthony. 1988. *Solitude: A Return to the Self*. New York: Free Press.

Strier, Karen B. 2000. *Primate Behavioral Ecology*. Boston: Allyn & Bacon.

Szymborska, Wislawa. 1995. *View with a Grain of Sand: Selected Poems*. New York: Harcourt Brace.

Tannen, Deborah. 1994. *Gender and Discourse*. New York: Oxford University.

Tiger, Lionel. 1999. *The Decline of Males*. New York: Golden Books.

Trevathen, Wenda R. 1987. *Human Birth: An Evolutionary Perspective*. New York: Aldine de Gruyter.

Trivers, R. L. 1985. *Social Evolution*. Menlo Park, Calif.: Benjamin Cummings.

Updike, John. 1997. *In the Beauty of the Lilies: A Novel*. New York: Fawcett Books.

Vaillant, George E. 1993. *The Wisdom of the Ego*. Cambridge: Harvard University.

Whiting, Beatrice, and Edwards, Carolyn Pope. 1988. *Children of Different Worlds: The Formation of Social Behavior*. Cambridge: Harvard University.

Wiessner, Polly, and Schiefenhövel, Wulf, eds. 1996. *Food and the Status Quest: An Interdisciplinary Perspective*. Providence: Berghahn Books.

Wiessner, Polly, and Tumu, Akii. 1998. *Historical Vines: Enga Networks of Exchange, Ritual, and Warfare in Papua New Guinea*. Washington, D.C.: Smithsonian Institution Press.

Wikan, Unni. 1990. *Managing Turbulent Hearts: A Balinese Formula for Living*. Chicago: University of Chicago.

Wilson, Edward O. 1998. *Consilience: The Unity of Knowledge*. New York: Alfred A. Knopf.

Wilson, James Q., and Herrnstein, Richard. 1985. *Crime and Human Nature*. New York: Simon & Schuster.

Wrangham, Richard W., and Peterson, Dale. 1996. *Demonic Males: Apes and the Origins of Human Violence*. Boston: Houghton Mifflin.

Wright, Robert. 1994. *The Moral Animal: The New Science of Evolutionary Psychology*. New York: Pantheon.

Index